Rainer Pöttgen, Thomas Jüstel, Cristian A. Strassert (Eds.)
Applied Inorganic Chemistry

Also of interest

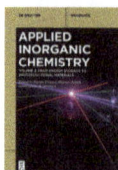

Applied Inorganic Chemistry
Volume 1: From Construction Materials to Technical Gases
Rainer Pöttgen, Thomas Jüstel and Cristian A. Strassert (Eds.) 2023
ISBN 978-3-11-073814-8, e-ISBN 978-3-11-073314-3

Applied Inorganic Chemistry
Volume 2: From Energy Storage to Photofunctional Materials
Rainer Pöttgen, Thomas Jüstel and Cristian A. Strassert (Eds.) 2023
ISBN 978-3-11-079878-4, e-ISBN 978-3-11-079889-0

Intermetallics
Synthesis, Structure, Function
Rainer Pöttgen, Dirk Johrendt 2019
ISBN 978-3-11-063580-5, e-ISBN 978-3-11-063672-7

Rare Earth Chemistry
Rainer Pöttgen, Thomas Jüstel and Cristian A. Strassert (Eds.) 2020
ISBN 978-3-11-065360-1, e-ISBN 978-3-11-065492-9

Zeitschrift für Kristallographie – Crystalline Materials
Rainer Pöttgen (Editor-in-Chief)
ISSN 2194-4946, e-ISSN 2196-7105

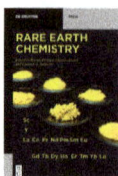

Applied Inorganic Chemistry

Volume 3: From Magnetic to Bioactive Materials

Edited by
Rainer Pöttgen, Thomas Jüstel, Cristian A. Strassert

DE GRUYTER

Editors

Prof. Dr. Rainer Pöttgen
Institut für Anorganische und Analytische Chemie
Westfälische Wilhelms-Universität Münster
Corrensstraße 30
48149 Münster
Germany
E-mail: pottgen@uni-muenster.de

Prof. Dr. Thomas Jüstel
Fachbereich Chemieingenieurwesen
Fachhochschule Münster
Stegerwaldstraße 39
48565 Steinfurt
Germany
E-mail: tj@fh-muenster.de

Prof. Dr. Cristian A. Strassert
Institut für Anorganische und Analytische Chemie
CiMIC – CeNTech – SoN
Westfälische Wilhelms-Universität Münster
Corrensstraße 28/30
48149 Münster
Germany
E-mail: ca.s@wwu.de

ISBN 978-3-11-073837-7
e-ISBN (PDF) 978-3-11-073347-1
e-ISBN (EPUB) 978-3-11-073357-0

Library of Congress Control Number: 2022935001

Bibliographic information published by the Deutsche Nationalbibliothek
The Deutsche Nationalbibliothek lists this publication in the Deutsche Nationalbibliografie; detailed bibliographic data are available on the Internet at http://dnb.dnb.de.

© 2023 Walter de Gruyter GmbH, Berlin/Boston
Cover image: Magnetfabrik Bonn
Typesetting: Integra Software Services Pvt. Ltd.
Printing and binding: CPI books GmbH, Leck

www.degruyter.com

Preface

The Periodic Table meanwhile lists 118 chemical elements, which leads to a vast number of inorganic compounds. Many of them have well-defined physicochemical properties, which are exploited for the realization of functional materials we all comfortably use in daily life without even thinking about it, including magnetic and optical materials, construction materials, materials for energy storage and conversion – just to name a few remarkable examples. The impact of inorganic chemistry in human evolution cannot be overstated, and is proven by the designation of historical ages, such as stone, copper, bronze or iron age (even golden ages and gold rush), or by geographical locations (such as the Silicon Valley and Argentina). While carbon-based organic chemistry has provided incredible breakthroughs in medicinal chemistry and plastic materials, there is no doubt that the solution of the most urgent problems currently faced by humanity will stem from inorganic chemistry providing high-density/high-stability materials for construction, information technologies, energy storage and conversion.

Chemical sciences and industries are often demonized, but the many indispensable materials we use in daily life impressively show how significantly they influence our society. Ecosystems, metabolic and pathophysiological processes, food production, construction in its broadest sense, mobility and energy conversion are determined by chemistry – these facts cannot simply be ignored! The present book summarizes the many basic examples of inorganic materials we use on a large scale in everyday life, but also niche products with thoroughly optimized properties. Many subchapters are written by experts from academia and industry. We tried to ensure a proper balance of topics, even though it is simply impossible to cover all aspects of applied inorganic chemistry. Nonetheless, we hope that we made a good compromise – if any topic is missing, this was unintentional. The final chapter focusses on energy flows and resources, which constitutes one of the most urgent topics. As a kind of appetizer for the following 16 chapters, we briefly summarize some applications for the elements of the first four rows of the Periodic Table. Several of these topics are picked up again in the following chapters:

Hydrogen: energy source; **helium:** low-temperature refrigerant; ballon gas, **lithium:** anode materials for lithium-ion batteries; **beryllium:** hardening component for light-weight alloys, non-spark alloys, X-ray windows; **boron:** hardening component for intermetallics; **carbon:** electrode materials, black pigment; **nitrogen:** source for ammonia and nitrate fertilizers, protective gas, low-temperature cooling; **oxygen:** medical gas, liquid oxygen for the Linz-Donauwitzer process in steel refinement; **fluorine:** uranium hexafluoride production; **neon:** helium-neon lasers; **sodium:** reducing agent; **magnesium:** alloying component and sacrificial anodes; **aluminum:** light-weight alloys, construction material; **silicon:** semiconductors; **phosphorus:** synthesis of phosphoric acid; matches; **sulfur:** vulcanization of rubber; **chlorine:** disinfection of water; **argon:** protective gas in chemical

https://doi.org/10.1515/9783110733471-202

synthesis and arc-welding; **potassium:** liquid sodium-potassium alloys as coolants in nuclear reactors; **calcium:** reducing agent in metallurgy; **scandium:** additive for aluminum-based alloys, component of electron emitters; **titanium:** steel additive, corrosion resistant alloys; **vanadium:** high-speed tool steels; **chromium:** stainless steel and chromium plating; **manganese:** ferromanganese, activator in LED phosphors; **iron:** steel and cast iron; **cobalt:** superalloys and samarium-cobalt magnets; **nickel:** catalysis and anti-corrosion coatings; **copper:** cables and water tubes; **zinc:** facade cladding, corrosion protection; **gallium:** gallium nitride, phosphide or arsenide semiconductors; **germanium:** semiconductors and detection technology; **arsenic:** doping of semiconductors; **selenium:** II-VI semiconductors and alloy additive for free cutting steel; **bromine:** special disinfection products and synthesis of flame retardants; **krypton:** excimer lasers, KrCl excimer discharge lamps. The reader might notice that transition metals and lanthanides are not even mentioned here; there would not be sufficient space in a preface to list their impact!

Such a book project is not realizable without the help of numerous colleagues and co-workers. We thank Gudrun Lübbering for continuous help with literature search and text processing and Thomas Fickenscher for providing with many photos of materials and devices. We are especially grateful to our colleagues for their immediate agreements to write up a subchapter. It is always challenging to compile a concise Table of Contents and find the right co-authors. We are indebted to the editorial and production staff of De Gruyter. Our particular thanks go to Kristin Berber-Nerlinger, Dr. Vivien Schubert and Melanie Götz for their continuous support during conception, writing and producing the present book.

Münster, Steinfurt, June 2022
Thomas Jüstel, Rainer Pöttgen, Cristian A. Strassert

This book contains two different tokens, pointing to:

📖　　　　list of references

📑　　　　recommended literature for further reading; i.e. relevant text books, review articles or important original articles

Contents

Volume 3 (From Magnetic to Bioactive Materials)

Volume 1 (From Construction Materials to Technical Gases)

List of contributors

Ackermann, Dr. habil. Lothar
Deutsche Stiftung Edelsteinforschung (DSEF)
Professor-Schlossmacher-Straße 1
55743 Idar-Oberstein
Germany
E-mail: lackermann@outlook.de

Agne, Dr. Matthias
Forschungszentrum Jülich GmbH
Helmholtz-Institut Münster (HI MS, IEK-12)
Corrensstraße 46
48149 Münster
Germany
E-mail: m.agne@fz-juelich.de

Apaydin, Dr. Dogukan H.
Institut für Materialchemie
TU Wien
Getreidemarkt 9/165
1060 Wien
Austria
E-mail: dogukan.apaydin@tuwien.ac.at

Arnault, Dr. Jean-Charles
Laboratoire des Edifices Nanométriques
Université Paris-Saclay, CEA, CNRS, NIMBE
91191 Gif sur Yvette
France
E-mail: jean-charles.arnault@cea.fr

Banik, Dr. Ananya
Institut für Anorganische und Analytische Chemie
Westfälische Wilhelms-Universität Münster
Corrensstraße 30
48149 Münster
Germany
E-mail: banik@uni-muenster.de

Bauer, Dr. Thomas
Deutsches Zentrum für Luft- und Raumfahrt (DLR)
Institut für Technische Thermodynamik
Thermische Prozesstechnik
Linder Höhe, Gebäude 26
51147 Köln
Germany
E-mail: thomas.bauer@dlr.de

Baur, Dr. Florian
Fachbereich Chemieingenieurwesen
Fachhochschule Münster
Stegerwaldstraße 39
48565 Steinfurt
Germany
E-mail: florian.baur@fh-muenster.de

Bayer, Dr. Bernhard
Institut für Materialchemie
TU Wien
Getreidemarkt 9/165
1060 Wien
Austria
E-mail: bernhard.bayer-skoff@tuwien.ac.at

Behrend, Prof. Dr.-Ing. habil. Detlef
Lehrstuhl Werkstoffe für die Medizintechnik
Fachbereich Maschinenbau und Schiffstechnik
Universität Rostock
Friedrich-Barnewitz-Straße 4
18119 Rostock
Germany
E-mail: detlef.behrend@uni-rostock.de

Behrens, Dr. Rainer
VDM Metals International GmbH
Kleffstraße 23
58762 Altena
Germany
E-mail: rainer.behrens@vdm-metals.com

Bertau, Prof. Dr. rer. nat. habil. Martin
Institut für Technische Chemie
TU Bergakademie Freiberg
Leipziger Straße 29
09599 Freiberg
Germany
and
Fraunhofer Technology Center for High-Performance Materials THM
Fraunhofer Institut for Ceramic Technologies and Systems IKTS
Am St.-Niclas-Schacht 13
09599 Freiberg
Germany
E-mail: Martin.Bertau@chemie.tu-freiberg.de

https://doi.org/10.1515/9783110733471-204

Binnewies, Prof. Dr. Michael
Institut für Anorganische Chemie
Naturwissenschaftliche Fakultät
Leibnitz Universität Hannover
Callinstraße 3-9
30167 Hannover
Germany
E-mail: michael.binnewies@aca.uni-hannover.de

Boos, Dr. Markus
Remmers GmbH
Bernhard-Remmers-Straße 13
49624 Löningen
Germany
E-mail: mboos@remmers.de

Boos, Dr. Peter
HeidelbergCement AG
Zur Anneliese 11
59320 Ennigerloh
Germany
E-mail: Peter.Boos@heidelbergcement.com

Bredol, Prof. Dr. Michael
Fachbereich Chemieingenieurwesen
Fachhochschule Münster
Stegerwaldstraße 39
48565 Steinfurt
Germany
E-mail: bredol@fh-muenster.de

Broll, Dr. Sascha
Broll-Buntpigmente GmbH & Co. KG
Karl-Winnacker-Straße 2-4
36396 Steinau
Germany
E-mail: drsascha@broll-buntpigmente.de

Buchner, Dr. Magnus R.
Anorganische Chemie, Fluorchemie
Philipps-Universität Marburg
Hans-Meerwein-Straße 4
35032 Marburg
Germany
E-mail: magnus.buchner@chemie.uni-marburg.de

Busch, Dr. Frank
Materialprüfungsamt Nordrhein-Westfalen
Marsbruchstraße 186
44287 Dortmund
Germany
E-mail: busch@mpanrw.de

Buttler, Dr.-Ing. Torben Alexander
-ISAF- Institut für Schweißtechnik und
Trennende Fertigungsverfahren
Technische Universität Clausthal
Agricolastraße 2
38678 Clausthal-Zellerfeld
Germany
E-mail: buttler@isaf.tu-clausthal.de

Dewalsky, Dr. Martin V.
Am Gemeindeholz 6
82205 Gilching
Germany
E-mail: martinvdew@gmail.com

Dorsch, Leonhard Yuuta
Institut für Anorganische Chemie
Universität Leipzig
Johannisallee 29
04103 Leipzig
Germany
E-mail: leonhard.dorsch@uni-leipzig.de

Dramicanin, Prof. Dr. Miroslav
Vinca Institute of Nuclear Sciences
University of Belgrade,
PO Box 522
11001 Belgrade
Serbia
E-mail: dramican@vinca.rs

Eckert, Prof. Dr. Hellmut
Institut für Physikalische Chemie
Westfälische Wilhelms-Universität Münster
Corrensstraße 30
48149 Münster
Germany
and
Instituto de Física de Sao Carlos
Universidade de Sao Paulo
Avenida Trabalhador Saocarlense 400
Sao Carlos, SP 13566-590
Brasil
E-mail: eckerth@uni-muenster.de

Eder, Prof. Dr. Dominik
Institut für Materialchemie
TU Wien
Getreidemarkt 9/165
1060 Wien
Austria
E-mail: dominik.eder@tuwien.ac.at

Engel, Stefan
Universität des Saarlandes
Anorganische Festkörperchemie
Campus C4 1
66123 Saarbrücken
Germany
E-mail: stefan.engel@uni-saarland.de

Engels, Ir. Marcel
Forschungsinstitut für Glas | Keramik (FGK)
Heinrich-Meister-Straße 2
56203 Höhr-Grenzhausen
Germany
E-mail: marcel.engels@fgk-keramik.de

Epple, Prof. Dr. Matthias
Anorganische Chemie
Fakultät für Chemie
Universität Duisburg-Essen
Universitätsstraße 7
45141 Essen
Germany
E-mail: matthias.epple@uni-due.de

Faust, PD Dr. Andreas
European Institute for Molecular Imaging (EIMI)
Waldeyerstraße 15
48149 Münster
Germany
E-Mail: faustan@uni-muenster.de

Feser, Prof. Dr.-Ing. Ralf
Fachbereich Informatik und
Naturwissenschaften
Fachhochschule Südwestfalen
Frauenstuhlweg 31
58644 Iserlohn
Germany
E-Mail: feser.ralf@fh-swf.de

Fickenscher, Thomas
Institut für Anorganische und Analytische
Chemie
Westfälische Wilhelms-Universität Münster
Corrensstraße 30
48149 Münster
Germany
E-mail: thomasfi@uni-muenster.de

Fröhlich, Dr. Peter
Institut für Technische Chemie
TU Bergakademie Freiberg
Leipziger Straße 29
09599 Freiberg
Germany
E-mail: peter.froehlich@chemie.tu-freiberg.de

Ghidiu, Dr. Michael
Institut für Anorganische und Analytische
Chemie
Westfälische Wilhelms-Universität Münster
Corrensstraße 30
48149 Münster
Germany
E-mail: ghidiu@uni-muenster.de

Glaum, Prof. Dr. Robert
Institut für Anorganische Chemie
Rheinische Friedrich-Wilhelms-Universität
Gerhard-Domagk-Straße 1
53121 Bonn
Germany
E-mail: rglaum@uni-bonn.de

Grönefeld, Dr. Martin
Magnetfabrik Bonn GmbH
Dorotheenstraße 215
53119 Bonn
Germany
E-Mail: Martin.Groenefeld@Magnetfabrik.de

Haberkamp, Prof. Dr.-Ing. Jens
Fachbereich Bauingenieurwesen
Fachhochschule Münster
Corrensstraße 25
48149 Münster
Germany
E-mail: haberkamp@fh-muenster.de

Haneklaus, Dr. Nils
Td Lab Sustainable Mineral Resources
Universität für Weiterbildung Krems
Dr.-Karl-Dorrek-Straße 30
3500 Krems an der Donau
Austria
E-mail: nils.haneklaus@donau-uni.ac.at
and
Institut für Technische Chemie
TU Bergakademie Freiberg
Leipziger Straße 29
09599 Freiberg
Germany

Hayen, Prof. Dr. Heiko
Institut für Anorganische und Analytische
Chemie
Westfälische Wilhelms-Universität Münster
Corrensstraße 48
48149 Münster
Germany
E-mail: heiko.hayen@uni-muenster.de

Hendriks, Dr. Theodoor
Forschungszentrum Jülich GmbH
Helmholtz-Institut Münster (HI MS, IEK-12)
Corrensstraße 46
48149 Münster
Germany
E-mail: t.hendriks@fz-juelich.de

Hermes, Dr. Wilfried
trinamiX GmbH
Industriestraße 35
67063 Ludwigshafen
Germany
E-mail: wilfried.hermes@trinamix.de

Herrmann, Dr. Fabian
Institut für Pharmazeutische Biologie und
Phytochemie
Westfälische Wilhelms-Universität Münster
Corrensstraße 48
48149 Münster
Germany
E-mail: fabian.herrmann@wwu.de

Hosono, Prof. Dr. Hideo
Materials Research Center for Element
Strategy (MCES)
Tokyo Institute of Technology
SE-1, 4259 Nagatsuta-cho
Midori-ku, Yokohama, Kanagawa, 226-8503
Japan
E-mail: hosono@mces.titech.ac.jp

Huppertz, Prof. Dr. Hubert
Institut für Allgemeine, Anorganische und
Theoretische Chemie
Universität Innsbruck
Innrain 80–82
6020 Innsbruck
Austria
E-mail: Hubert.Huppertz@uibk.ac.at

Janiak, Prof. Dr. Christoph
Institut für Anorganische Chemie und
Strukturchemie
Lehrstuhl für nanoporöse und nanoskalierte
Materialien
Heinrich-Heine-Universität Düsseldorf
Universitätsstraße 1
40225 Düsseldorf
Germany
E-mail: janiak@hhu.de

Janka, PD Dr. Oliver
Universität des Saarlandes
Anorganische Festkörperchemie
Campus C4 1
66123 Saarbrücken
Germany
E-mail: oliver.janka@uni-saarland.de

Jin, Wenqi
Xinjiang Technical Institute of Physics &
Chemistry
Chinese Academy of Sciences
40-1 South Beijing Road
830011 Urumqi
China
E-Mail: jwqineni@qq.com

Johrendt, Prof. Dr. Dirk
Department Chemie
Ludwig-Maximilians-Universität München
Butenandtstraße 5-13 (Haus D)
81377 München
Germany
E-mail: johrendt@lmu.de

Jüstel, Prof. Dr. Thomas
Fachbereich Chemieingenieurwesen
Fachhochschule Münster
Stegerwaldstraße 39
48565 Steinfurt
Germany
E-mail: tj@fh-muenster.de

Klapötke, Prof. Dr. Thomas M.
Department Chemie
Ludwig-Maximilians-Universität München
Butenandtstraße 5-13 (Haus D)
81377 München
Germany
E-mail: tmk@cup.uni-muenchen.de

Kohlmann, Prof. Dr. Holger
Institut für Anorganische Chemie
Universität Leipzig
Johannisallee 29
04103 Leipzig
Germany
E-mail: holger.kohlmann@uni-leipzig.de

Koller, PD Dr. Hubert
Institut für Physikalische Chemie
Westfälische Wilhelms-Universität Münster
Corrensstraße 30
48149 Münster
Germany
E-mail: hkoller@uni-muenster.de

Kratz, Dr. Nadja
Forschungsinstitut für Glas | Keramik (FGK)
Heinrich-Meister-Straße 2
56203 Höhr-Grenzhausen
Germany
E-mail: nadja.kratz@fgk-keramik.de

Kränkel, PD Dr. Christian
Leibniz-Institut für Kristallzüchtung (IKZ)
Max-Born-Straße 2
12489 Berlin
Germany
E-mail: christian.kraenkel@ikz-berlin.de

Krzywinski, Jacek
Magnetfabrik Bonn GmbH
Dorotheenstraße 215
53119 Bonn
Germany
E-Mail: Jacek.Krzywinski@Magnetfabrik.de

Langner, Dr. Bernd E.
Glockenheide 11
21423 Winsen
Germany
E-mail: langner@understanding-copper.com

Letz, Dr. Martin
SCHOTT AG
Research and Technology Development
Hattenbergstraße 10
55122 Mainz
Germany
E-mail: martin.letz@schott.com

Lider, Konstantin
Diener & Rapp GmbH & Co. KG Eloxalbetrieb
Junkersstraße 39
78056 Villingen-Schwenningen
Germany
E-mail: konstantin.lider@dienerrapp.de

Lox, Prof. Dr. Ir. Egbert S. J.
Am Laerchentor 8
36355 Grebenhain-Hochwaldhausen
Germany
E-mail: Egbert.Lox@gmail.com

Lovrincic, Dr. Robert
trinamiX GmbH
Industriestraße 35
67063 Ludwigshafen
Germany
E-mail: robert.lovrincic@trinamix.de

Maletz, Prof. Dr. Reinhard
VOCO GmbH
Anton-Flettner-Straße 1-3
27472 Cuxhaven
Germany
E-mail: r.maletz@voco.de

Matschke, Dr. Christian
BERLIN-CHEMIE AG
Glienicker Weg 125
12489 Berlin
Germany
E-mail: cmatschke@berlin-chemie.de

Mertens, Prof. Dr. Konrad
Fachbereich Elektrotechnik und Informatik
Fachhochschule Münster
Stegerwaldstraße 39
48565 Steinfurt
Germany
E-mail: mertens@fh-muenster.de

Mudryk, Dr. Yaroslav
Ames Laboratory, U.S. Department of Energy
Iowa State University
254 Spedding
Ames, IA 50011-2416
USA
E-mail: slavkomk@ameslab.gov

Niehaus, Dr. Oliver
Umicore AG & Co. KG
Rodenbacher Chaussee 4
63457 Hanau
Germany
E-mail: Oliver.Niehaus@eu.umicore.com

Pan, Prof. Dr. Shilie
Xinjiang Technical Insitute of Physics &
Chemistry
Chinese Academy of Sciences
40-1 South Beijing Road
830011 Urumqi
China
E-mail: slpan@ms.xjb.ac.cn

Pavón Regaña, Dr. Ing. Sandra
Fraunhofer-Institut für Keramische
Technologien und Systeme IKTS
Fraunhofer-Technologiezentrum
Hochleistungsmaterialien THM
Am St.-Niclas-Schacht 13
09599 Freiberg
Germany
and
Institut für Technische Chemie
TU Bergakademie Freiberg
Leipziger Straße 29
09599 Freiberg
Germany
E-mail: sandra.pavon.regana@ikts.
fraunhofer.de

Pecharsky, Prof. Dr. Vitalij K.
Ames Laboratory
Iowa State University
Ames, IA 50011-2416
USA
E-mail: vitkp@ameslab.gov

Piribauer, Dipl.-Ing. Dr. rer. nat. Christoph
Forschungsinstitut für Glas | Keramik (FGK)
Heinrich-Meister-Straße 2
56203 Höhr-Grenzhausen
Germany
E-mail: christoph.piribauer@fgk-keramik.de

Pöttgen, Prof. Dr. Rainer
Institut für Anorganische und Analytische
Chemie
Westfälische Wilhelms-Universität Münster
Corrensstraße 30
48149 Münster
Germany
E-mail: pottgen@uni-muenster.de

Quirmbach, Prof. Dr. rer. nat. Dr. h.c. Peter
Technische Chemie und
Korrosionswissenschaften
Universität Koblenz-Landau
Universitätsstraße 1
56070 Koblenz
Germany
E-mail: pquirmbach@uni-koblenz.de

Reiss, Prof. Dr. Günter
Physics Department
Center for Spinelectronic Materials and
Devices
Universitätsstraße 25
33615 Bielefeld
Germany
E-mail: guenter.reiss@uni-bielefeld.de

Riedel, Prof. Dr. Sebastian
Institut für Chemie und Biochemie
Anorganische Chemie
Freie Universität Berlin
Fabeckstraße 34/36
14195 Berlin
Germany
E-mail: s.riedel@fu-berlin.de

Rieger, Dr. Thorsten
VDM Metals International GmbH
Kleffstraße 23
58762 Altena
Germany
E-mail: Torsten.Rieger@vdm-metals.com

Salvermoser, Dr. Manfred
Riedel Filtertechnik GmbH
Westring 83
33818 Leopoldshöhe
Germany
E-mail: manfred.salvermoser@riedel-
filtertechnik.com

Sax, Dr.-Ing. Almuth
Technische Chemie und
Korrosionswissenschaften
Universität Koblenz-Landau
Universitätsstraße 1
56070 Koblenz
Germany
E-mail: asax@uni-koblenz.de

Schäferling, Prof. Dr. Michael
Fachbereich Chemieingenieurwesen
Fachhochschule Münster
Stegerwaldstraße 39
48565 Steinfurt
Germany
E-mail: michael.schaeferling@fh-muenster.de

Schmid, Jonas R.
Institut für Chemie und Biochemie
Anorganische Chemie
Freie Universität Berlin
Fabeckstraße 34/36
14195 Berlin
Germany
E-mail: jonas.schmid@fu-berlin.de

Schramm, Dr. Stefan
Merck KGaA
Frankfurter Straße 250
64293 Darmstadt
Germany
E-mail: stefan.schramm@merckgroup.com

Schupp, Prof. Dr. Thomas
Fachbereich Chemieingenieurwesen
Fachhochschule Münster
Stegerwaldstraße 39
48565 Steinfurt
Germany
E-mail: thomas.schupp@fh-muenster.de

Seifert, Dr. Markus
TU Dresden
Walther-Hempel-Bau
Mommsenstraße 4
01069 Dresden
Germany
E-mail: markus.seifert1@tu-dresden.de

Slabon, Prof. Dr. Adam
Chair of Inorganic Chemistry
University of Wuppertal
Gaußstraße 20
42119 Wuppertal
Germany
E-mail: slabon@uni-wuppertal.de

Staffel, Prof. Dr. Thomas
Research & Development, Phosphate
solutions
BK Giulini GmbH, ICL Group Ltd.
Dr.-Albert-Reimann-Straße 2
68526 Ladenburg
Germany
E-mail: thomas.staffel@icl-group.com

Stephan, Dr. Tom
Deutsche Gemmologische Gesellschaft e.V.
(DGemG)
Prof.-Schlossmacher-Straße 1
55743 Idar-Oberstein
Germany
E-mail: t.stephan@dgemg.com

Stengel, Dr. Ilona
Merck KGaA
Frankfurter Straße 250
64293 Darmstadt
Germany
E-mail: ilona.stengel@merckgroup.com

Stöwe, Prof. Dr. Klaus
Faculty of Natural Sciences
Institute of Chemistry, Chemical Technology
Technische Universität Chemnitz
09107 Chemnitz
Germany
E-mail: klaus.stoewe@chemie.tu-chemnitz.de

Strassert, Prof. Dr. Cristian A.
Institut für Anorganische und Analytische
Chemie
CiMIC – CeNTech – SoN
Westfälische Wilhelms-Universität Münster
Corrensstraße 28/30
48149 Münster
Germany
E-mail: ca.s@wwu.de

Teliban, Dr. Iulian
Magnetfabrik Bonn GmbH
Dorotheenstraße 215
53119 Bonn
Germany
E-Mail: Iulian.Teliban@Magnetfabrik.de

Termath, Dr. Andreas
Clariant Plastics & Coatings (Deutschland) GmbH
Chemiepark Knapsack
Industriestraße 149
Gebäude 2703, R. 128
50354 Hürth
Germany
E-mail: andreas.termath@clariant.com

Trodler, Dr. Jörg
Trodler-EAVT
Technische Beratung für die Aufbau und
Verbindungstechnik in der Elektronik
Grüner Weg 18/19
15712 Königs Wusterhausen
Germany
E-mail: joerg.trodler@trodler-eavt.de

Voigt, Dominik
Fachbereich Chemieingenieurwesen
Fachhochschule Münster
Stegerwaldstraße 39
48565 Steinfurt
Germany
E-mail: dv009200@fh-muenster.de

Voigt, Prof. Dr. Ingolf
Fraunhofer-Institut für Keramische
Technologien und Systeme IKTS
Michael-Faraday-Straße 1
07629 Hermsdorf
Germany
E-mail: ingolf.voigt@ikts.fraunhofer.de

**Warkentin, apl. Prof. Dr.-Ing. habil. Dr. rer.
nat. Mareike**
Fakultät für Maschinenbau und
Schiffstechnik
Universität Rostock
Friedrich-Barnewitz-Straße 4
18119 Rostock
Germany
E-mail: mareike.warkentin@uni-rostock.de

Weigand, Prof. Dr. Jan J.
TU Dresden
Walther-Hempel-Bau
Mommsenstraße 4
01069 Dresden
Germany
E-mail: jan.weigand@tu-dresden.de

Werner, Prof. Dr. Jan
Forschungsinstitut für Glas | Keramik (FGK)
Heinrich-Meister-Straße 2
56203 Höhr-Grenzhausen
Germany
E-mail: jan.werner@fgk-keramik.de

Wendel, Dr. Jörg
Wendel GmbH Email- und Glasurenfabrik
Am Güterbahnhof 30
35683 Dillenburg
Germany
E-mail: joerg.wendel@wendel-email.de

Wilhelm, Dr. Dominik
TYROLIT - Schleifmittelwerke Swarovski K.G.
Swarovskistraße 33
6130 Schwaz
Austria
E-mail: Dominik.Wilhelm@Tyrolit.com

Winter, Dr. Florian
Culimeta Textilglas-Technologie GmbH & Co.
KG
Werner-von-Siemens-Straße 9
49593 Bersenbrück
Germany
E-mail: fwinter@culimeta.de

Yang, Prof. Dr. Zhihua
Xinjiang Technical Institute of Physics &
Chemistry
Chinese Academy of Sciences
40-1 South Beijing Road
Urumqi 830011
China
E-Mail: zhyang@ms.xjb.ac.cn

Zeier, Prof. Dr. Wolfgang
Institut für Anorganische und Analytische
Chemie
Westfälische Wilhelms-Universität Münster
Corrensstraße 30
48149 Münster
Germany
E-mail: wzeier@uni-muenster.de

Ziegler, Raimund
Institut für Allgemeine, Anorganische und
Theoretische Chemie
Universität Innsbruck
Innrain 80–82
6020 Innsbruck
Austria
E-mail: Raimund.Ziegler@uibk.ac.at

Zumdick, Dr. Markus
H.C. Starck Tungsten GmbH
Im Schleeke 78-91
38642 Goslar
Germany
E-mail: markus.zumdick@hcstarck.com

9 Electronic and magnetic materials

9.1 Semiconductors

Michael Bredol, Dominik Voigt

9.1.1 General aspects

Semiconductors are found in many applications in which the transfer, the storage or the shuttling of electrons is key to the functionality: microelectronics, photovoltaics, light-emitting diodes, photocatalysts or electrocatalysts are prominent examples. The reason for this versatility is the presence of a so-called energy gap: in pure ideal semiconductors there is an energy interval above the highest occupied electronic state in which no further (unoccupied) states are found. Additional electronic states may be designed in this zone by using added impurities ("dopants") or point defects, leading to an enormous number of possible electronic properties and thus potential applications. Before discussing some inorganic semiconductors in more detail we will, therefore, have to inspect the necessary quantities and properties to understand electronic states in semiconductors.

The origin of energy bands and band gaps can only be understood on the background of the underlying quantum physics. We will not give a full account here, but rather compile the necessary concepts qualitatively (closely following an approach as laid out in [1], where also a somewhat deeper and more comprehensive discussion can be found). The basic idea is similar to molecular quantum chemistry: the electrons in a solid are approximated as individual negatively charged objects feeling the electrostatic attraction of the positively charged atomic nuclei. In a large crystal (say, of more than 50 nm particle size), however, in contrast to a molecule, there are many more of these charged nuclei – but regularly arranged on a crystal structure. This situation can be approximated by a potential function that combines a large and deep well (the full crystal size) with a local periodic modulation due to the crystalline periodicity.

As long as the periodically varying potential is disregarded, electrons will be moving quasi-freely in this large well – this is what we call the "electron gas" in a metal. For a truly free electron, according to the time-independent form of *Schrödinger's* equation the kinetic energy is not quantized and its wave function can be described by plane waves with arbitrary energy (for the sake of simplicity formulated here in one spatial dimension x only):

$$\text{classical: } E = \frac{p^2}{2m} \tag{9.1.1}$$

$$\text{quantum physical: } \Psi_k = A e^{ikx} \quad E = \frac{\hbar^2 k^2}{2m} \tag{9.1.2}$$

https://doi.org/10.1515/9783110733471-001

with p as the classical momentum, m as mass, $\hbar = h/2\pi$, k as so-called wave vector, Ψ as wave function and A as a constant depending on the boundary conditions. Modern DFT (density functional theory) methods for the modeling of solids are using this plane-wave approach in their "basis sets"; we will have a closer look into this in the following chapter.

The magnitude of the wave vector in the case of a free electron can vary continuously. In the expressions for the energy in eqs. (9.1.1) and (9.1.2) it can immediately be recognized that the classical momentum p is proportional to the magnitude of the wave vector $|k|$ – this analogy will help us very much with the interpretation of semiconductor properties. On the other hand, $|k|$ carries the unit of an inverse length and thus defines the inverse wavelength of the plane wave Ψ.

Equation (9.1.2) also tells us that for free electrons the dependence of the kinetic energy on the magnitude of the k-vector is a simple parabola. This somewhat oversimplified view does hold only as long as the wavelength associated with Ψ is much larger than the periodicity of the lattice – the wave function then simply will not notice variations in the potential (see Figure 9.1.1 for a sketch of the situation), and the average contribution of the potential is respected by exchanging the electron mass m_e against an effective mass m_e^*. However, as soon as the wavelength approaches the lattice parameter, the periodically varying potential has to be taken into account, and when the wavelength is in phase with the lattice parameter a (in other words, $1/|k| = a/n$ with n an integer value), we will expect *Bragg* reflection. This means that now backward and forward plane waves occur, leading to two distinct wave functions and consequently to two distinct charge densities according to $\Psi\Psi^*$: one being proportional to $\cos^2(\pi x/a)$, and the other to $\sin^2(\pi x/a)$, respectively. One will thus be centered at the (positively charged!) position of the atoms, the other just in between. The associated potential energy therefore must be different, leading to two states with different energy – this is the origin of energy gaps. The details of such a gap obviously depend on the form of the potential and thus on the nature of the atoms – the stronger the interaction, the larger the gap.

Figure 9.1.2 explains the approach: in an infinitely large crystal, the choice of the origin is arbitrary; therefore, it is always possible to displace and extend the bands periodically on the k-axis. This leads to the construction of a "reduced zone" in which the relations between the bands become visual: this is the core of "band structures".

Our discussion up to now was following eq. (9.1.2), and thus assuming, for the sake of simplicity, a simple one-dimensional sequence of evenly spaced atoms (the "lattice"). In three dimensions, however, we will find several independent periodic directions, depending on the crystal symmetry and structure, and for each of them the arguments laid out above apply: we will find bands and gaps for all these directions, and they need to be combined to the complete so-called "band structure" of the semiconductor (conveniently projected to a two-dimensional viewgraph). The symbols found on the k-axis then are crystal-symmetry dependent and each represents one

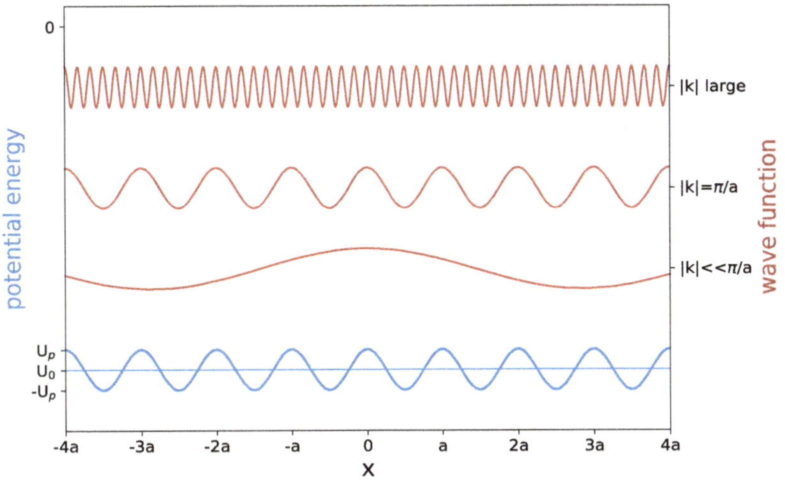

Figure 9.1.1: Periodic potential in one dimension and wavelength of the wave function, following [1]. Blue: potential energy; red: wave function.

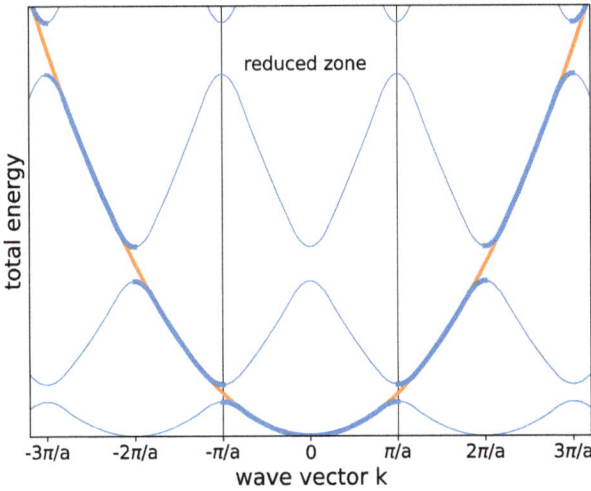

Figure 9.1.2: Development of band gaps and definition of a reduced zone in a one-dimensional crystal, following the treatment in [1]. Blue, strong: original scheme; orange: no periodic potential, no gap; blue, thin: periodically extended scheme.

of the typical directions within a three-dimensional periodic potential (see Figure 9.1.3 for two examples).

The electronic and optical properties of a semiconductor at low temperature are dominated by two specific bands and the gap in between: the "valence band" (VB) with the energetically highest lying yet fully occupied electronic states and the "conduction band" (CB) with the energetically lowest lying unoccupied ("empty") electronic states. The CB may be populated by optical or thermal excitation as well as by electrical carrier injection; removing electrons from states in the VB on the other hand will leave a "hole" there. Electrons in the CB and holes in the VB are mobile in the whole crystal (following eq. (9.1.2)) and are characterized by "effective masses" and mobilities. As a rule, holes in the VB have higher effective masses than electrons in the CB and thus are typically less mobile.

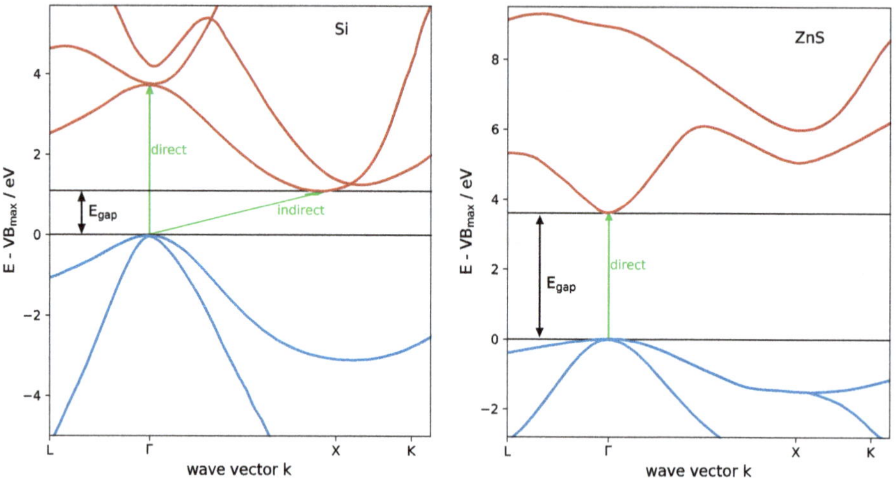

Figure 9.1.3: Band structures of cubic silicon and cubic ZnS (redrawn with data extracted from [2]). Blue: occupied bands; red: unoccupied bands.

Figure 9.1.3 shows another important feature of semiconductors: the energetic top level of the VB may be situated at the same k-value as the bottom level of the CB ("direct semiconductor", for example, ZnS) or at a different position ("indirect semiconductor", for example, Si). This has important consequences for applications: optical absorption as well as luminescence (optical de-excitation) are much more intense in a direct than in an indirect semiconductor; since momentum in these processes has to be conserved, the optical processes over the band gap in the indirect semiconductor have to couple with lattice vibrations ("phonons") in order to adjust the momentum difference.

One final important point to understand the nature of semiconductors can be recognized already in Figure 9.1.1: the value of U_0 depends on the chemical nature of the material, and thus the depth of the potential will vary strongly. When comparing different semiconductors, this will make a big difference – the absolute value of the band energy relative to the vacuum level of free electrons does indicate the chemical properties: the higher an electron is situated on this scale, the stronger reductive it is chemically, and the lower a hole is situated, the stronger its oxidative power is. We can even integrate the electrochemical standard reduction potentials here, since their positions on the energy scale relative to vacuum are well known. Figure 9.1.4 shows this effect for some common examples – we can immediately see that some semiconductors after excitation offer extremely reductive electrons in their CB (ZnTe e.g. is capable to reduce CO_2 to its radical anion, the first step of electrochemical CO_2 reduction), whereas others provide very oxidative holes in the VB capable to oxidize water to O_2. These capabilities are important parameters when designing photocatalysts e.g. for water splitting: both water reduction and water oxidation are necessary there, and after a look into Figure 9.1.4 it is obvious that this will typically need more than one kind of semiconductor and thus complex composites.

Now that we understand the origin and nature of band gaps, we may realize that the band gap can be populated with electronic states as soon as we disturb the strict periodicity, for instance, by suited defects or dopants. Depending on their redox activity (accepting or delivering electrons), these states will be closer either to the VB edge or the CB edge – in the former case, a hole conductor has been created ("p-doping"); in the latter one, an electron conductor is the result ("n-doping"). The *Fermi* energy E_F (chemical potential of the electrons) in an undoped semiconductor will be found approximately in the middle of the band gap – but after n-doping it is moving closer to the conduction band, and after p-doping closer to the valence band.

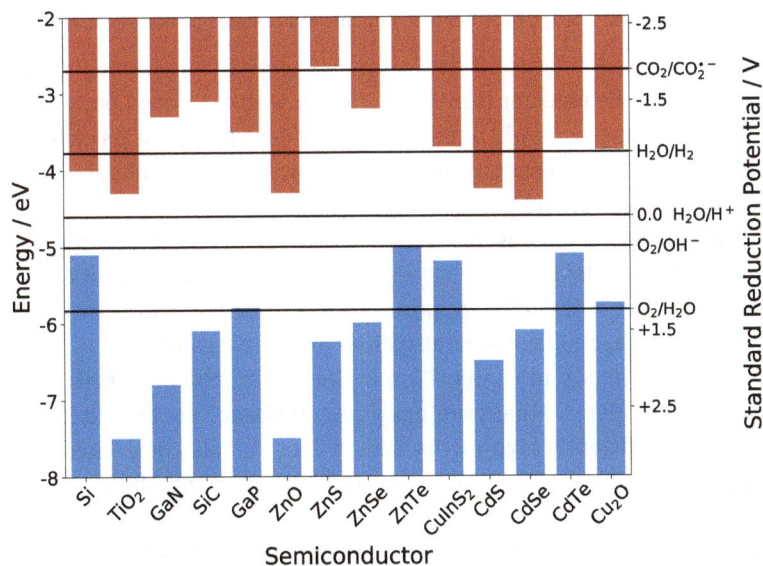

Figure 9.1.4: Positions of valence bands, conduction bands (redrawn, data extracted from [3–6]) and standard reduction potentials relative to the vacuum level for some common examples.

In our view of quasi-free electrons (eq. (9.1.2)) in a large (crystal) well every band does contain a large but limited number of states represented by a statistical quantity: the "density of states" in an energy interval, or DOS for short. It is largest at the (energetic) center of the band and fades to zero at the band edges – and of course inside the band gap. Figure 9.1.5 does show this schematically, together with two dopant states: those neighboring the conduction band (n-doping) are called "donors" and those close to the valence band (p-doping) are called "acceptors".

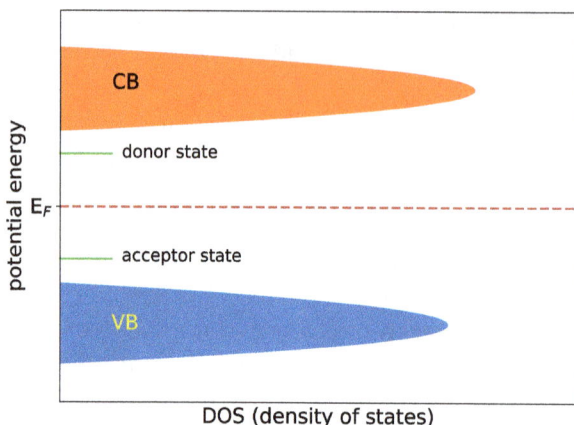

Figure 9.1.5: Density of states and energy, together with dopant/defect states (schematically).

If an n-doped part of a semiconductor is in direct contact with a p-doped part, electrons and holes in the contact zone will annihilate and a "depletion zone" is created. Such a "pn-junction" (or its combinations) is the very basic element of all microelectronics (diodes, transistors, etc.) and thus the foundation of all kinds of modern electronics. Photovoltaic elements also usually rely on pn-junctions: light absorbed in the depletion zone will generate pairs of electrons and holes, which are then first separated from each other due to the different positions of E_F relative to CB and VB in the two parts of the junction, and then are transported to external contacts.

Doping, defects or variation in composition under certain circumstances may lead to such a high carrier density that eventually new sub-bands are formed in the former energy gap. At sufficiently high carrier density they can also be excited collectively like in many metals – surface plasmon resonance can then be observed and utilized for near-field effects [7].

Keeping the general concepts of this chapter in mind, we will now step through some families of semiconductors with technical importance. It is impossible to cover all semiconductors here comprehensively, so we will focus on the most important ones. Some families with lesser importance (e.g. lead chalcogenides or copper oxides for electrocatalysis), therefore, will be completely omitted.

9.1.2 Workhorses Si and TiO$_2$

For silicon (Si) as the present major semiconductor for microelectronics and photo-voltaics, very accurate optical and electrical data are available. Using a critical compilation in [8], Figure 9.1.6 shows the optical data from the UV to the IR wavelength region.

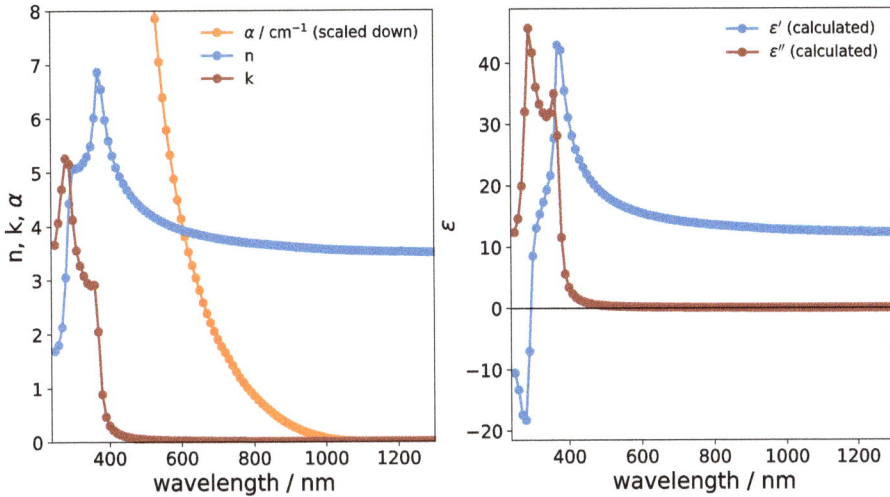

Figure 9.1.6: Optical properties and dielectric function of intrinsic silicon (drawn and computed using data taken from [8]).

The indirect nature of Si is obvious (compare also Figure 9.1.3): optical absorption α is starting to increase at around 1000 nm at the indirect band gap, but that is barely seen in the curves of the extinction coefficient k or the imaginary part of the dielectric function ϵ''. The direct gap in the near UV however leads to such a strong absorption that the real part of the dielectric function ϵ' turns negative in the UV: the material becomes reflective.

Figure 9.1.6 does also show another important feature present in all semiconductors: when approaching the wavelength of the direct band gap, relative permittivity and refractive index are increasing steeply – pigments under these conditions are scattering strongly and, therefore, have high "hiding power".

The exact location of the (indirect) optical band gap can be extracted from sufficiently accurate and complete experimental data by a so-called Tauc plot (originally designed for amorphous semiconductors [9], for a critical discussion of its general applicability see [10]), which has to be adjusted depending on the direct or indirect nature of the gap (E is the photon energy and B is an adjustable fit parameter):

$$\text{indirect: } (\alpha E)^{1/2} = B(E - E_G) \tag{9.1.3}$$

$$\text{direct: } (\alpha E)^2 = B(E - E_G) \tag{9.1.4}$$

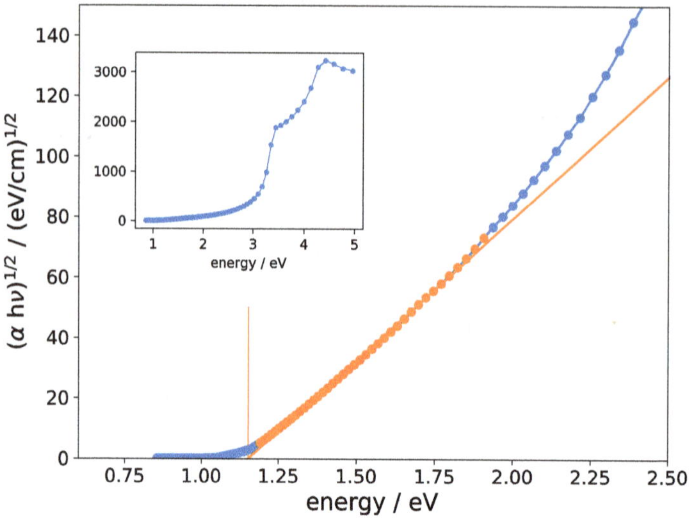

Figure 9.1.7: "Tauc" plot of absorption (data taken from [8]); inset: full *Tauc* transform up to 5 eV.

Figure 9.1.7 shows this for the indirect fundamental band gap of silicon, using the same data as in Figure 9.1.6: the extrapolation of the linear part in the plot to $\alpha E = 0$ leads to the correct optical band gap of $E_G = 1.15$ eV. The data belong to ultrapure crystalline silicon: there is basically no sub-band gap absorption caused by defects or impurities. In order to fabricate silicon on an industrial scale on such a high level of quality, enormous efforts are needed, but they are absolutely necessary and well established for present-day industrial applications in microelectronics and photovoltaics. "Tauc" plots in this context are an indispensable tool to report realistic optical band gaps and crystal quality for semiconducting materials in general.

The indirect nature of the band gap of silicon has an important practical consequence: optical absorption in the visible and near-IR part of the optical spectrum is comparably weak; therefore, photovoltaics based on silicon has to use very thick layers for complete absorption of the solar spectrum and thus needs huge material investment (in terms of primary energy, for details, see [11]). "Solar silicon" however does not need to be as pure as silicon for microelectronics; therefore, it is produced separately in order to avoid the energy-intensive purification processes as far as possible. On the other hand, there is also a variant of silicon with direct band gap: amorphous silicon saturated with hydrogen. This material allows to fabricate thin-film photovoltaic elements based on silicon, but unfortunately, they suffer from long-term chemical instability as well as from low conductivity so that the final efficiency is far below the level of crystalline silicon.

Silicon (and historically also germanium) are central for established microelectronics and photovoltaics, but for the more passive role as a (direct) semiconducting pigment TiO_2 is the most important material. These days, it is produced globally in large quantities, mostly as white pigment. However, TiO_2 is also an important photocatalyst [12, 13], mostly due to the very high oxidative power of its VB holes (see Figure 9.1.4). This property has advantageous and disadvantageous consequences: TiO_2 can be used for self-cleaning surfaces (surface holes decompose organic matter, including biochemical one) but will (unprotected) also oxidize organic binders and additives in coatings and plastics on the long run, when exposed to daylight. Pigment qualities of TiO_2, therefore, are always surface-passivated, typically by a coating with (hydrated) alumina and/or silica.

TiO_2 is applied in two different modifications: anatase (band gap at ca. 3.2 eV) and rutile (band gap at ca. 3.0 eV). At optical frequencies, rutile has the higher refractive index (in a powder on average 3.1), compared to an average of 2.5 in anatase, in accordance with the principle visible in Figure 9.1.6 and, therefore, is the preferred modification for use as a white pigment (with maximal hiding power at particle sizes of 200–300 nm). The use of such pigments is not limited to coatings and paints: under the code of "E171" it is allowed to be used in food, albeit under critical discussion for some time already. Anatase, on the other hand, is photocatalytically more active and thus is the preferred modification when it comes for instance to self-cleaning surfaces, which use the UV part of daylight to generate aggressively oxidizing holes on their surface.

TiO_2 is also produced commercially in nanoparticulate form, mostly from the gas phase, and under such conditions typically is a mixture of rutile and anatase, with particle sizes on the order of 20–40 nm. The scattering power of these small particles is low, but as a direct semiconductor they are still strongly absorbing UV radiation and thus find application in cosmetics (sun lotions) or body care (tooth paste). More important, however, is the photocatalytic activity of these materials typically branded as "P25", and their use as electron conductor: dye-sensitized solar cells for instance work best with nano-TiO_2 as dye carrier as well as (electron) current collector.

9.1.3 Established: binary and ternary semiconductors based on the II-VI and III-V families

Apart from silicon and titania, the semiconducting members of the II–VI (typical member: ZnS) and III–V (typical member: GaAs) families of materials have found wide application in electronics and optics. Quite striking is the fact that all these materials (like silicon) crystallize in tetrahedrally coordinated structures, mainly of the sphalerite and wurtzite type. The result is that extensive ranges of solid solutions are possible between the members of these families, opening an enormous width of possible compositions. In many cases, by mixing two or three members, band gaps and lattice parameters can be set independently (called "band gap engineering"), allowing for epitaxial growth of complex structures and layer packages. This ability has led to the development of laser diodes as well as high-efficiency LEDs, which dominate now lighting technology at least in the interior. The basic idea here is a double "heterojunction": in an epitaxially grown layer stack there are two wide band gap layers (one p-, the other n-doped) and a central thinner small band gap layer. Energetically, they are all "pinned" by a common *Fermi* energy: electrons injected into the conduction band system then will automatically be concentrated in the innermost layer; the same applies for holes injected into the valence band system. With electrons and holes concentrated in a small volume of only some dozens of nanometer thickness, radiative recombination becomes very effective – this is the basis for light output with high electrical efficiency. Figure 9.1.8 shows a sketch of this mechanism. Very obviously, the mechanism can only work if there are no additional barriers between the various layers – this requirement calls for perfect epitaxial growth and thus is only possible under the conditions of band gap engineering. Laser diodes follow a similar mechanism but usually contain a much more complex stack of layers.

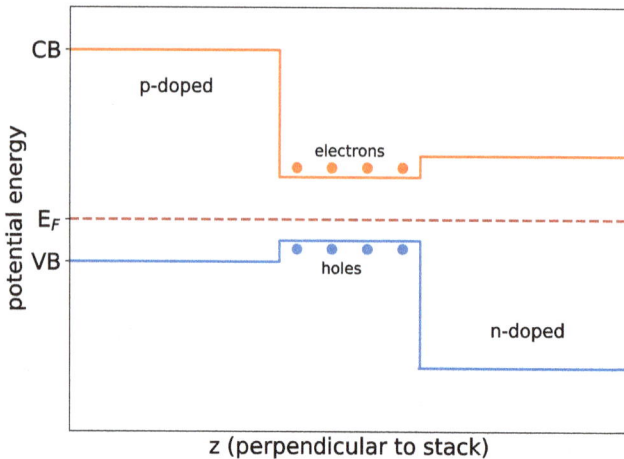

Figure 9.1.8: Cartoon of a double heterojunction in an LED (not to scale). Layers are "pinned" by a common Fermi energy.

However, the wide band gap members like GaN have not only led to a revolution in the lighting sector but also offer new approaches for high-voltage electronics needed, e.g. in high power applications. Another issue that might be resolved by III-V semiconductors is active electronic components for the THz region, which is not accessible by conventional materials: InP is one of the candidates for this purpose.

An important part of the success story of III-V semiconductors in the form of epitaxially grown stacks is the fact that all members of the family can be deposited with high crystal quality from the gas phase, using metal-organic precursors: Al, Ga, In all form very reactive liquid trimethyl compounds with high vapor pressure. Making (and switching) complex gaseous mixtures is thus very easy, facilitating epitaxial growth of complex heterogeneous stacks needed for laser diodes or high-efficiency LEDs, known as "metal organic chemical vapor deposition". To illustrate the chemistry used in this process family, a typical gas reaction for the deposition of solid GaAs is: $Ga(Me)_3 + AsH_3 \rightarrow GaAs + 3\ CH_4$.

Many members of the II-VI and III-V families are direct semiconductors and thus for long have been used as strongly absorbing pigments: ZnO is used as UV-protective additive in polymers as well as in cosmetic sunscreens, and ZnS with its comparatively high index of refraction was an important white pigment before this job was taken over mainly by TiO_2. CdS is a very brilliant yellow (or in solid solution with CdSe orange) pigment, which can also be found as optical filter when embedded in glass. But there are also applications requiring very small band gaps – a very prominent one is IR detection (using photoconductivity) at wavelengths around 10 μm, which allows "night vision" using blackbody emission, e.g. for military purposes. It turned out that solid solutions of the semimetals HgTe and HgSe with CdTe were excellent for this job, and (Hg,Cd)Te, therefore, has been investigated thoroughly. But there is an alternative to tune the band gap from semimetal to semiconductor: in nanocrystals, the band gap can be widened very precisely, just by setting the particle size. This approach allows to use pure mercury chalcogenides as narrow band gap materials – an excellent review of methods and properties has been published in 2021 [14]. We will have a more systematic look into the science of so-called quantum dots in the next chapter.

There are also technically interesting examples mixing the II-VI and III-V families to III-VI variants: compounds like Ga_2Se_3 or Ga_2S_3 tend to crystallize in defect blende or wurtzite types, but together with Cu_2S the defects may be filled leading to structures of the chalcopyrite-type (a superstructure of blende): $CuInS_2$, $CuGaS_2$ and their solid solutions, also (partially) selenized: $Cu(In,Ga)(S,Se)_2$. The latter one is known as "CIGS" and (as a suited direct semiconductor) has been investigated extensively in thin-film photovoltaics. The II-VI family does offer a candidate of its own for thin-film photovoltaics as well – CdTe, but suffering both from low abundance of Te and toxicity of Cd.

9.1.4 Semiconductor quantum dots and related nanostructures and nanocomposites

Semiconductor particles and structures with dimensions on the nanoscale have attracted enormous interest, because the optical and electrical data can be fine-tuned simply by control of the particle size, the particle surface and the internal structure.

This has offered unprecedented opportunities for novel applications not only in the field of optics and electronics but also for photo- and electrocatalysis. Enormous progress has been made in this field; therefore, we will have a closer look here into some examples, together with the physicochemical background and the theoretical understanding.

In nanoparticles, the periodic nature of a semiconductor crystal as outlined in this chapter is definitely disturbed by the limited number of atoms involved. Electrons and holes after excitation or injection will be confined to quite small volumes and thus may feel strong interaction. Fortunately, the dependence of the band gap energy on the particle size can be understood in a quite simple theoretical approach called "effective mass theory". It considers corrections to the bulk band gap energy by kinetic energies, *Coulomb* attraction of carriers and correlation of electrons and holes, respectively, as a function of particle radius r, as described by eq. (9.1.5) (m_e^* and m_h^* are the effective masses of electron and hole, respectively, all other symbols have the usual meaning) [15]:

$$E_G = E_G(\text{bulk}) + \frac{h^2}{8m_e r^2}\left(\frac{1}{m_e^*} + \frac{1}{m_h^*}\right) - 1.8\frac{e^2}{4\pi\epsilon_0\epsilon_r r} - 0.25\frac{e^4 m_e}{8\pi^2\epsilon_0^2\epsilon_r^2\hbar^2}\left(\frac{m_e^* m_h^*}{m_e^* + m_h^*}\right) \quad (9.1.5)$$

After optical excitation of a semiconductor, electrons and holes may be loosely bound in pairs, forming a mobile "exciton". The radius of such an exciton (so-called *Bohr* radius a_B) can be estimated in the same framework and marks the difference between "strong" and "weak" confinement of charge carriers (eq. (9.1.6)). The basic idea here is that so-called strong quantum confinement happens when the particle size is getting close to or even smaller than a_B; in slightly larger particles, "weak" confinement has to be considered, meaning that excitons still have limited mobility, as compared to the bulk case. In the case of "strong confinement" in spherical particles, the system is often called a "quantum dot" (QD). Equation (9.1.6) tells us that a_B is decreasing with the effective mass of the carriers; strong confinement thus happens on different length scales, depending on the nature of the material under investigation:

$$a_B = \frac{4\pi\epsilon_0\epsilon_r\hbar^2}{m_e e^2}\left(\frac{m_e^* + m_h^*}{m_e^* m_h^*}\right) \quad (9.1.6)$$

Since researchers first described the size-dependent properties of nanometer-sized semiconductor particles, a variety of different synthesis strategies emerged to produce them; an early review was published already back in 2001 [16]. They can be divided into two main approaches: the top-down approach which involves the breaking of bulk material down into nanosized structures or particles (e.g. by e-beam lithography or X-ray lithography), and the bottom-up approach which involves the building up of a material atom-by-atom (e.g. by self-assembly in solution following a chemical reaction producing precursors, or in the gas phase). The development of monodispersed colloidal QDs produced from the latter approach has been proven to be particularly interesting in the area of solution-processable optoelectronic devices. For example, well-established coating and printing methods like spin and blade coating as well as ink-jet and screen printing may pave the way for cost-efficient fabrication of large-area flexible devices.

Currently, most of the strategies to synthesize such monodispersed nanocrystals are based on the seminal work of LaMer and Dinegar, who explained how their formation is dependent on a rapid nucleation followed by the controlled growth of the existing nuclei, by means of a thermodynamical approach [17]. In classical nucleation theory, the change of the *Gibbs* free energy ΔG during the nucleation process arises from the formation of new volume (ΔG_V) and the energy required to create new surface area (ΔG_S), and is therefore related to the radius of the nuclei (r), the specific surface energy (σ), the density of the solid material (ρ) and the change in the chemical potential ($\Delta\mu$). $\Delta\mu$ in turn is basically a function of the supersaturation (S), which is the driving force for nucleation:

$$\Delta G = \Delta G_V + \Delta G_S = \frac{4}{3}\pi r^3 \rho \Delta\mu + 4\pi r^2 \sigma \tag{9.1.7}$$

$$r_c = \frac{2\sigma}{\rho|\Delta\mu|} \tag{9.1.8}$$

with

$$\Delta\mu = -kT\ln S \tag{9.1.9}$$

and

$$S = \frac{\text{ionic activity product } (a)}{\text{solubility product } (K)} \tag{9.1.10}$$

As shown in Figure 9.1.9, the curve of the cluster *Gibbs* free energy has a maximum at a radius r_c (critical radius), due to the negative (binding energy in the bulk material) and positive (surface energy) terms. Once nuclei reach sizes with $r > r_c$ they have overcome the activation barrier and will further grow and form stable entities, whereas for clusters smaller than r_c growth is unfavorable and dissolution is more probable.

The basic idea behind LaMer and Dinegar's concept for monodispersed nanoparticle formation now is to separate nucleation and growth processes, which ultimately enables control of the particle size as well as the particle size distribution, as displayed in Figure 9.1.9. The mechanism is as follows: first, the concentration (or more accurately the activity) of the precursors is increasing till it reaches the saturation concentration. Due to missing nucleation seeds/sites, the energy barrier for nucleation cannot be overcome yet. A supersaturated solution is the result and the concentration further increases up to a critical concentration where spontaneous nucleation

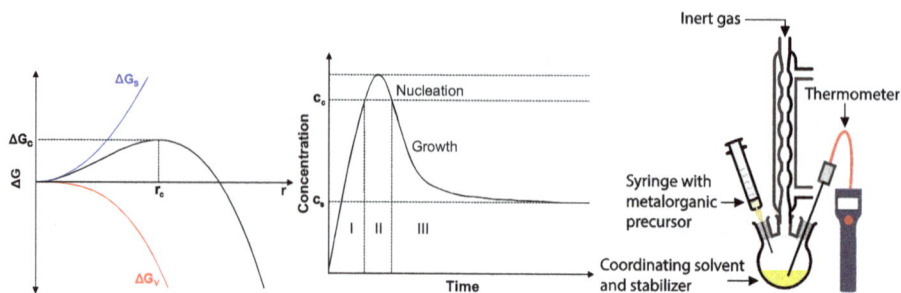

Figure 9.1.9: Left: Dependence of the particle free energy (ΔG) on the radius (r), with a maximum free energy (ΔG_c) at a critical particle size (r_c). Middle: LaMer and Dinegar's model used to describe time- and concentration-dependent nucleation (II) and growth (III) of nanoparticles. Right: Simple schematic representation of an apparatus used to prepare monodispersed nanoparticles by the "hot-injection" method.

without seeds is possible, leading to a rapid self-nucleation (so-called burst nucleation). This immediately lowers the supersaturation level below the critical concentration, which leads to crystal growth of the existing clusters without additional nucleation. Using this approach, many colloidal synthesis strategies were developed like the "hot-injection synthesis", "heat-up method", "cluster-assisted method" and "continuous-flow method" (for the interested reader we recommend [18]). The hot-injection synthesis remains till date the most common method to produce monodispersed QDs of high quality. By quickly injecting organometallic reagents into a hot solvent, burst nucleation occurs as described above in a short period of time. Further, nucleation is inhibited so that the width of the particle size distribution can be controlled by the injection rate (with faster injection rates leading to narrower size distributions and vice versa). The final or desired particle size of the nanocrystals can afterwards be altered simply by varying the growth time. It should also be noted that the reaction solution in hot-injection synthesis contains surfactant molecules (ligands), primarily to prevent the produced nanoparticles from agglomerating. Secondarily, these ligands (more accurately the terminal group of the ligands) play an essential role in determining the surface chemistry of the particles, ranging from specifying their hydrophilicity/hydrophobicity up to their applicability in optoelectronic devices (e.g. define the electronic coupling to other components like electrodes).

The surface-to-volume ratio in nanometer-sized materials like quantum dots is especially high, resulting in a high chemical potential and thus in a high reactivity, and secondly in an excessive population of surface atoms that can negatively influence the optoelectronic properties of the material (e.g. forming non-radiative decay channels, thus reducing the luminescence quantum yield). The most common solution to address this problem is by growing epitaxial layers of a suited inorganic material over the quantum dot so that the surface atoms are passivated and the nanoparticle becomes protected against environmental influence. The theoretical background for

the synthesis of such structures is the same as discussed above, with the only exception that the nanocrystals that are supposed to be passivated now act as nucleation
sites. So when new organometallic precursors are added to the synthesis solution,
epitaxial crystal growth on existing particles is observed rather than newly formed
nuclei. The result is referred to as a core-shell structure (see Figure 9.1.10, core =
quantum dot; shell = passivating layer around). Depending on the band gap and the
relative position of the valence and conduction band edges with respect to vacuum level, core-shell structures are classified into different categories as presented in Figure 9.1.10. Type I structures are obtained when the shell material has
a larger band gap than the core material with band edges of CB and VB higher
and lower in energy, allowing to confine excitons to the core region and shielding
them from the environment. Reverse type I structures are obtained when the band
gap of the shell is smaller than the core with band edges of CB and VB lower and
higher in energy, which leads in contrast to normal type I structures to a partial
delocalization of charge carriers within the shell. In type II configuration only one
charge carrier (electron or hole) is confined to the core region while the other is
primarily located in the shell. This is achieved when both the valence band maximum and conduction band minimum of the core material are both higher or lower
than the band edges of the shell material. Depending on the desired application
(e.g. in solar cells, as catalyst material and in light-emitting application), the confinement of charge carriers to the core region or vice versa the delocalization
within the shell material can be tremendously beneficial. However, the choice of
the inorganic coating material is limited to those that not only fulfill the configuration for the desired core-shell type but also have the same crystal structure as
the core material and a small lattice mismatch to enable epitaxial growth.

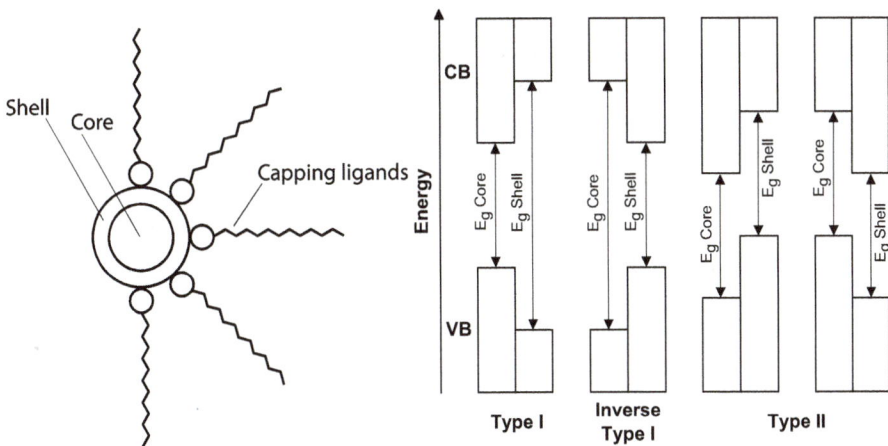

Figure 9.1.10: Simplistic representation of a core-shell particle capped with stabilizing ligands and
energetic band alignment for different types of core-shell structures.

Given the potential areas of application, the prediction of optical and electronic properties (like band gaps, band edges and density of states) is of particular interest. With decreasing costs and increasing power and availability of computational resources, ab initio approaches like the well-established DFT in principle are able to provide such valuable insights. For reviewing the foundations and approximations of DFT, we refer interested readers on that topic to [19, 20]. When determining the properties of nanomaterials theoretically with this approach, a few remarks should be considered. In DFT calculations due to the so-called self-interaction error (which arises from the overall mean-field approximation), band gap and electronic structure calculations often deviate massively from experimental findings. In standard DFT this error over-delocalizes occupied states, forcing them up in energy and leading to an underestimation of the band gap energy.

Table 9.1.1: Comparison of DFT, DFT + U and experimental results for ZnS and CuInS$_2$ (reprinted from [21] under CC-BY-NC-ND 4.0).

Compound	Method	E_g (eV)	Ed (eV)	a (nm)	c (nm)	EA (eV)	IP (eV)
ZnS	DFT	2.03	6.01	0.5435	–	3.33	5.36
ZnS	DFT + U	3.57	9.12	0.5423	–	3.13	6.70
ZnS	Exp.	3.6 [22]	9.10 [23]	0.542 [24]	–	3.28 [25]	6.82 [25]
CuInS$_2$	DFT	0.02	14.97 (In)	0.5575	1.1235	4.50	4.52
CuInS$_2$	DFT + U	1.56	18.08 (In)	0.5536	1.1205	4.18	5.74
CuInS$_2$	Exp.	1.53 [26]	18.2 (In) [27]	0.5523 [28]	1.112 [28]	3.55 [29]	5.05 [29]

Empirically it has been found that some hybrid functionals (mixing pure DFT with wave function theory results) may compensate the self-interaction error, thus correcting the band gap. While suited hybrid functionals may reproduce the band gap and electronic structure accurately enough, other (morphological) properties then may be predicted unreliably. Another, computationally even cheaper, method to correct for the self-interaction is to introduce a Hubbard U parameter (performing a DFT + U calculation). In short, the U parameter is a potential that favors localizing the states to which it is applied and thus lowering them in energy, correcting the electronic structure. In Table 9.1.1, such a correction and improvement from standard DFT towards more accurate results with DFT + U is shown for two important materials, namely ZnS and CuInS$_2$. Not only the electronic structure in form of the band gap (E_g) and energetic position of the d-orbitals with respect to the valence band maximum (Ed) is predicted well, also other properties like lattice parameters (a and c), electron affinities (EA) and ionization potentials (IP) are reproduced with reasonable accuracy. This shows that reasonably adjusted standard DFT methods are available these days to model and design complex semiconductor-based nanostructures.

For the modeling of nanostructures, a suitable (virtual) surface passivation method is needed; otherwise, states arising from the unsaturated dangling bonds of the surface atoms would dominate the electronic structure and lead to unphysical results. Especially easy to passivate are group IV semiconductors like Si or Ge. Crystallizing in the diamond structure, each atom is tetrahedrally coordinated and surrounded by four other atoms, so each atom provides one electron to form one Si–Si (or Ge–Ge) bond. Hydrogen can be used, both experimentally and theoretically, as a passivating agent because it provides also one electron per bond generating a natural structural and electronic termination. However, other technologically important semiconductors consisting of II-VI or I-III-V elements are not so easy to manage at their surface. Let us consider, for example, the above-mentioned compounds ZnS and $CuInS_2$ that crystallize in derivatives of the diamond structure: sphalerite (ZnS) and chalcopyrite ($CuInS_2$). Zn, Cu and In are all tetrahedrally surrounded by four S atoms, which means they provide (considering their formal valence) 0.5 (Zn), 0.25 (Cu) and 0.75 (In) electrons per bond with an S atom. To generate a suitable electronic termination, so-called fictitious hydrogen atoms are used with a formal valence of 1.5 (for termination of Zn bonds), 1.75 (for Cu) and 1.25 (In) to form bonding orbitals occupied by exactly two electrons. Obviously, hydrogen atoms with a fractional charge do not exist physically, but as it can be seen in Figure 9.1.11 that they are a helpful tool in DFT calculations as a passivating agent to predict size-dependent properties. In Figure 9.1.11, calculated band gaps of differently sized ZnS and $CuInS_2$ QDs are shown (DFT + U calculation plus the surface passivation with fictitious hydrogen atoms) and compared to experimental results, as well as to the size dependency calculated from the abovementioned "effective mass theory" equation. The viewgraph again highlights the small remaining mismatch between properly performed DFT calculations and experimental findings, and the suitability to predict such properties.

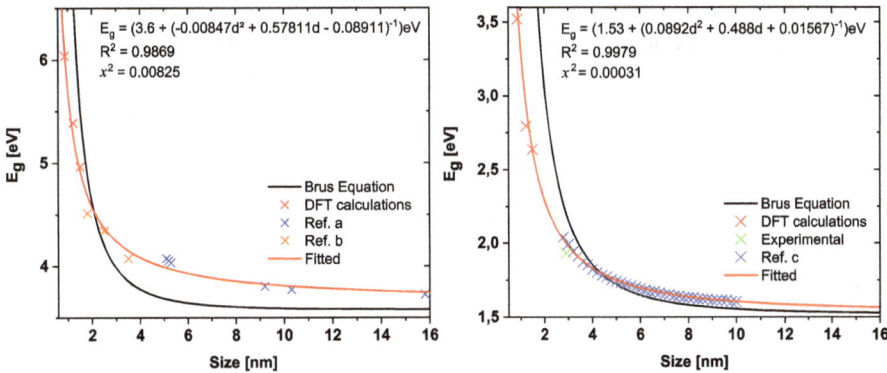

Figure 9.1.11: Band gap dependency with respect to the particle size obtained from DFT + U calculations, experimental values and calculated by means of the Brus equation for ZnS (left) and $CuInS_2$ (right) (reprinted from [21] under CC-BY-NC-ND 4.0).

To date, virtually all kinds of semiconductors have been prepared and described in the form of QDs, but especially CdSe has been fabricated with excellent control of particle size, surface chemistry, stoichiometry and reproducible electronic and optical properties and thus has grown into kind of a benchmark. The excellent control of properties and scalability of fabrication has also led to a discussion whether semiconductor QDs may be a platform for quantum information science [30], but this field is still in its infancy at the time of writing. Nevertheless, the excellent control of morphology has already led to another level of optical properties, namely in colloidal photonic crystals – uniform QDs in this case are spontaneously forming ordered superstructures in which the optical and electronic properties of the QDs can be combined with interference effects caused by the regular arrangement [31].

Semiconductor QDs have found their first commercial applications in some TV sets: using CdSe-based nanoparticles with precisely defined diameters, these dots are used as luminescence converters in order to push the light of blue (or UV)-emitting backlight LEDs into exactly those spectral ranges that match perfectly the color filters of an otherwise unmodified LCD screen, thus leading to improved brilliance, color gamut and contrast of the display. Quite recently, it became clear that the toxic CdSe particles can be substituted by InP without compromising the optical quality. The key for this is a complex system of shells around the core nanoparticle, modulating the optical properties as well as the colloidal behavior [32]. It even may be possible to get rid of the In component and use more earth-abundant components, namely, solid solutions of ZnSe and ZnTe, again tuned with multiple shells of other II-VI components [33].

Future applications of QDs are to be expected as a marker in medical applications (due to their potential rich surface chemistry and the ability to be excited by two- or three-photon absorption in the near infrared [34]) and most prominently in next-generation LEDs and displays: direct injection of electrical charge carriers into the particles appears to be feasible [35, 36], opening the route for filter-free brilliant and bright all-inorganic displays with high resolution – features that up to date are only possible with organic emitters (OLEDs), which (especially for the blue-emitting component) have problems with their service lifetime. Since surface-tuned QDs can be integrated into polymers, we may also expect to see novel approaches in device fabrication, e.g. by methods of additive manufacturing [37].

QDs based on II-VI materials may also mark their impact in photovoltaics: instead of using an organic or inorganic dye in sensitized TiO_2-based cells QDs have been explored as sensitizer and show high potential [38], at least once the use of Cd as one of the ingredients has been overcome.

9.1.5 Emerging: quaternary semiconductors and perovskites with organic cations

Many of the technologically most interesting compound semiconductors especially with band gaps in the optical region have a major disadvantage: one or more of their constituting elements are either toxic (Cd, As, Hg, etc.) or not very abundant (Ga, In, Te, etc.). The quest for earth-abundant and non-toxic semiconductors useful e.g. for cheap and efficient photovoltaics, therefore, is still open and has led to more and more complex compositions. Typical examples for such developments are Cu_2ZnSnS_4 and $Cu_2ZnSnSe_4$ – they crystallize in the kesterite or stannite structure (which again are superstructures of the zinc blende type), are direct semiconductors, have a tunable band gap around 1.5 eV and thus should be perfect candidates for thin-film inexpensive photovoltaics. However, the more components need to be used for preparation, the more demanding the process control becomes. Simultaneous control of stoichiometry, defects, crystal quality, etc. is mandatory for reproducible high-quality films, but difficult to achieve. The potential of this class of materials thus still has to be developed further.

Among all newly described and promising semiconductors of the last decade, there is one family that has been found basically by serendipity but with a very large impact: organic/inorganic perovskites. Initially reported in 2009 as potential dyes for dye-sensitized solar cells, it was soon discovered [39] that they constitute an extremely versatile family of semiconductors on their own with tunable optical band gaps in the visible region (see e.g. the compilation in [40]), very high carrier mobility and a striking ease of fabrication: processing from (organic) solution, followed by heat treatment appears to be sufficient to produce high-quality films of these materials. Stacked with an electron conductor and a (mostly organic) hole conductor to extract carriers, they form a simple yet extremely well-working photovoltaic cell. Soon they beat records on photovoltaic efficiency (about 23% as of 2019) and were also found to be interesting nanocrystalline light emitters. Nevertheless, there are two serious drawbacks: only the toxic Pb-containing variants offer high performance, and nearly all films still lack the necessary durability in humid and oxidizing atmosphere that is necessary for prolonged outdoor applications [40].

The archetypical member of the family is methylammonium triiodoplumbate, $(CH_3NH_3)PbI_3$, and with a direct band gap of about 1.55 eV it is an optimal absorber for photovoltaics (for a theoretical assessment of the electronic structure, see [41]). It is related to the general formula of simple perovskites (ABO_3) by using the organic cation as the (large) A-cation, and Pb^{2+} as the (small) B-cation. At high temperature (330 K and more), the crystal structure is cubic (original perovskite type), but around room temperature it is slightly distorted to a tetragonal variant [42]. This polymorphism leads to some dependency of the properties on the experimental conditions under which the materials are deposited as films. Instead of methylammonium, other even larger organic cations like formamidinium (or Cs^+ from the inorganic world) may be used to fine-tune the crystal structure and thus the electronic properties of these semiconductors. The anion may also be (partially) substituted by chloride or bromide, which not only allows fine-tuning of optical properties but also leads to somewhat higher chemical stability. Pb^{2+}, however, is much more difficult to replace; there was some success with Sn^{2+}, but unfortunately Pb seems to be crucial for high mobility of electrons as well as holes. The theoretical reasons for this behavior are not obvious, but it appears that relativistic effects need to be considered: high-level computations of the band structure show that the differences between Sn-based and Pb-based perovskites are based on strong spin-orbit coupling and lone-pair effects in the heavy Pb^{2+} ions [43]. As a result, the Pb-based perovskites are more stable against oxidation (due to a VB band edge at lower energy) and are good conductors for electrons as well as holes.

After the impact in photovoltaics it was soon realized that perovskites also have huge potential for light generation in LEDs or displays, especially with the Cs^+-containing variant, also in nanoparticulate form [44, 45]. It remains to be seen whether the organic/inorganic perovskites will have an impact on industrial application or not – the impressive performance increase shows the potential but the practical limitations at the time of writing (2021) were still immense.

9.1.6 Summary

Semiconductors are at the heart of modern technology – be it in microelectronics, light generation, photovoltaics, power generation, pigments or photocatalysis: the unsurpassed ability to interact both with optical radiation and electrical phenomena is lending itself to a broad variety of applications. From the point of view of materials science, the challenge in most cases is to prepare well-crystallized materials with low defect density in the first place (to establish a well-defined and empty band gap), followed by doping procedures or other measures of defect chemistry in order to generate exactly those states in the band gap that are needed for various applications. The morphology of applied semiconductors varies from large single crystals over complex stacks of epitaxially grown materials to nanometer-sized quantum dots, all with their specific fields of application. With suited state-of-the-art quantum chemical methods, morphology as well as electronic and optical properties of such systems can be predicted, at least qualitatively. Future developments will see more and more integration of semiconductors (especially in their nanoparticulate forms) into "soft" organic or even biochemical matter, making this class of materials even more versatile.

References

[1] Myers HP. Introductory Solid State Physics, Taylor & Francis, London, 1990.
[2] Cohen ML, Bergstresser TK. Band structures and pseudopotential form factors for fourteen semiconductors of the diamond and zinc-blende structures. Phys Rev, 1966, 141, 789–96.
[3] Grätzel M. Photoelectrochemical cells. Nature, 2001, 414, 338–44.
[4] Wöhrle D, Tausch MW, Stohrer W-D. Photochemie, Wiley-VCH, Weinheim, 1998.
[5] Van de Walle CG, Neugebauer J. Universal alignment of hydrogen levels in semiconductors, insulators and solutions. Nature, 2003, 423, 626–8.
[6] Luo J, Tilley SD, Steier L, Schreier M, Mayer MT, Fan HJ, Grätzel M. Solution transformation of Cu_2O into $CuInS_2$ for solar water splitting. Nano Lett, 2015, 15, 1395–402.
[7] Agrawal A, Cho SH, Zandi O, Ghosh S, Johns RW, Milliron DJ. Localized surface plasmon resonance in semiconductor nanocrystals. Chem Rev, 2018, 118, 3121–207.
[8] Green MA. Self-consistent optical parameters of intrinsic silicon at 300K including temperature coefficients. Sol Energ Mater Sol, 2008, 92, 1305–10.
[9] Tauc J. Optical properties and electronic structure of amorphous Ge and Si. Mater Res Bull, 1968, 3, 37–46.
[10] Makuła P, Pacia M, Macyk W. How to correctly determine the band gap energy of modified semiconductor photocatalysts based on UV–vis spectra. J Phys Chem Lett, 2018, 9, 6814–7.
[11] Mertens K. Photovoltaics – Fundamentals, Technology, and Practice, 2nd edition, Wiley, Weinheim, 2020.
[12] Schneider J, Matsuoka M, Takeuchi M, Zhang J, Horiuchi Y, Anpo M, Bahnemann DW. Understanding TiO_2 photocatalysis: Mechanisms and materials. Chem Rev, 2014, 114, 9919–86.
[13] Ma Y, Wang X, Jia Y, Chen X, Han H, Li C. Titanium dioxide-based nanomaterials for photocatalytic fuel generations. Chem Rev, 2014, 114, 9987–10043.
[14] Gréboval C, Chu A, Goubet N, Livache C, Ithurria S, Lhuillier E. Mercury chalcogenide quantum dots: Material perspective for device integration. Chem Rev, 2021, 121, 3627–700.
[15] Brus LE. Electron–electron and electron-hole interactions in small semiconductor crystallites: The size dependence of the lowest excited electronic state. J Chem Phys, 1984, 80, 4403–9.
[16] Trindade T, O'Brien P, Pickett NL. Nanocrystalline semiconductors: Synthesis, properties, and perspectives. Chem Mater, 2001, 13, 3843–58.
[17] LaMer VK, Dinegar RH. Theory, production and mechanism of formation of monodispersed hydrosols. J Am Chem Soc, 1950, 72, 4847–54.
[18] Murray CB, Norris DJ, Bawendi MG. Synthesis and characterization of nearly monodisperse CdE (E = sulfur, selenium, tellurium) semiconductor nanocrystallites. J Am Chem Soc, 1993, 115, 8706–15.
[19] Hohenberg P, Kohn W. Inhomogeneous electron gas. Phys Rev, 1964, 136, B864–71.
[20] Kohn W, Sham LJ. Self-consistent equations including exchange and correlation effects. Phys Rev, 1965, 140, A1133–8.
[21] Voigt D, Bredol M, Gonabadi A. A general strategy for $CuInS_2$ based quantum dots with adjustable surface chemistry. Opt Mater, 2021, 115, 110994.
[22] Ley L, Pollak RA, McFeely FR, Kowalczyk SP, Shirley DA. Total valence-band densities of states of III-V and II-VI compounds from x-ray photoemission spectroscopy. Phys Rev B, 1974, 9, 600–21.
[23] Chiang TC, Himpsel FJ. ZnS: Datasheet from Landolt-Börnstein, Springer-Verlag, Berlin, 1989, Chapter Subvolume A 2. 1.22, 77–80.

[24] McCloy JS, Tustison RW. Chemical Vapor Deposited Zinc Sulfide, Society of Photo-Optical Instrumentation Engineers, Cardiff, 2013, Chapter Physics and Chemistry of ZnS, 1–30. ISBN: 9780819495891.

[25] Swank RK. Surface properties of II-VI compounds. Phys Rev, 1967, 153, 844–9.

[26] Jaffe JE, Zunger A. Electronic structure of the ternary chalcopyrite semiconductors $CuAlS_2$, $CuGaS_2$, $CuInS_2$, $CuAlSe_2$, $CuGaSe_2$, and $CuInSe_2$. Phys Rev B, 1983, 28, 5822–47.

[27] Braun W, Goldmann A, Cardona M. Partial density of valence states of amorphous and crystalline $AgInTe_2$ and $CuInS_2$. Phys Rev B, 1974, 10, 5069–74.

[28] Spiess HW, Haeberlen U, Brandt G, Räuber A, Schneider J. Nuclear magnetic resonance in IB–III–VI2 semiconductors. Phys Status Solidi B, 1974, 62, 183–92.

[29] Lv M, Zhu J, Huang Y, Li Y, Shao Z, Xu Y, Dai S. Colloidal $CuInS_2$ quantum dots as inorganic hole-transporting material in perovskite solar cells. ACS Appl Mater Interf, 2015, 7, 17482–8.

[30] Kagan CR, Bassett LC, Murray CB, Thompson SM. Colloidal quantum dots as platforms for quantum information science. Chem Rev, 2020, 121, 3186–233.

[31] González-Urbina L, Baert K, Kolaric B, Pérez-Moreno J, Clays K. Linear and nonlinear optical properties of colloidal photonic crystals. Chem Rev, 2011, 112, 2268–85.

[32] Toufanian R, Chern M, Kong VH, Dennis AM. Engineering brightness-matched indium phosphide quantum dots. Chem Mater, 2021, 33, 1964–75.

[33] Lee S-H, Han C-Y, Song S-W, Jo D-Y, Jo J-H, Yoon S-Y, Kim H-M, Hong S, Hwang JY, Yang H. ZnSeTe quantum dots as an alternative to InP and their high-efficiency electroluminescence. Chem Mater, 2020, 32, 5768–75.

[34] Yu JH, Kwon S-H, Petrášek Z, Park OK, Jun SW, Shin K, Choi M, Park YI, Park K, Na HB, Lee N, Lee DW, Kim JH, Schwille P, Hyeon T. High-resolution three-photon biomedical imaging using doped ZnS nanocrystals. Nat Mater, 2013, 12, 359–66.

[35] Yang J, Choi MK, Yang UJ, Kim SY, Kim YS, Kim JH, Kim D-H, Hyeon T. Toward full-color electroluminescent quantum dot displays. Nano Lett, 2020, 21, 26–33.

[36] Kim T, Kim K-H, Kim S, Choi S-M, Jang H, Seo H-K, Lee H, Chung D-Y, Jang E. Efficient and stable blue quantum dot light-emitting diode. Nature, 2020, 586, 385–9.

[37] Wood V, Panzer MJ, Chen J, Bradley MS, Halpert JE, Bawendi MG, Bulovíc V. Inkjet-printed quantum dot-polymer composites for full-color AC-driven displays. Adv Mater, 2009, 21, 2151–5.

[38] Chebrolu VT, Kim H-J. Recent progress in quantum dot sensitized solar cells: An inclusive review of photoanode, sensitizer, electrolyte, and the counter electrode. J Mater Chem C, 2019, 7, 4911–33.

[39] Snaith HJ. Perovskites: The emergence of a new era for low-cost, high-efficiency solar cells. J Phys Chem Lett, 2013, 4, 3623–30.

[40] Fu R, Zhou W, Li Q, Zhao Y, Yu D, Zhao Q. Stability challenges for perovskite solar cells. ChemNanoMat, 2019, 5, 253–65.

[41] Yin W-J, Yang J-H, Kang J, Yan Y, Wei S-H. Halide perovskite materials for solar cells: A theoretical review. J Mater Chem A, 2015, 3, 8926–42.

[42] Zuo T, He X, Hu P, Jiang H. Organic-inorganic hybrid perovskite single crystals: Crystallization, molecular structures, and bandgap engineering. ChemNanoMat, 2019, 5, 278–89.

[43] Umari P, Mosconi E, Angelis FD. Relativistic GW calculations on $CH_3NH_3PbI_3$ and $CH_3NH_3SnI_3$ perovskites for solar cell applications. Sci Rep, 2014, 4, 4467.

[44] Akkerman QA, Rainò G, Kovalenko MV, Manna L. Genesis, challenges and opportunities for colloidal lead halide perovskite nanocrystals. Nat Mater, 2018, 17, 394–405.

[45] Liu X-K, Xu W, Bai S, Jin Y, Wang J, Friend RH, Gao F. Metal halide perovskites for light emitting diodes. Nat Mater, 2020, 20, 10–21.

9.2 Superconductors

Dirk Johrendt

The excellent electrical conductivity of metals is among the most fundamental properties of matter and indispensable for many technologies in our daily life. It is one of those material properties that seem so self-evident that we hardly give it a second thought. Stephen Gray was the first who divided matter into electrical conductors and non-conductors in 1731 [1], and even though hardly noticed and falsely considered trivial, he is the actual inventor of electrical communication. Today, after many decades with seminal findings by Joseph J. Thomson [2], Paul Drude [3], Arnold Sommerfeld [4] and further protagonists of quantum mechanics, we know that the metallic conductivity bases on a high density of easily movable charge carriers, viz. electrons, that move as waves almost freely through the periodic lattice of metal atoms pushed by an electric field. Quantum-mechanical laws [5] ensure that such electron waves do not interact with the atom lattice as long as this is perfectly periodic. However, the latter condition is never fulfilled in real metals, where lattice defects, impurities and the thermal movement of the atoms break the perfect periodicity of the lattice to a certain extent. Electrons collide with such impurities and transfer energy to the lattice, which leads to heating, viz. energy dissipation, known as electrical resistance. This is a self-reinforcing effect, since the increasing temperature causes even more lattice vibrations, which increases the resistivity further. This is well known from the lighting bulb, where an electric current flows through a thin metal filament which heats up to glow due to its resistance. If the current gets too large, the thin wire melts and the light bulb is broken. This shows the limitation of electronic transport even in highly conducting pure metals, which is quantified as the maximal current density J_{max}. Copper has a J_{max} of 1000 A/mm^2 at 25 °C, which decreases to 100 A/mm^2 at 125 °C; however, these values require cooling to the given temperature. Values of usual copper wirings are much smaller and range from 2 to 6 A/mm^2 depending on the insulation and environment. Modern long-range transmission lines lose on the average 5.6% of the electrical energy by the ohmic resistance, which is about 4.6 TWh per year in Germany. Similar limitations concern the generation of high magnetic fields necessary in many technologies like magnetic resonance imaging (MRI) in hospitals, particle accelerators or state-of-the-art wind power generators. As an example, typical MRI instruments operate with magnetic fields of 2–3 T, and the inner coil radius must be wide enough to accommodate a human body. The field generated by a conventional magnetic coil is mainly proportional to the number of windings and to the applied current, and inverse to the lengths of the magnet. It can easily be shown that such a magnet is hardly possible to construct with a coil of copper wires, finally limited by the ohmic resistance of the wire.

https://doi.org/10.1515/9783110733471-002

One way out of these problems would be a perfect metallic conductor with zero resistivity. Indeed, most metals undergo transitions to a superconducting state at very low temperatures. Therein the conduction electrons form pairs which do not feel any drag because quantum laws do not allow interaction of the pairs neither with the lattice nor with imperfections. The second intriguing property of the superconducting state is magnetic in nature and makes the superconductor a perfect diamagnet; thus, magnetic fields are completely expelled from the interior. However, most superconductors are nevertheless penetrated by the magnetic field in a unique way called flux vortices. Quantum laws again cause that these vortices are pinned and a change costs energy. This effect of flux pinning is responsible for the phenomenon of magnetic levitation, where a magnet levitates over the superconductors while its position remains fixed. This principle is already used in high-speed magnetic levitating trains (MAGLEV) which are already in regular service in Japan and China.

The electrical resistivity is zero and the magnetic fields are expelled as long as the superconducting state exists; however, unfortunately it becomes immediately destroyed above the so-called critical temperature T_c, which is below 10 K for all metallic elements at ambient pressure. Furthermore, zero resistivity of a superconductor does not imply an infinite current according to Ohm's law $I = U/R$ with $R = 0$ because a critical current density also exists for superconductors, which is, however, often several magnitudes higher than in the normal conducting metallic state. Just as the current, also the magnetic field penetrating a superconductor is limited to a critical field; thus, the superconducting state exists in between its critical parameters T_c, J_c and B_c, and immediately vanishes as soon as only one of these critical parameters is surpassed. The values of these critical parameters greatly vary in different materials and decide whether or not they are suitable for applications. The relationship between superconductivity and chemistry has ever been problematic. This is not too surprising given that superconductivity is in the end a not fully understood property of conduction electrons that we can hardly influence chemically, at least not in a concrete way. For this reason, many superconducting materials were discovered by chance, and predictions are still only possible to a very limited extent today. In fact, over the decades, many theories have been proposed and refuted mostly through discoveries of new superconductors that were not expected before. Until now, no theory of superconductivity has had a concrete relation to chemistry, making it still difficult to find new superconductors despite tremendous efforts.

This also caused the still slow spread of superconductors into industrial applications. Even though many application fields, especially in currently highly relevant energy technologies, would benefit from superconductors, in most cases, the materials still do not meet the requirements to be competitive with conventional conductors in terms of costs. Nevertheless, the global superconducting wire market grows at a rate of about 10% per year and reached more than 1 billion USD in 2021. The growing demand for superconductor-based MRI systems, superconducting magnets, and synergies of high-voltage transmission application and high efficiencies are major drivers of the market.

From the view of chemistry, of the thousands of superconductors known today, only three material classes have truly made it to the market so far, namely niobium metal, its compounds like Nb_3Sn and especially NbTi alloys, copper oxides [6, 7] and magnesium diboride [8]. The newest class of high T_c superconductor is the iron-based compounds, which have not yet overcome the threshold to applications despite intense research [9, 10].

9.2.1 Niobium and its compounds

The superconducting transition temperature of niobium is 9.465 K, which is the highest among the elements in the periodic system at normal pressure [11, 12]. Since the critical field is very small, pure niobium is not suitable for superconducting wires. It is however used for superconducting radio frequency cavities, which allow energy storage with an extremely low loss in the range of 10^{-10}. Such devices are used in a variety of applications and in high-performance particle accelerator components [13].

Many alloys of niobium with other transition metals have been studied in the search for superconductors with suitable properties, especially for the construction of high-field magnets [14, 15]. NbTi alloys finally prevailed and became the most widely used superconductors up to now. From a chemical point of view, the binary phase diagram appears simple and shows complete miscibility of the components in the solid phase. However, the intensively studied metallurgy of NbTi alloys in detail is complex and out of the scope of this book. In particular, the microstructure and possible precipitations of the pure metals are important for superconducting properties like the critical field. It is remarkable that the superconducting critical temperatures of NbTi alloys do not change strongly over a wide range of the composition as shown in Figure 9.2.1a. In contrast, the critical field (Figure 9.2.1b) has a distinct maximum above 12 T near 47 wt% Ti corresponding to the titanium-richer composition $Nb_{37}Ti_{63}$. However, the critical field is not a simple function of the composition but strongly depends on the heat treatment of the alloy. Current manufacturers use different compositions depending on the application area of the superconducting wire.

Niobium-titanium alloys emerged as the supermagnet workhorse in the early 1960s [17], mainly driven by the demand for high magnetic fields necessary for MRI in hospitals, particle accelerators, and laboratory magnets. The smart thing of superconducting magnets is that once the current flows, the coil can be short-circuited into a persistent mode where no further energy is needed, except for cooling. This allows not only very high but also extremely constant magnetic fields over long times. However, the up-to-now lasting era of NbTi-based superconducting wires began not at all promising because first measurements resulted in very poor critical

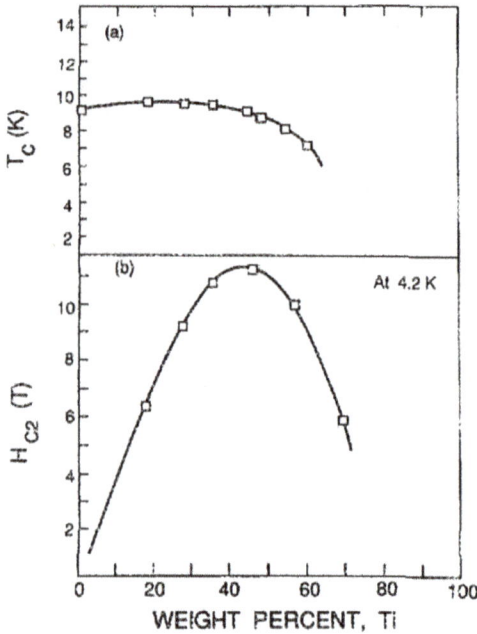

Figure 9.2.1: Critical temperature (T_c) and upper critical magnetic field (H_{c2}) of NbTi alloys (from [16] with permission).

currents. And at about the same time, a group from the Bell Telephone Laboratories published in 1961 superconductivity in Nb_3Sn at 8.8 T with a record current density of 10^5 A/cm^2, and moreover a critical temperature of 18 K, much higher compared with NbTi (4 K). But Nb_3Sn turned out to be a brittle material, from which it is anything but easy to make a wire. It was precisely the latter that was the great advantage of the easy-to-process NbTi alloys and motivated further experiments. In 1962, it was discovered that NbTi wires produced by cold forming allow high current densities and critical fields up to 14 T. This paved the way for niobium–titanium alloys for supermagnets capable of generating magnetic fields above 10 T. But there was still another obstacle to overcome. While such wires performed well as short samples in a magnetic field, they did not perform at all when wound to magnet coils. The reason is a physical effect called flux jumping. NbTi is a type-2 superconductor, which becomes penetrated by magnetic fields. Quantum rules enforce the formation of discrete flux vortices, which interact with the flowing current due to the Lorenz force and move perpendicular to the directions of the current and to the magnetic field. This movement generates resistivity and thus heat, and if this heat cannot escape rapidly, the temperature rises inside the superconductor, which itself causes further penetration of the field and a self-reinforcing situation. This electromagnetic thermal instability is called flux jump and, at worst, can cause the coil to become normally conducting and the electrical energy stored in it to be abruptly converted into heat.

This effect, also known as quenching, can destroy the entire superconducting magnet. However, the remedy for flux jumping is relatively simple and consists of dividing the wire into very fine filaments with diameters less than 50 µm and embedding them in a good heat-conducting material, usually copper. Magnetic coupling between the filaments is avoided by twisting them like a rope. A typical cross section of a multifilament strand is shown in Figure 9.2.2.

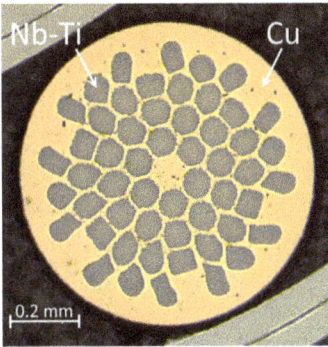

Figure 9.2.2: Cross section of a superconducting wire made of thin NbTi filaments embedded in copper (reproduced with permission from [18]).

At present, high-performance NbTi superconducting wires are manufactured as a commodity product for applications which require very high and uniform magnetic fields. MRI manufacturers are still the main customers (80%), while large amounts have been manufactured for particle accelerators like the LHC, which contains 7500 km of NbTi-based, so-called Rutherford cables, where the strands are arranged as a many-stranded helix that has been flattened into a rectangular cable. Even larger amounts require fusion reactors like ITER, which has revived the world market. ITER will use more than one-fifth of the annual production of niobium-titanium strands (275 tons), and the global production of niobium-tin strands (see below) has been multiplied by six to meet ITER's needs on the order of 600 tons. It is currently believed that the development of fusion technology will have a significant impact on the future superconductor market [19].

As mentioned above, another technically important superconducting compound is Nb_3Sn. Its critical temperature (18 K) and upper critical field (30 T) almost doubles the accessible field-temperature regime with respect to NbTi. Nb_3Sn is therefore used for extreme high-field laboratory magnets, NMR instruments, and large-scale applications in particle accelerators and fusion facilities. Nb_3Sn is a member of the large family of compounds with the A15- or better V_3Si-type structure, shown in Figure 9.2.3 [20].

The tin atoms are in a body-centered cubic arrangement and the niobium atoms form regular chains with two atoms on each face of the unit cell. These chains are separated from each other and run in all spatial directions. Nb_3Sn written as $Nb_{1-\beta}Sn_\beta$ ($\beta = 0.25$) has a homogeneity range of $0.05 \leq \beta \leq 0.25$. The solid solution of tin

Figure 9.2.3: Crystal structure of cubic Nb_3Sn.

in niobium at low concentrations ($\beta < 0.05$) reduces the critical temperature of bcc Nb from about 9.2 K to about 4 K at $\beta = 0.05$ [21].

Structural phase transitions around 32–45 K from cubic (space group $Pm\bar{3}n$) to tetragonal symmetry ($P4_2/mmc$, $c/a = 1.006$ at 4 K) has long been a subject of controversial discussions with regard to the influence on the critical field H_{c2} [21, 22]. Calculations [23] and some experimental data [21] indicate that tetragonal strain reduces H_{c2}, and other authors conclude that H_{c2} does not depend on the tetragonal distortion [24]. Figure 9.2.4 shows how the critical temperatures and upper critical fields change with the composition. The best superconducting performance occurs near the stoichiometric Nb_3Sn phase or slightly below, where also the structural transitions have been observed.

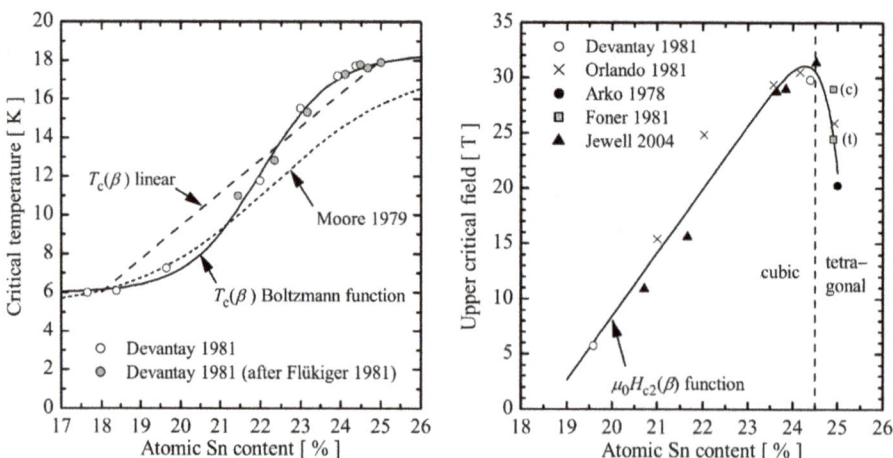

Figure 9.2.4: Critical temperature and upper critical field as a function of the tin concentration (reproduced with permission from [21]).

In contrast to the very ductile NbTi alloys, Nb_3Sn is a brittle material and its fabrication into woundable wires has been and still is a big challenge. However, the first 7–15 T superconducting magnets with Nb_3Sn coils have been commercialized already in the early 1960s [25]. These early magnets made from thin Nb_3Sn tapes applied by chemical vapor deposition (CVD) on ductile metallic substrates suffered from magnetic instabilities (flux jumping). They were shortly afterwards replaced by multifilament Nb_3Sn wires in a metallic matrix with high thermal conductivity, similar to the simultaneous development of NbTi wires. After the discovery that copper improves the Nb_3Sn formation at lower temperatures, multifilamentary wires were manufactured by the so-called bronze process [26]: 800 mm long bronze rods with a diameter of 200 mm and a weight of 250 kg are used. First, a hexagonal arrangement of axial holes was drilled (usually 19 or 37), into which niobium rods were inserted. This composite is then processed to hexagonal rods by extrusion and drawing, and several of these rods are combined in a copper cylinder. These bronze billets become hot extruded, drawn to a multifilament wire and wound to a coil. Nb_3Sn forms only during the final heat treatment at 675 °C for 200 h as layers around the original niobium filament. A cross section of a typical strand produced by the bronze process is shown in Figure 9.2.5.

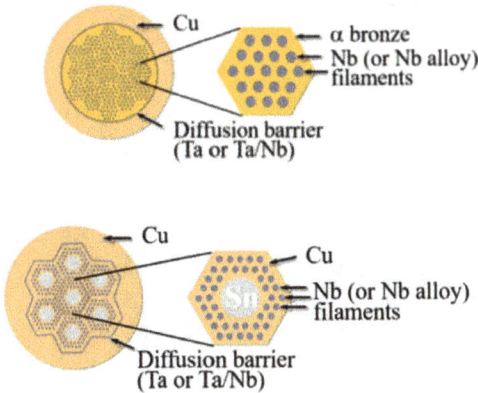

Figure 9.2.5: Schemes of Nb_3Sn composite wire based on the bronze route (top) and on the internal tin process (bottom) (reproduced from [27]).

A further development of the bronze process was the internal tin approach: Niobium and tin rods were first inserted in the holes of a copper billet. During heat treatment, Cu and Sn form Cu-Sn alloys which react with Nb to form the Nb_3Sn phase (Figure 9.2.5). Both the bronze and the internal tin process require large Cu/ Nb ratios and lead to small filament sizes, which yield relatively low critical currents J_c. This was overcome by continuous development of the production processes based on the internal tin method and driven by requirements of large-scale facilities like ITER and LHC [27]. Nb_3Sn multifilament strands finally reached J_c of 3000 A/mm^2 at 4.2 K in a 12 T field [28]. Details of these complex processes are beyond the scope of this book; therefore, we refer to the literature [26].

9.2.2 Copper oxides

The year 1986 marked a milestone in superconductivity research. Georg Bednorz and Karl Alexander Müller discovered superconductivity in the copper oxides $(La_{1-x}Ba_x)_2CuO_4$ and $(La_{1-x}Sr_x)_2CuO_4$ with critical temperatures up to 38 K and won the Nobel Prize in physics in 1987 [29, 30]. Shortly afterwards, Chu et al. discovered the compound $YBa_2Cu_3O_{7-x}$ with $T_c = 93$ K [31], higher than the boiling point of liquid nitrogen (77 K). Figure 9.2.6 shows the crystal structures of the cuprate superconductors, which are relatives of the perovskite-type structure. However, in contrast to the cubic perovskite, these cuprates form layered structures, and it is mainly this strong anisotropy which makes it difficult to process them into superconducting wires.

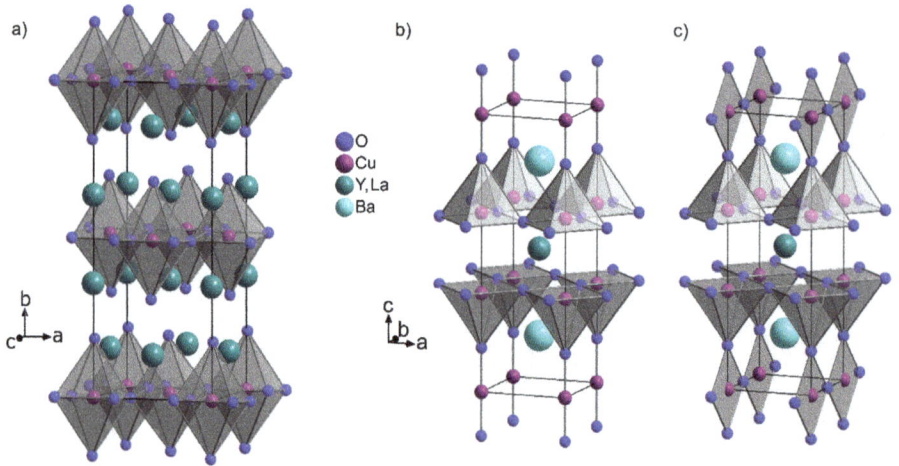

Figure 9.2.6: Crystal structures of (a) La_2CuO_4 (*Cmce*), (b) $YBa_2Cu_3O_6$ (*P4/mmm*) and (c) $YBa_2Cu_3O_7$ (*Pmmm*).

Unlike NbTi and Nb_3Sn, the cuprates require doping to become metallic and superconducting at low temperatures. In La_2CuO_4, lanthanum has been substituted by barium or strontium yielding $(La_{1-x}Ba_x)_2CuO_4$, ($x \sim 0.2$), and $YBa_2Cu_3O_{7-x}$ is superconducting at oxygen contents between 6.5 and 7. Figure 9.2.7 shows the variation of the critical temperatures T_c with the oxygen content in detail. The maximum is very close to but not at $x = 7$. This shows how important it is to maintain the exact oxygen content, which is a big challenge for the industrial productions of kilometer-long wires.

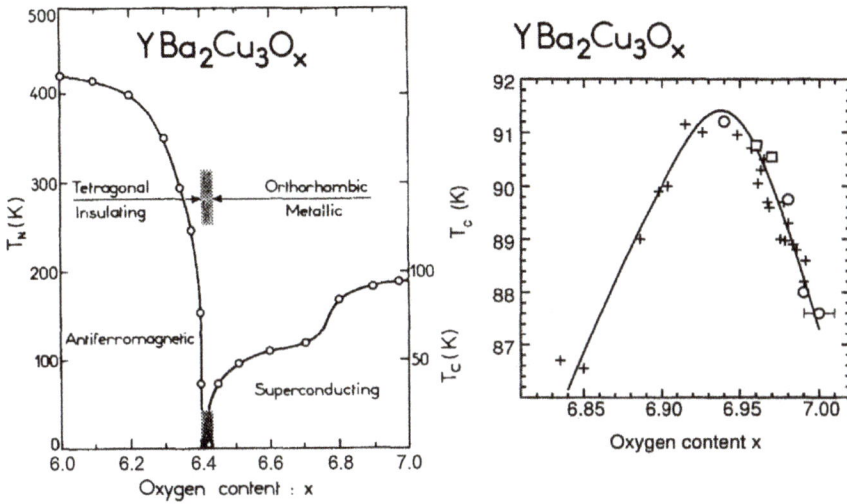

The advantages over the NbTi and Nb_3Sn materials are striking [6]. The much higher transition temperatures (>77 K) considerably reduce the cooling effort, and the superconductivity is quite robust in terms of the upper critical field H_{c2} in the range of 140 T in the low-temperature limit and around 40 T at 77 K [34]. The critical current densities exceed 10^4 and 10^6 A/cm^2 at 77 and 4.2 K, respectively. However, although the euphoria after the discovery of superconductivity in cuprates was enormous, it took almost 20 years before they were technically applied in superconducting wires. It turned out to be extremely complicated to get large-scale products, since these brittle materials must be synthesized at high temperatures (600–800 °C), while already small deviations from the stoichiometry like $YBa_2Cu_3O_{7-x}$ reduce T_c substantially. At the same time, defects had to be introduced in a controlled manner at average distances of about 20 Å acting as magnetic flux pinning centers; otherwise, the critical currents are low. Furthermore, the strong anisotropy due to the layered structures requires highly textured materials, since the critical current decreases exponentially with the grain misalignment angle [35].

In spite of many obstacles, the first generation of high-temperature superconducting wires (1 G-HTS) based on the bismuth-cuprate $Bi_2Sr_2Ca_2Cu_3O_{10}$ (Bi2223, $T_c = 110$ K) became commercially available since the 1990s [36]. They consist of thin superconducting filaments produced by the oxide-powder-in-tube technique. A ductile metal tube is filled with a stoichiometric mixture of the binary oxides and shaped to a thin filament by drawing. Many of these filaments are combined to bundles and placed into a silver cladding tube until the required thickness of the wire is reached. The starting oxide mixture reacts to the desired superconducting material during several heat treatments at 900 °C under noble gas atmosphere in order to avoid oxidation of the cladding metal. The alignment of the superconductive crystallites required for a high critical current density is achieved by several rolling steps between the heat treatments (Figure 9.2.8). However, only the CuO layers and not the other crystal axes became aligned parallel by this method. This enabled current densities of up to 10^4–10^5 A/cm^2, which are about a factor of 10 lower than in fully oriented material [7].

Figure 9.2.8: Main steps of the powder-in-tube fabrication of 1 G Bi2223 multifilament wires. Bundles of filaments in the cladding tube (left), shaping of the cladding tube (middle) and heat treatment to react the oxide mixture to the superconducting material (right) (reproduced with permission from [7]).

It turned out that these BSCCO-based wires have low irreversibility fields H_{irr} at higher temperatures. H_{irr} marks a transition in the vortex lattice, which "melts" at $H > H_{irr}$, and the consequence is that the 1 G-BSCCO wires are not suitable for applications in high fields at temperatures >50 K. To overcome this problem, wires based on the even more brittle YBa$_2$Cu$_3$O$_{7-x}$ (YBCO, $T_c = 93$ K) were developed [37]. These so-called coated conductors or second generation high-temperature superconducting wires (2 G-HTS) based on thin films of the superconducting material deposited on metal substrate strips with an intermediate ceramic buffer layer. Artificial texture of the buffer layer enabled the biaxial alignment of the grains (Figure 9.2.9).

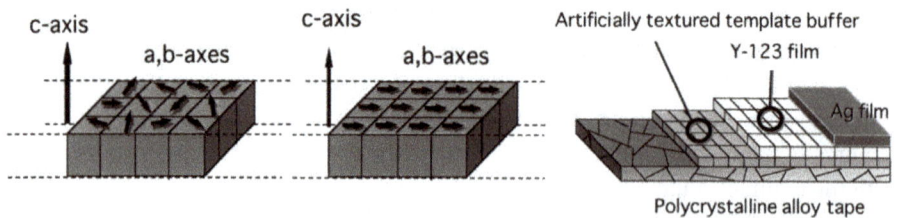

Figure 9.2.9: Left: Uniaxially aligned layer; middle: biaxial aligned layer; right: scheme of a biaxially aligned YBCO (Y-123)-coated conductor with artificially textured template buffer on an alloy tape (with permission from [37]).

The rolling-assisted biaxially textured substrate (RABiTS) technique [38] paved the way to the 2 G-HTS tapes with high critical currents exceeding 10^6 A/cm^2. The starting metal strip is hot rolled and annealed to promote a textured surface. The best texture quality is achieved using Ni/W(5%) alloys. Another approach to textured substrates is the ion beam-assisted deposition (IBAD) technique. Here, the first buffer layer deposited on the metallic substrate is forced to have a preferred texture. It is achieved using an ion gun that orients the growing oxide buffer layer while it is being deposited on the polycrystalline metallic substrate. Early studies used yttrium-stabilized zirconia (YSZ), and other buffer materials are $RE_2Zr_2O_7$ (RE: Gd, La), $AZrO_3$ (A = Ba, Sr), CeO_2 and MgO. A continuous reel-to-reel system has been realized using MgO deposited on a Hastelloy tape. Disadvantages are that the IBAD method is slow and requires expensive high vacuum technique [39].

A series of methods have been developed for the deposition of YBCO or $REBCO$ (RE = Y or Gd) layers on textured substrates [40]. Physical methods are pulsed laser deposition, electron beam-based deposition, magnetron beam-based deposition and thermal evaporation [41]. These methods are well established but require expensive equipment like high-power UV lasers and high vacuum conditions, which makes them hardly suitable for low-cost production.

CVD and metal-organic CVD (MOCVD) make use of deposition after thermal decomposition of mixtures of volatile inorganic or metal-organic precursors and require only moderate vacuum. MOCVD also enabled the metal strips to be coated with YBCO on both sides, which increased the superconducting cross section and thus the current-carrying capability [42]. Figure 9.2.10 shows the scheme of a corresponding reel-to-reel system. The metal-organic precursors are mixtures of the yttrium, barium and copper complexes of 2,2,6,6-tetramethyl-3,5-heptanedionate (tmhd), respectively, dissolved in tetrahydrofuran (see Figure 9.2.10). About 2 μm thick YBCO films have been deposited at 850 °C and 5 mbar pressure on LAO (LaAlO$_3$) buffer layers with a rate up to 500 nm/min.

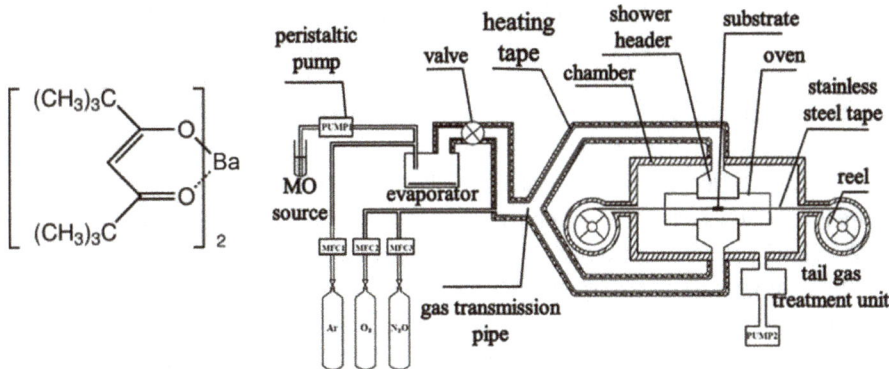

Figure 9.2.10: Scheme of a reel-to-reel MOCVD system. The formula of one of the metal-organic precursors (Ba(tmhd)$_2$) is shown on the left (with permission from [42]).

The further development of new methods should enable deposition under atmospheric environment and from aqueous precursor materials, which is a big challenge for solid-state chemists. Interestingly, as early as 1987, when $YBa_2Cu_3O_{7-x}$ was discovered, chemical solution deposition was already being considered. In the first works [43], aqueous solutions of Y-, Ba- and Cu-acetates [40] or -nitrates [44] were spun onto ceramic substrates like MgO, $SrTiO_3$ or YSZ, pyrolyzed at 400 °C, heat treated at 800–900 °C for 2 min and post-annealed for 1 h at 400 °C in an O_2 atmosphere to ensure a high oxygen content. However, carbon contamination during decomposition of carboxylates to $BaCO_3$ and BaO result in the formation of films with broad superconducting transitions. Another approach used metal trifluoroacetates (TFA) in organic solvents like methanol [45]. The spun-on films decompose at 400 °C and form amorphous films of BaF_2, YF_3, CuF_2 and CuO. Subsequent heating steps at 800–900 °C in a helium/H_2O and O_2 atmospheres form ~1 µm thick $YBa_3Cu_3O_{7-x}$ films. It turned out that water vapor is essential for converting the fluorides to oxides through formation of hydrogen fluoride and that the control of the microstructural factors which may influence the superconducting properties are a complex issue [46]. However, the TFA process, also known as TFA-MOD, was further developed in combination with IBAD and RABiTS which produce the ceramic buffer layers which enable the growth of biaxial aligned layers. Disadvantages of the original TFA-MOD method were long pyrolysis times (10 h) and defects in the films due to escaping HF gas. Subsequently, methods with lower fluorine contents were developed, in which the trifluoroacetates were increasingly replaced by acetates starting from the conventional low fluorine method with 54% F-content over the super low fluorine (31% F) up to the extremely low fluorine with only 7% F-content [47–49]. However, it is still not possible without fluorine completely, because this would cause the abovementioned problems with $BaCO_3$ and BaO. It turned out that $REBa_2CuO_{7-x}$ compounds with RE = Sm-Tm have about 5 K higher critical temperatures and are partly better processable than YBa_2CuO_{7-x}. Since gadolinium and yttrium do not much differ in price, some manufacturers use $GdBa_2CuO_{7-x}$ (GBCO or generally REBCO) for coated conductors [50]. During the last years, the production of 2 G-HTS wires and tapes has made great strides, and several companies produce coated conductors using different combinations and flavors of the methods described above. Figure 9.2.11 shows a typical example of a coated conductor architecture.

2G-REBCO superconducting films with $BaZrO_3$ (BZO) nanocolumns have been shown to be very effective for raising the vortex pinning. These conductors have reached a record of critical current densities of 10^7 A/cm^2 in magnetic fields up to 30 T [51], however, the tapes currently produced commercially by several companies worldwide already achieve similar values up to 6×10^6 A/cm^2 [52]. Note that these values do not directly reflect the actual current capacity of the whole wire. This is better described by the so-called transport current I_c. I_c is the amount of current in ampere that produces a voltage of 1 µV/cm along the conductor. Typical values are in the range of 300–600 A in a 1 cm wide tape with 1–2 µm REBCO. To carry

Figure 9.2.11: Architecture of a typical commercial 2 G-HTS REBCO-coated conductor (with permission from [52]).

the same amount of current with a copper wire, it would have to have a diameter of about 20 mm. The German company Theva recently announced more than 1350 A of transport current in tapes with 12 mm width and 75 µm total thickness [53].

9.2.3 Magnesium diboride

Superconductivity in MgB_2 was discovered in 2001 [54], although the compound was known since 1954 [55]. The critical temperature is 39 K, much lower than in the cuprates, but considerably higher than in Nb_3Sn (18 K). Magnesium and boron are light elements and abundant, therefore, MgB_2 was quickly considered to produce superconducting wires. Figure 9.2.12 shows the crystal structure, which is the AlB_2 type with honeycomb nets of boron and magnesium in between them.

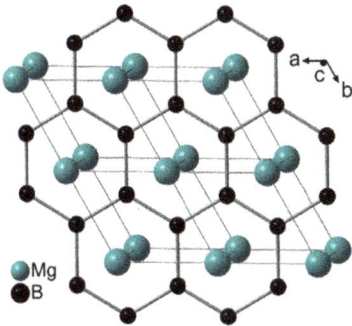

Figure 9.2.12: Crystal structure of MgB_2 with graphite-like nets of boron.

MgB_2 can be synthesized from the pure elements at high temperatures, and superconducting wires were fabricated by different methods [56]. The ex situ method uses prereacted MgB$_2$ powders in niobium tubes with a surrounding copper tube. Several wires are inserted in nickel or Monel sheath tubes. The whole composite is rolled and drawn to a round wire, which is twisted and then rolled to a tape or a round wire of about 1 mm followed by heat treatment at 965 °C for 1 h to recrystallize. The critical currents in the range of 3–5 × 10^2 A/mm^2 in 3–5 T fields are not very high, but this method yields good homogeneity over long lengths. The in situ method starts with a mixture of the powdered elements, which has the advantage that the reaction temperatures are significantly lower at 650 °C and reduces the unwanted reaction with the sheath. The critical currents are similar to those of the ex situ processed wires. It turned out that minor substitution of boron by carbon improves the critical current in higher fields. The in situ and ex situ methods were also combined [57, 58]. A third method is the internal magnesium diffusion technique. A 2 mm thick magnesium rod is inserted in the center of a niobium or tantalum tube, and the hollow space is filled with boron powder and a small amount of carbon. The rods are bundled in a Cu-Ni tube, cold deformed to a wire and heated to 650 °C. The magnesium diffuses into the boron and reacts to an MgB$_2$ layer of 10–30 μm thickness along the inner wall of the sheath tube. The critical currents reach 3 × 10^6 A/cm^2 at 20 K and a 4 T field.

Generally, the self-field (zero external field) critical currents of MgB$_2$ are up to 5 × 10^6 A/cm^2 even at 20 K and thus quite high, but it falls off quickly to higher magnetic fields as shown in Figure 9.2.13. This is a general property of MgB$_2$-based superconductors, and the main reason for the limited usability of MgB$_2$ in applications with high magnetic fields. Nevertheless, due to the abundance of the light elements magnesium and boron and the good processability, MgB$_2$ has a high potential for lower

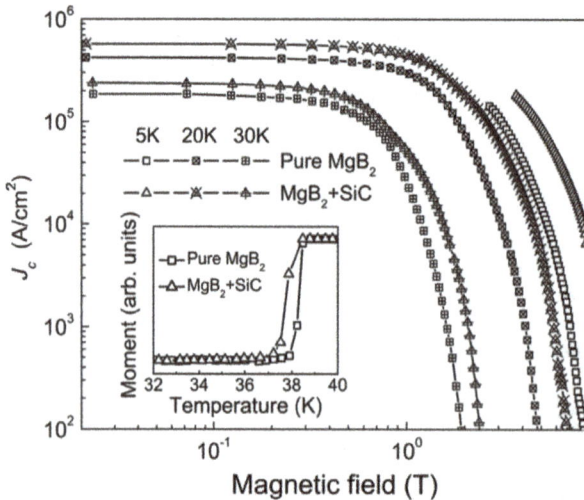

Figure 9.2.13: Critical current J_c versus field at 5, 20 and 30 K for pure and nano-SiC-doped samples. The inset shows the superconducting transition of the two samples [57].

field applications and temperatures up to 20 K and is already in use for a new generation of MRI instruments which do not require liquid helium [56, 59, 60]. As for the cuprates, the development is actively going on, and MgB_2 is certainly one of the superconducting materials which made it to the market. Many details of this development go beyond the scope of this book, for the interested reader we refer to the comprehensive literature [8, 61].

References

[1] Gray S. Phil Trans Royal Soc, 1731, 37, 397–407.
[2] Thomson JJ. Nobel Lectures: Physics 1901–1921, Elsevier, Amsterdam, 1967, 145–53.
[3] Drude P. Ann Phys, 1900, 306, 566–613.
[4] Sommerfeld A, Bethe H. Elektronentheorie der Metalle, Springer, Berlin, Heidelberg, 1933, 333–622.
[5] Bloch F. Z Phys, 1929, 57, 545–55.
[6] Hackl R. Z Kristallogr, 2011, 226, 323–42.
[7] Bäcker M. Z Kristallogr, 2011, 226, 343–51.
[8] Flükiger R. MgB_2 Superconducting Wires, World Scientific, Singapore, 2016.
[9] Hosono H, Yamamoto A, Hiramatsu H, Ma Y. Mater Today, 2018, 31, 278–302.
[10] Yao C, Ma Y. Supercond Sci Technol, 2019, 32, 023002.
[11] Finnemore DK, Stromberg TF, Swenson CA. Phys Rev, 1966, 149, 231–43.
[12] Eisenstein J. Rev Mod Phys, 1954, 26, 277–91.
[13] Padamsee HS. Ann Rev Nucl Particle Sci, 2014, 64, 175–96.
[14] Berlincourt TG. Brit J Appl Phys, 1963, 14, 749–58.
[15] Berlincourt TG, Hake RR. Phys Rev Lett, 1962, 9, 293–5.
[16] Meingast C, Lee PJ, Larbalestier DC. J Appl Phys, 1989, 66, 5962–70.
[17] Berlincourt TG. Cryogenics, 1987, 27, 283–9.
[18] Patel D, Kim S-H, Qiu W, Maeda M, Matsumoto A, Nishijima G, Kumakura H, Choi S, Kim JH. Sci Rep, 2019, 9, 14287.
[19] Mitchell N, Zheng JX, Vorpahl C, Corato V, Sanabria C, Segal M, Sorbom B, Slade R, Brittles G, Bateman R, Miyoshi Y, Banno N, Saito K, Kario A, Ten Kate H, Bruzzone P, Wesche R, Schild T, Bykovskiy N, Dudarev A, Mentink M, Mangiarotti FJ, Sedlak K, Evans D, Van Der Laan DC, Weiss JD, Liao M, Liu G. Supercond Sci Technol, 2021, 34, 10301.
[20] Geller S, Matthias BT, Goldstein R. J Am Chem Soc, 1955, 77, 1502–4.
[21] Godeke A. Supercond Sci Technol, 2006, 19, R68–80.
[22] Shirane G, Axe JD. Phys Rev B, 1971, 4, 2957–63.
[23] De Marzi G, Morici L, Muzzi L, Della Corte A, Buongiorno Nardelli M. J Phys: Condens Matter, 2013, 25, 135702.
[24] Zhou J, Jo Y, Sung ZH, Zhou H, Lee PJ, Larbalestier DC. Appl Phys Lett, 2011, 99, 122507.
[25] Rosner HC. IEEE/CSC & ESAS Eur Supercond News Forum, 2012, 19, 1–29.
[26] Xu X. Supercond Sci Technol, 2017, 30, 093001.
[27] Barzi E, Zlobin AV. Nb_3Sn wires and cables for high-field accelerator magnets. In: Schoerling D, Zlobin AV (Eds). Springer Nature, Cham, Switzerland, 2019, Chapter 2, 27.
[28] Field MB, Parrell JA, Zhang Y, Meinesz M, Hong S. AIP Conf Proc, 2008, 986, 237–43.
[29] Bednorz JG, Müller KA. Z Phys B: Condens Matter, 1986, 64, 189–93.
[30] Bednorz JG, Mueller KA, Takashige M. Science, 1987, 236, 73–5.

[31] Wu MK, Ashburn JR, Torng CJ, Hor PH, Meng RL, Gao L, Huang ZJ, Wang YQ, Chu CW. Phys Rev Lett, 1987, 58, 908–10.

[32] Rossat-Mignod J, Regnault LP, Vettier C, Burlet P, Henry JY, Lapertot G. Phys B: Condens Matter, 1991, 169, 58–65.

[33] Breit V, Schweiss P, Hauff R, Wühl H, Claus H, Rietschel H, Erb A, Müller-Vogt G. Phys Rev B, 1995, 52, R15727–30.

[34] Smith JL, Brooks JS, Fowler CM, Freeman BL, Goettee JD, Hults WL, King JC, Mankiewich PM, De Obaldia EI, O'Malley ML, Rickel DG, Skocpol WJ. J Supercond, 1994, 7, 269–70.

[35] Hilgenkamp H, Mannhart J. Rev Mod Phys, 2002, 74, 485–549.

[36] Fischer B, Arndt T, Gierl J, Munz M, Szulczyk A, Thöner M. Adv Solid State Phys 40, Springer, Berlin, Heidelberg, 2000, 741–51.

[37] Iijima Y, Matsumoto K. Supercond Sci Technol, 2000, 13, 68–81.

[38] Goyal A, List FA, Mathis J, Paranthaman M, Specht ED, Norton DP, Park C, Lee DF, Kroeger DM, Christen DK, Budai JD, Martin PM. J Supercond, 1998, 11, 481–7.

[39] Sheth A, Schmidt H, Lasrado V. Appl Supercond, 1999, 6, 855–73.

[40] Van Driessche I, Schoofs B, Penneman G, Bruneel E, Hoste S. Measure Sci Rev, 2005, 5, 19–29.

[41] Wördenweber R. Supercond Sci Technol, 1999, 12, R86–102.

[42] Zhang F, Xiong J, Liu X, Zhao R, Zhao X, Tao B, Li Y. J Vac Sci Technol A, 2014, 32, 041512.

[43] Rice CE, V Dover RB, Fisanick GJ. Appl Phys Lett, 1987, 51, 1842–4.

[44] Kawai M, Kawai T, Masuhira H, Takahasi M. Jpn J Appl Phys, 1987, 26, L1740–2.

[45] Gupta A, Jagannathan R, Cooper EI, Giess EA, Landman JI, Hussey BW. Appl Phys Lett, 1988, 52, 2077–9.

[46] Castaño O, Cavallaro A, Palau A, González JC, Rossell M, Puig T, Sandiumenge F, Mestres N, Piñol S, Pomar A, Obradors X. Supercond Sci Technol, 2002, 16, 45–53.

[47] Li MJ, Cayado P, Erbe M, Jung AL, Hanisch J, Holzapfel B, Liu ZY, Cai CB. Coatings, 2020, 10, 31.

[48] Chen Y, Wu C, Zhao G, You C. Supercond Sci Technol, 2012, 25, 062001.

[49] Palmer X, Pop C, Eloussifi H, Villarejo B, Roura P, Farjas J, Calleja A, Palau A, Obradors X, Puig T, Ricart S. Supercond Sci Techno, 2015, 29, 024002.

[50] Erbe M, Cayado P, Freitag W, Ackermann K, Langer M, Meledin A, Hänisch J, Holzapfel B. Supercond Sci Technol, 2020, 33, 094002.

[51] Xu A, Delgado L, Khatri N, Liu Y, Selvamanickam V, Abraimov D, Jaroszynski J, Kametani F, Larbalestier DC. APL Mater, 2014, 2, 046111.

[52] Namburi DK, Shi Y, Cardwell DA. Supercond Sci Technol, 2021, 34, 053002.

[53] THEVA Dünnschichttechnik GmbH (31.07.2020). THEVA reaches record performance in HTS tapes. Retrieved from https://www.va.com/theva-erreicht-rekord-performance-bei-hts-bandleitern/?noredirect=en_GB, accessed November 19th, 2021.

[54] Nagamatsu J, Nakagawa N, Muranaka T, Zenitani Y, Akimitsu J. Nature, 2001, 410, 63.

[55] Jones ME, Marsh RE. J Am Chem Soc, 1954, 76, 1434–6.

[56] Ballarino A, Flükiger R. J Phys: Conf Ser, 2017, 871, 012098.

[57] Li WX, Zeng R, Lu L, Dou SX. J Appl Phys, 2011, 109, 07E108.

[58] Li W, Dou S-X. Superconductors – new developments. In: Gabovich A (Ed). IntechOpen Book Series, Rijeka, Croatia, 2015.

[59] Patel D, Matsumoto A, Kumakura H, Maeda M, Kim S-H, Al Hossain MS, Choi S, Kim JH. J Mater Chem C, 2020, 8, 2507–16.

[60] Yao W, Bascunan J, Hahn S, Iwasa Y. IEEE Trans Appl Supercond, 2010, 20, 756–9.

[61] Flükiger R, Suo HL, Musolino N, Beneduce C, Toulemonde P, Lezza P. Phys C, 2003, 385, 286–305.

9.3 Inorganic materials for chemical sensors

Michael Schäferling

The development of chemical sensors meets the increasing demand for easily accessible analytical data, characterizing the status of biological systems or industrial processes. For example, the physiological state of the human body can be monitored by means of physical, chemical and biochemical parameters used for medical diagnosis, e.g. blood gas and electrolyte values. The quality of the ambient and natural environment can be supervised by measuring the concentration of noxious chemical species in short intervals. The monitoring of nitrate pollution in ground and water is one prominent example. No less important is the automated continuous control of industrial chemical or biotechnological processes that are affected by specific parameters (pH, temperature, concentration of reactants, oxygen partial pressure, etc.).

The IUPAC defines a chemical sensor as a device that transforms chemical information, usually the concentration of a specific sample component, into an analytically useful signal. Chemical sensors are characterized by two basic components connected in series: a chemical (molecular) recognition system (receptor or selective membrane) and a physicochemical transducer [1]. The transducer converts the chemical information into an optical or electrical signal (Figure 9.3.1). Thus, chemical sensors can be classified into two main categories: optical and electrical/electrochemical sensors. In addition, also acoustic sensors are in use, e.g. sensor chips based on surface acoustic waves, or piezoelectric resonators (oscillating quartz microbalance). A practically useful sensor should provide the following characteristics:

- applicability on-site and in direct contact to the sample,
- integrated in a small and handheld device,
- fast response time,
- cheap,
- good sensitivity and selectivity for a certain analyte.

Ideally, the response of chemical sensors is reversible, i.e. they are capable for continuous online monitoring of an analyte in a sample matrix [2].

Optical sensors are often a combination of optical fibers or waveguides and sensitive layers which consist of organic or metal-organic dyes incorporated in a polymer or silica film which change their absorbance or photoluminescence (fluorescence or phosphorescence) due to interaction with the analyte molecules. The focus of this chapter is on the description of inorganic materials used in electrochemical sensors because these found widespread applications in gas sensors and ion-selective electrodes (ISEs). The response of such sensors can be due to a change of inherent properties of the sensing material (conductivity, capacitance or permittivity) or a change of the measured current or voltage in an electrochemical cell (amperometric or potentiometric sensors).

https://doi.org/10.1515/9783110733471-003

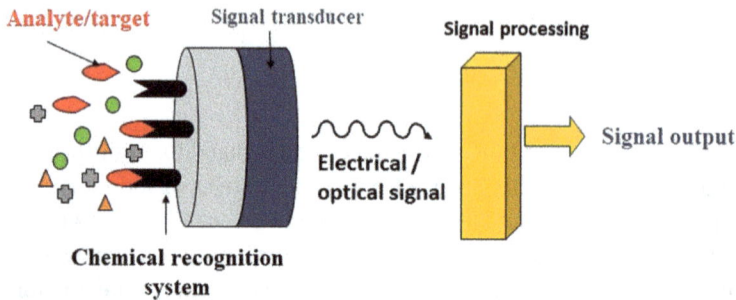

Analyte/target   Signal transducer

Signal processing

Electrical /
optical signal

Signal output

Chemical recognition
system

Figure 9.3.1: The general configuration of a chemical sensor consists of two components: an immobilized chemical recognition system (e.g. a molecular receptor or selective membrane) and a signal transducer converting chemical information (concentration of an analyte) in an electrical or optical signal which can be processed by further instrumentation.

This overlook is not focused on the synthesis or the characterization of the different crystalline structures, but to the general chemical composition and the mode of operation of the sensing materials.

9.3.1 Ceramic materials for electric gas sensors

The determination of the electric resistance of a sensor layer is a typical signal transduction method used in electrical chemical sensors. Resistance changes are usually analyzed by alternating current (AC) measurements. The AC resistance is termed as impedance and is a complex number. Electrical conductivity or specific conductance G is the reciprocal of electrical resistivity (or specific electrical resistance) ρ of a material.

The relation is given by

$$\rho = R \, A \, l^{-1} \tag{9.3.1}$$

where R is the electrical resistance of a uniform specimen of the material, l is the length of the specimen and A is the cross-sectional area of the specimen.

In electrical sensors, a layer of a semiconducting solid or gel serves as a sensing layer whose resistance changes as a result of interaction with a liquid or gas sample.

One of the oldest and most widely used chemical resistance sensors, which are also termed as chemoresistors, is the Taguchi sensor (developed by Naoyoshi Taguchi in 1962). Taguchi sensors are ceramic bodies made by pressing and sintering of polycrystalline metal oxides such as SnO_2 or ZnO. These are n-type semiconductors in which oxygen vacancies in the crystal structure act as electron donors. Millions of such sensors are used to indicate traces of reducing gases in air, e.g. for monitoring gasoline vapors, for the detection of leaks in gas pipes, for measurement of alcohol vapor in the breathing air or for determination of carbon monoxide. The sensors are operated at elevated temperatures (>200 °C). It is assumed that on the individual grains of the sinter pills, oxygen molecules are chemisorbed. These accept electrons from the outer layer of the grains and oxygen anions are formed on the surface by reduction. As a result, the charge carrier concentration inside the grains drops and a depletion region is formed, increasing the potential barrier at the grain boundaries. Overall, the conductivity of the material is decreased according to the extent of oxygen adsorption. The molecules of a reducing analyte gas in turn react with the adsorbed oxygen anions, lower the potential barrier and increase the conductivity of the sensing layer by transferring electrons back into the semiconducting material. This change in conductivity is reversible and can be measured with a very simple device. The sensors are highly sensitive robust and cheap, but suffer from low selectivity, aging and a signal drift over time.

Numerous reducing gases such as hydrogen, methane, carbon monoxide, ethanol or hydrogen sulfide can be determined in an air atmosphere. The concentration dependency of the conductance G of the sensors is non-linear and roughly obeys the following formula:

$$G = kc_i^n$$

where c_i is the analyte concentration. The constants k and n have to be determined by calibration. The sensing principle is shown in Figure 9.3.2.

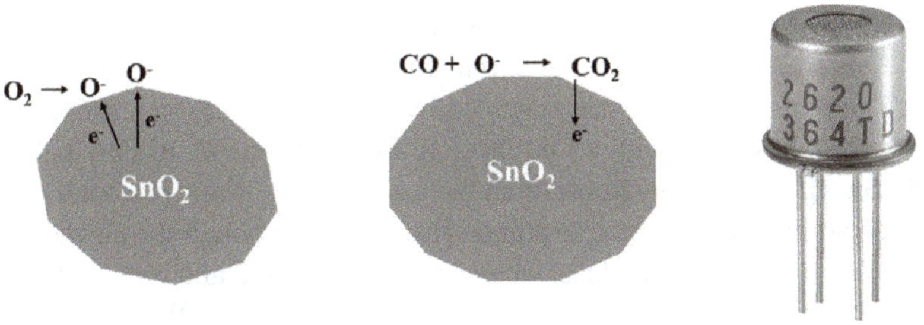

Figure 9.3.2: Left: Electron exchange after chemisorption of oxygen from the surface of an n-type oxide semiconductor. Middle: Interaction with a reducing gas, e.g. carbon monoxide increasing the conductivity of the sensor layer. Right: Commercial Taguchi-type gas sensor (commercialized by the company Figaro Engineering).

An increasing interest arises in planar sensor film structures, whose design is shown in Figure 9.3.3. On a relatively thick plate made of alumina (Al_2O_3), interdigitated metal strips are deposited by photolithography, mostly made from noble metals such as platinum. On top of these structures, the semiconducting oxide is grown in the form of a thin layer. At the underside of the ceramic plate a heating layer made of an inert metal is located [3]. For the fabrication of the semiconducting oxide layer, established techniques such as spin coating, vacuum evaporation or chemical vapor deposition are used.

Figure 9.3.3: Design of a planar film gas sensor. The sensing layer (SnO_2) is deposited on the interdigitated Pt electrode structure which is used to measure the change of conductance in the sensing layer (reproduced with permission from Wiley-VCH [3]).

9.3.2 Schottky-diode-based gas sensors

A Schottky diode represents an alternative type of an electrical sensor which is made from a metal film deposited on a semiconductor surface, forming a heterogeneous metal-semiconductor junction. Typically, Schottky diodes show low resistance under forward bias and very high resistance under reverse bias conditions. The response of such sensors is based on the variation of the work function of the metal interacting with the gas. This leads to a shift of the current-voltage characteristic curve of the diode device. The principle is here demonstrated by means of a hydrogen sensor. A semiconductor oxide is coated with a thin layer of a catalytic metal, usually palladium or platinum. This absorbs hydrogen from the gaseous sample and promotes dissociation of the hydrogen molecules into atoms. These diffuse through the metal and reach the metal/semiconductor interface, where they get polarized and form a dipole layer (Figure 9.3.4). This layer reduces the work function of the metal; thus, electrons can be transferred more easily from the metal to the semiconductor [4].

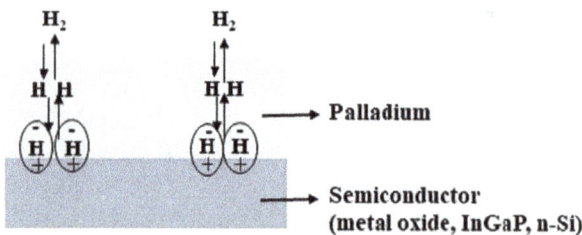

Figure 9.3.4: Principle of a Schottky-diode hydrogen sensor.

9.3.3 Electrode coatings for amperometric sensors

The mode of operation of amperometric sensors is based on an electrolytic cell. A defined voltage is applied between working and counter electrode, and the measured current is a function of the analyte concentration. This implies an electron transfer between the electrode and a redox-active species in the electrolyte. Thus, amperometric sensors are applicable to analytes (ions, gases or organic molecules) which can be oxidized or reduced at potentials within the electrochemical window of water. The measured current I is proportional to the reaction rate at the electrode surface. The flow of current is the generated or consumed charge q involved in the electrode reaction per time t:

$$I = dq/dt \tag{9.3.2}$$

The reaction rate depends on the amount of redox-active species that can reach the electrode surface per time. Hence, not only the concentration of redox-active species but also their diffusion rate determines the reaction rate. In case of a constant diffusion rate, the correlation between analyte concentration and the diffusion limited current I_d is given by the Cottrell equation

$$I_d = z \, F \, A \, c_0 \sqrt{\frac{D}{\pi t}} \tag{9.3.3}$$

where D is the diffusion constant, A is an electrode area, F is the Faraday constant and z is the number of electrons involved in oxidation/reduction reaction.

The typical setup of amperometric sensors is either a two-electrode configuration with a wire of a noble metal (Pt, Au) or carbon (graphite or glassy carbon) as working electrode and a reference electrode (Ag/AgCl or saturated calomel Hg/Hg$_2$Cl$_2$) as counter electrode. Alternatively, a potentiostatic three-electrode configuration can be used. This consists of a working electrode and a counter electrode (both Pt or Au) and a reference electrode (Ag/AgCl) with high impedance; thus, the current flows exclusively between working and counter electrodes, which enables to create a time-constant potential between working and reference electrodes.

The most prominent amperometric sensor is the Clark electrode for determination of dissolved oxygen. The setup is shown in Figure 9.3.5. A polytetrafluoroethylene (Teflon) membrane protects the surface of the platinum working electrode and increases the selectivity since it lets only gaseous components pass into the internal electrolyte and not ions, which could be also reduced at the working electrode. Furthermore, it forms a diffusion barrier layer, enabling constant diffusion of oxygen molecules to the working electrode. This is polarized with a negative voltage of –700 mV versus the Ag counter electrode so that every oxygen molecule reaching the cathode surface is reduced and the diffusion-limiting current conditions are achieved.

The cathodic electrode net reaction is

$$O_2 + 4\,e^- + 4H^+ \rightarrow 2\,H_2O$$

Clark electrodes are also the basis for many enzyme electrodes, e.g. blood glucose sensors using the enzyme glucose oxidase immobilized on the membrane surface, which consumes oxygen for the oxidation of glucose. Thus, the glucose concentration can be measured by means of the amount of oxygen reduced at the working electrode.

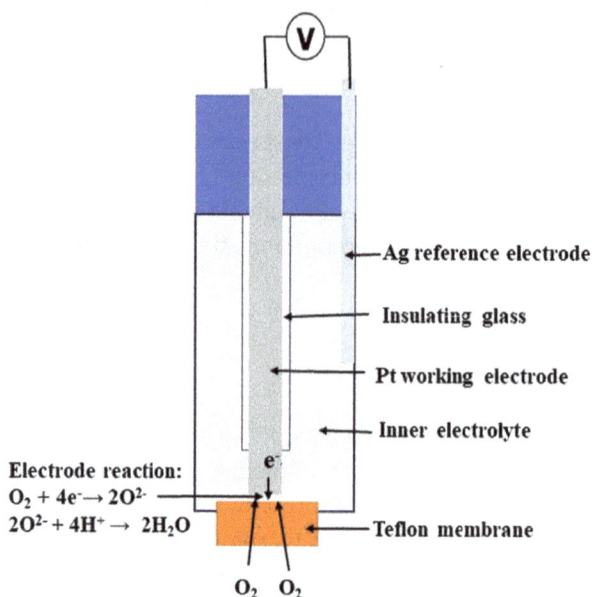

Electrode reaction:
$O_2 + 4e^- \rightarrow 2O^{2-}$
$2O^{2-} + 4H^+ \rightarrow 2H_2O$

—Ag reference electrode

— Insulating glass

— Pt working electrode

— Inner electrolyte

—Teflon membrane

$O_2 \quad O_2$

Figure 9.3.5: Schematic representation of a Clark oxygen electrode.

A major challenge for inorganic material chemistry is the development of electrocatalytically active electrode coatings. Such surface modifications can increase the selectivity of amperometric sensors, as every catalyst reduces the activation energy required for the redox reaction at the electrode surface and increases the reaction rate, which consequently increases the current and therefore makes the sensor response more sensitive. The catalytic coating mediates the electron transfer from the analyte molecules to the electrode or vice versa. Such mediators are also required for an efficient electron transfer between electrode and the active site of immobilized enzymes in enzymatic sensors [5].

Efficient electron transfer mediators are metal-ligand complexes such as ferrocene or osmium(II) polypyridyl complexes [6] as well as ruthenium(II) polypyridyl complexes [7].

Other electrocatalytically active electrode coatings can be prepared from:
- Carbon nanotubes
- Graphene
- Doped diamond
- Prussian blue $Fe_4[Fe(CN)_6]_3$ and other metal hexacyanoferrates
- IrO_2 and RuO_2 nanoparticles
- Gold, platinum or palladium nanoparticles

Amperometric sensors for oxygen, hydrogen peroxide, sulfite (e.g. for determination of sulfites in wine) or nitric oxide (NO) are commercially available. The latter is a molecule of vast physiological significance as it is involved in the regulation of blood pressure. Therefore, a large amount of catalytically active electrode modifications for the selective electrooxidation of NO in physiological milieu have been developed and tested in the past years [8]. Besides the several types of nanoparticles listed above, metalloporphyrins and phthalocyanines (mainly with Fe, Co or Ni as metal center) found an increased interest because they can specifically interact with NO. The design of electrocatalytic electrode coatings, particularly for oxygen reduction, is also of major interest in the development of efficient energy conversion systems such as fuel cells.

9.3.4 Solid membrane potentiometric sensors

The mode of operation of potentiometric sensors is based on a galvanic cell. A typical galvanic cell is the Daniell element which consists of copper and zinc electrodes immersed in solutions of their respective metal sulfates and separated by a porous diaphragm. According to the Nernst equation, the corresponding electrode potentials are a function of the activity of the metal ions in the electrolyte solution. This principle can be used for measuring the concentration of ions in aqueous solutions. In case of ISEs, the potential difference between a working electrode and a reference half-cell, e.g. an Ag/AgCl or an Hg/Hg_2Cl_2 electrode in a saturated KCl solution, is measured. Various types of potentiometric ion sensors are commercialized, which cannot be discussed here in detail. Usually, the measured value is the boundary potential formed at membrane interfaces, where a distribution of ion activities between the sample solution (aq) and the membrane (m) is generated. The phase boundary potential E_{PB} at this interface can be derived according to the Nernst equation:

$$E_{PB} = \phi_m - \phi_{aq} = E^o{}_{PB} + (RT/z_iF) \ln a_{i(aq)} \qquad (9.3.4)$$

if the activity of ion i in the solid membrane $a_{i(m)}$ is held constant [9].

With ϕ being the electric potential and z being the charge of the respective ion.

This approach can be used to determine the concentrations of metal ions in solution as well as for measuring the pH. The solid membrane is in this case a sodium and calcium silicate glass with cation-exchanger properties which provides an outstanding selectivity for protons in solution. This semipermeable membrane is connected to an internal reference electrode (Ag/AgCl) via an inner buffer electrolyte solution (KCl/HCl). Furthermore, pH electrodes in combination with additional gas-permeable membranes are known as Severinghaus electrodes. The assembly can be used for determination of acidic or basic gases such as CO_2, SO_2 and NH_3 which, after passing the membrane, undergo a hydrolysis reaction and therefore change the pH in an internal solution between the gas-permeable membrane and the pH-sensitive glass membrane. The dissociation equilibrium for CO_2 is

$$CO_2 + H_2O \rightleftharpoons H_2CO_3 \rightleftharpoons HCO_3^- + H^+$$

Commonly, ISEs consist of a solid diaphragma which can be electrically contacted to a wire directly or via an internal electrolyte. Crystal membrane electrodes are widely used types of solid membrane electrodes using an internal electrolyte. The membrane is made of an ion-conducting crystalline material which is the basis for the specific response of the electrode. A prominent example is the silver/silver chloride electrode (Ag/AgCl) which cannot only be used as a reference electrode, but can also be applied as a working electrode for the measurement of chloride anions, e.g. for determination of salt concentrations in seawater. Many solid-state membranes are prepared using low-soluble silver sulfide (Ag_2S) as a crystalline matrix. This can be easily pressed to pills and has a high ionic conductivity due to disorders in doped crystal structures. Pure Ag_2S can be used for sensing of S^{2-}. More often it is used as a blend with silver halides (AgX) for halide-sensitive electrodes (X = Cl, Br, I). It is also possible to prepare membranes in which the highly insoluble Ag_2S is used as a matrix and a more soluble sulfide as an additive, e.g. CuS, PbS or CdS. In this way, solid membrane ISEs for determination of Cu^{2+}, Pb^{2+} or Cd^{2+} can be configured, respectively [10].

A widely used single-crystal electrode is the fluoride electrode because of its good performance and high selectivity. The ion-conductive crystalline material consists of LaF_3 doped with Eu^{2+}. The doping with divalent Eu^{2+} induces anion-conducting properties due to anionic defects in the structure which are selective for F^-. Usually, an internal electrolyte is incorporated between metal wire and LaF_3 crystal (Figure 9.3.6). Thus, the crystalline membrane is in contact with the internal electrolyte with constant activity $a_{i,2}$ of F^- and also to the sample solution with the unknown activity $a_{i,1}$ of F^-. The potential difference between these two interfaces is affected by the equilibrium at the interface between crystal and water:

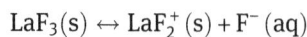

$$LaF_3(s) \leftrightarrow LaF_2^+(s) + F^-(aq)$$

The dissolved fluoride anions leave positive charged vacancies on the surface. Fluoride anions in the sample solution will affect this equilibrium. The same arises at the opposite side, where the membrane is in contact with the internal electrolyte. In this case, the transmembrane potential ΔE_m is

$$\Delta E_m = RT/zF \ ln(a_{i,1}/a_{i,2}) \tag{9.3.5}$$

All other phase boundary potentials in the electrochemical cell have to be kept constant.

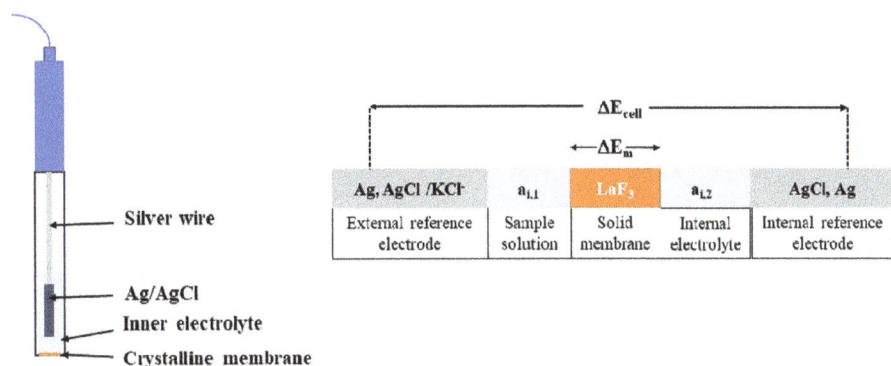

Figure 9.3.6: Basic design of an ISE with a crystalline solid membrane. This is connected to an external reference electrode (left). In the case of fluoride electrodes, the crystalline membrane consists of LaF$_3$ doped with Eu^{2+}. The formation of the corresponding transmembrane potential ΔE_m is illustrated on the right, with $a_{i,1}$ as the activity of fluoride in the sample solution and $a_{i,2}$ as the activity of fluoride in the internal electrolyte.

9.3.5 Solid electrolyte membranes: the lambda sensor

The lambda probe is one of the mostly used and robust electrochemical sensors, incorporated in every car with fuel engine. It was commercialized in 1977 by Bosch and is based on the principle of a redox electrode. It monitors the residual oxygen content in exhaust gas of cars compared to the oxygen content in air (~21%) as reference. The sensor regulates the air/fuel (A/F) ratio in combustion engines to enable maximum efficiency of the catalytic converter. The working principle is shown in Figure 9.3.7. The sensor consists of two porous platinum electrodes which enclose a solid ceramic electrolyte layer (ZrO$_2$ blended with Y$_2$O$_3$). The sensor is placed between two inlets, one containing the reference air and the other the exhaust gas. Usually, the oxygen concentration in air is higher than in the exhaust gas; thus, oxygen molecules will diffuse through the permeable layer to the side with the exhaust gas, thereby O$_2$ molecules are reduced by the platinum electrode to two O^{2-} anions at high temperature. These can move through the doped zirconia matrix at temperatures higher than 350 °C, at which the material gains a high ion conductivity. The doping with Y^{3+} creates oxygen defects in the structure due to charge neutralization, and O^{2-} anions can hop between these holes and be transported through the material in an electrical field. This so-called yttrium-stabilized zirconia is also the basis for solid oxide fuel cells, often used with a co-dopant.

In lambda sensors, the voltage between the two platinum electrodes is measured, which is generated by the potential difference caused by the different partial oxygen pressures pO_2 on both sides. Due to the reduction of O_2 at the inner side of the sensor (reference air side), an electron deficiency is created at this electrode, whereas at the second electrode (exhaust gas side), where the arriving O^{2-} ions are oxidized, an electron excess is generated.

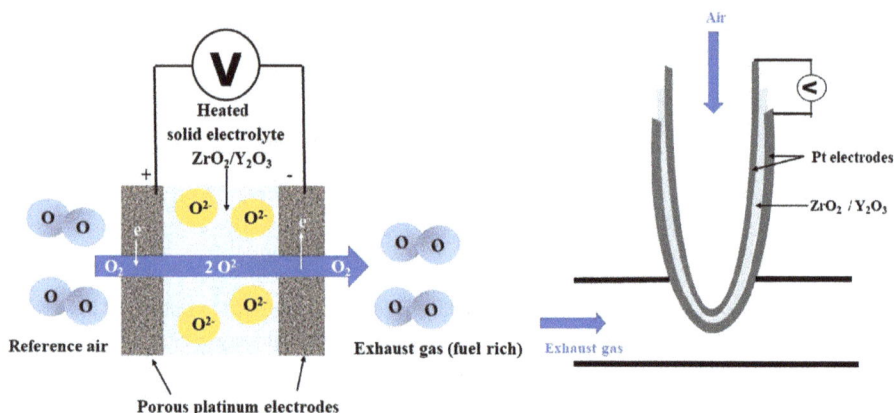

Figure 9.3.7: Scheme of the mode of operation of a lambda sensor (left) and general design of the sensor placed in the exhaust gas flow (right).

9.3.6 Chemical-sensitive field effect transistors

Miniaturization and multiplexing of chemical sensors lead to the development of sensor technologies integrated in electronic semiconductor devices. The most prominent example is the metal oxide-semiconductor field effect transistor (MOSFET). The basic setup can be used in combination with chemoresistors, gas-sensitive layers or ion-selective membranes. The common feature is that the metal gate of the transistor device is removed and the insulating metal oxide is either directly exposed to the sample solution or in contact with an ion-selective membrane, the latter are known as ion-selective field effect transistors (ISFET) as outlined in Figure 9.3.8. Though the gate can also be a metal which can interact with gas molecules, e.g. palladium for hydrogen detection as shown above in case of Schottky diodes. The different approaches are nowadays summarized as "CHEMFETs" [11]. Direct contact of the gate insulator, e.g. silica (SiO_2) leads to a pH-sensitive FET, which was developed by P. Bergveld in 1970 [12], and is nowadays commercialized by different companies. In this assembly, the surface hydrolysis of Si–OH groups of the gate oxide is affected by the pH value. The hydroxyl groups at the silica surface can donate and accept protons according to the following pH-dependent equilibria:

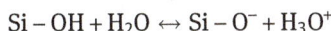

$$Si - OH + H_2O \leftrightarrow Si - O^- + H_3O^+$$

$$Si - OH + H_3O^+ \leftrightarrow Si - OH_2^+ + H_2O$$

This approach can also be used for sensing of acidic or basic gases in combination with gas-permeable membranes in analogy to the Severinghaus electrodes. For the use of ISFETs and related devices, miniaturized reference electrodes are required. In an ideal setup, the reference electrode would be not a separated element but integrated in the sensing system on a single chip device. Sensing layers and integrated reference electrodes are a challenge for inorganic material chemistry and nanochemistry, forming catalytically active structures, e.g. discontinuous or porous metal films and suspended metal gates forming a mesh for gas sensing [13]. Advanced inorganic materials such as carbon nanotubes or graphene are also gaining interest as sensitive layers in CHEMFETs. In this case, the nanotube material forms the channel between source and drain. The current flowing across this channel is very sensitive to the adsorption of gas molecules.

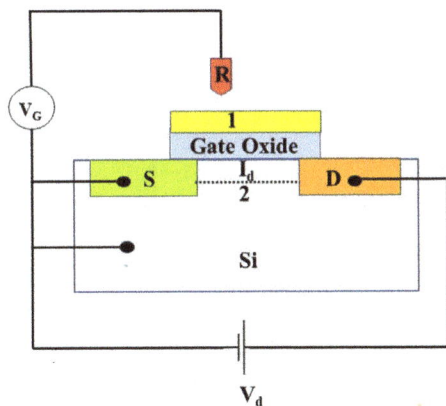

Figure 9.3.8: Schematic view of an ISFET. S: source; D: drain; R: reference electrode; V_g: gate voltage; V_d: drain voltage; I_d: drain current; 1: ion-selective membrane or electrolyte; 2: channel. The electron flow in the channel between drain and source I_d depends on V_d, which is fixed and on the potential difference over the gate oxide. Therefore, I_d is influenced by the interface potential between gate oxide and sample solution or between membrane and sample solution, which again changes the gate voltage V_g.

9.3.7 Conclusion and outlook

In this chapter, selected examples of electrical and electrochemical sensors were discussed which are based on inorganic chemical recognition and signal transduction elements. This overview cannot claim to be complete. Nevertheless, for the sake of completeness, it should be mentioned that inorganic materials are also applied in optical chemical sensors, mainly fluorescent nanoparticles based on gold, lanthanide-doped oxides or fluorides (e.g. photon upconversion nanoparticles) or quantum dots (nanoscale semiconductors).

The improvement of the selectivity and sensitivity of the recognition system, the modification of electrode surfaces, e.g. with electrocatalytic layers, as well as further miniaturization of the sensor elements and their integration in multisensor arrays and systems are further challenges for inorganic materials chemistry. Their combination with online and data processing tools helps to manage a large amount of analytical data which accrue for example in industrial process control. Comprehensive concepts for process analytical technologies require the integration of smart chemical sensors, particularly to control complex processes in the chemical industry and to realize what is aimed for as "Industry 4.0" [14]. In this regard, a smart sensor system is able to measure multiple components, does not require extensive calibration, can easily be integrated in the process environment and can be operated autonomously. Thus, further progress in this field can only be achieved by interdisciplinary cooperation of inorganic chemistry, nano- and material chemistry, engineering and data processing.

References

[1]   Thevenot DR, Tóth K, Durst RA, Wilson GS. Pure Appl Chem, 1999, 71, 2333–48.
[2]   Gründler P. Chemical Sensors, Springer, Berlin, Heidelberg, 2007.
[3]   Tiemann M. Chem Eur J, 2007, 13, 8376.
[4]   Lee SP. Sensors, 2017, 17, 683.
[5]   Banica F-G. Chemical Sensors and Biosensors, Wiley, Chichester, 2014.
[6]   Heller A. Acc Chem Res, 1990, 23, 128.
[7]   Kosela E, Elzanowska H, Kutner W. Anal Bioanal Chem, 2002, 373, 724.
[8]   Brown MD, Schoenfisch MH. Chem Rev, 2019, 119, 11551.
[9]   Bakker E, Bühlmann P, Pretsch E. Talanta, 2004, 63, 3.
[10] Janata J. Principles of Chemical Sensors, Springer, Dordrecht, 2009.
[11] Bergveld P. Sens Actuators, 1985, 8, 109.
[12] Bergveld P. IEEE Trans Biomed Eng, 1970, BME-17, 70.
[13] Yoshizumi T, Miyahira Y. Field-effect transistors for gas sensing. In: Pejovic MM (Eds). Different Types of Field-Effect Transistors – Theory and Applications, IntechOpen, 2017. DOI: 10.5772/intechopen.68481.
[14] Eifert T, Eisen K, Maiwald M, Herwig C. Anal Bioanal Chem, 2020, 412, 2037–45.

9.4 Transparent conducting oxides

Hideo Hosono

Non-transition metal oxides are optically transparent but mostly electrically insulating in general. However, some class of oxides has both high transparency and conductivity. These materials are called transparent conducting oxides (TCOs) and are widely used as transparent metals in displays and solar cell applications. A TCO is a heavily carrier-doped transparent oxide semiconductor (TOS). In this chapter, TCO and TOS are outlined along with materials' design concepts.

9.4.1 Transparent conducting oxides and transparent oxide semiconductors

Transparent oxides are mostly electrical insulators such as Al_2O_3 and SiO_2. However, some of the transparent oxides have metallic conductivity or tunable conductivity by impurity doping or applying an external voltage. The former and the latter are called TCO and transparent conductive semiconductors (TOS), respectively.

TCO is used as a transparent metal for applications in which both optical transparency and high electrical conductivity are required. The major application concerns transparent electrodes for electronics represented by flat panel displays and solar cells. Figure 9.4.1 shows the constitution of organic light-emitting displays (OLEDs). Since an OLED emits light by flowing current through the organic LEDs, emitted light passes through the electrode to forward. Thus, in addition to low resistivity, high transparency is required. This is the reason why a TCO thin film is inevitable for the displays. TOS is a wide-gap semiconductor like GaN and SiC, and its research has rapidly grown during the last three decades [1, 2]. The essential difference between TCO and TOS is controllability of the Fermi level. Since conduction carriers in oxide materials are practically restricted to electrons, both TCO and TOS are n-type materials. In TCO, the Fermi level is located above the conduction band minimum (CBM). Thus, the Fermi level is uncontrollable by applying voltage because biasing voltage is impossible due to high carrier concentration, i.e., TCO works as a persistent transparent metal. Meanwhile, the Fermi level of TOS is controllable by impurity doping or applying voltage. When the Fermi level is upshifted beyond the CBM, the resulting TOS functions as TCO.

A thin-film transistor (TFT) composed of semiconductors with three terminals (source, drain and gate) is a fundamental building device and functions as a switch driven by gate voltage in electronic circuits. In flat panel displays, a TFT array works as the backplane to make an image as shown in Figure 9.4.2. Amorphous hydrogenated Si (a-Si:H) is exclusively used as the semiconductor in TFTs for drive pixels of displays so far. However, the mobility of a-Si:H is insufficient to drive OLEDs and high-resolution liquid panel displays. Amorphous $InGaZnO_x$ (a-IGZO), a

https://doi.org/10.1515/9783110733471-004

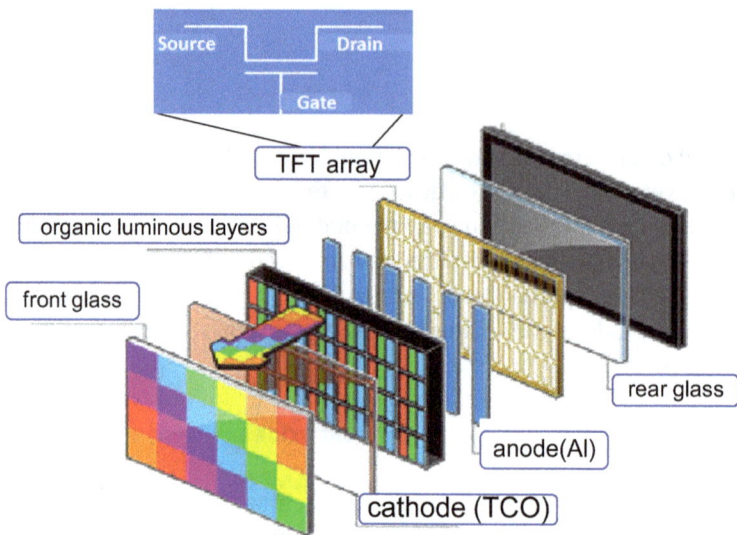

Figure 9.4.1: Constitution of an OLED-TV. TCO and TOS are used as cathode and TFT, respectively.

transparent amorphous oxide semiconductor (TAOS), is now widely applied for state-of-the-art displays [3]. Large-sized OLED TVs are realized by adopting a-IGZO-TFTs as the backplane utilizing higher mobility and easy deposition [4].

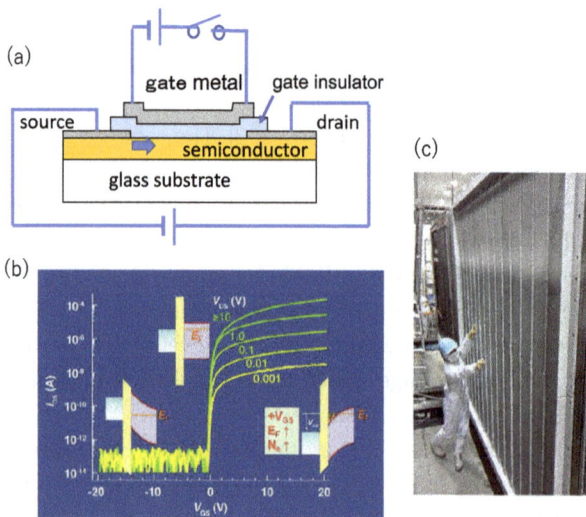

Figure 9.4.2: Structure of a TFT (a), performance (b) and photo of ceramic target for sputtering of IGZO (c). Facile synthesis of large-sized and dense ceramic targets is an advantage of IGZO. V_{GS}: voltage applied to gate-source; I_{DS}: current between source and drain.

9.4.2 Outline of TCOs [5]

9.4.2.1 Electrical conductivity

Figure 9.4.3(a) shows the electrical conductivities of several classes of conductive materials. Metals such as Ag and Al possess high electrical conductivities of the order of 10^5 S/cm. Semiconductors such as Si and Ge have medium conductivities. Impurity doping to pure semiconductors markedly increases the conductivities up to the order of 10^2 S/cm. The typical conductivities of TCOs such as In_2O_{3-x} and Sn-doped In_2O_3 are located in the middle of metals and doped semiconductors. This would be natural because TCOs are originally doped TOSs. The electrical conductivity, $\sigma = ne\mu$, is a product of carrier concentration n, elementary electric charge e and mobility μ.

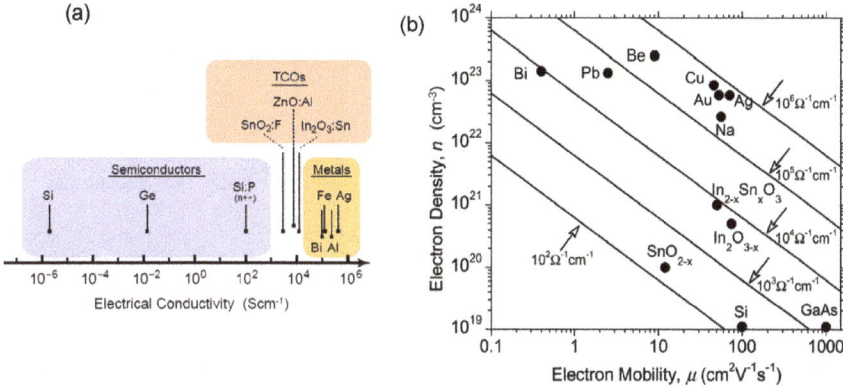

Figure 9.4.3: Electric conductivity (a) and relation between electron density and mobility (b) in different classes of conducting materials.

Figure 9.4.3(b) shows the relation between n and μ in the conducting materials. Ag metal exhibits high electrical conductivity because of high carrier concentration in the order of approximately 10^{23} cm^{-3}. The mobility of Ag metal is not markedly high among these conductors. P-doped Si shows the lowest electrical conductivity among them because the carrier concentration is low. It can become an electrical conductor simply because the carrier mobility of Si is much higher than that of Ag metal. Even if Si is heavily doped with the carrier concentration of up to 10^{19} cm^{-3}, the mobility decreases by carrier scattering and the conductivity does not increase as high as Ag metal. Sn-doped In_2O_3 (ITO) is the most widely used TCO. Its conductivity is 1/10 smaller than that of Ag metal. Because the carrier mobility of ITO does not differ largely from that of Ag metal, the carrier concentration of ITO, which is 1/10 smaller than that of Ag metal, is responsible for its one-tenth smaller conductivity. These features can be extended to other conducting materials, and their carrier concentrations and mobilities are plotted in the figure. Metals are located at the top left region in

Figure 9.4.3(b), which indicates that the metals have extremely high carrier concentration but their mobilities are not so high. In contrast, doped semiconductors are located at the right bottom region because of low carrier concentration and high mobility. It is obvious from this figure that TCOs are located between metals and doped semiconductors. Because the mobilities of TCOs are not largely different from those of typical metals such as Ag, each TCO can be regarded as a metal with small carrier concentration, which is called a degenerate semiconductor. The schematic electronic structure and the temperature dependence of the conductivity of a degenerate semiconductor, a metal and a semiconductor are shown in Figure 9.4.4. Although the thermal activation energy is necessary for semiconductors to generate carriers, it is unnecessary for metals and degenerate semiconductors, indicating that the Fermi level is located beyond the CBM.

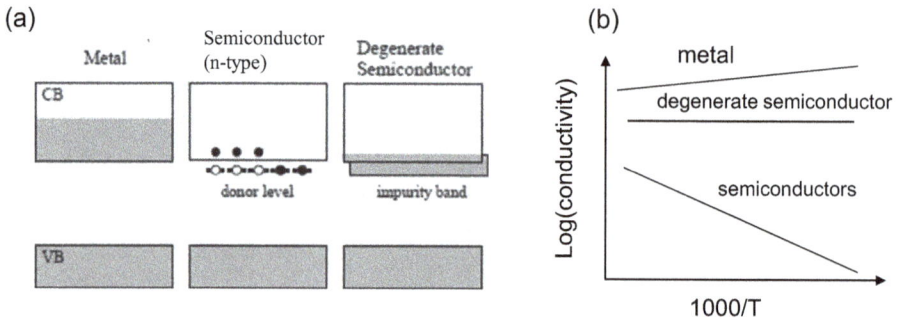

Figure 9.4.4: Schematic energy structure (a) and temperature dependence of conductivity (b) of conductive materials.

9.4.2.2 Optical transparency

The main group metal oxides such as MgO and Al_2O_3 are electrically insulating but optically transparent in the visible wavelength range from 400 to 800 nm. Therefore, oxide single crystals or glasses are frequently used as windows in optical application. Typical oxides are transparent in the range approximately from 300 to 10 μm including the visible region. On the other hand, conventional semiconductors such as Si and Ge have a window range only in the infrared region. Therefore, it is not surprising that TCOs are transparent in the visible region. In insulators, the absorption edge in the infrared region is limited by the lattice vibration and that in the ultraviolet region by the fundamental absorption caused by the electronic transition from the valence band to the conduction band.

When free carriers are introduced into the parent (undoped) material of TCOs, optical absorption and reflection are induced by the carriers. The collective motion of the conducting carriers behaves as a kind of plasma in materials. When light is irradiated to the conductors, the carriers oscillate at the frequency of the light, which is called plasma oscillation. This motion gives rise to the reflection of the light at the surface of the conductors. However, against the light with higher frequency, the carriers cannot catch up with the fast electric field oscillation of the light. Then, light transmits through the conductors without causing the plasma oscillation (i.e., reflection) at the surface. This maximum threshold frequency is called plasma frequency μ_p and can be expressed by eq. (9.4.1). The plasma frequency ω_p can be converted to the wavelength λ_p as shown in eq. (9.4.2):

$$\omega_p = \frac{ne^2}{\varepsilon_0 m} \tag{9.4.1}$$

$$\lambda_p = \frac{2\pi c}{e} \sqrt{\frac{\varepsilon_0 m}{n}} \tag{9.4.2}$$

Here n, m, ε_0 and c denote carrier concentration, electron rest mass, permittivity and speed of light in vacuum, respectively. For Al metal with high carrier concentration over 10^{22} cm^{-3}, the threshold λ_p is located in vacuum UV region (below 200 nm) as shown in Figure 9.4.5. The visible light is totally reflected at the surface. This is the origin of the metallic cluster of the Al metal, and the wide-ranged reflection of Al is often used as a mirror. On the other hand, since Sn-doped In_2O_3 (ITO), a representative TCO, has a carrier concentration lower by an order of magnitude than Al metal, the λ_p moves to the near-infrared region, which results in the transparency in the visible region because ITO has a sufficient band gap (>3 eV) to transmit visible light. Consequently, the maximum carrier concentration of TCOs should be less than ~10^{21} cm^{-3} because the carrier concentration beyond this value makes the transparent windows narrower in the visible region.

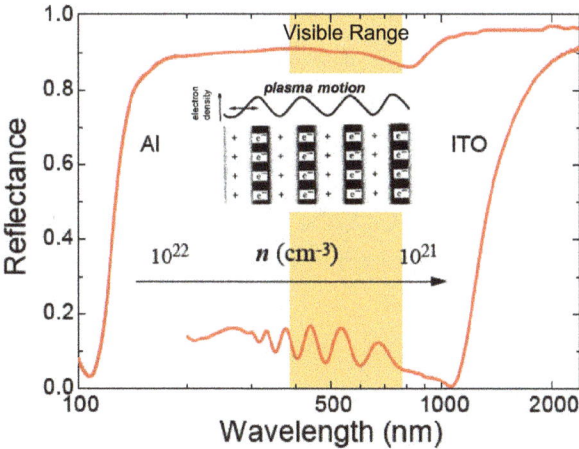

Figure 9.4.5: Optical reflection spectra of Al metal and Sn-doped In_2O_3 (ITO) thin films. Reflection is due to plasma oscillation of free electron.

9.4.3 Materials design for TCOs

The essence of the materials design for TCOs is how we can dope carrier into oxides, because the metal oxides except the transition metal (with an open d-shell) systems are basically transparent due to the large band gap. There are two approaches to increase the electrical conductivity, an increase of carrier concentration and/or mobility. Because the carrier concentration is controllable, the carrier concentration is in many cases independent from the nature of materials. Therefore, the mobility is the material-dependent parameter to be designed. The mobility is a velocity of carriers under unit electric field strength and inversely proportional to the effective mass m^* of carriers as follows:

$$\mu = \frac{v}{E} = \frac{e\tau}{m^*} \tag{9.4.3}$$

The effective mass m^* is evaluated by band calculations because m^* is inversely proportional to the curvature of an energy band in the energy diagram as a function of the wavenumber k as follows:

$$\frac{1}{m^*} = \left(\frac{2\pi}{h}\right)^2 \frac{\partial^2 E}{\partial k^2} \tag{9.4.4}$$

In a simple one-dimensional crystal with the lattice parameter a, a simple energy band is formed and composed of a one-dimensionally aligned atomic orbital ϕ_m as shown in Figure 9.4.6.

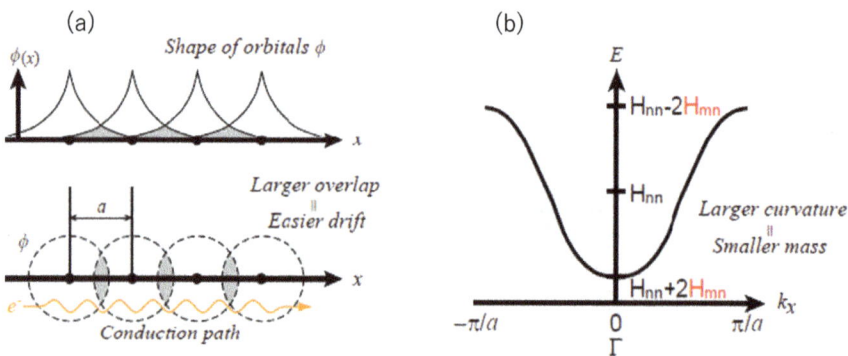

Figure 9.4.6: Overlap of post-transition metal vacant s-orbitals (a) and the conduction band composed of these orbitals (b).

The energy E of the band is expressed in eqs. (9.4.5) and (9.4.6) by using the orbital interaction H_{mn} between the neighboring orbitals:

$$H_{mn} = \int \phi_m^* \hat{H} \phi_n dx \tag{9.4.5}$$

$$E = H_{nn} + 2H_{mn} \cos(ka) \approx H_{nn} + 2H_{mn} - H_{mn}(ka)^2 \quad (x \approx 0) \tag{9.4.6}$$

From eqs. (9.4.3) to (9.4.6), a simple conclusion (9.4.7) that the mobility is proportional to the orbital interaction H_{mn} is obtained:

$$\mu \propto \frac{\partial^2 E}{\partial k^2} = -2H_{mn}a^2 \tag{9.4.7}$$

where H_{mn} becomes large when the overlap of the atomic orbitals between ϕ_m and ϕ_n is large because H_{mn} is calculated from a product of the atomic orbitals as seen in eq. (9.4.5). Making large overlap of the orbitals just corresponds to enhancing a probability of carrier transport in the crystal. Then, carriers may become mobile between the orbitals through their overlapped areas. Therefore, a smooth conduction path for carriers must be constructed in wide-gap oxides by making sufficient overlap of component orbitals to provide them with large carrier mobility [5].

9.4.4 Examples of n-type TCOs

From the concept of the materials design for n-type TCOs, oxides which are composed of the post-transition metal (PTM) cations with $d^{10}ns^0$ ($n = 4$ or 5) configuration and have the crystal structures with edge-sharing octahedra that can be candidates [6]. Actually, such materials often show transparent n-type conductive properties. The transparent conductors $CdIn_2O_4$ and $MgIn_2O_4$ with spinel structure meet these requirements. Here, SnO_2 is taken as an example of a representative TCO material. This material has a rutile-type structure, composed of edge-sharing octahedra. The inter-atomic distance is 3.0 Å in Sn metal, while the inter-cationic separation is 3.2 Å in SnO_2. Such a close separation between Sn in the oxide and the metal suggests that Sn^{4+} $5s^0$-orbitals, the lowest unoccupied atomic orbital, in the oxides can largely overlap with each other like in the metal.

Figure 9.4.7 shows the energy bands and density of the states of SnO_2 [7] along with the unit cell structure. The conduction band bottom (CBM) is primarily composed of Sn 5s-orbitals and has large energy dispersion, which means light electron mass. This large dispersion originates from large overlap of Sn 5s-orbitals in this crystal structure, i.e., separation between Sn^{4+} ions in the rutile-type structure composed of edge-sharing SnO_6 octahedra is rather short compared with that in the perovskite structure with corner-sharing MO_6 octahedra. Closely connected oxygen octahedra of PTM cations with $d^{10}ns^0$ ($n = 4$ or 5) configuration work as a good electron path. These oxides having edge-sharing octahedra of PTM are favorable as TCOs as depicted in Figure 9.4.8(a).

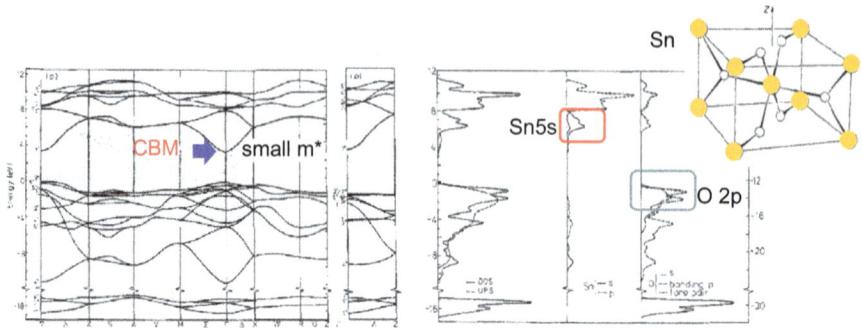

Figure 9.4.7: Band structure and density of states of SnO_2.

(a)   (b)

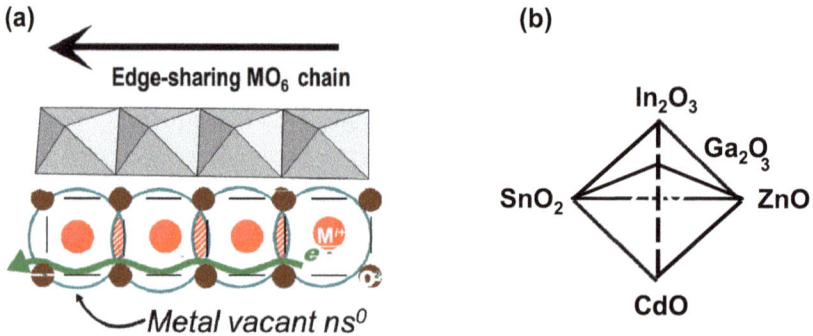

Figure 9.4.8: Connection of the SnO_6 octahedra and the conduction pathway in SnO_2 with rutile-type structure (a) and *polyhedral* representation of transparent conductive oxides (b). The chemical compositions within this polyhedron are candidates for TCOs. This *polyhedron* gives transparent conductivity.

Typical TCO materials are SnO_2, In_2O_3, ZnO, Ga_2O_3 and CdO, and each of them meets the above requirements. The solid solutions and complex oxides are good candidates for TCO materials as illustrated in Figure 9.4.8(b). For practical applications as transparent metals, these oxides appropriately doped are used to enhance the conductivity like F-doping to SnO_2 (abbreviated as FTO), Sn-doped In_2O_3 (ITO) and Al-doped ZnO (AZO).

9.4.5 Amorphous TCO and TOS

What happens when TCO materials become amorphous? In an amorphous state, structural disorder concentrates on energetically weak structural units. In most amorphous materials, structural disorder appears prominently as the bond angle distribution. When the bond angle has a large distribution, how the effective mass

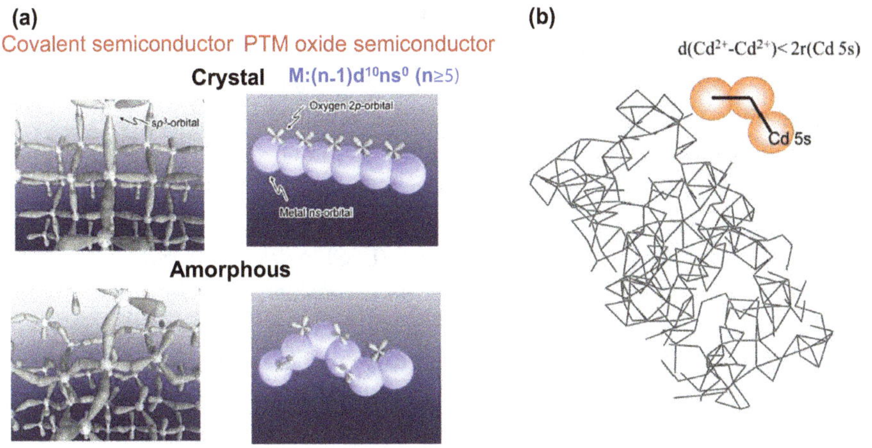

(a)
Covalent semiconductor PTM oxide semiconductor

Crystal M:(n-1)d¹⁰ns⁰ (n≥5)

(b)

$d(Cd^{2+}\text{-}Cd^{2+}) < 2r(Cd\ 5s)$

Amorphous

Figure 9.4.9: Schematic orbital drawing of the conduction band bottom (CBM) in crystalline and amorphous semiconductors (a). PTM: post-transition metal cation with electronic configuration $(n-1)\,d^{10}ns^0$, where n is the principal quantum number. Connectivity of PTM s-orbitals at the CBM in amorphous TOS ($2CdO$-GeO_2). When the separation d of two neighboring metal ns-orbitals is smaller than the double of the metal ns-orbital radius, two metal cations are connected by a line. The atomic position of the cations was determined by simulating the experimental radial distribution function using the reverse Monte Carlo method.

(in other words, the transfer rate between neighboring cation s-orbitals or overlap integrals) is modified for carrier electrons? The following two cases are considered: (i) covalent semiconductors and (ii) ionic semiconductors. In the former case, the magnitude of the overlap between the unoccupied orbitals of the neighboring atoms is very sensitive to the variation in the bond angle. As a consequence, rather deep localized states would be created at somewhat high concentrations; therefore, the drift mobility would be largely degraded [8].

On the other hand, the magnitude of the overlap in the latter case is critically different depending on the choice of the metal cations; when the spatial spread of the vacant s-orbital of the metal cation is larger than half the nearest inter-cation separation, the magnitude of overlap between the metal vacant s-orbitals constituting the CBM is large and should be insensitive to the bond angle variation because the s-orbital is isotropic in shape. As a consequence, these ionic amorphous materials have large electron mobility comparable to that in the corresponding crystalline phase. In the case that the spatial spread of the metal s-orbital is small, such a favorable situation cannot be expected. The spatial spread of the s-orbital of a metal cation is primarily determined by the principal quantum number (n) and is modified by the charge state of the cation as discussed for the crystalline TCOs. Thus, candidates for high-mobility TAOSs are found in oxides of PTM cations with an electronic configuration $(n-1)d^{10}ns^0$, where $n \geq 5$ [3, 8]. Please note that for crystalline oxide semiconductors, this requirement is relaxed to $n \geq 4$ as exemplified by ZnO with the $(3d)^{10}(4s)^0$

configuration. Figure 9.4.9(a) draws the difference in orbitals between Si and PTM oxides for crystalline and amorphous states. The drastic reduction of the electron mobility in the amorphous state from c-Si may be understood intuitively from the figure, whereas mobility in the c-PTM oxides is reserved even in the amorphous state. In a sense, the situation of CBM in PTM oxides is like that in amorphous metal alloys (the conductivity is slightly lower than that in crystalline phase) in the aspect that metal orbitals dominantly constitute the electron pathways. This simple idea is demonstrated quantitatively by observing the density of states (DOS) by inverse photoelectron spectroscopy and analyzing the computed DOS on atomic positions determined by a combination of the X-ray radial distribution function with reverse Monte Carlo simulations. Figure 9.4.9(b) illustrates the connectivity of Cd $5s$-orbitals at the CBM of amorphous 2CdO-GeO$_2$ as an example of TAOS. Here, Cd^{2+} with $(4d)^{10}(5s)^0$ meets the requirement for a post TM cation for TAOS. Here, two Cd^{2+} ions are connected by a line for visualization of orbital overlap when the $2 \times$Cd $5s$-orbital radius (Slater type) is larger than the interatomic separation. It is clearly observed that the lines are three-dimensionally connected throughout the sample forming a percolated electron pathway [9].

TAOS has several common and unique properties, which are not seen in conventional amorphous semiconductors. The first point concerns their large electron mobilities >10 cm^2/(V s), which is higher by an order of magnitude than that in a-Si: H. Second is that a degenerate state, i.e., TCO, can be realized. This is totally different from the other amorphous semiconductors. For instance, c-Si is easily changed to the degenerate state by impurity doping (~10^{16} cm^{-3}), but no such state is attained in a-Si:H, that is, carrier conduction takes place by hopping through localized tail states in conventional amorphous semiconductors. This is the reason why mobility in the amorphous state is so small as compared with that in the crystalline state. On the other hand, in TAOS, the Fermi level (E_F) can exceed the mobility gap easily by carrier doping, leading to the situation that the band conduction occurs. It is considered that this striking difference originates from the nature in chemical bonding between the materials, i.e., strong ionic bonding with spherical potential is much favorable to form a shallow tail state having small DOS. A representative TAOS material is amorphous InGaZnO$_x$ (IGZO), which is widely used as the semiconductor for TFTs to drive the pixel of flat panel displays as described in Figure 9.4.1. Amorphous TCO is applied as the transparent metal in flexible electronics taking advantages of excellent homogeneity and bending toughness over polycrystalline analogs although their conductivity is inferior to that of polycrystalline thin films. Amorphous In$_2$O$_3$, Sn-doped In$_2$O$_3$ and In$_2$O$_3$-ZnO are well-known amorphous TCOs [10].

9.4.6 p-Type TOS [6, 11]

n-Type TOSs can be designed by selecting metal cations with spatially spread s-orbitals that constitute the CBM and a crystalline structure with a smaller separation between metal cations as illustrated in Figure 9.4.10(a). However, no guidelines for designing p-type semiconductors were presented until 1997. Needless to say, pn-junctions are the origin of various semiconductor functions; therefore, high-quality p-type transparent semiconductors are essential for not only transparent oxide electronics but also all-solid dye-sensitized solar cells. For wide-gap oxides, the VBM, which serves as the conduction path of holes, is mainly composed of oxygen $2p$-orbitals and the contribution of the orbitals of metal cations is generally small. Therefore, the VBM is little dispersed (i.e., the effective mass of holes is large), and the energy level is deep. This is why p-type TOSs are difficult to realize. Resolving this problem is a key strategy in realizing p-type TOSs. Three previously proposed approaches are given below:

(i) Oxides of transition metal cations with electronic configuration of $(n - 1)d^{10}ns^0$
     $(n = 4$ or $5)$ [12]:
Although transition metal cations have orbitals with energy levels close to that of oxygen $2p$-orbitals, most of them absorb visible light owing to a d-d transition. Therefore, oxides of cations with closed-shell d-orbitals, such as Cu^+ and Ag^+, are considered to have the potential to exhibit p-type conductivity. As shown in Figure 9.4.10(b) for such oxides, the antibonding orbital component of the bond composed of metal d-orbitals and oxygen $2p$-orbitals constitutes the VBM, and the holes doped into the VBM are delocalized to realize p-type conductivity. A typical example is delafossite $CuMO_2$ ($M = Al^{3+}$, Ga^{3+}, In^{3+}) with dumbbell-type O–Cu–O bonds as the building block. Although several p-type transparent semiconductors have been found to date, no TFT operation based on Cu^+ has been reported to date like the Cu_2O case by Shockley. The high concentration of hole traps at the surface associated with oxidation of Cu^+ would be the most plausible cause of these results.

(ii) Oxides of metal cations with electronic configuration of ns^2 [13]:
Cations with an electronic configuration of the form ns^2 have lone pairs similar to those of anions. For oxides of such metal cations, the VBM is mostly occupied by s-orbitals, as shown in Figure 9.4.10(c). Lone pairs occupy the spatially dispersed s-orbitals, which overlap with the s-orbitals of adjacent cations via the oxygen, forming a largely dispersed band above the oxygen $2p$ band. Oxides of Sn^{2+} with an electronic configuration of $5s^2$ are typical and exhibit p-type conductivity in Hall effect measurements. In addition, the first-ever oxide TFT that can operate as a p-channel TFT was realized using SnO for the active layers.

Realization of CMOS based on an oxide semiconductor had been a milestone in oxide electronics. This objective was first attained in 2011 using SnO [14]. Although much improvement was reported [15], the CMOS performance is still insufficient for

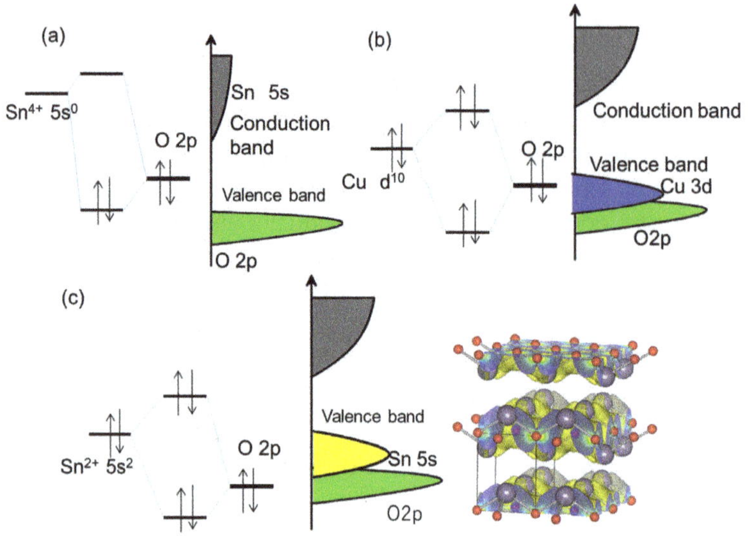

Figure 9.4.10: Schematic energy diagram: (a) n-type SnO$_2$, (b) p-type Cu$_2$O and (c) p-type SnO. The right is orbital drawing of VBM. Sn 5s electrons with large spread occupy the VBM.

application, unfortunately. Low-temperature processing as well as improvement of mobility is strongly required for display application.

It has been reported that chalcogenides and oxides of Pb^{2+} and Bi^{3+}, both with an electronic configuration of $6s^2$, cannot be used to enhance hole transport properties because the energy level of the $6s$ electrons is much deeper than the VBM. Although oxides of Sb^{3+} with an electronic configuration of $5s^2$, similar to that of Sn^{2+}, are expected to exhibit p-type conductivity, and no examples of such oxides have been reported to date.

(iii) Oxides of metal cations with electronic configuration of nd^6 [16]:
For oxides of Rh^{3+} and Ir^{3+}, both with an electronic configuration of $4d^6$ or $5d^6$, these cations stabilize in the low-spin-state octahedral configuration, where electrons occupy the three orbitals, i.e., d_{xy}, d_{yz} and d_{xz}. The $4d$- and $5d$-orbitals are spatially spread, and the state in which two electrons occupy each of the three d-orbitals is similar to that in (2), i.e., a pseudo s-orbital with large spread. When doped with holes, these oxides are expected to exhibit p-type conductivity. A typical example is ZnRh$_2$O$_4$, which has a normal spinel structure. Similar to the case of ZnO, the Zn^{2+} ions in this material are coordinated in tetrahedra, which are not continuously connected, and do not exhibit n-type conductivity. ZnRh$_2$O$_4$ is the only oxide that is known to exhibit p-type conductivity even in the amorphous state and has been reported to form pn-diodes on plastic substrates when combined with TAOS. Subsequently, p-type conductivity was reported for ZnCo$_2$O$_4$ as an extension of this series to $3d^6$ system.

These materials work as *p*-type semiconductors for *pn*-junction but do not work as *p*-channel material in TFTs because of high concentrations of vacant *d* levels which serve as hole trapping in the band gap.

9.4.7 Doping limit in TOS

Predicting the possibility of carrier doping and *p/n* directivity is a central issue in semiconductors. Here we consider this doping limit from the band lineup. In the band lineup, the levels of the CBM and VBM for various materials are aligned to the vacuum level (E_{vac}) [i, 17, 18].

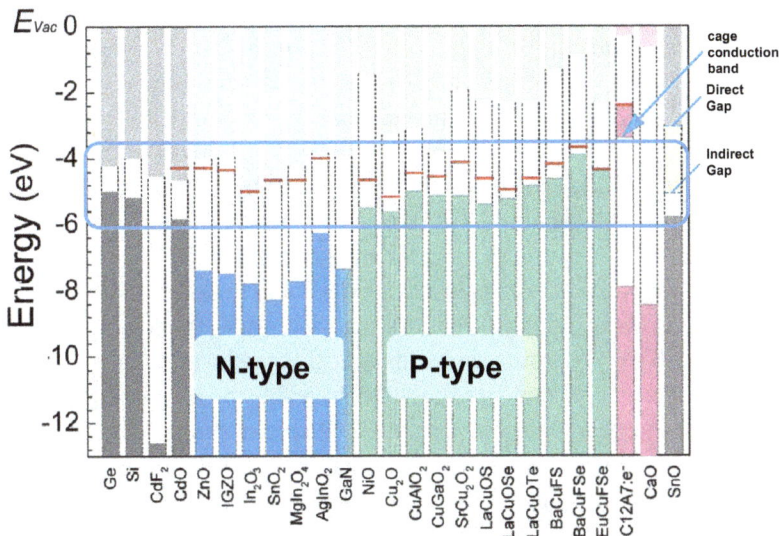

Figure 9.4.11: Energy band alignment of various semiconductors. The empirical carrier-doping window is shown by a dotted line [i].

Figure 9.4.11 shows a band lineup of oxide semiconductors and related materials [1]. The following findings are noticed from the band lineup: (1) The CBM of oxides that can be doped with electrons is deeper than ~4 eV. (2) The VBM of oxides that can be doped with holes is shallower than ~6 eV. These findings mean that the *pn*-orientation is determined by the stability of the carriers. This empirical rule that wide-gap materials are difficult to dope with carriers mostly holds when the type of anion is fixed, because the energy level of the VBM is almost determined by the highest occupied atomic orbitals of the anions. However, this rule does not generally hold because the level of the VBM strongly depends on the type of anion. A typical example is CdF_2, which has a wide band gap of 7 eV but can be easily doped with electrons. The level of 2p-orbitals of fluorine, which has a higher electronega-tivity than oxygen, is 4–6 eV deeper than the level of 2p-orbitals of oxygen. How-ever, it is reasonable that CdF_2 can be easily doped with electrons because the level of the CBM is as deep as 5 eV. It is thus evident that the carrier doping ability is not judged from the band gap in general. $12CaO \cdot 7Al_2O_3$ (C12A7) is a major constituent of alumina cement and was known as a typical insulator until 2003. However, we found this material could be converted to an electronic conductor by successful electron doping. The RT electrical conductivity can be controlled from below 10^{-10} to $>10^3$ S/cm by varying doped electron concentrations [19]. There are several com-pounds in the $CaO-Al_2O_3$ system but such a successful electron doping was possible only for C12A7. The crystal structure of C12A7 is composed of sub-nanometer-sized cages and each cage is positively charged and is connected by sharing monolayer of Ca-O-Al. The free oxygen ion O^{2-} is entrapped as the counter anion to keep elec-troneutrality. This O^{2-} anion can be replaced by electrons and such electrons can pass through the very thin (monolayered) cage wall by electron tunneling. This is the reason why C12A7 can be converted to an electronic conductor [20]. These cages form a conduction band which differs from that primarily composed of the vacant 4s-orbitals of Ca^{2+} constituting the cage wall. This conduction band is called cage conduction band, which is located at ~2 eV below the cage wall CBM whose loca-tion is similar to that of CaO or other calcium aluminate compounds.

The p/n orientation of oxide materials may be generally understood by the cri-teria described here. However, some exceptions are seen. An example is CdO with CBM <4 eV and VBM >6 eV. Following the criteria, CdO is predicted to be a bipolar semiconductor, i.e., p- or n-type doping is possible. In reality, p-type doping to CdO has been unsuccessful to date. Since the formation of oxygen vacancies giving rise to electron generation occurs easily, the obtained CdO is restricted to n-type. Please note that the p/n orientation proposed here is an empirical guide based on experi-mental works.

9.4.8 Further reading

Research on TCOs has a long history since the report of conductivity in oxidized in-dium thin films in the 1930s. Attention for these oxide thin films is driven by focused research on solar cell and display applications, which need high transparency and metallic conductivity. Active research on TOS rose as a wide-gap semiconductor in the 1990s, especially on ZnO as a II–VI-type semiconductor, and a TAOS, IGZO, is now widely used as the TFT to drive flat panel displays. Recently, Ga_2O_3 is attracting attention as a semiconductor for power electronics controlling large voltage and cur-rent because of its very large band gap (~4.8 eV). Thus, so many papers and reviews have been published to date. The following is a list of monographs and books rele-vant to this chapter.

(i)   Ginley DS, Hosono H, Paine D. Handbook of Transparent Conductors, Springer, Heidelberg, 2010.
(ii)   Ellmer K, Klein A, Rech B. Transparent Conductive Zinc Oxide, Springer, Heidelberg, 2008.
(iii) Higashiwaki M, Fujita S. Gallium Oxide, Springer, Heidelberg, 2020.
(iv)   Hosono H, Kumomi H. Amorphous Oxide Semiconductor: IGZO and Related Materials for Display and Memory, Wiley, Weinheim, 2022.

References

[1]   Hosono H. Jpn J Appl Phys, 2013, 52, 090001-1-14.
[2]   Ginley DS, Hosono H, Paine D. Handbook of Transparent Conductors, Springer, Heidelberg, 2010.
[3]   Nomura K, Ohta H, Takagi A, Kamiya T, Hirano M, Hosono H. Nature, 2004, 432, 488–92.
[4]   Hosono H. Nat Electron, 2018, 1, 428.
[5]   Hosono H, Ueda K. Handbook of Electronic and Photonic Materials. In: Kasap S, Capper P (Eds). Springer, Heidelberg, 2017, Chapter 58.
[6]   Hosono H. Thin Solid Films, 2007, 515(15), 6000–14.
[7]   Falabretti B, Robertson J. J Appl Phys, 2007, 102(12), 123703.
[8]   Hosono H. J Non-Cryst Solids, 2006, 352, 851–8.
[9]   Narushima S, Mizoguchi H, Shimizu K, Ueda K, Ohta H, Hirano M, Hosono H. Adv Mater, 2003, 15, 1409–13.
[10]   Leenheer A, Perkins J, Van Hest F, Berry J, O'Hayre P, Ginley D. Phys Rev B, 2008, 77, 115215.
[11]   Hosono H. In: Ginley DS, Hosono H, Paine D (Ed). Handbook of Transparent Conductors, Springer, Heidelberg, 2010, Chapter 10.
[12]   Kawazoe H, Yasukawa M, Hyodo H, Kurita M, Yanagi H, Hosono H. Nature, 1997, 389, 939–42.
[13]   Ogo Y, Hiramatsu H, Nomura K, Yanagi H, Kamiya T, Hirano M, Hosono H. Appl Phys Lett, 2008, 93, 032113.
[14]   Nomura K, Kamiya T, Hosono H. Adv Mater, 2011, 23, 3431–4.
[15]   Nomura K. J Info Disp, 2021, 22, 211–29.
[16]   Mizoguchi H, Hirano M, Fujitsu S, Takeuchi T, Ueda K, Hosono H. Appl Phys Lett, 2002, 80, 1207–9.
[17]   Zunger A. Appl Phys Lett, 2003, 83, 57–9.
[18]   Robertson J. J Non-Cryst Solids, 2012, 358, 2437–42.
[19]   Matsuishi S, Toda Y, Miyakawa M, Hayashi K, Kamiya T, Hirano M, Hosono H. Science, 2003, 301, 626–9.
[20]   Hosono H, Kitano M. Chem Rev, 2021, 121, 3121–85.

9.5 Inorganic detector materials

Wilfried Hermes, Robert Lovrincic, Thomas Jüstel, Rainer Pöttgen,
Cristian A. Strassert, Frank Busch

Detector materials for radiation detection are directly coupled to the wavelength of the process that should be monitored. The electromagnetic spectrum (Figure 9.5.1) covers a broad wavelength range of more than 17 orders of magnitude from long radio waves to energy-rich gamma rays. While microwaves and especially radio waves can easily be detected by antennas, detection of the IR and high-energy radiation requires tailored solid-state materials.

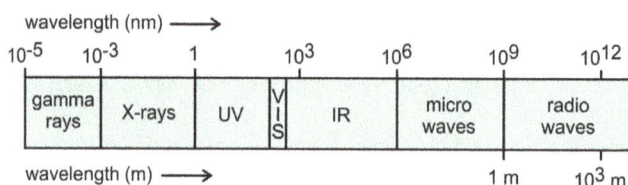

wavelength (nm) ⟶

10^{-5}	10^{-3}	1		10^3	10^6	10^9	10^{12}

gamma rays	X-rays	UV	V I S	IR	micro waves	radio waves

wavelength (m) ⟶   1 m 10^3 m

Figure 9.5.1: The electromagnetic spectrum.

This chapter shortly summarizes the many functional inorganic materials that find application for detectors. We purely focus on the material aspects. For the physics behind the different radiation sources and detectors, we refer to [1]. The field of detector materials is so broad [2, 3] that only some representative examples along with their applications are summarized herein.

9.5.1 IR sensors

The infrared range can be separated into two distinct regimes of high technological importance: near-infrared (NIR) from 800 to 2500 nm (12000–4000 cm^{-1}), and mid-infrared (MIR) from 2500 nm to 20 μm (4000–500 cm^{-1}). Infrared detectors for both MIR and NIR fall into two main categories: thermal and photon. The earliest detectors of IR were thermal in nature, e.g. a thermometer. The subsequent developments of these detectors, such as resistance bolometers, pyroelectric detectors, or thermopiles, can operate at ambient temperature but have disadvantages of insensitivity and slowness.

Photodiodes have a simple structure, high stability and broad bandwidth. Thus, photodiodes are among the technologically most important detector materials for the VIS and IR spectral range. The detection principle is shown in Figure 9.5.2.

The diodes are composed of a semiconductor material. Whenever the energy of an incident photon is higher than the bandgap of the semiconductor used, the photon is absorbed. It transfers its energy to an electron that is subsequently excited to

https://doi.org/10.1515/9783110733471-005

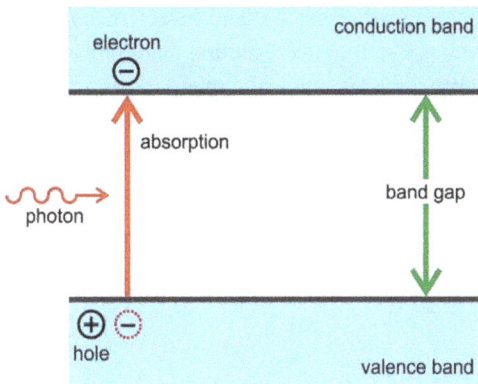

Figure 9.5.2: Detection mechanism of a semiconductor detector.

the conduction band. If one can neglect recombination, in an ideal photodiode, one photon can generate one pair of carriers. To create a measurable photocurrent and photovoltage, the electron-hole pair (exciton) needs to be separated, overcoming its exciton binding energy. The driving force for this separation is the built-in electric field in the depletion region of the diode. Different types of photodiodes exist, each with its own advantages: PIN diodes, p-n diodes, and Schottky diodes to name just the most common ones. We refer to more specialized literature for details on operational principle and structure [4].

Since the photon energy decreases with increasing wavelength, each detector material has a cut-off value for the detectable wavelength that directly depends on the value of the bandgap. Typical materials for the VIS and IR detection range along with their cut-off value are summarized in Table 9.5.1.

Table 9.5.1: Basis semiconductors for detector materials and their cut-off values (µm) for VIS and IR detection.

Si	1.1 µm	$(In_{1-x}Ga_x)As$	1.6 µm
Ge	1.4 µm	$(In_{1-x}Ga_x)(As_{1-x}P_x)$	1.4 µm
PtSi	5 µm	InSb	5 µm
$(Hg_{1-x}Cd_x)Te$	14 µm	PbS	3.3 µm
		PbSe	5.2 µm

$In_{1-x}Ga_xAs$ PIN photodiodes are photovoltaic detectors just like Si photodiodes. Since $In_{1-x}Ga_xAs$ photodiodes have a smaller energy gap than Si, they have sensitivity in a longer wavelength range than Si. These solid solutions are referred to as III–V compounds and have properties intermediate between those of GaAs and InAs. $In_{1-x}Ga_xAs$

has a lattice parameter that increases almost linearly from GaAs to InAs. The bandgap increases with decreasing In content (Figure 9.5.3). Single crystalline materials in form of thin films can be grown by MO-CVD (metal-organic chemical vapor deposition) or by MBE (molecular beam epitaxy). In$_{1-x}$Ga$_x$As devices are grown on indium phosphide (InP) substrates. In order to match the lattice parameter of InP and avoid mechanical strain, In$_{0.53}$Ga$_{0.47}$As is used. This composition has an optical absorption edge at 0.75 eV, corresponding to a cut-off wavelength of $\lambda = 1.68$ µm at 295 K.

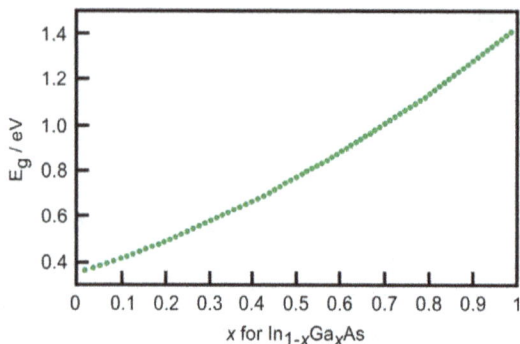

Figure 9.5.3: Course of the energy bandgap in the solid solution In$_{1-x}$Ga$_x$As (bulk material) (redrawn and adapted based on [5]).

Figure 9.5.4 summarizes some crystallographic and electronic properties of the family of III–V compound semiconductors. The III–V pnictides crystallize either with the hexagonal wurtzite type (space group $P6_3mc$) or the cubic sphalerite type (space group $F\bar{4}3m$). The size of the atoms determines the lattice parameters and consequently the value of the bandgap. As emphasized at the right-hand part of Figure 9.5.4, the III–V pnictides or their solid solutions can be used as tailored semiconductor detector materials in the IR-to-UV range.

A semiconductor-based alternative to photodiodes are photoconductors. The absorption mechanism is identical to photodiodes, and the incoming light can excite free carriers if the photon energy is larger than the bandgap. However, without a built-in electrical field, no photovoltage is created. Charge separation is therefore reliant on an external applied bias. Photoconductors utilize the photoconductive effect where exposure to infrared radiation causes a decrease of the resistance of the active area as a function of the radiation intensity. Depending on the application, photoresistors are a viable alternative to photodiodes both in terms of performance and costs.

Polycrystalline lead sulfide (PbS) and lead selenide (PbSe) detectors are thin-film photoconducting devices (Figure 9.5.5) that have been in use as infrared (IR) detectors for almost a century owing to their versatility and wide wavelength coverage. PbS detectors respond to light with wavelengths between 1 and 3 µm and are applicable for NIR detection. PbSe detectors extend into the MIR, covering applications from 1 to 5 µm.

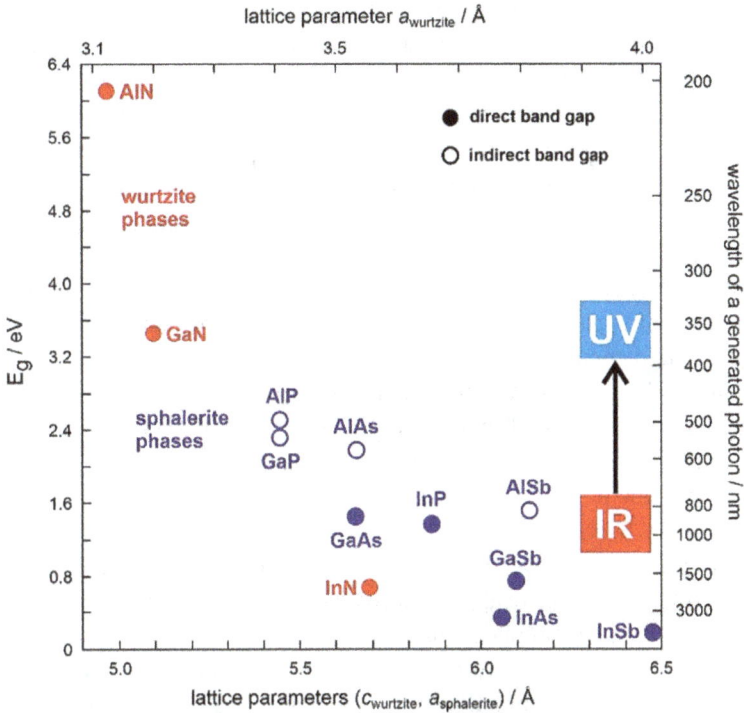

Figure 9.5.4: Course of the energy bandgap as a function of the lattice parameters of III–V semiconductors with wurtzite- or sphalerite-type structure (redrawn and adapted based on [6]).

Figure 9.5.5: Lead sulfide detectors with different sizes from trinamiX.

The standard manufacturing method is the chemical bath deposition (CBD). Thin films (several hundreds of nm) are deposited from an aqueous solution. The process is fast and scalable. As the resulting films must be polycrystalline for a sufficiently high photoresponse, no lattice match to the substrate is required. Therefore, suitable substrates for PbS and PbSe detectors are quartz glass or silicon wafers.

Further important thermal detectors concern bolometers, the pyroelectric effect or thermopiles. Such thermal detectors use IR radiation energy as heat source. Their response is a function of the incident thermal energy and is independent of the wavelength.

A bolometer consists of an absorptive element, such as a thin layer of a metal or a semiconductor, connected to a thermal reservoir through a thermal link. A microbolometer consists of a grid of vanadium oxide (VO_x) or amorphous silicon. IR radiation is absorbed and changes the electrical resistance.

Lithium tantalate, $LiTaO_3$, or lead zirconate titanate $PbZr_xTi_{1-x}O_3$ (PZT) are widely used pyroelectric detector materials. The detector releases an electrical signal induced by temperature changes. The latter occurs via light absorption. Pyroelectric crystals have a rare asymmetry due to their single polar axis. This causes their polarization to change with temperature. Such materials are discussed in detail in Chapter 9.9.

A thermopile is composed of several thermocouples connected usually in series. It converts thermal gradients into electrical potential differences, i.e. electrical energy.

Finally, we turn to the many applications of IR detectors. IR is used in a number of applications in the fields of agriculture, medicine, science, communications, remote sensing, food and beverage, animal nutrition, chemical and pharmaceutical industry and more. (N)IR spectroscopy is mainly used in those industries, for example, for quality control. The IR radiation energy is related to the vibrational or rotational energy of molecules. This phenomenon makes it possible to identify molecules and therefore can be used for spectroscopy. Further typical applications of IR sensors are night vision devices and thermal cameras. Two typical applications from the food and recycling sectors are discussed in detail below.

Organic molecules can absorb light via the excitation of roto-vibrational modes. The energy of such a motion depends on the reduced mass or moment of inertia of the involved moieties and the bond strength between them. These absorption bands result in a molecular "fingerprint", located in the MIR spectral region (wavelength range from ca. 3 to 10 μm).

NIR spectroscopy covers the wavelength range from 0.9 to 2.5 μm (or 11000 to 4000 cm^{-1}). This spectral range allows to measure overtone (0.9–2 μm) and combination bands (2–2.5 μm) of molecular vibrations in organic materials. For decades, NIR spectroscopy has been a working horse in food, agriculture and pharmaceutical industries for the following reasons: (i) The higher penetration depth of light in the NIR region compared to MIR makes sample preparation much easier and thereby sample throughput higher. (ii) Highly sensitive semiconductor-based detectors for the MIR have to be operated at cryogenic temperatures due to their small bandgaps, whereas NIR detectors can operate efficiently at room temperature, greatly reducing costs.

The overtone and combination bands in the NIR range carry very detailed information on the samples under test, e.g. the fatty acid composition in oilseeds or the amino acid profile of proteins in cereals. This is exemplified in Figure 9.5.6, where NIR absorption spectra of important macronutrients are shown. Hence, spectrometers capable of measuring between 900 and 2500 nm are widely used in agriculture for quality control and pricing. To extract detailed information on the composition of a complex sample such as canola, soy or wheat, multivariate statistical methods and machine learning algorithms are typically used. This scientific discipline is commonly referred to as chemometrics.

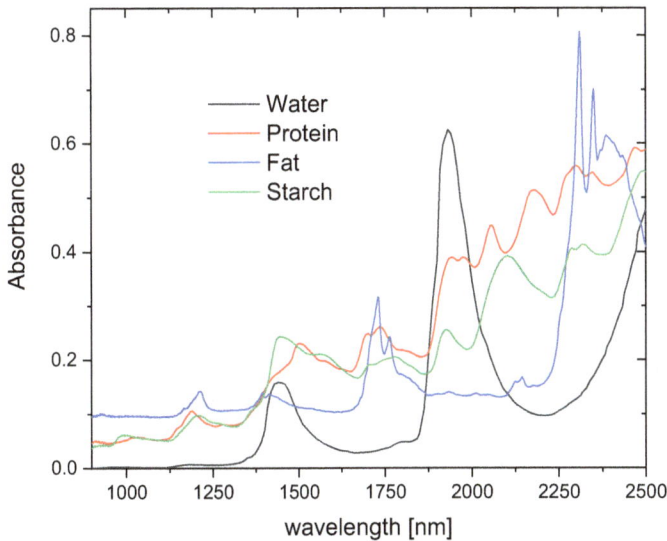

Figure 9.5.6: NIR absorption spectra of macronutrients (courtesy of trinamiX). Each nutrient exhibits a distinct absorption pattern which allows for quantitative analysis of food and feed compositions via chemometric analysis.

Fairly recently, NIR spectrometers have managed to break out of their well-controlled laboratory environment by becoming smaller, lighter and scalable. Main applications for such portable devices (Figure 9.5.7) are currently in the fields of on-site analysis for smart farming and plastics identification for circular economy [7].

Additionally, there is a strong commercial interest in the utilization of NIR spectroscopy for health and well-being applications.

Figure 9.5.7: Mobile NIR spectroscopy solution (courtesy of trinamiX).

9.5.2 X-ray detection

The classical detection of X-rays in research (single crystal and powder diffraction) and medicine was similar to light photography. Silver halide-coated polyester foils are irradiated, developed with quinone/hydroquinone-based redox chemicals followed by fixation with thiosulfate solution. This technique is meanwhile completely out-of-date. The use of image plate systems, CCD detectors or photodiodes allows for faster data collection, higher resolution and electronic data processing – without the material costs and waste production of the classical film technology. For medical applications, the main advantage is the strong reduction of X-ray exposure for the patient!

9.5.2.1 Image plate detectors

Image plate detectors [8, 9] rely on photostimulated luminescence. The standard image plate phosphors are $BaFBr:Eu^{2+}$ and $BaF(Br_{0.85}I_{0.15}):Eu^{2+}$ as ca. 150 µm thick layer (with a grain size of ca. 5 µm) on a rigid or flexible plastic plate. The coating is mostly fixed by an organic binder, and Mylar foil can be used as a protective layer.

The incident X-ray photons oxidize the Eu^{2+} cations to Eu^{3+}, and the liberated electron (a so-called photoelectron) is transferred to the conduction band. This electron is trapped in a lattice defect (color center, F center), resulting from bromide or fluoride vacancies that were introduced into the detector material within the manufacturing process. The resulting states are metastable. When exposing the exposed plate to HeNe laser radiation (the reader unit), the laser radiation excites the photoelectrons and recombination with the Eu^{3+} ions occurs under emission of a photon with a wavelength of 390 nm. Signal intensification is realized with a photomultiplier tube (readout unit of the detector). A large advantage of the image plate detectors is their thousand times reusability. In a final step, the residual image can be completely erased by visible light exposure and the plate is ready for a new exposure experiment.

As an example, we present the image plate detection unit of an X-ray single crystal (Stoe IPDS-II) diffractometer in Figure 9.5.8.

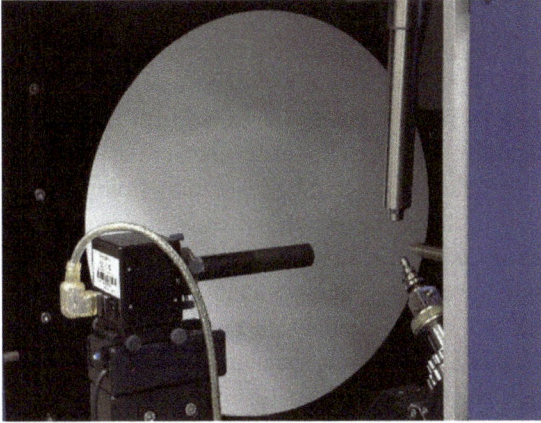

Figure 9.5.8: BaFBr:Eu^{2+}-coated image plate used for X-ray detection in a Stoe IPDS-II single crystal diffractometer (photo by Thomas Fickenscher).

The use of highly sensitive image plate detectors was a first decisive step in early protein crystallography. A reduced X-ray dosage and much shorter counting times enormously helped to reduce the radiation damage of protein crystals during data collection.

Although a huge number of single crystal and powder X-ray diffractometers worldwide are equipped with BaFBr:Eu^{2+}-based image plate detectors, this use may be considered as niche application. Image plates are nowadays broadly used as a detection unit for medical X-ray imaging (computed radiography) [10]. Classical X-ray films have meanwhile completely been replaced. Image plate technology allows for better resolution, much lower radiation exposure for the patient and reusability.

9.5.2.2 CCD image sensors

CCD (charge-coupled device) sensors are meanwhile standard detector materials in the field of X-ray diffractometry and medicine (imaging application). These detectors are larger arrays with a manifold of light-sensitive photodiodes (pixels) with edge length ranging from ca. 1.4 to 20 µm. The pixel size determines the light sensitivity, the image resolution and the detector dynamics. The CCD devices are so-called MIS structures, i.e. metal-insulator-semiconductor stacks. The stacking sequence is metal contact–insulator (mostly SiO$_2$)–semiconductor (p-doped silicon)–metal contact. The incoming light transfers its energy to the electrons of the semiconductor via the inner photoelectric effect. The CCD devices are coupled with phosphors for converting the X-rays. The complete detailed CCD technique is far beyond the scope of this chapter. We refer to an excellent review by Gruner et al. [11] for further reading.

9.5.2.3 Silicon photodiodes

Silicon photodiodes have the same working principle as the semiconducting IR detectors discussed above. The peculiarity for the detection of X-ray concerns the coupling with a scintillator material, typically CsI(Tl), Gd_2O_2S:Pr (so-called GOS ceramic) or $CdWO_4$ (so-called CWO; no broad application due to the heavy metal toxicity of cadmium). The scintillator (coated with a reflector to avoid light escaping from the photosensitive area) converts the X-ray into light that is subsequently detected by the photodiode. The photoelectric effect or Compton scattering as dominating effects in a silicon photodiode depend on the energy of the X-ray/gamma rays as well as on the substrate (silicon film) thickness [12]. Scintillator-coupled silicon photodiodes can be arranged in larger arrays and find broad application for examining the shape of materials, most prominently in baggage inspection in airports (Figure 9.5.9).

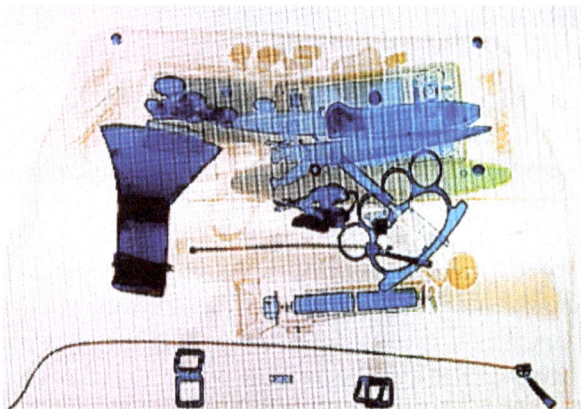

Figure 9.5.9: Image of a baggage inspection scan carried out with X-rays.

9.5.2.4 Silicon drift detector

A special semiconductor detector material is lithium-doped silicon, Si(Li), the so-called lithium-drifted silicon detector. The Si(Li) detectors find broad application in X-ray spectrometers for both energy- and wavelength-dispersive analyses [13]. These detectors have materials' stacking gold–p-type Si–Si(Li)–n-type Si–gold. The advantage of Si(Li) detectors is their broad spectral range and a comparatively high quantum yield.

9.5.3 Scintillation materials

Scintillators are basic components for a variety of applications in high-energy physics and medical imaging since more than 100 years, while the first materials in application were $CaWO_4$ and ZnS:Ag [14]. Figure 9.5.10 gives a historical overview on the applied scintillator materials.

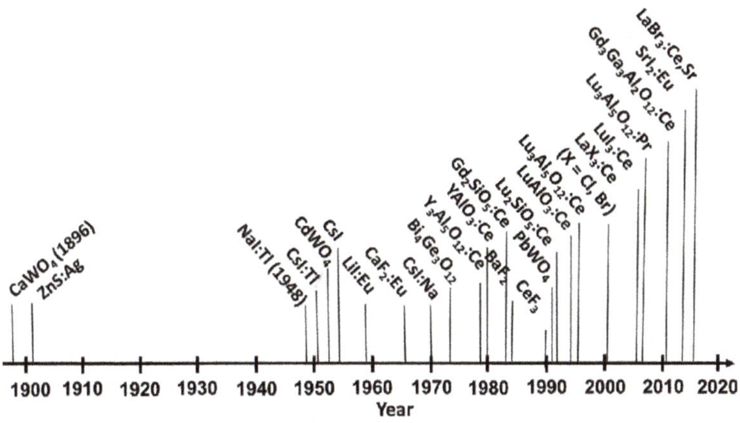

Figure 9.5.10: Brief history of scintillator material discovery (modified after [14]).

They are used as converter materials in all sorts of detectors for ionizing radiation, film-screen systems, image intensifiers, solid-state flat panel detectors (FPD), radiation therapy detectors (portal imaging), computed tomography detectors (CT), gamma cameras, single photon emission computed tomography (SPECT) or positron emission tomography (PET).

Instead of scintillators, direct converters (X-ray-sensitive semiconductors) could be used. There are a few materials like amorphous selenium (alpha-Se) or cadmium zinc telluride (CZT) that are widely applied, alpha-Se in static X-ray detectors and CZT in small gamma ray detectors. Direct converter materials are, however, not as far developed as scintillators, which have been used for many decades in X-ray imaging.

Different classes of scintillators or direct converters are used for different applications, characterized by:
- high signal, slow decay (µs range) (FPD, CT)
- high signal, fast decay (ns range) (nuclear medicine)
- low signal, ultrafast decay (<3 ns) (particle physics)

Fast and ultrafast scintillators were mainly developed for particle physics research. Counting detectors using fast scintillators are already applied in nuclear medicine, e.g. in PET machines.

9.5.3.1 Some general considerations

There are four different mechanisms that can give rise to fast scintillation:
- exciton-like luminescence, e.g. CsI
- core-valence luminescence or cross-luminescence (CL), e.g. BaF_2
- luminescence quenching, e.g. in $PbWO_4$
- parity-allowed $5d$–$4f$ transitions of di- or trivalent rare earth ions

Only a few "classical" scintillators (like CsI, CsI:Tl or NaI:Tl) exist that show a fast scintillation decay. CL and quenched luminescence give a very fast decay, but the light yield is generally moderate to low. Rare earth doping is most promising for rather fast decay combined with high light yield.

The scintillation process, as discussed in the literature [15], can be described in wide gap materials as follows: A high-energy photon/particle generates an excitation cascade due to a series of inelastic electron-electron scattering events and core-hole Auger decays occurring at a timescale of 10^{-15} s. The subsequent thermalization (I) stage ($<10^{-12}$ s) is defined by an electron-phonon scattering of charge carriers, whose energy is too small for inelastic processes (see Figure 9.5.11). The next stage is the migration (II) of thermalized electrons and holes to recombination centers. Finally, the recombination (III) stage occurs, where most of the useable light is produced by conventional scintillators as luminescence. Stages II and III are also very important for the processes of energy transfer to rare earth luminescence centers used in doped scintillators and these set the limits for decay time range.

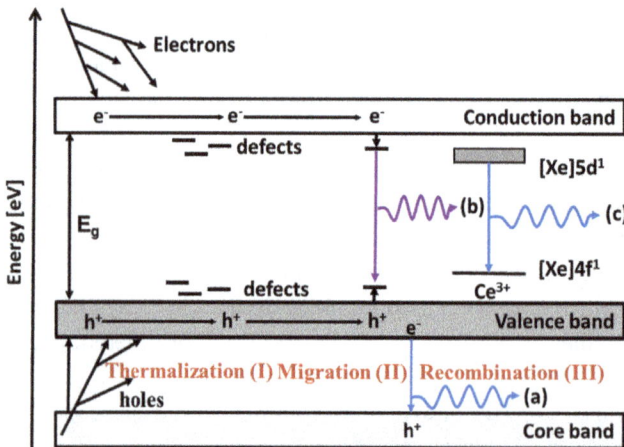

Figure 9.5.11: Simplified sketch to illustrate the scintillation process in wide bandgap materials, whereby the curved arrows are related to core (a), exciton (b) and activator (c) luminescence.

The timescale and efficiency of these stages may vary widely determining the performance of a particular scintillator. Therefore, in order to advance performance of scintillators, some other approaches have to be used to overcome limitations set by the migration stage (including the rise time) and the recombination stage responsible for luminescence decay time. In particular, one can improve time resolution using ultrafast intrinsic emission processes, providing prompt photons in the very beginning of the scintillation process.

There are materials with specific band structures, e.g. BaF_2, CsCl or $KMgF_3$, where the Auger decay of the holes at the uppermost core level is energy-forbidden. In such cases, CL is observed due to radiative recombination of electrons from the valence band with these holes [16]. The rise time of CL is determined by the duration of the cascade stage and subsequent self-trapping of core holes, which occurs on the 100 fs scale [17]. Core hole phonon relaxation is prevented by the large energy distance to the valence band, while the CL decay time is determined by the probability of dipole-allowed radiative transitions ($\leq 10^{-9}$ s). Subsequently, the dependence of CL on non-radiative processes is rather weak and the light yield is reasonably high, e.g. about 2000 ph/MeV in BaF_2 [18]. This brands CL materials as perfect candidates for ultrafast scintillation with an acceptable light yield.

As mentioned above, rare earth ions are excellent activators for scintillators with a high light yield. Those which have allowed interconfigurational $5d$–$4f$ transitions, i.e. Ce^{3+}, Pr^{3+} and Nd^{3+}, show fast luminescence processes with decay times in the tens of nanoseconds regime. The decay time increases as the square of the emission wavelength ($\tau \sim \lambda^2$). For trivalent cerium, the decay time varies between 20 ns in the ultraviolet and 80 ns in the visible range. Pr^{3+} and Nd^{3+} exhibit shorter decay times at the expense of short wavelength emission in the deep ultraviolet, which requires detectors with improved sensitivity in the UV range. Additionally, the light yield is usually smaller than with cerium doping so that Ce^{3+} ions remain as the most interesting dopants. Bivalent or tetravalent states of the cerium ion normally act as quenching centers so that a host structure with trivalent sites is preferred for cerium doping.

It has been shown that the scintillation efficiency depends on the type of rare-earth host and increases according to the order of fluorides < chlorides < bromides < iodides < oxides < sulfides. Fluorides, therefore, generally show a low light yield. Chlorides, bromides and iodides are often hygroscopic so that there is not much effort on these compounds, and oxysulfides or sulfides are required through control of the synthesis conditions. Therefore, most of the host crystals used for rare earth doping are oxygen-based.

Scintillators can only have a high light output if the electronic absorption bands of the activator are in resonance with the intrinsic (exciton) luminescence bands of the host crystal. Analysis of spectroscopic properties of many oxide crystals has shown that ideally Y, Ce, Gd and Lu can be used as host structure components, except for some La- and Yb-based compounds. However, Gd^{3+}-comprising crystals show a slow scintillation component due to the intraconfigurational ^6P_J–$^8S_{7/2}$ transition related to the [Xe]$4f^7$ configuration so that Y^{3+}- and Lu^{3+}-based host materials are mostly preferred for cerium-activated scintillators.

9.5.3.2 SPECT and PET detector materials

In single photon emission computed tomography (SPECT), gamma-emitters (e.g., ^{99m}Tc or ^{123}I) are used as diagnostic imaging tools (for further details, see Section 13.2.1). When they accumulate in a targeted organ or tissue, the randomly oriented (i.e., non-coherent) emission of gamma-photons is detected with suitable scintillators coupled to photodiodes. In its most simple version, two-dimensional planar-static images are taken if patient and detector are not moved while gamma-photons are counted and integrated (e.g., in a thyroid gland scintigraphy). If the patient is slowly scanned at a constant angle, an extended 2D image of a larger area can be generated. Typically, the gamma-camera is composed of two detectors placed facing each other at 180° (or at 90° for heart imaging). If the images are taken without moving the patient or the camera but at different time points, a time-dependent activity profile of a defined area can be recovered (i.e., a planar-dynamic image). If 3D resolution is needed, the gamma-camera is rotated around the patient in a circular or elliptical trajectory, which enables the generation of a tomographic array of images from different angles that are computed into a single image, from which 2D planes can be recovered. In any case, the spatial resolution can only be achieved by collimating the gamma-photons that reach the camera; this is achieved by employing an LED plate with holes absorbing non-perpendicular quanta. For SPECT, the spatial resolution is significantly lower than for PET (*vide infra*).

Positron emission tomography is another important technique, which relies on the use of ionizing radiation. It is also one of the most recently developed and most advanced medical imaging techniques. There are many requirements which have to be met by the scintillating material, such as high density, radiation hardness, mechanical strength, high light yield, an emission spectrum aligned to the detector sensitivity curve and short decay time, which makes the material selection rather delicate (see Table 9.5.2).

However, before a PET scan can take place, the patient is injected with a short-lifetime radioactive tracer, which accumulates especially in tissues of increased metabolic activity. Mostly, the compound is a modified form of glucose with one of the hydroxyl groups replaced by the radioactive fluorine isotope ^{18}F. The isotope undergoes ß-decay and produces a positron as a result. Almost immediately afterwards, the positron annihilates with an electron from one of the surrounding atoms, two γ-photons are emitted in opposite directions, due to the momentum conservation principle. These 511 keV photons are weakly absorbed by the human tissue and are able to reach the detector ring surrounding the patient placed in the PET machine (Figure 9.5.12).

Table 9.5.2: Selected scintillator materials in application.

	NaI:Tl	CsI:Tl	$Bi_4Ge_3O_{12}$	Lu_2SiO_5:Ce	$Lu_2Si_2O_7$:Ce	Gd_2SiO_5:Ce	$LaCl_3$:Ce	$LaBr_3$:Ce	LuI_3:Ce
Density (g/cm³)	3.67	4.51	7.13	7.4	6.2	6.7	3.86	5.29	5.60
Z_{eff}	51	52	75	66	64	59	48.3	46.8	61
Light yield (photons/MeV)	40000	52000	9000	30.000	26.300	8000	50.000	70.000	95000
Emission peak at (nm)	415	550	480	420	380	440	337	358	474
Decay time (ns)	230	1000	300	42	30	60	24	16	24
Energy resolution at 662 keV	7%	10%	> 10%	9%	10%	9.2%	3.5%	2.6%	3.3%
Hygroscopic?	Yes	Yes	No	No	No	No	Yes	Yes	Yes

In the detector ring, scintillators are placed, which convert the high-energy radiation into short visible light pulses. These pulses are subsequently detected by photomultipliers and recorded as electric signals by the connected computer. Data analysis and processing are based on the fact that the γ-photons are always emitted at $\alpha = 180°$ angle, and, therefore, by the simultaneous detection of both of them, it is possible to identify precisely the location of the annihilation event at the crossing point of the so-called lines-of-response (LOR). This, therefore, means that the distribution of the labeled tracer in the patient's body can be determined. It might, however, occur that a γ-photon, which was produced from the annihilation event, undergoes Compton scattering and, consequently, its trajectory is deflected from 180°. The Compton scattered photon is characterized by lower energy (different wavelength) than the non-scattered photon and this can be diminished. Other tracers are also used for PET, as further discussed in Section 13.2.1.

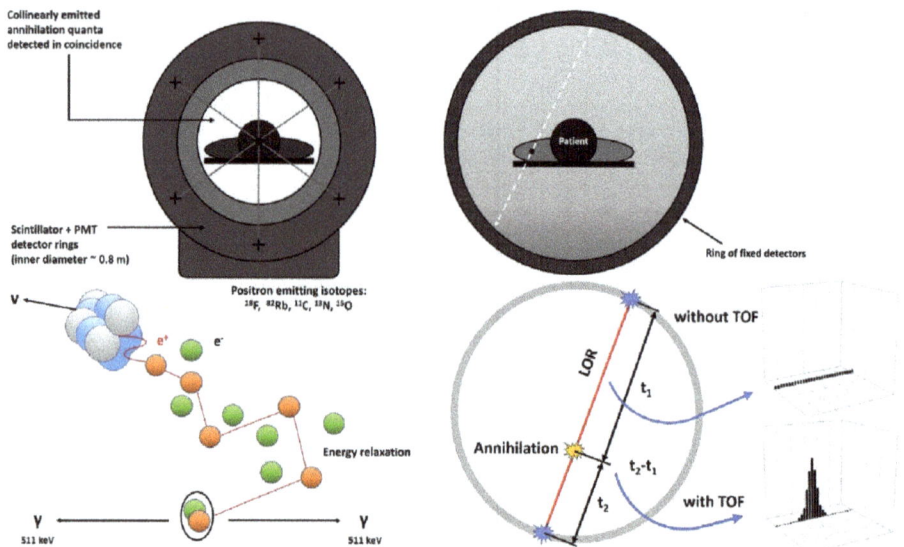

Figure 9.5.12: Cross section and principle of operation of a positron emission tomograph (PET). The scintillator crystals (several thousands) are mounted to form an external ring. The line of response can be used for the detection without time of flight (TOF) of the photons or with TOF resulting in the so-called coincidence analysis enabling a higher image resolution (adapted from Philips Medical Systems).

State-of-the-art time-of-flight (TOF) PET data processing is based on the measurement of differences in arrival times of the two γ-quanta coming from the same annihilation event. In such a way, it is possible to obtain very high spatial resolution (2 mm). However, since γ-rays travel at the speed of light, a very fast response of the detection system is needed which requires the use of scintillators with a fast luminescence build-up (in the order of sub-ns) and a short decay time (<40 ns).

In conclusion, there are many requirements which have to be met by the scintill-ating material for PET, whereby a short decay time is a must for the coincidence analysis in TOF PET.

A very high light yield combined with a rather short decay time of the $5d–4f$ luminescence was demonstrated for LuI_3:Ce [19]. This halide exhibits excellent scin-tillation properties including a short decay time, little afterglow and a high energy resolution, resulting in a remarkably linear spatial resolution between 8 and 14 µm compared to the commercially used $Lu_3Al_5O_{12}$:Ce transparent oxide ceramics.

9.5.3.3 Outlook on scintillators

Although scintillation as a process has been investigated and used for a long time in very different areas of technology, including medical diagnostics, homeland se-curity, high-energy physics at synchrotrons, such as BESSY, CERN, DELTA or DESY, and controlled nuclear fusion research, there are always new challenges.

A number of these applications have specific demands, e.g. as improved time resolution. In particular, ultrafast scintillation processes will be useful in high count rate applications like new generation of particle colliders, such as the high luminosity-Large Hadron Collider (HL-LHC) or recently announced the Future Circu-lar Collider (FCC). They can also be used in the development of gamma-cameras with GHz frame rates for hard X-ray imaging, which will allow a detailed study of fast evolving processes, such as laser-driven implosion and dynamics of matter-radiation interaction at extreme conditions.

An improved time resolution will enable enhancement of TOF scintillation tech-niques, which is a strong motivation for the development of new materials with bet-ter time performance. The fundamental PET resolution is limited by the mean free path of a positron in living tissue, which is about 1.5 mm and corresponds to the coincidence time resolution of 10 ps. The present TOF-PET systems in the market have 215 ps of TOF resolution, mainly limited by the rise and decay time of the scin-tillator single crystals in the ring.

A better scintillator material, which is operating using a concept different from classical rare-earth-doped Lu_2SiO_5:Ce crystals, would allow a direct reconstruction less real-time imaging with an ultimate spatial resolution, leading to much lower patient doses. Thus, this diagnostic method can be applied for pediatric, prenatal and neonatal examinations. Progress towards a 10-ps PET would open a door for more compact tomograph designs without sacrificing spatial resolution, which means that economically affordable machines can be installed in all major hospi-tals. The development of new multifunctional scintillation materials will result in further advancement of high-level medical diagnostic technologies.

By considering the scale of healthcare costs in modern societies, any improve-ment of medical diagnostics will help to reduce these costs substantially and even more important to rise the life quality of patients through more personalized treatment strategies prohibiting unneeded inspection or surgery.

9.5.4 Gas ionization detectors

The detection of radiation (particularly alpha and beta particles as well as X-ray and gamma photons, whereas neutrons pose a special problem) can be also carried out with gaseous ionization detectors. In general, a gas-filled tube (grounded or negatively biased) surrounds one (or more) wire(s), to which a (typically positive) potential is applied (in ionization chambers, two parallel plates are preferred, *vide infra*). The incident radiation causes the formation of charge carriers, which in turn enhances the electrical conductivity of the otherwise insulating gas while leading to a drastic change in current intensity. Depending on their operation mode, they are classified as ionization chambers, proportional counters or Geiger-Müller tubes (i.e., sorting the devices by the increasing voltage applied between the electrodes, respectively) [1].

With the relatively high potential applied in Geiger-Müller tubes, every incident particle produces primary charges that in turn cause a massive current peak involving a cascaded avalanche of secondary charge carriers (due to high-voltage-related acceleration and concomitant emission of ionizing UV radiation). Single pulses are counted, and the resulting current peak for every detection event is independent from the energy of the radiation type. It provides a relatively high detection sensitivity yet without requiring sophisticated electronics. Nonetheless, the dead-time (and the risk of pile-up effect) after every ionization-avalanche-cascade incident limits its applicability to the detection of comparatively lower activities (i.e., less detection events per unit time).

On the other hand, ionization chambers are run at relatively low voltages, meaning that every ionizing particle causes a number of primary conductive charges that is proportional to its energy (which can be determined for a single ionization event in the pulsed current detection mode). Since no avalanche-like discharge occurs, the dead-time is drastically shortened while enabling the detection of higher activities. If run at constant voltage, the measured current is proportional to the number of primary charge carriers per unit time, i.e. to the activity (and to the energy dose upon integration over time; with proper calibration, it constitutes the basis for dosimetry, even though the energy resolution of single particles is then lost). However, the sensitivity is significantly lower than for a Geiger-Müller tube, while requiring rather sophisticated amplification electronics.

Proportional counters (Figure 9.5.13) are devices run at intermediate voltages (between ionization chambers and Geiger-Müller tubes), where the peak current of a pulse is proportional to the applied voltage and to the energy of the incident particle. Since every ionization event also involves a voltage-dependent gas amplification

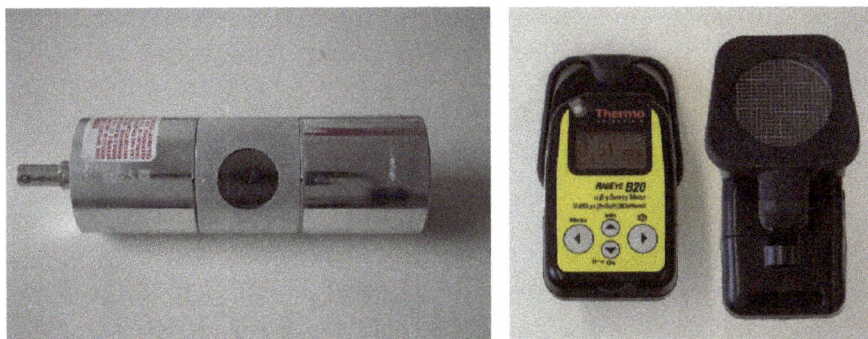

Figure 9.5.13: Depiction of a proportional counter tube (LND, Inc.) from a Mössbauer spectrometer (left) and of a portable contamination detector (Thermo Fisher Scientific) based on a "pancake" Geiger-Müller counter (right) (photos taken by Thomas Fickenscher).

(limited avalanche), the sensitivity is significantly higher than for ionization chambers; however, as the applied voltage is lower than in a Geiger-Müller tube, the cascaded production of secondary charge carriers is limited, meaning that the peak current detected for a single ionizing particle is proportional to its energy (provided that single events are properly resolved, since the dead-time between ionization events limits the detected activity, but not as severely as in the case of Geiger-Müller tubes). In a way, proportional tubes constitute a compromise between the high-resolution of ionizing chambers and the high-sensitivity of Geiger-Müller tubes and are generally operated in the pulsed detection mode.

By moving from plain air as a detector gas towards optimized devices, noble gases at different pressures are used, due to their comparatively lower ionization potentials and electron affinities as well as their high chemical inertness. The noble gas (or mixture of gases) is combined with a quencher to limit the duration of the ionization avalanche (e.g., 90% argon, 10% methane). Alternatively, butane, ethanol or halogens (such as Cl_2 or Br_2) can also be added (acting as polyatomic species able to neutralize cations along with a concomitant thermalization while suppressing ionizing UV radiation) to shorten the dead-time. Gamma and X-ray photons as well as high-energy beta particles are not as efficiently absorbed as alpha or low-energy beta particles, causing a lower sensitivity for their detection. Hence, longer tubes with high-pressurized heavier noble gases can be used to improve the sensitivity. Another challenge is represented by neutrons: having no charge or significant electric dipole moment, they do not cause any ionization event on the electron cloud of any gaseous species (also their inherent magnetic dipole moment cannot be harvested in any meaningful way). However, certain isotopes possess high neutron absorption cross sections and are able to trap these uncharged particles, which is followed by the emission of ionizing entities (e.g., alpha particles). If the (thermal) neutrons to be detected are slow enough and a proper gaseous isotope is used, proportional counters can be realized to detect these otherwise elusive particles. Typically, ^{3}He or ^{10}B-enriched BF_3 are used [20], which undergo the following nuclear reactions, respectively:

$$^3He + {}^1n \rightarrow {}^3H + {}^1H$$

$$^{10}B + {}^1n \rightarrow {}^7Li + {}^4He$$

High-pressure ^{3}He can be used to increase the sensitivity by increasing the efficiency of neutron capture (this is not possible with BF_3, due to its higher reactivity). However, ^{3}He can be prohibitively expensive, since it is only produced by the radioactive decay of tritium, a synthetic nuclide. On the other hand, fast neutrons can be detected with high-pressure hydrogen-filled proportional counters, where their elastic scattering on protons causes the ionization of hydrogen atoms.

9.5.5 Dosimetry

Radiation protection is a central task with respect to health and safety of workers. Personal dosimetry as passive/indirect protection plays an important role in all medical X-ray facilities and nuclear medical labs as well as in all research and production facilities dealing with any kind of ionizing radiation, e.g. neutron research and synchrotron facilities.

In research and medical X-ray labs, passive radiation detectors are either classical X-ray films (Figure 9.5.14) or solid-state detectors (SSD), with thermoluminescence detectors (TLD) and optical stimulated detectors (OSL) being most commonly used in radiation protection. TLDs are typically based on LiF:Mg,Ti or CaF_2:Mn (Figure 9.5.15), while OSL utilizes BeO or AlO. The passive personal dosimeters are usually evaluated monthly. In case of a film dosimeter, radiation exposure gives rise to a blackening on the film, which is evaluated by comparison to exposed films with standardized irradiation doses. Using metal filter elements during exposure allows to widen the usable range beyond X-rays to higher photon energies.

Solid-state detectors generally rely on radiation energy being transferred to electrons, which then are lifted to the conduction band similar to a number of active detectors mentioned above. In contrast, SSDs are capable of storing a fraction of the radiation energy in traps located in the bandgap between valence and conduction bands. LiF:Mg,Ti has a thermal fading (20–25 °C rage) of approximately 5% within 3 months [21]. The dose evaluation is then performed by stimulating the trapped electrons with external energy to induce emission of luminescence when the electrons are returning to the valence band. The radiation dose is then determined by measuring the amount of light.

Different types of SSDs are distinguished by the stimulating process: With TLDs, energy is thermally delivered by heating up to 573 K and OSL uses light

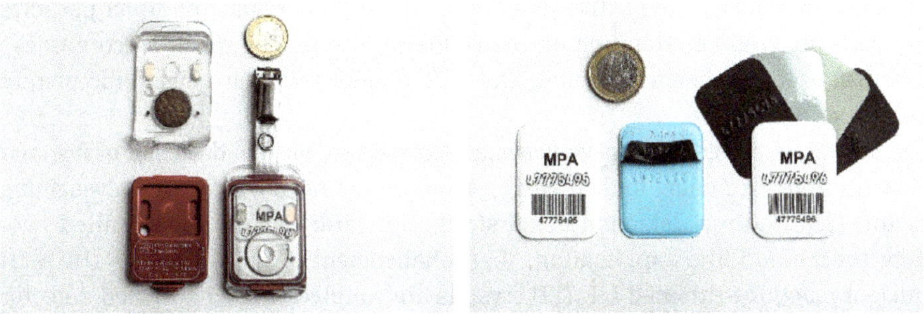

Figure 9.5.14: Film dosimeters for personal dosimetry. Complete batch (left) and sealed film package with two detector films with different sensitivity (right). The images were kindly supplied by MPA NRW, Dortmund.

pulses in the visible range. The widely used thermoluminescence detector material LiF:Mg,Ti delivers luminescence with an emission maximum at about 400 nm during the heating process.

These solid-state detectors work for X-rays, gamma rays and beta radiation. However, beta radiation requires thin entrance windows and thin detector design in order to deliver reliable dose results. Thermoluminescence detectors are commercialized for most personal dosimetry purposes: whole body, finger, eye lens and skin monitoring (Figure 9.5.15).

Figure 9.5.15: Thermoluminescence dosimeters: whole body batch (left), finger ring with single detector chip (middle) and band-mount dosimeters for wrist, head or ankle mounting (right). The images were kindly supplied by MPA NRW, Dortmund.

Radon exposure (alpha radiation of Rn-222 and decay products) can typically occur in active mining, in visitor mines, in radon spas or waterworks. Passive radiation protection monitoring proceeds via small diffusion chambers using so-called radon dosimeters. Radon diffuses into the diffusion chamber through a fine glass fiber filter and then alpha particles that form during the radioactive decay hit a cellulose nitrate film, which is the active detector material. This impact of alpha particles causes a material deterioration on the cellulose nitrate. The resulting core tracks are visualized by chemical etching and quantified by standard electronic picture analyses.

The last example for passive personal dosimeters concerns detection of fast and slow neutrons (according to their energy, neutrons are rated by a radiation-weighting factor) [22]. Neutron detection is not straightforward: Usually the so-called two-detector method finds application. ^{6}Li (enhancement of the ^{6}Li + ^{1}n → ^{4}He + ^{3}H nuclear reaction)-enriched LiF TLD crystals are coupled to a ^{7}Li-enriched detector presenting a very low cross section for the reaction above. A specific field calibration of these neutron dosimeters is often necessary due to the large neutron energy dependence of the weighting factor. An improved version of this dosimeter uses two pairs of ^{6}Li–^{7}Li detectors to monitor neutron radiation from the source as well as backscattered from the monitored person's body, in order to derive neutron energy information from the ratio of primary and moderated neutrons. This type is called Albedo dosimeter. Further developments of this method exploiting higher temperatures during readout yield improved neutron sensitivity [23].

References

[1] Cerrito L. Radiation and Detectors – Introduction to the Physics of Radiation and Detection Devices, Springer, Cham, Switzerland, 2017.

[2] Iniewski K (Ed). Advanced Materials for Radiation Detection, Springer, Cham, Switzerland, 2022.

[3] Milbrath BD, Peurrung AJ, Weber WJ. J Mater Res, 2008, 23, 2561–81.

[4] Rogalski A, Adamiec K, Rutkowski J. Narrow-Gap Semiconductor Photodiodes, SPIE Press, Bellingham, 2000. ISBN: 9780819436191.

[5] Guisbiers G, Abudukelimu G, Wautelet M, Buchaillot L. J Phys Chem C, 2008, 112, 17889–92.

[6] https://de.m.wikipedia.org/wiki/Datei:III-V-Halbleiter.png. Accessed on December 23rd, 2021.

[7] Schmidt F, Christiansen N, Lovrincic R. Photonics Rev, 2020, 17, 56–9. DOI: doi.org/10.1002/phvs.202000036.

[8] Amemiya Y. J Synchr Rad, 1995, 2, 13–21.

[9] Gales SG, Bentley CD. Rev Sci Instr, 2004, 75, 4001–3.

[10] Hoheisel M. Nucl Instr Meth Phys Res A, 2006, 563, 215–24.

[11] Gruner SM, Tate MW, Eikenberry EF. Rev Sci Instr, 2002, 73, 2815–42.

[12] Tietze U, Schenk C, Gamm E. Halbleiterschaltungstechnik, Vol. 16, Auflage, Springer, Berlin, Heidelberg, 2019. ISBN: 978-3-662-48553-8.

[13] Beckhoff B, Kanngießer B, Langhoff N, Wedell R, Wolff HH. Handbook of Practical X-ray Fluorescence Analysis, Springer, Berlin, Heidelberg, 2006. ISBN 978-3-540-36722-2.

[14] Dorenbos P. Opt Mater X, 2019, 1, 100021. DOI: doi.org/10.1016/j.omx.2019.100021.

[15] Rodnyi PA, Dorenbos P, van Eijk CWE. Phys Stat Sol (B), 1995, 187, 15–29. DOI: doi.org/10.1002/pssb.2221870102.

[16] Makhov VN. Phys Scr, 2014, 89, 044010. DOI: doi.org/10.1088/0031-8949/89/04/044010.

[17]   Sekikawa T, Yamazaki T, Nabekawa Y, Watanabe S. J Opt Soc Am B, 2002, 19, 1941–5. DOI: doi.org/10.1364/JOSAB.19.001941.

[18]   Laval M, Moszyński M, Allemand R, Cormoreche E, Guinet P, Odru R, Vacher J. Nucl Instr Meth, 1983, 206, 169–76. DOI: doi.org/10.1016/0167-5087(83)91254-1.

[19]   Marton Z, Miller SR, Ovechkina E, Kenesei P, Moore MD, Woods R, Almer JD, Miceli A, Singh B, Nagarkar VV. AIP Conf Proc, 2016, 040035, 1741–4. DOI: doi.org/10.1063/1.4952907.

[20]   Kouzes RT, Ely JH, Erikson LE, Kernan WJ, Lintereur AT, Siciliano ER, Stephens DL, Stromswold DC, Van Ginhoven RM, Woodring ML. Nucl Instr Meth Phys Res A, 2010, 623, 1035–45.

[21]   Krieger H. Strahlungsmessung und Dosimetrie, Springer Fachmedien Wiesbaden, 2013. DOI: 10.1007/978-3-658-00386-9_14.

[22]   d'Errico F, Bos AJJ. Rad Prot Dosimetry, 2004, 110, 195–200.

[23]   Heiny M, Busch F, Kröninger K, Theinert R, Walbersloh J. Rad Prot Dosimetry, 2020, 188, 8–12.

9.6 Soft magnets

Torsten Rieger

9.6.1 Fundamentals of magnetic alloys

9.6.1.1 History of magnetic alloys

The first mention of magnetic materials refers to "magnetic stones" that got stuck on the shoes of a Greek shepherd, described by Pliny the Elder in the *Naturalis Historia*. The magnetic stones are loadstones, a naturally magnetized form of magnetite (Fe_3O_4). There are different theories about the origin of the term "magnet": either it is based on the region where the loadstones were found, Magnesia, or the name of the shepherd mentioned by Pliny, Magnes.

In China, a first form of a compass made of a magnetic spoon of lodestone placed on a brass/bronze plate was used around 200 BC. The spoon pointed to the south and was therefore called "si nan" (south pointer) (Figure 9.6.1). However, this compass was not used for navigation, but for prophecy. The first compasses for navigation were developed in China around AD 1100, and around AD 1200 compasses were also used in Europe.

Figure 9.6.1: Replica of a "Si nan" compass.

The first scientific approach to describe magnetic phenomena was given by William Gilbert in 1600 with his book *De Magnete, Magnetisque Corporibus, et de Magno Magnete Tellure* which led to the many later physicists like Ørsted, Ampère, Faraday and finally Maxwell to come to a comprehensive understanding of the relation between electricity and magnetism.

https://doi.org/10.1515/9783110733471-006

Magnetic materials became an important issue with industrialization and the first applications in the generation and electromechanical use of electric currents. The first application of an electromagnet was the horseshoe magnet developed by William Sturgeon in 1824, in which an electrical conductor was wound around a horseshoe-shaped piece of iron. As a result of the ongoing industrialization, the systematic development of magnetic materials began in the nineteenth century. Today, there are two extreme groups of magnetic materials: soft magnetic materials with very low coercive force and hard magnetic materials with very high coercive force.

9.6.1.2 Types of magnetism

Five types of magnetism are distinguished: diamagnetism, paramagnetism, ferromagnetism, ferrimagnetism and antiferromagnetism.

In diamagnetic materials, the density of magnetic field lines in the material is lower than in the environment, and the material migrates out of the magnetic field. Diamagnetic effects occur in every material, but they are much smaller than, e.g., influences of para- or ferromagnetism. If materials or their atoms have no unpaired electrons, they are diamagnetic.

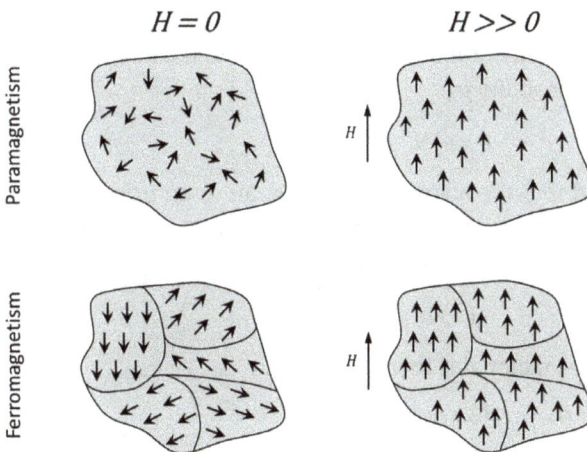

$H = 0$ $H >> 0$

Figure 9.6.2: Orientation of magnetic moments in a paramagnetic and a ferromagnetic material without and with external magnetic field.

If the materials or atoms have unpaired electrons, there is a magnetic moment which can be influenced by an external magnetic field. In a paramagnet, the individual magnetic moments are isolated from each other. In an external magnetic field, the moments align parallel to the magnetic field and the magnetic field is amplified in the material (see Figure 9.6.2). When the external magnetic field is removed, the parallel alignment of the magnetic moments is lost again due to the thermal movement of the atoms.

In some materials, the magnetic moments are coupled to each other by an exchange interaction, and they align parallel to each other within certain areas. This is called ferromagnetism. Areas with parallel alignment of the magnetic moments are called Weiss domains. Without an external magnetic field, the magnetizations of the individual domains cancel each other out and the material is not macroscopically magnetized. In an external magnetic field, the magnetic moments align parallel to the external magnetic field (see Figure 9.6.2) and significantly amplify it inside the material. When the external magnetic field is switched off, in ferromagnetism, in contrast to paramagnetism, a residual magnetization remains, the so-called remanence. At room temperature, Fe, Co, Ni and Gd are ferromagnetic.

In antiferromagnetism, the magnetic moments are aligned antiparallel and each has the same magnitude. Antiferromagnetic materials are macroscopically non-magnetic.

In ferrimagnetic materials, there are antiparallel aligned magnetic moments of different magnitudes. Macroscopically, the magnetic moments thus do not cancel each other out, in contrast to antiferromagnetism. Ferrimagnetic materials behave similarly to ferromagnetic materials.

9.6.1.3 Soft magnetic hysteresis

In vacuum, the magnetic flux density or induction B is related to the magnetic field strength H via the vacuum permeability μ_0:

$$B = \mu_0 H \qquad (9.6.1)$$

When a material is introduced into the magnetic field, e.g. air or an iron core, it is magnetized. The proportionality factor between the magnetization M and the magnetic field strength H is the magnetic susceptibility χ:

$$M = \chi H \qquad (9.6.2)$$

The magnetic flux density in the material is then

$$B = \mu_0(H + M) = \mu_0 H + J \qquad (9.6.3)$$

The polarization J is the intrinsic magnetic flux density of the material, i.e. the magnetic flux density minus the flux density in vacuum:

$$J = B - B_0 = \mu_0 M \qquad (9.6.4)$$

Substituting eq. (9.6.2) into eq. (9.6.3) yields

$$B = \mu_0(1 + \chi)H = \mu_0 \mu_r H \qquad (9.6.5)$$

$\mu_r = 1 + \chi$ is the relative permeability of the material and a measure for the "magnetic conductivity" of a material. For soft magnetic materials, $\mu_r \gg 1$ and thus $B \approx J$ is valid.

The relative permeability is thus given by

$$\mu_r = \frac{B}{\mu_0 H}$$

However, the ratio B/H of a material is not constant but depends on the magnetic field strength H as well as on the history of the material. This behavior is called hysteresis and is shown schematically in Figure 9.6.3.

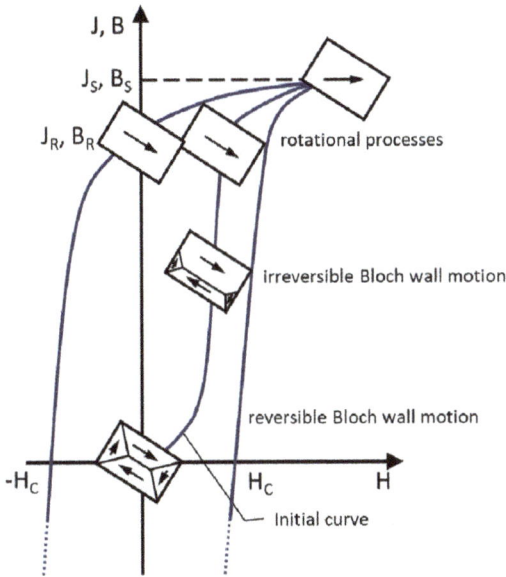

Figure 9.6.3: Section of the hysteresis curve of a soft magnetic material also showing the elementary remagnetization processes (adopted from [1]).

In the initial state, the magnetic domains are randomly distributed and the material is therefore not macroscopically magnetized. This initial state typically exists, for example, after manufacturing or after a demagnetization process. By applying an external magnetic field strength H, initially domains which are favorably oriented to the external field strength grow by Bloch wall motion. The individual domains are separated from each other by Bloch walls. The Bloch walls are pinned by impurities, stresses or other inhomogeneities. At low field strengths, the Bloch walls move reversibly, i.e. they can return to their initial state when the external magnetic field is switched off. At higher field strengths, the Bloch walls detach from the pinning centers. This abrupt process is called Barkhausen jump and leads to the fact that the magnetization does not change continuously but, with appropriate resolution, discrete jumps can be seen in the B–H curve. The Barkhausen jumps are irreversible and thus contribute significantly to the formation of the hysteresis curve.

With a further increase of the magnetic field strength, the domains have grown to maximum size due to the irreversible wall displacements, and a further increase of the magnetization of the material is achieved by rotational processes. In this process, the magnetization vectors of the domains rotate from a preferred orientation in to the direction of the external field. When this process is complete, the material is in saturation. This is described by the saturation flux density B_S or saturation polarization J_S. The curve traversed during magnetization starting from a macroscopically non-magnetic material is called the initial curve or commutation curve. This curve is usually traversed only once.

When the external field strength is reduced again, a macroscopic magnetization remains at $H = 0$. Here, the magnetization vectors within the individual domains have turned back to their preferred directions. The magnetization of the material at $H = 0$ is called remanence B_R or remanent polarization J_R.

The magnetic flux density becomes zero only at negative field strengths, i.e. fields with reversed orientation. The field strength required for this is the coercive force $-H_c$. The coercive force describes how easily a material can be remagnetized and is usually used to classify materials into hard magnet and soft magnet.

When the negative field strength is further increased, the saturation flux density $-B_s$ is reached in the material. When the external field is reversed again, the remanence $-B_R$ remains at $H = 0$, the coercive force H_c at $B = 0$ and finally the saturation flux density B_s again.

Relative permeability is not a material constant but rather individual, specific permeabilities are typically considered. The most important permeabilities are:

μ_i: initial permeability, permeability at $H = 0$
μ_4:   permeability at a field strength of 4 mA/cm
μ_{max}: maximum permeability

Depending on the material, the manufacturing route and the final annealing, different shapes of the $B–H$ curve and thus of the permeabilities can be achieved, which are useful depending on the application. Three different principal forms of the $B–H$ curve shown in Figure 9.6.4 can be distinguished here:

Flat hysteresis (also called F-loop): the relationship between B and H is linear over large field strengths, and the remanence is small compared to the saturation magnetization. The permeability is almost independent of the field strength; only at high field strengths the permeability decreases, because the material is in saturation. In a flat hysteresis loop, the rotational processes dominate the remagnetization process. This is the case when there is a preferred direction perpendicular to the later flux direction so that most Weiss domains are magnetized perpendicular to the later magnetization direction. How a preferred direction can be imprinted in a material is described in Section 9.6.2.2.2.

Rectangular hysteresis (also called Z-loop): The Bloch wall motions contribute most to the remagnetization, and rotational processes hardly occur. This is the case

Flat hysteresis	Round hysteresis	Rectangular hysteresis

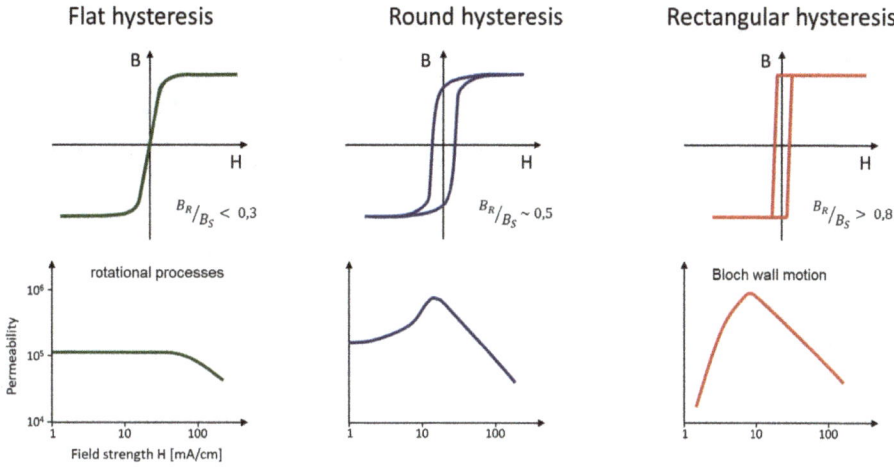

Figure 9.6.4: Loop shape and permeability curve for the flat, round and rectangular hysteresis loop.

when there is a preferred direction parallel to the subsequent flux direction. μ_4 is low for rectangular hysteresis loops, while μ_{max} is very high.

Round hysteresis (also called R-loop): In a round hysteresis loop, both Bloch wall motions and rotational processes are involved in the remagnetization, both μ_i, μ_4 and μ_{max} reach high values.

Another characteristic of magnetic materials is the Curie temperature. The magnetic moments are coupled with each other via the exchange interaction and have the same orientation within the Weiss domains. The thermal motion of the magnetic moments counteracts the coupling (see Figure 9.6.5). At low temperatures, the exchange interaction is much larger than the thermal motion. However, as the temperature increases, the thermal motion increases and the coupled alignment of the magnetic moments becomes progressively weaker. Above the Curie temperature T_C, the thermal motion is higher than the exchange interaction and the material loses its ferromagnetic properties and is paramagnetic. The Curie temperature of iron is 769 °C, nickel has 358 °C and cobalt has a Curie temperature of 1127 °C.

Soft magnetic materials are characterized by low coercive forces and high permeability. This is achieved by an adapted chemical composition, which leads to the low magnetostriction and crystalline anisotropy.

The crystalline anisotropy K_1 indicates whether there is a preferred direction for the magnetic moment within the unit cell. A distinction is made between a preferred direction along the <100> direction and a preferred direction along the <111> direction. With a preferred orientation along <100> as shown in Figure 9.6.6, $K_1 > 0$. This is the case, for example, with pure Fe or Ni-Fe alloys with ~50 wt% Ni. $K_1 < 0$ means that the preferred orientation is along the <111> direction. This is true for Ni, for example. Deflection of the magnetic moment from the preferred direction requires energy,

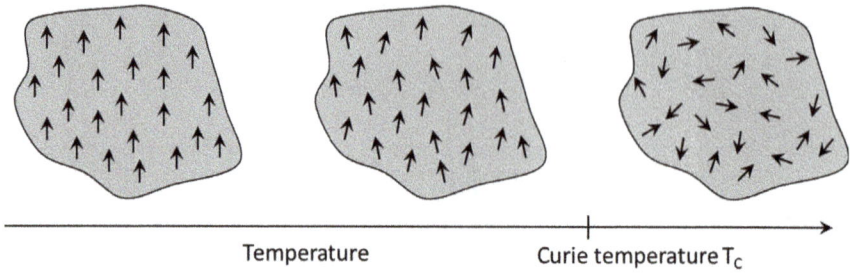

Figure 9.6.5: Effect of temperature on the alignment of magnetic moments in a ferromagnetic material. Above the Curie temperature T_C, the alignment is lost.

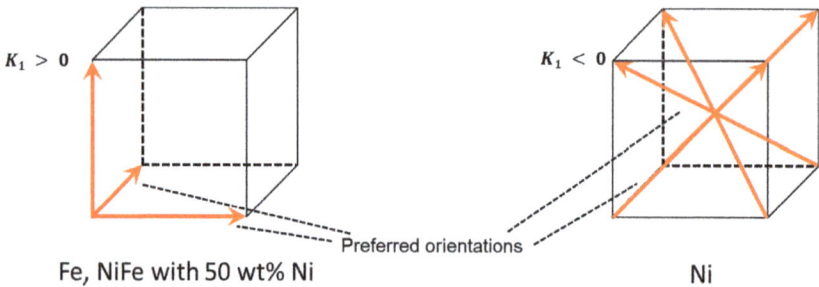

Figure 9.6.6: Diagram showing the preferred orientation of magnetic moments in materials with a crystalline anisotropy $K_1 > 0$ and a crystalline anisotropy $K_1 < 0$.

which is greater the larger the magnitude of K_1. At $K_1 = 0$, there is no pronounced preferred orientation; this is the desired state for the highest possible permeabilities. In some cases, however, a specific preferred orientation is desired, which can then be adjusted by rolling and annealing processes or magnetic field annealing.

Magnetostriction describes how a material is elastically deformed by an external magnetic field (see Figure 9.6.7). If the material becomes longer in the direction of the magnetic field, the magnetostriction is $\lambda > 0$, which is the case for Ni-Fe alloys with ~50 wt% Ni, for example. Pure nickel, on the other hand, becomes shorter, i.e. $\lambda < 0$. Practically, one distinguishes the magnetostriction along the lattice direction. In a cubic lattice, λ_{100} and λ_{111} are relevant for this. For polycrystalline material with a statistical grain distribution, the "mean magnetostriction" λ_s is sometimes also used. For most applications, deformation should be avoided to a large extent, especially since it has a negative effect on the magnetic properties. High permeabilities and low coercive forces are obtained at $\lambda \approx 0$.

A mechanical stress on a material also affects the orientation of the magnetic moments: In a material with positive magnetostriction, the direction of tensile stress becomes the preferred direction for the magnetic moments; with negative magnetostriction, it becomes a direction of difficult magnetizability. The directions

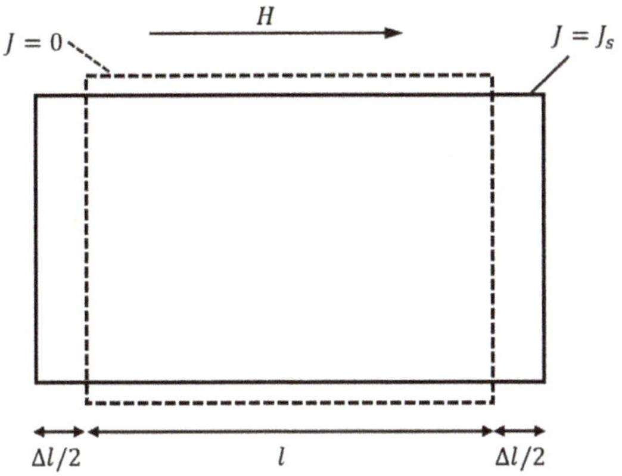

Figure 9.6.7: Schematic representation of the change in length of a ferromagnetic material when magnetized into saturation due to magnetostriction $\lambda > 0$.

perpendicular to the tensile stress direction are then preferred orientations for the magnetic moments.

As described, good soft magnetic properties are only achieved when K_1 and λ are small. In addition, a final annealing under highly controlled conditions is usually required to achieve good properties. During this annealing, stresses are relieved, grain growth takes place (exceptions are the amorphous and nanocrystalline alloys) and an additional purification effect occurs. The material condition obtained after annealing is sensitive to mechanical stress, even small impacts can negatively affect the properties. For this reason, sensitive components are protected, e.g. by soaking them in lacquer, powder coating them or fixing them in trays.

9.6.2 Soft magnetic materials

In general, soft magnetic materials are those with a coercive field strength of less than 1000 A/m or 10 A/cm. A large number of materials fall into this range. Figure 9.6.8 gives an overview of the most important soft magnetic materials with typical data for the coercivity and saturation magnetization. 3% SiFe (Fe with 3 wt% Si) represents by far the largest amount of soft magnetic materials produced.

9.6.2.1 Electrical steel

High-iron soft magnetic materials have by far the largest production volume of the soft magnetic materials. The most important representative of this group is Fe with about 3 wt% Si, the so called electrical steel.

Pure Fe has a high saturation flux density (2.15 T) and a high Curie temperature (769 °C). However, the magnetocrystalline anisotropy is too high, so good soft

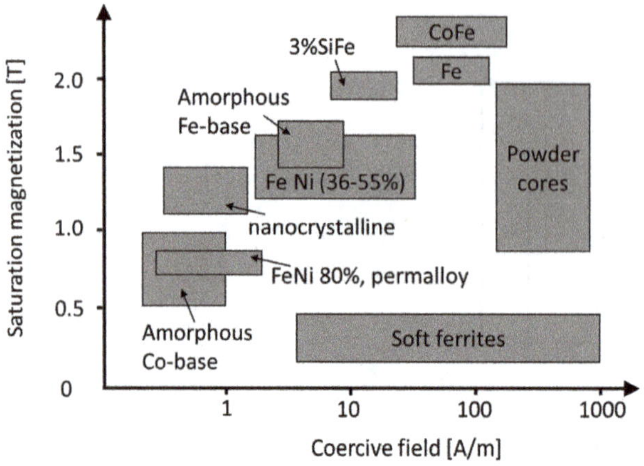

Figure 9.6.8: "Map" of soft magnetic materials (based on [2, 3]).

magnetic properties are not obtained. At the same time, the electrical resistivity of Fe is relatively low, so remagnetization losses (eddy current losses) are high. However, eddy current losses can be reduced by using smaller strip thicknesses and a soft magnetic material with a higher electrical resistivity.

Adding Si reduces the crystal line anisotropy of Fe and increases the electrical resistivity. At the same time, however, the saturation flux density and the Curie temperature decrease. Si additions higher than about 3.5 wt%, however, lead to severe processing difficulties because the material becomes brittle. The technically relevant composition is about 3.2 wt% Si. At 3.2 wt% Si, the electrical resistance of Fe is increased by a factor of about 4 [4, 5].

In general, a distinction is made between two types of electrical steel with different grain orientation: non-grain-oriented (NGO) and grain-oriented (GO).

In NGO electrical steel, the grains are randomly distributed (see Figure 9.6.9), and there is no preferred orientation. The magnetic properties are identical in all directions (isotropic). NGO electrical steel is mainly used in rotating components, e.g. electric motors.

GO electrical sheet has a preferred orientation for magnetization, the so-called Goss texture [6]. In the Goss texture, the directions of easy magnetizability (easy axis), i.e., the <100> directions, are oriented in the strip direction. The {110} plane is in the band plane, as shown in Figure 9.6.9. Therefore, the Goss texture is also indicated as {110} <100> texture. The production of GO electrical steel requires close analytical limits and very accurate process control in terms of rolling and annealing steps. The actual Goss texture is formed by secondary recrystallization during high-temperature annealing.

Since the easy axis of GO electrical steel is in the strip direction, the magnetic properties are also particularly good there. Remagnetization processes take place

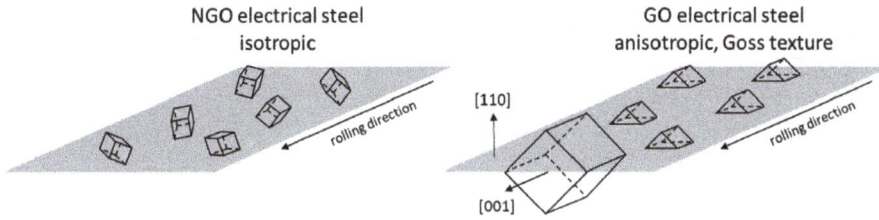

Figure 9.6.9: Grain orientations in NGO and GO electrical steel.

mainly through Bloch wall motions; rotation processes play only a minor role. Perpendicular to the rolling direction, however, the magnetic properties are worse than those of NGO. GO electrical steel is used, for example, in transformers.

9.6.2.2 Ni-Fe alloys

Soft magnetic Ni-Fe alloys are a very versatile family of alloys where the soft magnetic properties can be precisely tailored and varied over a wide range by composition, annealing and rolling treatments.

Ni-Fe alloys, like the other crystalline soft magnetic materials electrical steel and CoFe alloys, are melted on an industrial scale of several tons and transformed to the required dimensions and geometry (foil, strip, wire, rod or sheet) by hot and, if necessary, cold forming. Depending on the material and application, melting can take place in a vacuum induction furnace to achieve high purity.

Figure 9.6.10 shows the variation of crystalline anisotropy, magnetostriction, saturation flux density and Curie temperature as a function of Ni content [7–11]. For the magnetostriction, λ_{100} and λ_{111} are distinguished, and for the crystalline anisotropy, the diagram distinguishes between a quenched state and a state with slow cooling after final magnetic annealing. The final magnetic annealing is necessary to achieve ideal soft magnetic properties. For Ni-Fe alloys, the final annealing ideally takes place under pure hydrogen with a dew point $<-60\,°$C or in a vacuum; typical temperatures are in the range of 1050–1200 °C with an annealing time of several hours. Of particular importance is the subsequent cooling phase. As shown in Figure 9.6.10, the crystal line anisotropy K_1 depends on the cooling rate. Highest permeabilities and smallest coercive field strengths are achieved at crystalline anisotropy $K_1 = 0$ and magnetostriction $\lambda = 0$.

In principle, three relevant groups of Ni-Fe alloys are distinguished:
- High Ni content (approx. 80 wt%): highest permeabilities and lowest coercive field strengths.
- Medium Ni content (approx. 50 wt%): high saturation flux density, texture formation possible
- Low Ni content (approx. 36 wt%): high electrical resistance, low thermal expansion

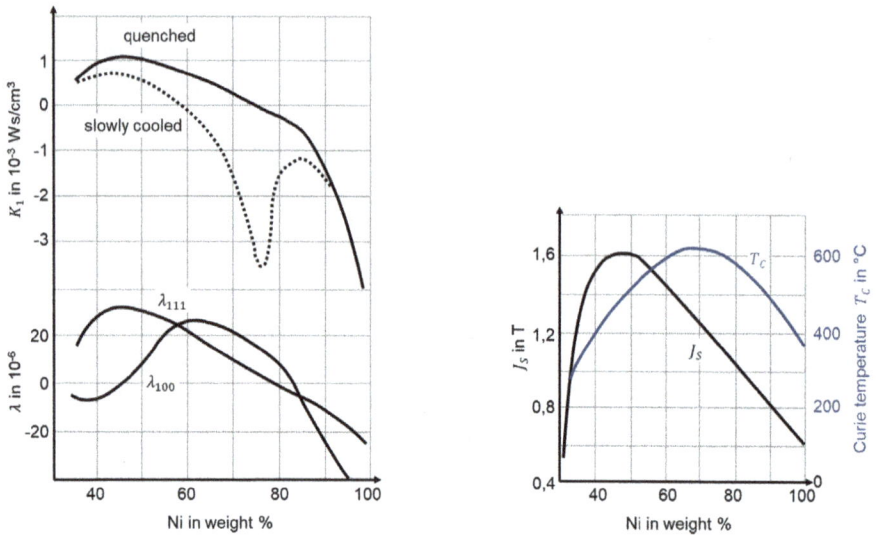

Figure 9.6.10: Crystalline anisotropy K_1, magnetostriction λ, saturation flux density J_s and Curie temperature T_c as a function of Ni content.

9.6.2.2.1 Ni-Fe alloys with high Ni content

At Ni contents in the range of 80 wt%, both crystalline anisotropy and magnetostriction are low, but K_1 and λ never become zero simultaneously. This is only achieved by adding further elements such as Cu, Mo or Cr [12–15].

The variant NiFe15Mo (2.4545, DIN 17745) has about 80 wt% Ni, 5 wt% Mo, remainder Fe and thus achieves $K_1 = 0$ and $\lambda = 0$. This can also be achieved with the variant NiFe16CuCr (2.4501, DIN 17745) with about 77 wt% Ni, 5 wt% Cu, 2 wt% Cr and remainder Fe. Both variants are also referred to as Mu-metal, permalloy or supermalloy. $K_1 = 0$ is achieved in each case during slow cooling after final magnetic annealing to approx. 500 °C and subsequent air cooling.

The saturation flux density in both cases is about 0.8 T, and the Curie temperature is about 400 °C.

Applications are typically either in the static or low-frequency range, e.g. for shielding magnetic interference fields, or at mains frequency.

In the static and low-frequency range, the static hysteresis curve and the characteristic values derived from it are particularly relevant, i.e. μ_4, μ_{max} and H_C. In the case of 2.4545, initial permeabilities $\mu_4 > 600000$ and maximum permeabilities $\mu_{max} > 700000$ can be achieved at 0.35 mm strip thickness after suitable annealing. The coercivity is $H_C < 0.5\,\text{A/m}$.

At mains frequency or higher frequencies, the remagnetization losses (eddy current losses) become relevant. Therefore, either laminated stacks or toroidal tape-wound cores are used, as shown in Figure 9.6.11. The individual layers are electri-

Figure 9.6.11: Laminated stack of stamped soft magnetic sheets and toroidal tape-wound core.

cally insulated from each other. The tape thicknesses are significantly reduced in order to minimize eddy current losses. For toroidal tape-wound cores, typical tape thicknesses are in the range of 60–200 μm. At 50 Hz, initial permeabilities μ_4 of approx. 300,000 and maximum permeabilities μ_{max} of approx. 400,000 are then achieved with NiFe15Mo.

9.6.2.2.2 Ni-Fe alloys with medium Ni content

Ni-Fe materials with Ni content in the range of 45–55 wt% such as 1.3922 have the highest saturation flux density achievable with Ni-Fe alloys (about 1.55 T) and achieve high DC permeabilities ($\mu_{max} > 160,000$) and low coercive field strengths ($H_C < 5\,\text{A/m}$) at static excitation.

An important property of Ni-Fe alloys with medium Ni content is that they can be textured by suitable rolling and annealing processes, similar to electrical steel. Two possible textures are distinguished for Ni-Fe alloys: the cube texture and the (210)[001] texture [16].

In the cube texture ((100)[001] texture), one cube facet (the (100) facet) of the unit cell lies in the band plane, the [001] direction is parallel to the rolling direction (see Figure 9.6.12). Since the crystalline anisotropy is greater than zero for Ni-Fe alloys with ~50 wt% Ni, the [100] direction is the preferred direction (easy axis). Remagnetization processes are thus mainly accomplished by Bloch wall motions, and the hysteresis loop is correspondingly rectangular with a high remanence. However, the grain size is small and thus the permeabilities are relatively low.

The orientation of the unit cell in the (210)[001] texture is shown in Figure 9.6.12. The [100] direction, i.e. the magnetically preferred direction, is still in the band direction. In contrast to the cube texture, significantly larger grains are obtained with the (210)[001] texture, the grain size being in the range of mm. This leads to significantly higher permeabilities than with the cube texture.

The formation of the textures depends on the rolling and annealing conditions (see Figure 9.6.13). If the strip receives less than about 85% cold working, an isotro-

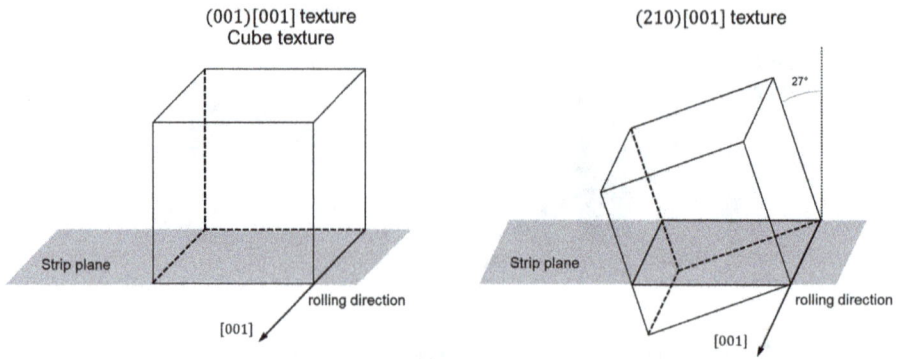

Figure 9.6.12: Textures of Ni-Fe alloys with Ni content of about 50 wt%. Left: cube texture. Right: (210)[001] texture.

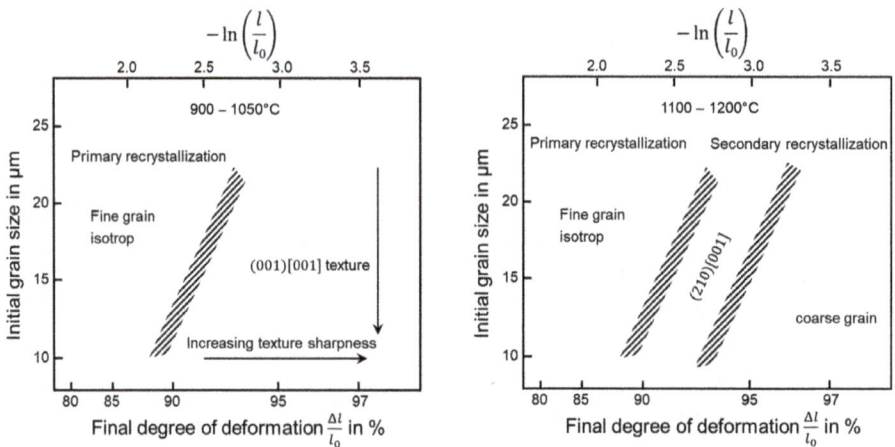

Figure 9.6.13: Diagram showing the conditions (degree of cold deformation, temperature, etc.) where the cube texture (left) and the (210)[001] texture (right) are obtained in Ni-Fe alloys with medium Ni content (adapted from [1]).

pic microstructure forms during final annealing. The hysteresis curve for isotropic microstructures is flatter. At more than 90% cold working and relatively low annealing temperatures in the range of about 900–1050 °C, the cube texture is formed. The higher the degree of deformation and the finer the initial microstructure, the sharper the cube texture is formed.

At temperatures of about 1100–1200 °C and degrees of deformation of about 90–95%, a secondary recrystallization takes place after the primary recrystallization, during which the (210)[001] texture is formed (Figure 9.6.13, right). The grain size is then in the range of millimeters. At higher degrees of deformation, a coarse-grained microstructure with grain sizes in the centimeter range is formed instead of the (210)[001] texture. The coarse-grained structure does not provide good magnetic properties because the grains are not oriented in the preferred direction.

Similar to electrical steel, the isotropic texture is preferably used for rotating machines, and the (210)[001] texture is used when the magnetic properties in strip direction are crucial (e.g. for transformers).

Due to the high degrees of cold working required, the cube texture and the (210)[001] texture are only available for thin strips with strip thicknesses <0.35 mm.

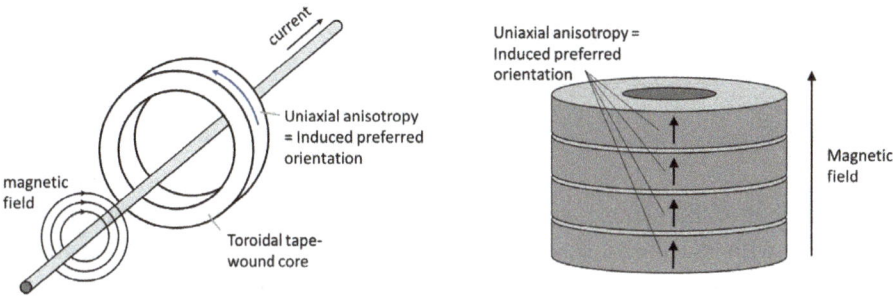

Figure 9.6.14: Left: longitudinal magnetic field annealing. Right: transversal magnetic field annealing.

The shape of the hysteresis loop can also be influenced by annealing in a magnetic field, which defines a preferred direction either perpendicular or parallel to the later flux direction. For this purpose, the toroidal core is heated to a temperature below the Curie temperature and subjected to a magnetic field for several hours. In this case, an induced uniaxial anisotropy K_U is formed which is larger than the crystalline anisotropy K_1. By annealing in a longitudinal field (see Figure 9.6.14), the preferred direction is in the band direction, and remagnetization processes occur almost exclusively via Bloch wall motions. Accordingly, the hysteresis loop is rectangular. For annealing in the transverse field, the preferred direction is perpendicular to the strip direction, and mainly rotational processes contribute to remagnetization. The hysteresis loop thus becomes flat.

9.6.2.2.3 Ni-Fe alloys with low Ni content

Ni-Fe alloys with about 36 wt% Ni have a resistivity of about 75 µΩcm and a saturation flux density of 1.3 T. The permeabilities are lower than Ni-Fe alloys with high or medium Ni content. Due to the higher electrical resistivity, the remagnetization losses are reduced.

A special property of Ni-Fe materials with about 36 wt% Ni is the low thermal expansion, the so-called invar effect. Here, a negative volume magnetostriction compensates for the normal thermal expansion, resulting in very low expansion coefficients up to the Curie temperature of about 230 °C [17]. Figure 9.6.15 shows

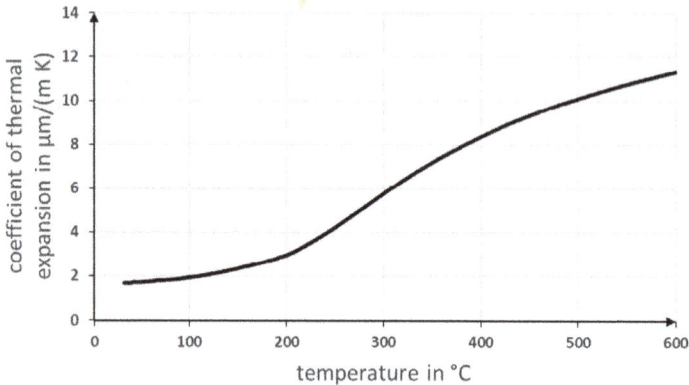

Figure 9.6.15: Coefficient of thermal expansion of Ni-Fe alloy with 36 wt% Ni.

the coefficient of thermal expansion as a function of the temperature for a typical Ni-Fe alloy with 36 wt% Ni. By adjusting the chemical composition, the expansion behavior can be precisely adapted to other materials such as glasses, ceramics, semiconductors or carbon fiber-reinforced plastics, or even reduced to zero. Invar materials are used, for example, in mold tooling for the aerospace industry, in the production of OLED screens and as passive components in bi-metals.

9.6.2.3 Co-Fe alloys
The highest saturation flux densities of up to 2.43 T are obtained in Co-Fe alloys, and these have very high Curie temperatures of up to 980 °C. However, in the Co-Fe phase diagram, there is a phase transformation from an austenitic phase at high temperatures to a ferritic phase at low temperatures at about 900 °C. This limits the possible temperature for a final magnetic annealing. The best magnetic properties are obtained at about 50 wt% Co, where the crystalline anisotropy has a zero crossing. However, at this composition, an abrupt ordering process takes place below 730 °C, which leads to embrittlement of the material. This must be taken into account during hot forming and requires accurate process control and rapid cooling. The addition of about 2 wt% V slows down the ordering process and significantly increases the electrical resistivity [4, 18].

Co-Fe alloys are used where high saturation flux density or an elevated application temperature is required. The high saturation flux density makes it possible, for example, to design smaller components and thus save weight. The most common soft magnetic Co-Fe alloy is Fe49Co49V2.

9.6.2.4 Amorphous and nanocrystalline materials
A relatively young class of soft magnetic materials are the amorphous and nanocrystalline materials. It is not the long-range crystalline order that defines whe-

ther a material is ferromagnetic or not, but the magnetic coupling between neighboring atoms. In amorphous and nanocrystalline materials, this short-range magnetic coupling is present, while no long-range crystalline order is present. Amorphous materials are often referred to as metallic glasses.

In order to produce amorphous soft magnetic materials, the liquid state must be frozen, which requires very high cooling rates in the range 10^5–10^6 K/s. In addition, glass formers, i.e. atoms with small atomic radii such as Si and B, are helpful to avoid crystallization. At the same time, the band thickness is limited. Figure 9.6.16 shows schematically the process for the continuous production of amorphous films, the so-called melt spinning. The molten metal is supplied via a ceramic nozzle onto a rotating, cooled wheel made of a material with very good thermal conductivity, usually Cu. The metal cools very quickly and has no time to crystallize [19–21]. The amorphous strip is then released from the wheel and wound into coils. Typical strip thicknesses are in the range of 20 μm.

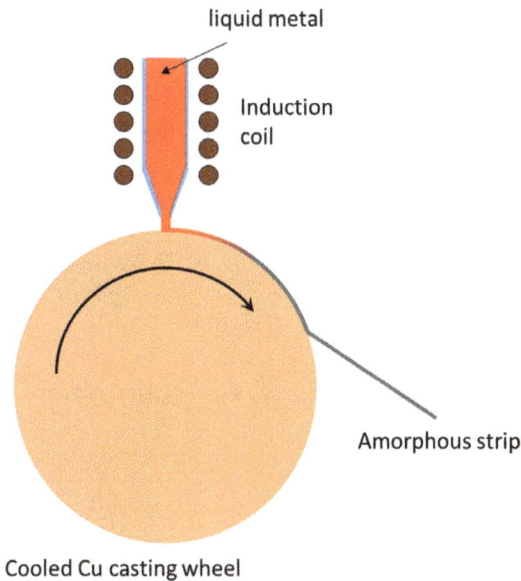

Figure 9.6.16: "Melt spinning" of amorphous alloys.

Due to the lack of crystalline order, the crystal anisotropy K_1 in amorphous alloys is zero, and very high magnetic values can be achieved if the magnetostriction λ_s is sufficiently small. Typical representatives of amorphous soft magnetic alloys are Fe-based and Co-based alloys. Both types of alloys have about 15–30 atom% of glass formers, the rest being ferromagnetic elements such as Fe, Co and/or Ni and, depending on the application, other elements such as Mn. In the amorphous Fe-based alloys, magnetostriction is larger than 25 ppm, and saturation inductions of

1.4–1.8 T are achieved. A vanishing magnetostriction can be achieved with the Co-based materials, and the saturation induction is then between 0.4 and 1.2 T [22].

In order to achieve the desired soft magnetic properties, the materials are wound, e.g. as toroidal cores, and annealed to relieve stresses. The annealing temperature for the amorphous materials must be selected so that no crystallization takes place, but should still be as high as possible. Depending on the application, a subsequent annealing below the Curie temperature in a magnetic field can take place, similar to the Ni-Fe materials, i.e. either as longitudinal or transverse field annealing.

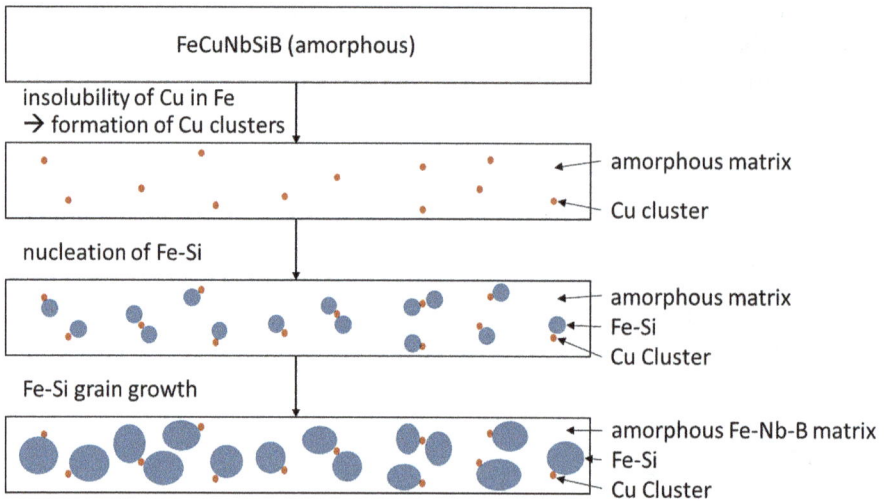

Figure 9.6.17: Transformation from an amorphous FeCuNbSiB matrix to a nanocrystalline alloy during annealing (based on [23–25]).

If amorphous materials are annealed at a too high temperature, crystallization takes place and crystalline anisotropy becomes relevant. In this case, the soft magnetic properties deteriorate. However, if grain sizes in the range of 10–15 nm are achieved, excellent soft magnetic properties are obtained. Such materials are called nanocrystalline soft magnetic materials and are typically Fe-based (e.g. FeCuNbSiB). The production of the nanocrystalline materials is displayed in Figure 9.6.17. Initially, an amorphous tape is produced via the melt-spinning process. Subsequently, annealing at 500–600 °C takes place, and small Cu clusters with a high density precipitate from the matrix. In the further annealing process, these clusters serve as nucleation points for Fe-Si grains. The remaining amorphous matrix is then enriched with Nb and B. In the further annealing process, the Fe-Si grains grow to a size of 10–15 nm. Finally, the amorphous matrix consists of Fe-Nb-B [23–25]. Subsequently, annealing in a magnetic field can also take place to selectively modify the hysteresis loop. The grain size of the nanocrystalline materials is 10–15 nm, which is smaller than the range of the ferromagnetic exchange interaction, averaging out the crystalline aniso-tropy K_1.

Both amorphous and nanocrystalline soft magnetic materials achieve high per-meabilities and low coercivity, with amorphous Co-based materials and nanocrystal-line Fe-based materials having the main technical importance. The low strip thickness and the high electrical resistance of >1 µΩm compared to crystalline materials are also advantageous for keeping eddy current losses low and thus also enable use at high application frequencies.

9.6.2.5 Soft ferrites

In contrast to the previously mentioned materials, the soft magnetic ferrites are not ferromagnetic but ferrimagnetic materials. The most important ferrite materials are the Ni-Zn ferrites and the Mn-Zn ferrites, described by the chemical formula $MeOFe_2O_3$. Here Me is a divalent metal ion (e.g. Fe, Mn, Ni and Zn). The production of ferrites also differs from the previously mentioned materials: The basic materials are first mixed, then pre-sintered, milled and organic additives are added. The material is then granu-lated, followed by final sintering. This process allows a wide variety of geometries to be produced.

Compared with other soft magnetic materials, the permeability of ferrites is low. The highest permeability ferrite has just an initial permeability of about 20,000, and the saturation induction is less than 0.55 T. The decisive factor in ferrites, however, is their high electrical resistivity: Mn-Zn ferrites have an electrical resistivity of about 1 Ωm, Ni-Zn ferrites even of more than 10^5 Ωm. This enables their use at very high frequencies (e.g. in the range of GHz), since eddy currents only become relevant at high frequencies due to the high electrical resistance [26, 27].

9.6.3 Applications of soft magnetic materials

In the following, some exemplary applications of soft magnetic materials are descri-bed, which demonstrate their great technical importance.

9.6.3.1 Transformers

Transformers are needed to transform voltages, e.g. in power distribution networks, from high voltages (e.g. 110 kV) to medium voltages (e.g. 20 kV) or low voltages (e.g. 230 V).

A transformer consists of a soft magnetic core, a primary-side winding and a secondary-side winding, as shown in Figure 9.6.18. The primary winding consists of N_1 turns and the secondary winding consists of N_2 turns. An AC voltage V_1 is app-lied to the primary side, which induces a magnetic flux in the core, which in turn

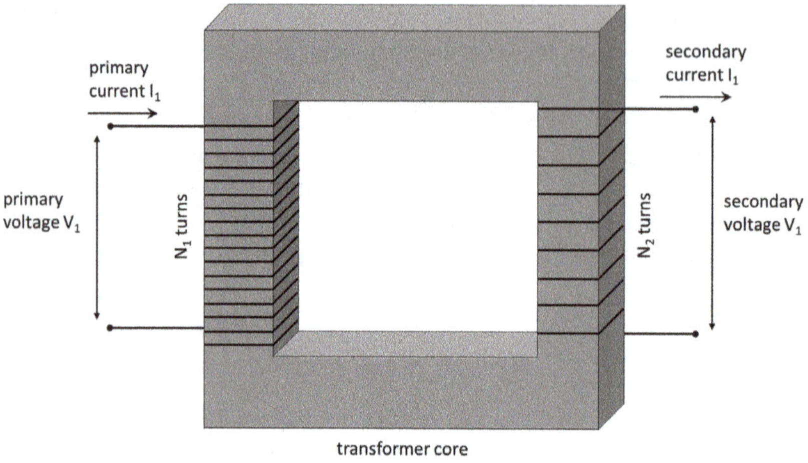

Figure 9.6.18: Schematic of a transformer.

induces an electric voltage V_2 in the secondary side. In the case of an ideal transformer, the ratio of the two voltages is equal to the ratio of the number of turns:

$$\frac{V_2}{V_1} = \frac{N_2}{N_1}$$

Furthermore, since there are no losses in an ideal transformer, the power on the primary side is equal to the power on the secondary side and thus

$$\frac{I_1}{I_2} = \frac{N_2}{N_1}$$

In a real transformer, however, these ratios are influenced by, among other things, electrical resistances of the windings, remagnetization losses in the soft magnetic core and nonlinear hysteresis curves.

Transformers operating at mains frequency usually have a core of GO electrical steel, e.g. an EI core. In special cases, Ni-Fe alloys or amorphous Fe-based materials are also used, and ferrite cores are used at higher frequencies.

Another transformer is the current transformer: Here, the primary winding often consists of only one turn, and the electrical conductor is only inserted through the transformer. A measuring device for determining the induced current, e.g. an ampere meter, is attached to the secondary side. The current on the secondary side is then

$$I_2 = I_1/N_2$$

Current transformers are typically made of toroidal cores with the highest possible permeabilities, large saturation flux densities and linear hysteresis curves. Depending on the application, electrical steel, ferrites, Ni-Fe materials (with 80 wt% Ni or ~50 wt% Ni) or amorphous or nanocrystalline alloys are used.

A special case of the current transformer is the differential current transformer, which is used in a residual current circuit breaker.

9.6.3.2 Residual current device

A residual current circuit breaker is a central, safety-related element in electric circuits. It consists of a differential current transformer, a tripping relay and a latch. Figure 9.6.19 shows an example of the structure of a residual current circuit breaker. All current-carrying lines pass through the differential current transformer. In fault-free operation, the sum of the currents flowing to and from the consumer is zero, the magnetic fields of the individual lines cancel each other out and there is no induction in the differential current transformer. However, if current flows out through the ground due to a fault in the electrical system, the currents flowing to and from the load no longer cancel each other out. Due to the net current effectively present, the differential current transformer becomes magnetized and a current is induced in the winding wound on the differential current transformer. This induced current is used to operate a relay, which mechanically interrupts the current flow via a latch.

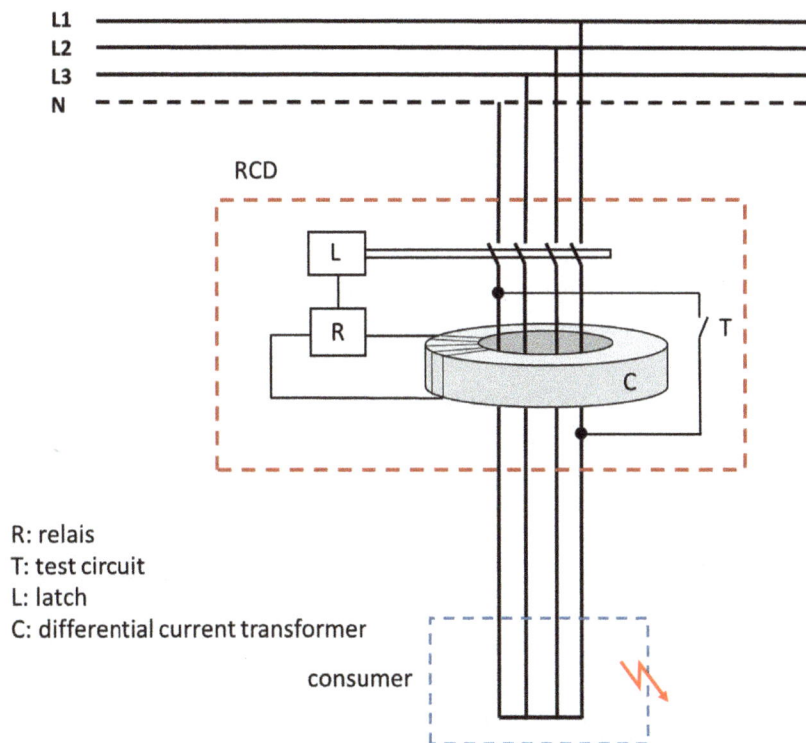

R: relais
T: test circuit
L: latch
C: differential current transformer

Figure 9.6.19: Basic principle of residual current circuit breaker (RCD).

The two central components of a residual current circuit breaker are thus the differential current transformer and the relay.

The differential current transformer must pass on as high a tripping current as possible to the relay even with small residual currents (i.e. with a small H field). This means that the induction swing, i.e. the difference between B at $H = 0$ and B at the H field generated by the residual current, should be as large as possible. In the case of sinusoidal currents, this is achieved, for example, with the Ni-Fe alloy 2.4545 (NiFe15Mo) with a round hysteresis loop. In the case of pulsed DC currents, however, the available induction swing with a round hysteresis loop is only small due to the comparatively high remanence. Therefore, for pulsed DC currents, a soft magnetic material with low remanence but high permeability is required. This can be achieved by an annealing in the transverse field, which modifies the hysteresis loop to a flat loop as the rotational processes dominate. Alternatively, amorphous or nanocrystalline materials with a flat hysteresis loop can be used here.

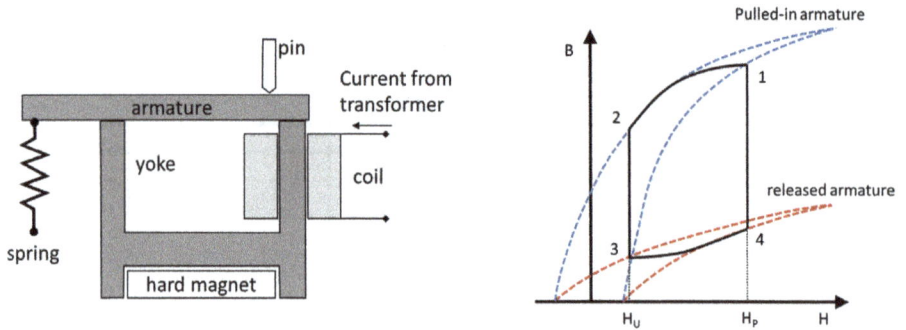

Figure 9.6.20: Left: Schematic of a tripping relay. Right: Working principle of a pre-polarized relay.

The relay is used to interrupt the current flow in the faulty circuit when a fault current has been detected via the differential current transformer. In many applications, relays are made of electrical steel sheet, but in the case of RCDs, Ni-Fe materials are used because the relay operates independently of the power supply and with low tripping currents, requiring high permeabilities and low coercive field strength. Most often, pre-polarized relays are used, as shown in Figure 9.6.20 (left). In this case, the yoke is biased by a hard magnet so that the armature is in position. When a fault current is present, a tripping current is applied to the coil through the differential current transformer. This current reduces the magnetic field of the hard magnet, which means that the armature no longer adheres on the yoke. This is additionally amplified by the spring.

Figure 9.6.20 (right) shows the sequence in the hysteresis curve: At point 1 the armature is attracted, H_P is the field of the hard magnet. When a sufficiently large fault current is detected, the field is reduced to H_U by the coil current (point 2). In this condition, the spring force is higher than the holding force by the hard magnet and the armature detaches from the yoke (point 3). This interrupts the current flow and thus also the fault current (point 4). To close the circuit again, the armature must be mechanically pressed back onto the yoke. If there is still a residual current, the RCD will trip again directly.

To ensure that the relay trips even at small fault currents but not due to slight vibrations, both the flux density B and the difference between the flux density with the armature attracted and the flux density at the fault current must be as high as possible. At the same time, a high difference should be achieved with small tripping currents, i.e. a small difference between H_P and H_U. This requires high permeabilities but also very precise manufacturing of the yoke and armature so that air gaps are minimal.

9.6.3.3 Current sensor

Current measurement sensors can be used to measure the currents in cables or busbars contact-free, i.e. galvanically isolated. The current is measured via the magnetic field of the current-carrying cable. For this purpose, the magnetic field is focused onto a Hall probe by means of a soft magnetic core. Two setups can be distinguished: open-loop and closed-loop current sensors.

Open-loop current sensors represent the simplest design. The structure is shown in Figure 9.6.21 (left): the current-carrying conductor is guided through a soft magnetic core with an air gap, and the Hall element is located in the air gap. The soft magnetic core concentrates the magnetic field of the current-carrying conductor. For a core with a linear hysteresis curve, B is proportional to H and thus to I_1. Thus, the Hall voltage V_h is also proportional to I_1. Due to the air gap, there is a shearing of the hysteresis curve, the $B–H$ curve flattens and is linearized. Open-loop current sensors can measure both DC and AC currents, have low current consumption, and are very compact. Compared to closed-loop sensors, open-loop sensors have lower bandwidth and higher response time. The measurement error of the open-loop sensors is higher than that of closed-loop sensors.

Closed-loop current sensors are similar in design to open-loop current sensors, but have an additional compensation winding. Therefore, they are also called compensation current sensors. Figure 9.6.21 (right) shows the structure of the closed-loop current sensors. A voltage proportional to the magnetic field is generated in the Hall probe. The Hall voltage is used to control an electronic amplifier, which drives a current into an additional winding on the core. This current compensates the magnetic field generated by the current-carrying conductor to zero. Then $I_2 = n_1 I_1 / n_2$ is satisfied, so the secondary current I_2 is proportional to the current I_1 to be measured. Since the flux density is compensated to zero, the core operates near the coercive field strength causing an offset. For this reason, high-permeability cores made of

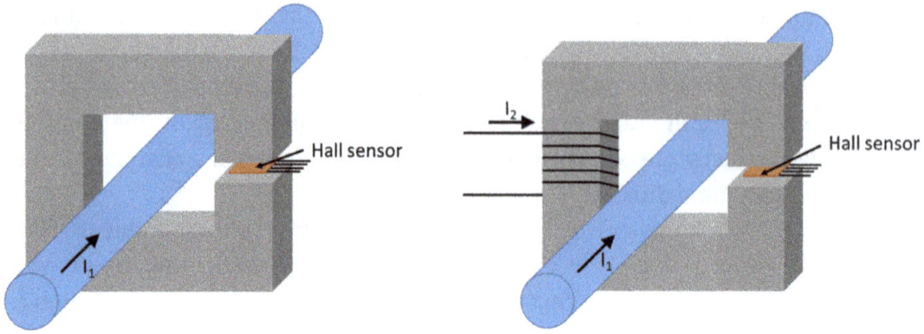

Figure 9.6.21: Left: open-loop current sensor. Right: closed-loop current sensor.

high-nickel Ni-Fe alloys with low coercive field strength or amorphous and nanocrystalline materials are particularly suitable for this application. Closed-loop current sensors offer excellent accuracy, high linearity and high bandwidth. Compared to open-loop sensors, the dimensions are larger and the construction is more cost-intensive.

9.6.3.4 Shielding

Sensitive electronic devices often need to be shielded from magnetic interference fields. Likewise, sources of high magnetic fields are shielded to prevent them from affecting the environment or humans. The size of a magnetically shielded area can range from a few cubic millimeters to entire shielded rooms with a volume of several cubic meters.

The shielding factor S describes the strength of the shielding and is the ratio of the field strength without shielding H_E to the field strength with shielding H_I:

$$S = H_E/H_I$$

Two extreme cases can be considered for magnetic shielding: high-frequency fields and low frequencies/DC fields. The principle of a magnetic shield for the two extreme cases is illustrated in Figure 9.6.22 using the example of an infinitely long round cylinder.

In the case of high frequencies (Figure 9.6.22, left), eddy currents are induced in the shielding which forces the external field H_E out of the shielding. Decisive for the shielding effect is then the penetration depth of the external field into the shielding, described by the skin depth δ. The higher the electrical conductivity, the better the shielding at high frequencies. For this reason, copper or aluminum are often used for this case.

At low frequencies or DC fields (Figure 9.6.22, right), the shielding effect is based on magnetic permeability. In this case, the magnetic flux lines are preferentially located in the shielding material. Since the permeability is $\mu \gg 1$, the field lines

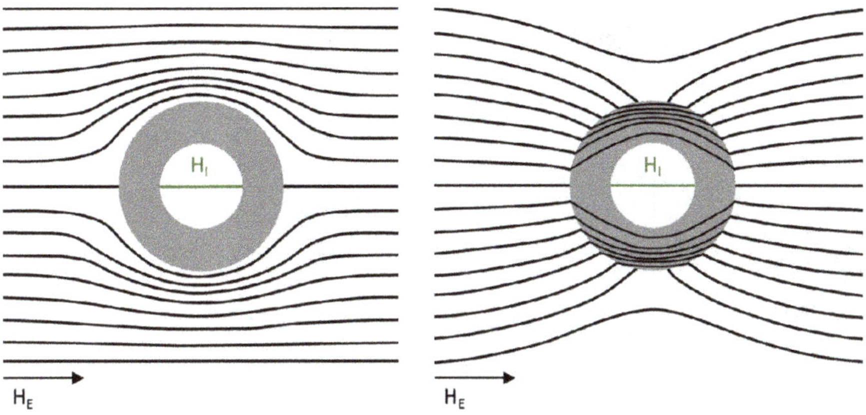

Figure 9.6.22: Working principle of magnetic shielding. Left: high-frequency shields. Right: low-frequency shields.

enter the shielding almost perpendicularly and run parallel to the surface inside the shielding. The field inside the shield is significantly reduced. The shielding factor increases with permeability. Therefore, when high shielding factors are required, Ni-Fe alloys such as NiFe15Mo are used. If very high field strengths have to be shielded, materials with higher saturation flux density are also used (e.g. Ni-Fe alloys with 50 wt% Ni (1.3922) or electrical steel).

Magnetic shielding is used, for example, as room shielding to protect sensitive equipment such as transmission electron microscopes, electron beam lithography systems or magnetic resonance tomographs from external fields. Figure 9.6.23 shows a room magnetically shielded with sheets of NiFe15Mo. In the further process, the Ni-Fe sheets are covered so that the shielding is not visible anymore.

Areas with extremely high shielding factors, so-called zero Gauss chambers, are achieved by several layers of shielding material. Here, each new layer shields the remaining field from the outer layer. The shielding factor for the ideal case of two hollow spheres inside each other is then given by

$$S = S_1 + S_2 + S_1 S_2 \left(1 - \frac{D_2^3}{D_1^3}\right)$$

Here S_1 is the shielding factor of the outer sphere, S_2 is the shielding factor of the inner sphere and D_1 and D_2 are the diameters of the two hollow spheres. By using several layers of shielding material, the shielding factor can thus be increased much more effectively than by increasing the sheet thickness.

Figure 9.6.23: Magnetically shielded room (courtesy of Systron EMV GmbH).

📖 References

[1]   VDM Metals, VDM Report No. 27 – Soft magnetic Ni-Fe base alloys.
[2]   Collocott S. Magnetic materials: Domestic applications. In: Reference Module in Materials Science and Materials Engineering, Elsevier, 2016.
[3]   Tumanski S. Handbook of Magnetic Measurements, CRC Press, 2019.
[4]   Chin GY, Wernick JH. Chapter 2 soft magnetic metallic materials. In: Handbook of Ferromagnetic Materials, 2 Hrsg, Elsevier, 1980, 55–188.
[5]   Boll R. Magnettechnik – Technik und Anwendung der Weichmagnetischen Werkstoffe, expert Verlag, 1980.
[6]   Goss NP. U.S. Patent 1.965.559, 1934.
[7]   Hegg F. Etude Thermomagnétique sur les Ferro-nickels, 1910.
[8]   Volk KE. Nickel- Und Nickel-Legierungen, Springer Verlag, Berlin, Heidelberg, New York, 1970.
[9]   Bozorth RM, Walker JG. Magnetic crystal anisotropy and magnetostriction of iron-nickel alloys. Phys Rev, 1953, Nr. 89, 624.
[10] Puzei IM. Dependence of energy of magnetic anisotropy in Invar on temperature and magnetic field. Fizika Metallov I Metallovedenie, 1961, Nr. 11, 525–8.
[11] Hall RC. Single crystal anisotropy and magnetostriction constants of several ferromagnetic materials including alloys of NiFe, SiFe, AlFe, CoNi, and CoFe. J Appl Phys, 1959, Nr. 30, 816.
[12] Smith WS, Garnett HJ. Magnetic Alloy. U.S. Patent 1582353, 1924.
[13] Smith WS, Garnett HJ. Magnetic Alloy. U.S. Patent 1552769, 1924.
[14] Supermalloy': A new magnetic alloy. Nature, 1948, Nr. 161, 554.
[15] Boothby OL, Bozorth RM. A new magnetic material of high permeability. J Appl Phys, 1947, Nr. 18, 173–6.
[16] Pfeifer F, Radeloff C. Soft magnetic Ni-Fe and Co-Fe alloys – some physical and metallurgical aspects. J Magn Magn Mater, 1980, Nr. 19, 190–207.
[17] Guillaume CE. The anomaly of the nickel-steels. Proc Phys Soc London, 1919, Nr. 32, 374–404.

[18] Kawahara K. Effect of additive elements on cold workability in FeCo alloys. J Mater Sci, 1983, Nr. 18, 1709–18.

[19] Strange EH, Pim CA. Process of manufacturing thin sheets, foil, strips, or ribbons of zinc, lead, or other metal or alloy. U.S. Patent 905,758, 1908.

[20] Pond RJ, Maddin RA. Method of producing rapidly solidified filamentary castings. Trans AIME, 1969, Nr. 11, 2457–76.

[21] Liebermann H, Graham C. Production of amorphous alloy ribbons and effects of apparatus parameters on ribbon dimensions. IEEE Trans Magn, 1976, Bd. 12(Nr. 6), 921–3.

[22] Hilzinger R, Rodewald W. Magnetic Materials: Fundamentals, Products, Properties, Applications, Publicis, 2013.

[23] Hono K, Ohnuma K. Microstructure and properties of nanocrystalline and nanogranular magnetic materials. In: Mag Nanostruc, American Scientific Publishers, 2002, 327–58.

[24] Hono K, Hiraga K, Wang Q, Inoue A. The microstructure evolution of a Fe73.5Si13.5B9Nb3Cu1 nanocrystalline soft magnetic material. Acta Metal Mater, 1992, Nr. 40, 2137–47.

[25] Hono K, Ping DH, Ohnuma M, Onodera H. Cu clustering and Si partitioning in the early crystallization stage of a Fe73.5Si13.5B9Nb3Cu1 amorphous alloy. Acta Metal Mater, 1999, Nr. 47, 997–1006.

[26] Michalowsky L, Schneider J. Magnettechnik: Grundlagen, Werkstoffe, Anwendungen, Vulkan-Verlag GmbH, 2005.

[27] Goldman A. Modern Ferrite Technology, 2 Hrsg, Springer-Verlag New York Inc., 2010.

9.7 Permanent magnets

Martin Grönefeld, Iulian Teliban, Jacek Krzywinski

9.7.1 Fundamentals of permanent magnets

Permanent magnetic materials have been already observed in the ancient Greece in the form of "loadstones". A development of permanent magnets as we know them today started about a hundred years ago. Since then, many new groups of materials have been developed and implied new areas of application. Knowing that simple iron already used in the ancient time is one of the best materials with respect to the magnetic saturation strength, it is important to understand that the development of better materials is mainly focused on the coercivity. Based on that, some materials such as hard ferrites for low class and SmCo- and NdFeB-based materials for higher class have been established for a wide usage. Today, permanent magnetic fields in their applications cover about two decades of magnitude in a spectrum of about 30 decades of magnetic field strength observed in nature (Figure 9.7.1).

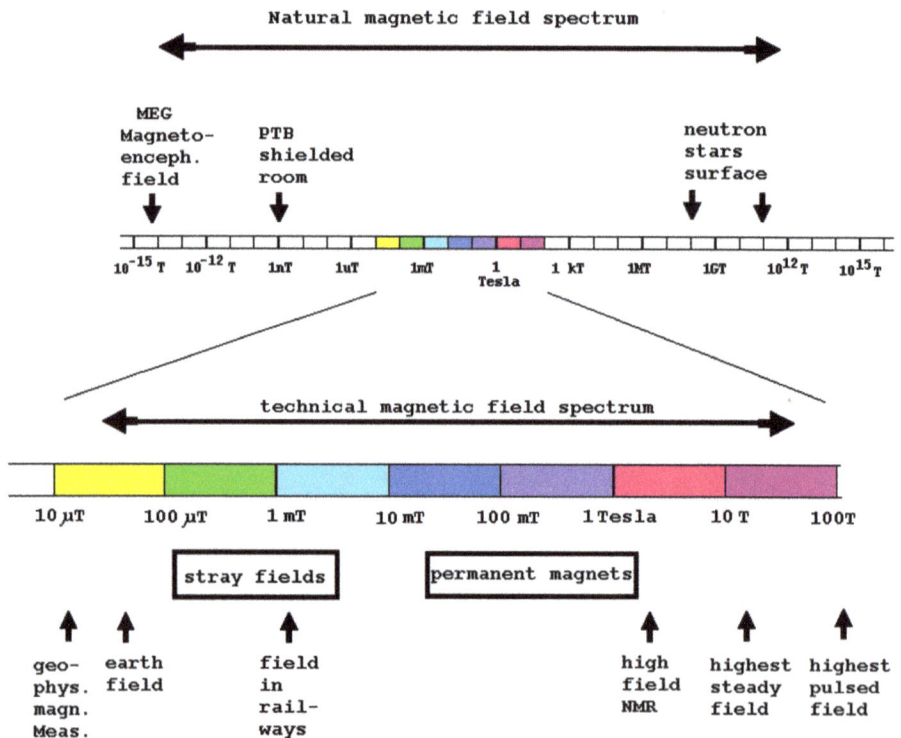

Figure 9.7.1: Magnetic field value's spectrum.

https://doi.org/10.1515/9783110733471-007

9.7.1.1 Magnetic properties of permanent magnets: hysteresis curve

The properties of magnetic materials are represented by their magnetization curve or hysteresis loop which shows the induction $J(H)$, respectively, the magnetic flux density $B(H)$ in the magnet on application of an external field, H. The diagram in Figure 9.7.2 shows a schematic hysteresis curve of a hard magnetic material [1]. When a non-magnetized magnet is exposed to an external field H, the polarization J will increase proportional to H along the "initial curve" extending to saturation. With saturation the slope of $J(H)$ vanishes, and the slope of $B(H)$ is equal to μ_0 when the material is fully magnetized along the direction of the external field H.

Gradually reducing the external field H from the saturation state to zero decreases the magnetization in the material to a value B_r, known as residual induction or remanence. If the demagnetizing field $-H$ is now increased in the negative direction, the resulting flux density B in the material is reduced to zero at $-H$ equal to H_{cB} called the coercive force. Even if there is no net flux, the magnet is not completely demagnetized at this point. The value H_{cJ} is the intrinsic coercivity when half of the magnetic domains have been reversed resulting in no net magnetic induction J in the magnet. At this point, the magnet is completely demagnetized.

Increasing the demagnetizing field H further into the negative direction will magnetize the material in the opposite polarity to the saturation. Reducing H again to zero produces $-B_r$ and by reversing the external field direction the material is gradually remagnetized again to saturation completing the hysteresis loop.

The hysteresis loop measured by $B(H)$ in the material is called the normal curve. The flux density B inside the magnetic material is the superposition of the applied field H and the magnetization M, respectively, the polarization $J = \mu_0 M$, thus $B = \mu_0 \cdot H + J = \mu_0 (H + M)$.

The properties of permanent magnets are usually represented only by their demagnetization curve (the second quadrant of the full hysteresis loop) which shows the falling induction B of the fully magnetized magnet on application of an increasing demagnetizing field H.

Even without an external applied field, the magnetized state of the magnet produces a demagnetizing field inside the magnet and the shape of the magnet defines the ratio between B and H. From this, the energy product $B \cdot H$ with the unit of an energy density (J/m^3) can be deduced and the maximum of this value on the demagnetization curve $(BH)_{max}$ is an important material characteristic. The maximum of the energy product is roughly quadratically dependent on the remanence for most materials, $(BH)_{max} \sim B_r^2$. The demagnetization curves are generally represented by their characteristic remanence B_r, coercivities H_{cB} and H_{cJ} and maximum energy product $(BH)_{max}$.

When the magnet is removed completely from the external field, the magnetic state in the magnet will drop to its open circuit values H_m, B_m. In a closed circuit (steel part short cutting north and south pole) the operating line is close to the B-axis and $B_m = B_r$. The open-circuit induction B_m is dependent on the geometry of

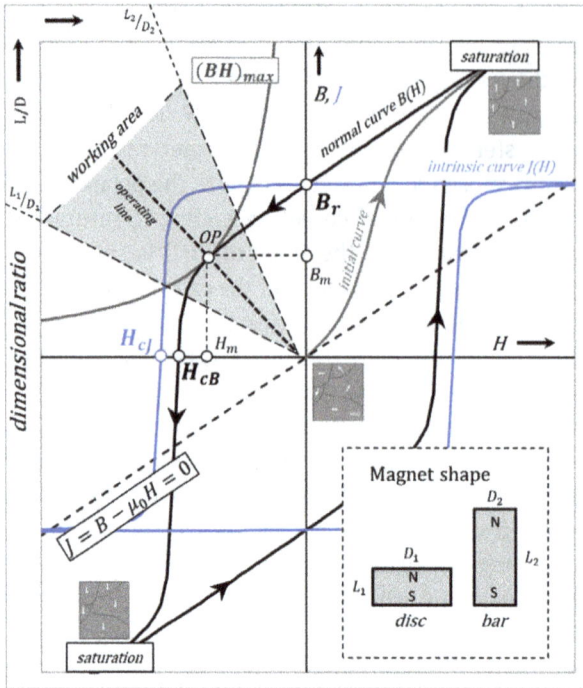

Figure 9.7.2: Hysteresis curve of a permanent magnet [1].

the magnet which can be represented by operating line with slope, B/H called permeance coefficient. Figure 9.7.3 (left) shows the values of permeance coefficient B/H for axial magnetized cylinder [2]. For a given magnet geometry the operating point $OP(B_m, H_m)$ is located by the crossing point of the demagnetization curve and B/H slope.

Only in an ellipsoidal-shaped magnet, the induction B_m and the field strength H_m within the magnet is a constant within the whole volume. The induction in cylindrical or rectangular magnets decreases towards the polar ends as the internal demagnetizing field increases. In addition to the shape of the magnet, the permeability of the magnetic material is also of significance. The AlNiCo materials, due to their high permeability, are less homogeneous in respect of induction towards the ends than ferrite materials, with permeability close to 1. In a magnet which has no ellipsoid shape, the individual areas are in various states of magnetization such that there is no working point but a working area on the demagnetization curve. B_m and H_m are average values of this area from which the approximate effect of the magnet in an external situation can be evaluated. The average value of induction B_m and the field strength H_m can be measured with a fluxmeter and a surrounding coil and potential coil, respectively.

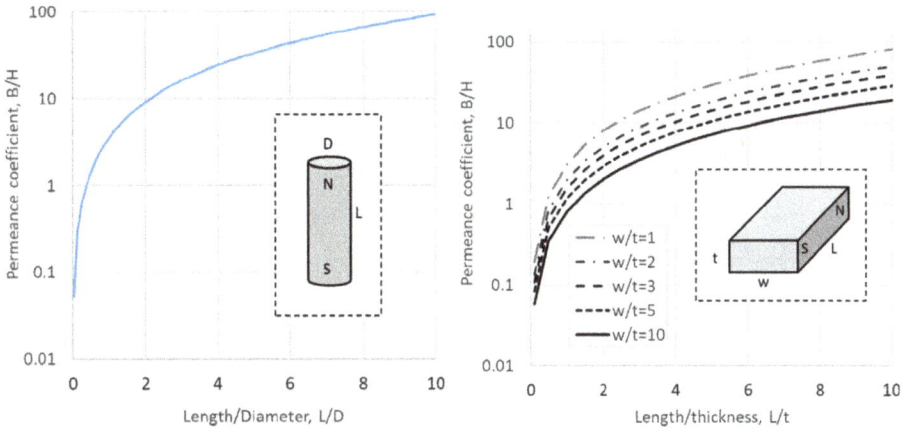

Figure 9.7.3: (left) Permeance coefficient for an axial magnetized cylinder magnet. (right) Permeance coefficient (B/H) for square magnet magnetized over the length of the magnet [2].

In most of the applications, it is required to design the magnet with an optimal operational slope. The operation point can be calculated using numerical methods, e.g. finite element method (FEM) simulations or for simple geometries using analytical approximations [2].

9.7.1.2 Magnetic property losses due to aging, temperature and external fields

The properties of all magnetic materials change with ambient temperature. In physical terms, the effects of temperature are fully described by the demagnetization curves under temperature (Figure 9.7.4). Because the demagnetization curves are generally reduced to the characteristic's remanence B_r, coercivities H_{cB} and H_{cJ} and maximum energy product $(BH)_{max}$, the material manufacturers generally document the corresponding coefficients for linear change over temperature. As the energy product is roughly quadratically dependent on the remanence for most materials and coercivities are interdependent often only the linear coefficients $\alpha(B_r)$ and β (H_{cJ}) are specified in the material characteristic tables. Both values indicate the percentage change of the characteristic between 20 and 120 °C, referred to the value at 20 °C. The linearization is an approximation that can lead to incorrect results, particularly when extrapolating for values above 120 °C.

As far as the user is concerned, the physical description of a material often fails to provide practical information for evaluating their application. In this context, a far more important question is often how the magnetic field of a permanent magnet at a fixed position changes as a function of time and temperature. Qualitatively different effects occur here, and these are described briefly below. All three effects occur together, and their sum effect must therefore be considered.

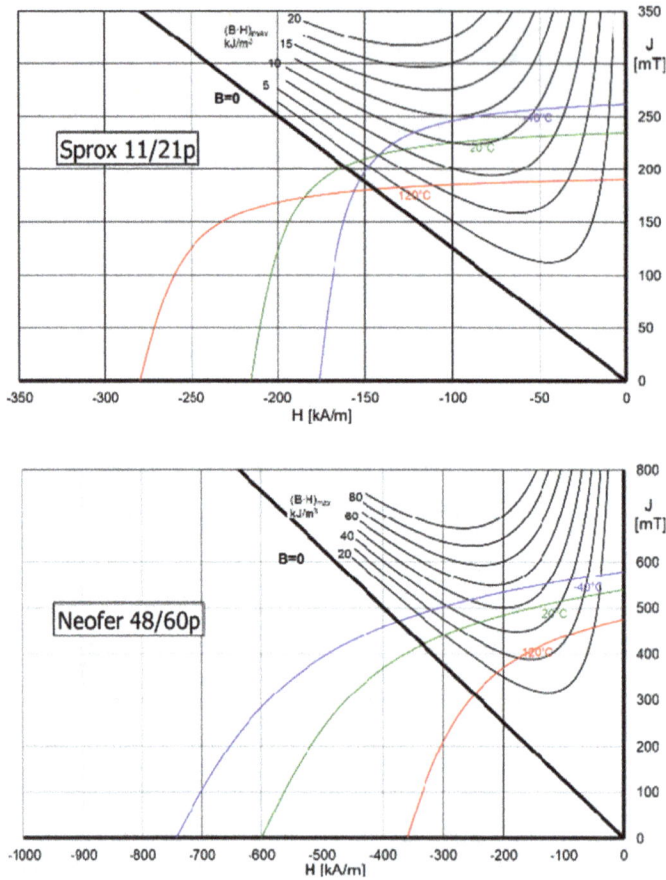

Figure 9.7.4: Temperature dependence of hysteresis curves, exemplarily for two polymer-bonded magnetic materials from the portfolio of Magnetfabrik Bonn GmbH.

9.7.1.2.1 Reversible changes over temperature

The magnetic field changes reversibly with temperature. As a good approximation, this is proportional to the remanence and is not dependent on the type of magnetization and the shape of the magnet. Typical values of the major materials are listed in Table 9.7.1.

Since a first approximation of this reversible change is linear, i.e. it shows a constant increase or decrease per degree Centigrade, the coefficient $\alpha(B_r)$ is sufficient to describe it. In the case of rare earth magnet materials (samarium and neodymium based), the coercive field also decreases as temperature increases as well as the remanence, i.e. the two coefficients $\alpha(B_r)$ and $\beta(H_{cJ})$ are negative. In the case of hard ferrites, on the other hand, the coercive field decreases at low temperatures, i.e. $\beta(H_{cJ})$ is positive.

Table 9.7.1: Typical temperature coefficients for most used material groups.

Material called	Temperature	$\alpha(B_r)$, %	$\beta(H_{cJ})$, %
AlNiCo	−270–400 °C	−0.02	−0.01
Sr-ferrite	−50–350 °C	−0.2	0.27
Neodymium iron boron	−50–200 °C	−0.12	−0.5
Samarium cobalt	−50–200 °C	−0.04	−0.4

9.7.1.2.2 Irreversible loss

The coercive field describes a magnet's stability with respect to demagnetization, i.e. partial demagnetization can occur with rare earth magnets at high temperatures and with hard ferrite magnets at low temperatures. This results in a change of the magnetic field when the relevant temperature is reached the first time. This loss is not compensated when the temperature goes back to its original value and is thus called irreversible loss.

This weakening stabilizes itself as a result of the loss of magnetization and the self-demagnetizing field it generates. There is no further decrease or only a slight decrease when the magnet is heated or cooled to the same temperature again. Compared with the reversible change to the field, the irreversible losses are more complex to describe and depend not only on the magnet material but also on the shape of the magnet and the type of magnetization as well as on any external fields applied.

9.7.1.2.3 Irreversible loss over time

In the event of repeated temperature cycles or long aging times, rare earth materials exhibit progressive irreversible demagnetization losses over very long periods (Figure 9.7.5). These arise partly as a result of delayed thermal demagnetization and as a result of chemical changes in the material.

The delayed thermal demagnetization is determined by an Arrhenius law, i.e. there is a logarithmic progression over the aging time. This logarithmic dependency means that the loss can be demonstrated almost entirely after a period of a few minutes or hours. Further change over days and months is comparably low.

In addition, however, some materials exhibit a considerable progressive loss of magnetic field strength over long time periods. This loss is caused by different physical processes and can, for instance, be traced back to a slow but continuous chemical decomposition at high temperatures.

No irreversible losses were observed on the ferrite materials investigated in a temperature range above room temperature up to 160 °C. In the case of rare earth magnet materials, depending on coercivity and chemical stability of the magnetic material, up to 35% loses at 130 °C has been measured [27].

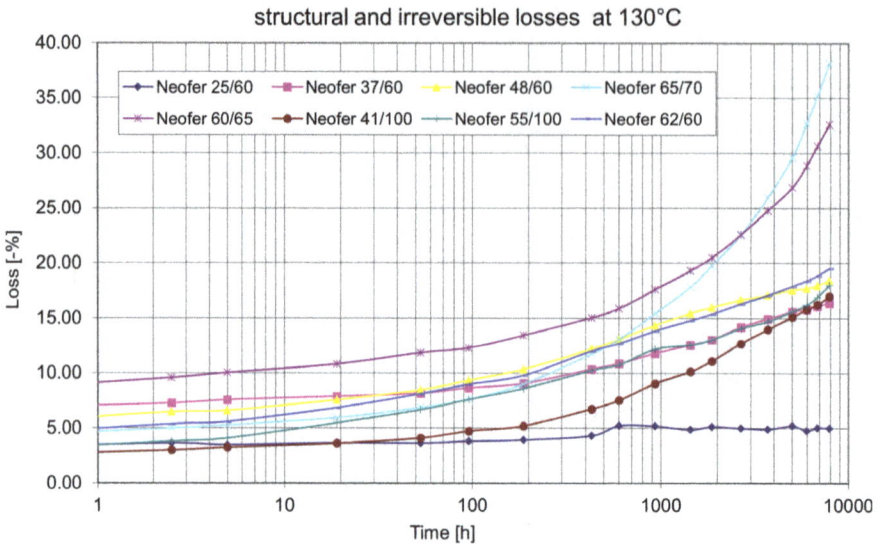

Figure 9.7.5: Magnetic losses for rare earth polymer bound materials (Neofer® registered by Magnetfabrik Bonn GmbH) for long time exposure to 130 °C for 10000 h [27].

The fundamental trends for the magnetic losses are well known for most common magnetic materials [8–10]; however, a reliable estimation for a given application requires a testing procedure designed to suit the application. The long-term losses also depend on the environmental conditions and possibly also on any electrochemical contact with other metals.

9.7.1.2.4 Losses caused by external fields

The above-discussed magnetic property losses apply to magnets when the operating slope remains constant. If an external field impacts on the magnet, the operating slope shifts with the amplitude of the external field. As shown in Figure 9.7.6, an "unfavorable" working point (A) or a large external field (H_a) will decrease the flux density in the magnet.

With an external field H_a is applied to the magnet, the flux density B_m of the magnet will decrease from the working point A to the point B. If the working point for the magnet materials shifts beyond the "knee" in the demagnetization curve, the material will be partly demagnetized. Thus, when the external demagnetizing field is removed, the flux density does not return to the point A, but instead will return to the point C. The slope of line is the slope of recoil permeability represented by BC. Recoil permeability is a material characteristic. When the external field is removed, the magnet's operating slope returns on the operating line to a new flux density, B_C.

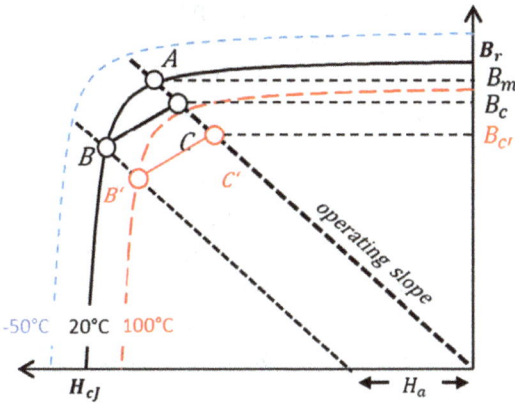

Figure 9.7.6: Influence of externally applied field, H_a, on magnetic properties.

Simultaneous temperature and external field applied to the magnet will increase the flux loss due to the change of coercivity with temperature shifting the "knee" on the demagnetization curve. On the example in Figure 9.7.6, the red-lined demagnetization curve under temperature increases the impact of the demagnetization field H_a and the field drops to $B_{c'}$.

For materials that exhibit a straight-shaped demagnetization curve like samarium-cobalt or neodymium iron boron with the knee far away from the working point by an optimal design, the flux loss associated with the presence of an external field is greatly reduced or not present. These materials are also more resistant to demagnetization by reverse fields.

9.7.2 Permanent magnetic materials and fabrication

While many steels, alloys, intermetallic phases and even ceramics show magnetic ordering and are magnetized by a magnetic field, permanent magnets require that a magnetic state is sustained even after the disappearance of a transient magnetizing field. Therefore, a certain coercivity is required resulting from a phenomenon called magnetic hardening mechanism.

Depending on the shape with respect to the magnetization direction, a magnetized body produces an internal self-demagnetizing field that can range between less than 1% of its magnetization for a horseshoe magnet or a few percent for a needle-shaped magnet magnetized along its long axis but can go up to 40–100% for a flat magnet magnetized in the short direction.

Speaking of permanent magnetic materials would thus ideally mean a material with coercivity higher than the remanent magnetization. Looking on steels that are technologically used longer than magnetism is known, we find many alloys with a polarization J above 1 T and up to more than 2 T even for pure iron [3]. The magnetization $M = J/\mu_0$ is thus in the range of megaampere/meter. It took a long time that materials appeared with coercivity in this range. In the beginning of the twentieth century, engineers had to live with steels that were especially alloyed with tungsten or later with cobalt to give some coercivity at all [4–7]. Therefore, a longtime historic permanent magnetic materials would be called half-hard materials today, and there are still some of them in use.

However, today we have a few material groups that fulfill the requirement of showing coercivity comparable or even higher than their magnetization at room temperature [3, 8–10]. The mainly used materials are the ceramic hard ferrites based on an $SrFe_{12}O_{19}$ structure, the SmCo magnets based on an $SmCo_5$ or an Sm_2Co_{17} structure [11], and the NdFeB magnets based on an $Nd_2Fe_{14}B$ structure [12, 17].

Table 9.7.2 gives an overview of all materials, but we will concentrate in this chapter on the ferrites and on the rare earth magnets.

Table 9.7.2: Standard magnetic materials with their magnitude of remanence B_r and coercivity H_{cj}.

Material called	Constituents	B_r (mT)	H_{cj} (kA/m)	Advantage
Hardened steels	Fe + V, Cr, Co, W	600–1200	5–20	Historic
Iron cobalt chromium	FeCoCr(V)	800–1300	40–70	Ductile
AlNiCo	Fe/Al/Ni/Co	500–1350	40–160	Temperature
Hard ferrite	$SrFe_{12}O_{19}$	200–450	120–350	Cheap
Platinum cobalt	PtCo	650	400	Ductile, strong
Manganese aluminum carbon	MnAlC	300–600	200	Cheap constituents
Neodymium iron boron	NdFeB	1150–1450	800–2800	Strong
Samarium cobalt	SmCo	800–1100	600–3000	Strong, temperature

Even if not mentioned in the table, the main figure of merit of a permanent magnet material is the energy product $(BH)_{max}$, which gives the field energy per volume that the material is able to deliver in maximum. This value is more or less proportional to the magnetic force on a soft magnetic body attracted by a permanent magnet.

In Figure 9.7.7, a schematic view on the historical development of materials, with respect to the energy product is shown.

Coercivity, also known as magnetic hardness, is linked to the mechanic hardness as we find similar mechanisms in the microstructure that prevents ductile deformation and growth of reversed magnetic domains [13]. Thus, we only find a few workable materials in the table and many magnetic parts are of a simple shape and have to be grinded to give narrow tolerances. As an alternative, complex shape of magnets and narrow tolerances without expensive machining are available with the group of polymer-bonded permanent magnets. The basic idea is to use permanent magnetic powders or even just crushed permanent magnets embedded in a polymer to process the composite material on machines that are also used for conventional polymers.

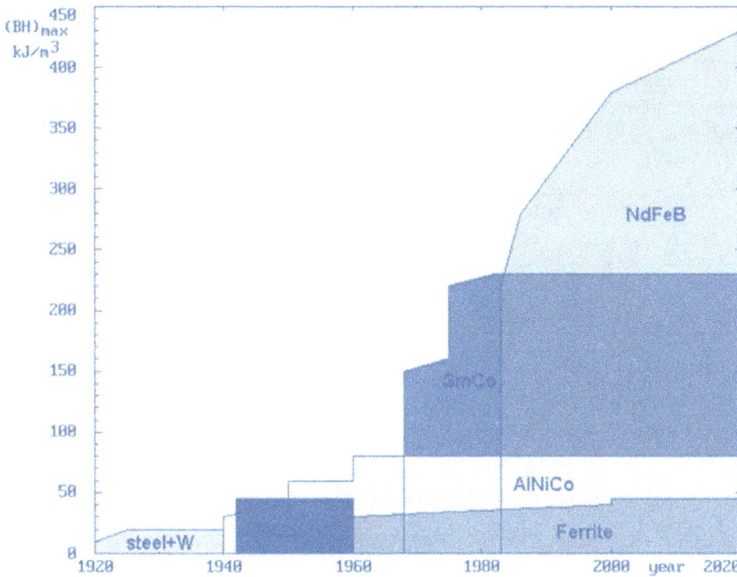

Figure 9.7.7: First appearance of material families with development of the energy product [8–10].

The shape of the magnet is defined by forming tools during an injection molding or compression molding process. Nowadays, magnets with new technologies such as 3D printer are also possible [23–26]. In Table 9.7.3, the bonded magnetic materials are listed. It can be recognized that magnetic properties, especially the remanent polarization, is considerably reduced through the dilution of magnetic effective material by the polymer.

Polymer-bonded magnets are mainly used where shaping and mechanic function stands beside the magnetic requirement and where not the highest magnetic field is needed. A large application area is the challenging field of electronic touchless sensing of mechanic movements. In this application, a small sensing semiconductor device is fixed on one side and a permanent magnet on the other side of two parts that are in change of the relative position to each other [14]. The electronic device detects a movement of the magnet by the change of the magnetic field and as there is no mechanic contact between the two sides the wearless detection can stand for much more than 10^9 cycles, which are unimaginable for a mechanic system with contact.

Microdevices are another interesting and growing field of applications for bonded magnets where the cost of machining or the advantage of complexity of the magnet shape dominates against the magnetic requirements [15].

Table 9.7.3: Polymer-bonded magnetic materials.

Material	B_r (mT)	j_{Hc} (kA/m)
Isotropic bonded ferrites	80–150	150–220
Anisotropic bonded ferrites	160–270	180–300
Isotropic bonded NdFeB	400–900	600–1200
Anisotropic bonded NdFeB	600–1000	800–1200
Anisotropic bonded SmCo	400–840	800–2000

9.7.2.1 Hard ferrite materials

Hard ferrites are produced from iron oxide and strontium carbonate that are wet milled together, dried and heated to 1000–1350 °C in air to react and form the ceramic raw material. This material is ground to a fine powder. The size of particles has to reach values as low as below 1 μm [13].

Sintering compressed green parts of this powder give the final material with permanent magnetic properties. Historically, besides strontium, barium was also used for the hard ferrites, but barium is used today only in small amounts as compared to strontium ferrites, due to the better environmental compatibility [6, 7].

The basic $SrFe_{12}O_{19}$ structure is hexagonal. On the atomic length scale, the magnetic moments are fixed by a large crystalline anisotropy along the c-axis. To carry this anisotropy to reveal coercivity on the macroscopic scale, a special microstructure is achieved by the powder sintering process.

The raw materials are widely available and cheap, and the production process is scaled up to thousands of tons of materials resulting in cost in the range of US$ <1 per kg.

9.7.2.2 Rare earth materials

9.7.2.2.1 SmCo magnets

In the end of the 1960s and mid of the 1970s, magnets based on samarium and cobalt were developed [11, 16]. The arc-melted samples are ground to fine powder, compressed and sintered. The first material family of nearly pure $SmCo_5$ was extended by the so-called Sm_2Co_{17} materials which contain much less of the expensive and strategic cobalt metal as one would expect from the basic formula as a considerable part of the cobalt is replaced by iron within a solid solution $Sm_2Co_{17-x}Fe_x$. While the raw materials are thus less expensive and more abundant, the magnetic strength of the 2–17 magnets is even a little bit higher. The pretentious heat treatment of the 2–17s and very large coercivity of the 1–5s allow that also the 1–5 magnets are of large importance even today.

As for all modern magnetic materials, the coercivity results from the high crystalline anisotropy of the basic cell and from the microstructure which comes from the powder sintering route for the 1–5 magnets and from a multiphase system after heat treatment for the 2–17 magnets.

Note that SmCo and NdFeB are abbreviations that are used for a larger class of magnetic materials in the following.

9.7.2.2.2 NdFeB sintered materials

Finally in the year 1983/1984, it was recognized in parallel from two groups that the intermetallic compound $Nd_2Fe_{14}B$ exhibits excellent prerequisites for a permanent magnetic material and that it can be processed to a microstructure to reveal good magnetic properties by two different ways [12, 17, 18].

The first way is comparable to the production route of samarium cobalt magnets where an arc-melted sample is ground to a fine powder, compacted and sintered. As the neodymium alloy is even more reactive with oxygen than the SmCo one, the process has to be performed in inert gas.

NdFeB magnets contain no or only a small amount of cobalt, and rare earth ores contain more Nd than Sm, so that the NdFeB magnets are in many cases more economic when compared to SmCo. However, for high temperature, SmCo is much more suitable. In electric motors, the elevated temperature and the electromagnetic counter field requires that part of the neodymium has to be replaced by much less abundant heavy rare earths like dysprosium or terbium [19–22], but mainly NdFeB magnets play the key role in many arising applications like wind energy and electric mobility.

9.7.2.2.3 NdFeB powders for bonded magnets or for thermally densified magnets

In parallel to the development of sintered NdFeB magnets it turned out that a fine microstructure of the NdFeB alloy can also be achieved by extreme quenching of the material from the melt state. This technique that is also used for metallic glasses is called melt spinning. The melt is spurt on a rotating copper wheel and cools down with 1000 K in a millisecond. The result is a thin ribbon that already exhibits the good magnetic properties of the NdFeB magnet. As compared to sintered magnets, the alloy composition is much closer to the stoichiometry of $Nd_2Fe_{14}B$. Sintered magnets need a certain excess of neodymium to give a grain-separating nonmagnetic phase. However, the ribbon is isotropic while sintered magnets are produced by concentrating the crystalline easy directions in the same line as the later magnetization. This makes sintered magnets much stronger from their final magnetization.

The metallic ribbon is crushed to a coarse powder and for the forming process, the material is transferred to the route of polymer-bonding line injection molding or compression molding. As an alternative, it is also possible to densify the powder under temperature to give to it the full bulk properties. In these routes, densification with a flow called "die up setting" can even give anisotropy to the final part. These kinds of magnets developed by Magnequench are called MQP and MQA magnets.

9.7.3 Applications of permanent magnets

Permanent magnets are used in many applications like generators, motors, electronics, hard drives, televisions, telephones, headphones, speakers, sensors and transmitters [28–32]. In the last decades, the production on permanent magnets has drastically increased due to their various applications. The total production of permanent magnets in 2020 (Figure 9.7.8), valued at around US $15,000 million, is estimated at 1000 to 1200 kT [33].

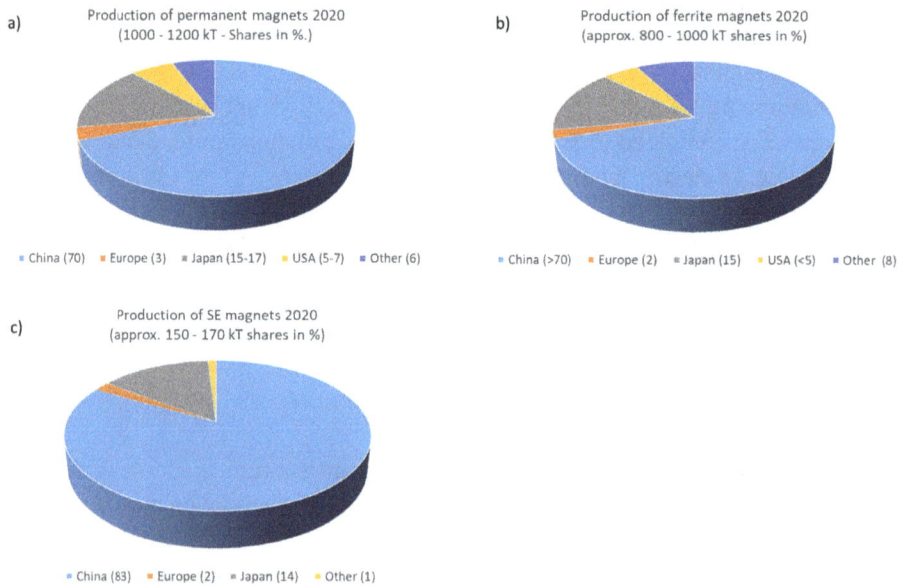

a) Production of permanent magnets 2020
(1000 - 1200 kT - Shares in %.)

b) Production of ferrite magnets 2020
(approx. 800 - 1000 kT shares in %)

■ China (70) ■ Europe (3) ■ Japan (15-17) ■ USA (5-7) ■ Other (6)

■ China (>70) ■ Europe (2) ■ Japan (15) ■ USA (<5) ■ Other (8)

c) Production of SE magnets 2020
(approx. 150 - 170 kT shares in %)

■ China (83) ■ Europe (2) ■ Japan (14) ■ Other (1)

Figure 9.7.8: (a) Sintered and bonded ferrites, SmCo, NdFeB and others (AlNiCo) with a total of US $15,000 million. (b) Sintered and bonded hard ferrites (US $7,000 million). (c) Sintered and bonded rare earth magnets (US $8,000 million) [33].

Ferrite magnets (approx. 800–1000 kT), followed by rare earth magnets (approx. 150–170 kT), make up the largest share of production. Due to the environmental challenges and the need of energy efficiency, the demand for rare earth magnets, especially neodymium-iron-boron magnets, which are used in many modern applications, is increasing. An area of application with a great growth potential is hybrid and fully electric cars with permanently excited synchronous motors, e-bikes, industrial motors such as servo motors with permanently excited synchronous motors, electronic devices, generators for wind turbines and nuclear magnetic resonance tomographs (Figure 9.7.9) [28, 30].

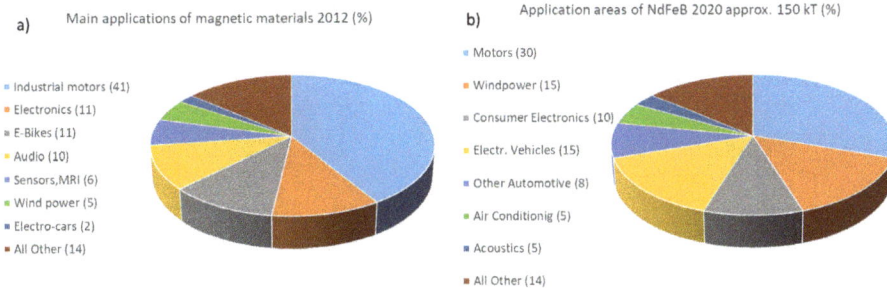

a) Main applications of magnetic materials 2012 (%)
- Industrial motors (41)
- Electronics (11)
- E-Bikes (11)
- Audio (10)
- Sensors,MRI (6)
- Wind power (5)
- Electro-cars (2)
- All Other (14)

b) Application areas of NdFeB 2020 approx. 150 kT (%)
- Motors (30)
- Windpower (15)
- Consumer Electronics (10)
- Electr. Vehicles (15)
- Other Automotive (8)
- Air Conditionig (5)
- Acoustics (5)
- All Other (14)

Figure 9.7.9: (a) Main areas of application of magnetic materials (2012) [34]. (b) Applications of only neodymium-iron-boron magnets [33].

From the physical point of view, the numerous applications of permanent magnets can be divided into five groups:
– Devices providing a mechanical force
– Energy conversion mechanical energy ⇔ electrical energy
– Acoustic applications
– Contactless information exchange in combination with sensors
– Physical devices using magnetic field in laboratory, NMR, etc.

9.7.3.1 Mechanical devices

The oldest application of permanent magnets is based on the force between two or more magnets or between magnets and solid iron pieces. A most common application of magnetic forces is the construction of various types of holding magnets, magnetic tapes and office holding magnets. Holding magnets are used for holding iron objects and for fixing objects to iron surfaces, e.g. holding measurement instruments or tools. They are constructed in such a way that the fully magnetic holding force is effective on one or two holding surfaces depending on the system (Figure 9.7.10). Even with relatively small magnets, high mechanical forces can be achieved. Holding stripes are used for lifting larger iron pieces, for building magnetic chucks, for separating iron particles from nonmagnetic materials and for similar purposes.

Other applications using the magnetic force by permanent magnetic fields include roles in printers transporting toner and keeping it in shape for applying it to the paper as well as magnetic separators removing ferromagnetic parts or impurities from bulk goods [29]. In the metal processing industry, magnets are often used to remove iron particles from rising fluids and lubricants in the same way as well-established magnetic grids, workpiece holders and other tools incorporating magnets. In the furniture

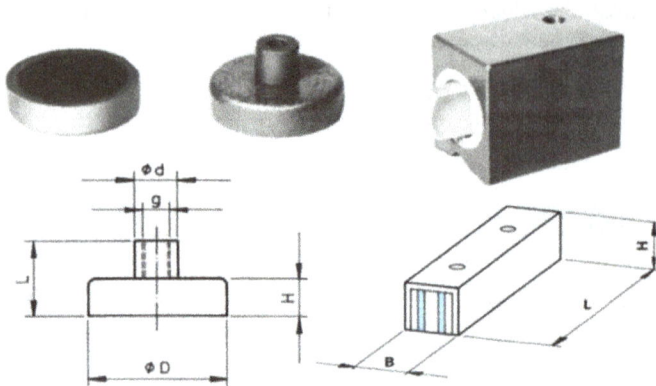

Figure 9.7.10: Magnetic holding systems with or without fastening threat. All sides are shielded except the holding surface. Holding stripe for technical purposes. Mechanical switching magnet (up right) used for holding instruments and tools.

and appliance industries, magnets are implied in door fastening systems that do not require a spring mechanism [29, 37]. The magnetic separators are often used for refuse processing (separation of iron and steel scrap), old car recycling at scrap yards, processing of cores for the blast furnace process by separating the gangue (rock) from ferromagnetic iron cores.

These forces diminish at least with the fourth power of the distance to the magnet so that with distances of more than a few centimeters very strong or large magnets are required to preserve any notable strength.

Thanks to contactless power transmission by magnetic fields, permanent magnets are used as components of magnetic couplings (Figure 9.7.11) and eddy current or hysteresis brakes [35, 36, 40, 42]. Magnetic couplings are used to transmit mechanical motions through solid walls, depending on the installation space and application in frontal or central rotary constructions [29].

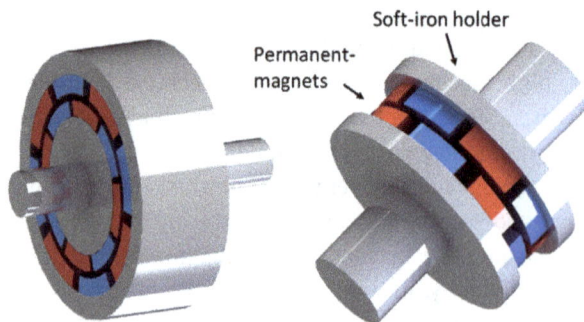

Figure 9.7.11: Magnetic coupling systems with radial and front coupling, respectively.

Finally, the compass needle is the most prominent and oldest example of mechanical forces or to put it more precisely a mechanic torque generated by a magnetic field.

9.7.3.2 Application of permanent magnets for motors and generators

There is a constantly increasing number of applications for permanent magnets in construction of motors and generators. For small motors with performance capacity up to a few watts, permanent magnets are used almost exclusively [29, 31, 32]. At the other end of the scale, permanent magnets are used for producing energy in generators of 100 kW or above; they simplify the construction of machines and provide a high degree of rotability [38].

Permanent magnets used in these applications can be arranged in a variety of ways in the magnet system. This involves either block magnets with soft iron pole pieces which direct the flux to the armature of the machine (AlNiCo, hard ferrite magnets), or an arrangement where magnets themselves border the air gap which is suitable with rare earth materials due to their high coercivity.

9.7.3.2.1 Energy generators

The voltage created by the movement of a magnet relative to an electrical conductor is mainly used in electrical generators. Historically, the first generator known as "alternator" was constructed by Hippolyte Pixii & Ampere in 1832. The generator consisted of two coils and a horseshoe magnet. At the same time, Faraday built a unipolar machine that generates a direct current when the cylindrical permanent magnet rotates on its axis [39]. Conversion of mechanical energy into electrical energy was well known and further developed but not efficient to be implied in mass production and to replace electrically excited machines. The discovery of cheap ferrites and finally of rare earth magnets beginning in 1965 made it possible to build systems with permanently excited synchronous generators. The environmental and energy debate in the 1970s, two oil crises and finally the CO_2 atmospheric problems raised a renaissance of wind energy use.

Permanently excited generators using permanent magnets instead of electric coils for the prime field generation (excitation) have several advantages over electrically excited generators. In addition to a slightly higher degree of efficiency due to the elimination of the excitation power, they can be built more compact and lighter due to their higher field strength. The use of permanent magnet generators is particularly advantageous in gearless offshore wind turbines. Permanently excited generators are also used in gear systems with compact generators. The nominal output of newly installed wind turbines on land (Figure 9.7.12) is usually in the range of 2 to approx. 5 MW, while the largest offshore systems developed to date reach up to 15 MW [34].

Along the way, synchronous generators of small powers are built based on permanent magnets in alternators of motorcycles and bicycle dynamos. From larger units generating several kVA down to small generators with performance figures

Figure 9.7.12: Wind power plants in NRW, Germany.

measured in fraction of a watt, permanent magnets offer economic advantages as a result of a fall-off of electrical field excitation.

An important application is in electroacoustic instruments used for the conversion of sound into electric energy: microphones and pickups. An interesting application is magnetic damping, where a solid disk with good electric conductivity rotates or moves in a strong magnetic field. The eddy currents induced in the disk dissipate the kinetic energy and damp vibrations [40–42].

9.7.3.2.2 Electric motors
Motors are electrical machines working inverse to the generators. An electric current impacts a mechanical force when it moves inside a magnetic field excited by current or permanent magnets, an effect that is called "Lorenz force"; this is the basis of applications of permanent magnets in the conversion of electrical energy into mechanical energy for electrical motors and mobility. The first practical application of an electric motor for mobility was demonstrated in 1835, by Sibrandus Stratingh and Christopher Becker who used an electric motor that powered a small model vehicle [43]. In the first half of the nineteenth century, the first electromechanical energy converters were DC machines. In the low-power range and in comfort applications, the permanent-magnet DC motor is particularly common because of its simple design. In typical DC motors, permanent magnets made of hard ferrite, samarium-cobalt or neodymium-iron-boron are built into the stator. These shell magnets are largely standardized in terms of dimensions, which leads to a very

advantageous price–performance ratio. Nowadays, the DC machine lost its importance in large machine construction as they use brushes liable to wear to alternate
current and due to lower efficiency. In particular, the synchronous machines and,
for low-maintenance drive systems, the asynchronous motor replaced the direct
current machine in many areas of application [29, 38].

In AC and three-phase motors, especially stepper motors (Figure 9.7.13) and
synchronous motors, permanent magnet rotors made of hard ferrite, samarium cobalt and, increasingly, neodymium iron boron are used. Permanent magnets are
also used in conventional linear drives [44]. Linear motors are increasingly being
built into small household electrical appliances because of their manyfold design
advantages. Magnets made of neodymium-iron-boron are mainly used as stator.

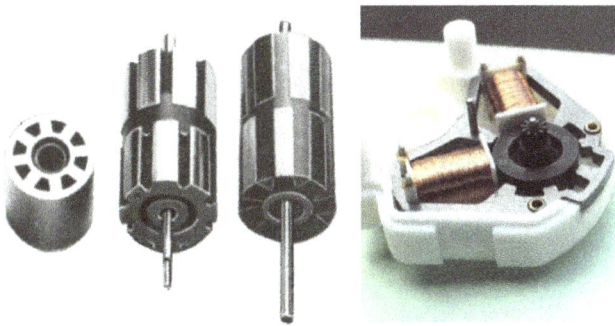

Figure 9.7.13: Micromagnets of hard ferrite for stepper and electric clocks.

In the last decades, permanent magnets are playing a crucial role in e-mobility. E-
scooters (Figure 9.7.14), pedelecs and e-bikes are driven by electronically controlled
multiphase DC motors [45, 46]. Neodymium iron boron magnets are used because
of their high energy density.

9.7.3.3 Acoustic applications

The development of the loudspeaker is directly linked to the invention of the telephone. Werner von Siemens received the patent for an electrodynamic loudspeaker
as early as 1878. The first electrodynamically powered loudspeaker was presented
to the public at the Berlin radio exhibition in 1925. In the same year, the American
company General Electric developed the dynamic moving coil loudspeaker, as it is
basically used in most sound-emitting systems until today [47]. Cylindrical air gaps
with radial fields are mainly used in electro-acoustic industry for loudspeakers, microphones, etc. In these applications, the permanent magnet in ring form can encir-
cle a soft iron pole piece (Figure 9.7.15a) or as a block located in the interior of a
soft iron cup or tube (Figure 9.7.15b, c). Particularly flat systems can be created by

Figure 9.7.14: Nowadays e-scooter.

the appropriate design and location of pole pieces (Figure 9.7.15d) as well as selection of a magnetic material.

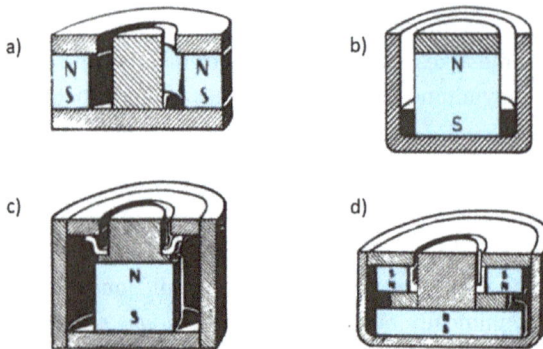

Figure 9.7.15: Magnet systems for loudspeakers.

In modern acoustics, moving coil loudspeakers, ribbon loudspeakers and, predominantly in the live area, dynamic moving coil microphones and ribbon microphones that are driven by permanent magnets are used. In the large loudspeakers and PA systems, hard ferrites are preferred because of their low price. Neodymium-iron-boron magnets have a right to exist in today's electronic devices (flat screen TVs, headphones, laptops, smartphones, etc.). Sound-emitting systems based on permanent magnets are also used in a variety of infrasound and ultrasound applications (cleaning devices, analysis devices for measuring vibration sensitivity, sonographs, etc.) [29, 31].

9.7.3.4 Magnets for sensors

Sensors for detecting magnetic fields in conjunction with a permanent magnet (actuator) are used in a variety of applications to measure the relative position or angle between mechanical parts. In general, these applications are divided into switching (e.g. limit switch of a linear unit) and analog or incremental continuous measurements. A common example of the latter group is any type of speed or position control in electrical machines (Figure 9.7.16).

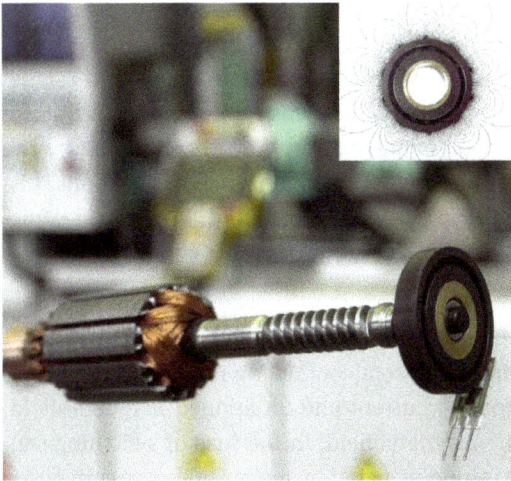

Figure 9.7.16: Multipole magnet for automotive window control (with courtesy of Magnetfabrik Bonn GmbH) [48, 49].

Magnetic switches and sensors (Figure 9.7.17) have no moving parts (except reed switches) comparably but more robust than mechanical microswitches. Their main advantage is therefore their nearly infinite lifetime. For continuous position or rotation measurement, magnetic sensors compete with optical methods. While the latter provide more accurate position measurement, magnetic systems are less sensitive to dust and external influences.

Sensor
Applications

for indirect Detection of ...   Position   Mat. Prop.,
Temp, etc

Output   Digital   Analog   coin validation

Evaluation   Counting   switching   Field strength   Field Angle

Magnetization   Multipole   multi-track   one track   one transition   XMR   Hall Array

Examples   Speed Control   lock systems   Electronic
Commutation   seatbelt lock
recognition   linear position
detection   steer by wire

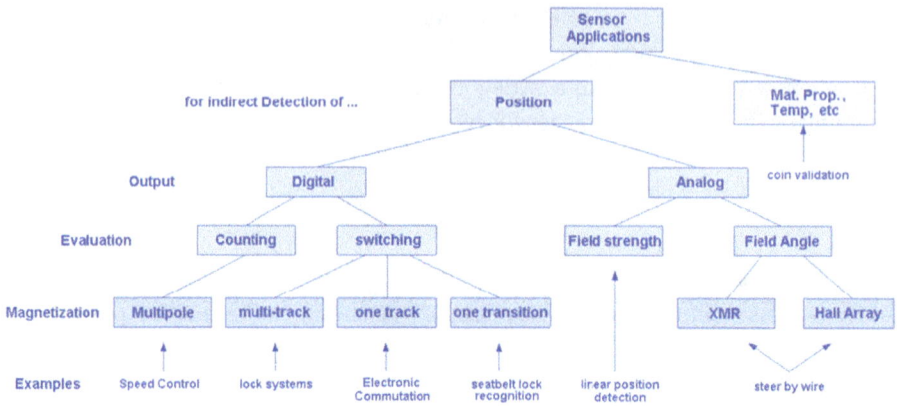

Figure 9.7.17: The most common current sensor applications with associated sensor types and magnetization types.

9.7.3.4.1 Reed switches

The direct counterpart to mechanical microswitches are reed switches. In these devices, two ferromagnetic contacts are enclosed in a glass tube in an inert atmosphere. An external field magnetizes the contacts, and they are attracted by the magnetic poles induced there, closing the contact. In reed switch applications, historically AlNiCo magnets have been used, but today often hard ferrites and sometimes rare earth magnets are the first technical solution.

9.7.3.4.2 Hall probes, CMOS Hall sensors, CMOS Hall array

Hall effect sensors are based on a thin film of semiconducting material producing a small voltage perpendicular to an applied current and an applied magnetic field. This voltage is a direct measure of the magnetic field. In the case of switching sensors, the analog signal is converted to a digital signal by means of a post-converting Schmitt trigger. With analog sensors, compensation for temperature drift and nonlinearity effects is necessary. The main part of permanent magnets in this field belongs to plastic-bonded multipole magnets with different patterns of magnetization and irregularly magnetized permanent magnets which are used for detection of velocity and position.

9.7.3.4.3 XMR sensors

Magnetoresistive (MR) sensors exploit the field dependence of electrical resistance in a conducting or semiconducting probe. While traditional MR effects in most semiconductors tend to be small (about 1–3%), there has been a renaissance of this type of sensor in recent years with extremely large MR effects (XMR sensors) called AMR (anisotropic MR), GMR (giant MR), and CMR (colossal MR). Although the traditional MR effect has been used to measure field strength, MR and XMR sensors are now and will be increasingly used in the near future to measure field angles by evaluating the change in resistance in two directions.

In on-axis applications, neodymium-iron-boron magnets are mainly used. Due to the shortage of rare earths in 2010 and for economic reasons, injection-molded ferrite magnets have been developed for these applications [48, 49]. For off-axis detection of a rotor in permanently excited synchronous motors, multipole magnetized plastic-bonded hard ferrites are commonly used.

9.7.3.4.4 Induction coils
In dynamical devices rotating with a constant speed, the induced voltage in a coil can be evaluated to determine the angular position of a spindle. Typically, multipole polymer-bonded hard ferrite magnets are fixed to a spindle, and the rotation speed can be detected by either the amplitude of the induced AC voltage or by counting the periods of this voltage. This method is used for speed control in tool machines and in washing machines.

9.7.3.5 Permanent magnets in field systems
A defined magnetic field is often required to show physical effects, for example, for calibration purposes or for the treatment of materials in a magnetic field, for guiding electron beams or for NMR spectroscopic measurements. For these applications, large and complex systems with high field strength and field stability are designed. For large dimensions, the yoke construction (Figure 9.7.18a) has proved very suitable. This involves either plates or bar magnets arranged in an iron frame, or configurations where permanent magnets form a part of the outer frame so that there is sufficient magnet length available to produce a high field strength in a large airgap.

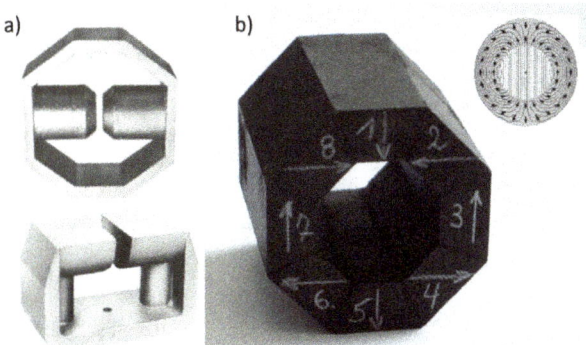

Figure 9.7.18: (a) Yoke-shaped and C-shaped magnetic field systems and (b) Halbach magnets arrangement for homogeneous field.

In modern designs, a homogeneous field with low stray field is provided by a so-called "Halbach arrangement" of magnets without iron parts (Figure 9.7.18b) [50]. While the permanent magnet material needed is comparable to the yoke systems with iron, the complete weight can be essentially reduced.

9.7.4 Magnetization and measuring of permanent magnets

9.7.4.1 Magnetization techniques

The final production step of a permanent magnet is magnetization when the structure or pattern of magnetization is defined and that the magnetization in each volume element is oriented along the required direction by a strong magnetic field. There are generally two methods to magnetize permanent magnets: magnetization by a static or by a pulsed field [51–53].

9.7.4.1.1 Magnetization by a static field

In order to achieve the maximum magnetic potential, it is necessary to magnetize the material to its saturation state. A field of at least 2–3 times the coercivity is typically necessary for complete saturation. Low coercive materials like ferrites or AlNiCo can be magnetized with the field of a permanent magnet or DC-based electromagnetic coils. Magnetization fields up to the range of 1000 kA/m can be achieved using DC coil systems as well as with strong permanent magnet systems as a homogeneous field.

9.7.4.1.2 Magnetization by a pulsed field

High coercive materials like NdFeB or SmCo have to be magnetized in electromagnetic coils (Figure 9.7.19) that are pulsed with a current peak of typically several thousand amps. Magnetization fields in the range of 3000–6000 kA/m are applied for a short time. Magnetization is achieved in less than a millisecond; thus, short pulses are sufficient. In an isotropic material, the magnetization is fixed exactly to the direction of the local magnetizing field. Anisotropic materials have a preferred direction of the magnetization and have to be magnetized along this axis.

The preferred orientation is defined in the previous forming of the magnet, e.g. by an orientation process while pressing a green part before sintering. This has to be kept in mind if a magnet should be magnetized in a special manner as discussed in the following.

9.7.4.1.3 Magnetization types

Depending on the application besides the shape and the material of the magnet, different types of magnetization have to be chosen in a proper way. The most important types of magnetization which can be used with permanent magnets are

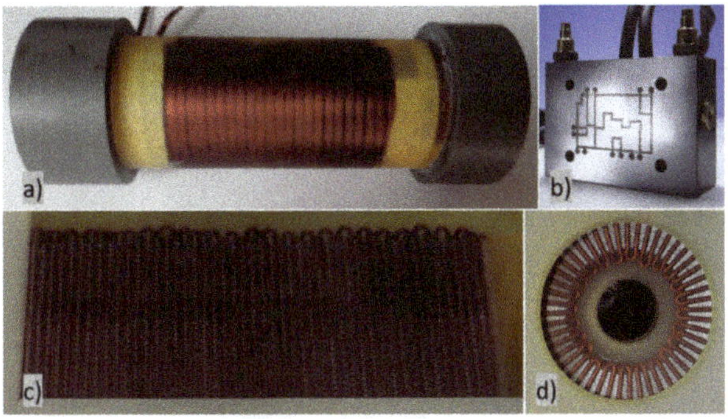

Figure 9.7.19: Magnetization coils. (a) Axial coil, (b) irregular and (c) and (d) regular multipole coil for one-sided surface magnetization.

shown in Table 9.7.4. By using different magnetization types with identical magnet shapes, different magnetic properties/qualities can be achieved. As an example, the magnetic adhesion force can be tuned to a high force over small gaps between the magnet surface and a steel part by many small poles on the surface or to a lower force reaching over longer gaps with only one pole facing to the steel part. One property of the used magnet material plays an essential role for the kind of magnetization. Isotropic materials have the same demagnetization curve in all directions, and they thus can be magnetized in any direction.

In an anisotropic material, the crystallites have their preferred magnetization direction called "easy axis" all in parallel, and thus magnetization is efficient only along this preferred direction. The magnitude of magnetization is nearly doubled compared to the isotropic version of the same material. Table 9.7.4 shows the most common regular magnetization types.

9.7.4.1.4 Demagnetization

Sometimes magnets have to be stored, handled or mounted in the non-magnetized state as mechanical forces and the attraction of pollution might bother. Because magnets are partly already magnetized as a result of the production process, e.g. during orientation of anisotropic materials, the complete demagnetization is an additional production step which can be challenging because of internal domain coupling.

The simplest way for the complete canceling of magnetization is to heat the magnet above a certain temperature defined for each material and called Curie point. Ferrites, for example, completely lose their magnetization at around 450 °C when thermal energy exceeds the energy of exchange interactions. However, most materials suffer by heating them to their Curie temperature.

Another way is using an electromagnetic coil with a decaying AC field [52]. By this way, the bulk mean value of magnetization is reduced to nearly zero but microscopically loops of magnetization may remain, and the demagnetized state is different

Table 9.7.4: Types of magnetization.

Axial	
Diametral for isotropic if diametral anisotropic	
In all directions H, L, B for anisotropic materials along orientation	
Multipolar on both sides (defined by the number of pole pairs) for anisotropic materials only if orientation along height	
Multipolar on one side (defined by the number of pole pairs) for anisotropic materials only if orientation along height	
External multipole (defined by the number of pole pairs) for isotropic or anisotropic with radial orientation	
Internal multipole (defined by the number of pole pairs) for isotropic or anisotropic with radial orientation	
Radial for isotropic or anisotropic with radial orientation	

compared to thermal demagnetization. Thus, AC demagnetized magnets might require higher fields in a later second magnetization.

9.7.4.2 Magnetic measurement

Quality control of a magnet has to guarantee that the magnetic material exhibits at least the minimum values characterized in the data sheet but also to prove that the magnetization is done in a proper way.

9.7.4.2.1 Measurement of hysteresis curve

The characterization of the magnetic material is usually represented only by their de-magnetization curve (the second quadrant of the full hysteresis loop) which shows the falling induction B of the fully magnetized magnet on application of an increasing de-magnetizing field H. The measurement may be done in an open circuit or in a closed circuit, according to how the external field is applied on the magnetic sample. In the open-circuit measurement (vibrating-sample magnetometer), the sample is kept vibrating in the field generated by an electromagnet. Using lock-in technique, the induced voltage in a measurement coil by the sample can be quantified as the magnetic moment of the magnet. Depending on the geometry of the sample, the self-demagnetizing field of the sample itself has to be considered to generate the exact $B(H)$ curve. In a closed-circuit measurement called Permagraph measurement (Figure 9.7.20) [54], the poles of the sample are in a shortcut with the iron poles of an electromagnetic yoke. Using highly permeable magnetic poles, the self-demagnetizing field is eliminated, and the normal $B(H)$ hysteresis can be directly measured.

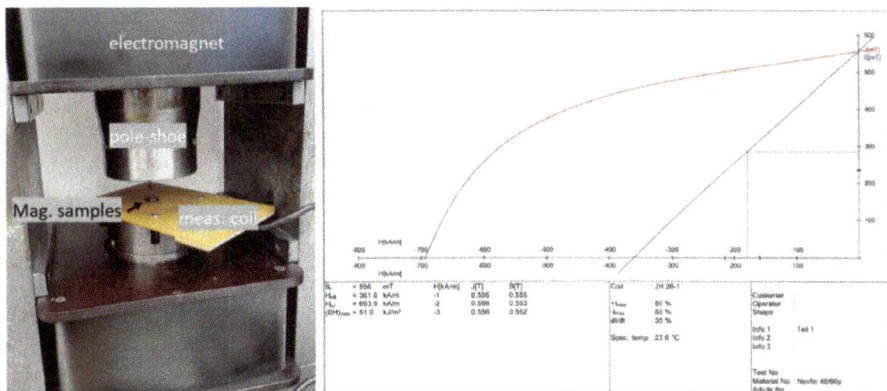

Figure 9.7.20: Permagraph device and $B(H)$ measurement. Protocol with the main magnetic characteristics such as remanence B_r, coercivities H_{cB}, and energy product $(BH)_{max}$.

9.7.4.2.2 Measurement of magnetic moment and field strength

The values of remanence B_r, coercivity H_{cJ} and the energy product $(BH)_{max}$ characterize the magnetic material, and an actual magnet is characterized by the values of polarization J_m and field strength H_m at the working point (OP in Figure 9.7.2). This point depends on the geometry of the magnet and can be designed in a way to guarantee the best aging and temperature stability for the magnet. A fast and very precise way of characterization of the material in the magnetic working point as well as a correct magnetization can be performed by a so-called Helmholtz coil in combination with an electronic voltage integrator called flux meter. The measurement principle is based on the induction law [55]. The flux change in the measurement coil induces a voltage which gives, after integration, one value total magnetization called the magnetic moment M, which is the product of polarization, J_m and volume of the magnet. The internal H_m field of the magnet can be estimated by magnetic potential differences at both ends of the magnet. Measuring of the magnetic potential is done by special long solenoidal coils as in Figure 9.7.21c. Punctual field coils (Figure 9.7.21b) with an integrator measure locally the resulting field, e.g. in the proximity of the magnet.

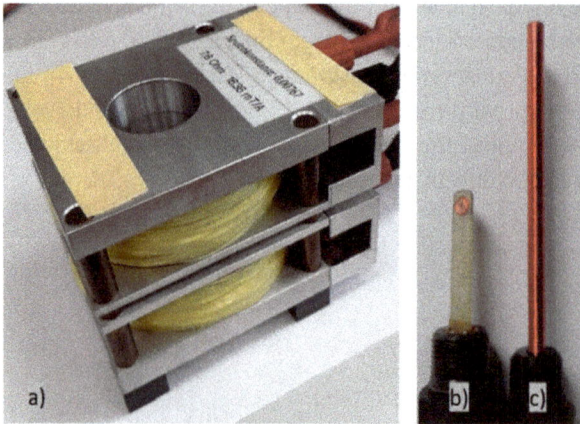

Figure 9.7.21: (a) Helmholtz coil for magnetic moment measurement, *M*. (b) Field coil and (c) potential coil for measurement of *H* field in the magnet.

9.7.4.2.3 Magnetic field mapping

To verify the right magnetization pattern, a complete mapping of the magnetic field in the free space around it is essential. Depending on the magnet geometry and application, 1D circular or linear traces are sufficient, or the field must be mapped into two or three dimensions for product characterization and qualification. Sensors for the field may only be sensitive for one direction of the field but more and more a complete vectorial field mapping is required.

The challenge of magnetic field measurement has pushed measurement techniques to an impressive range of fields from several Tesla down to fT (SQUIDS). For permanent magnets with a field above about 10 mT, the most familiar sensor technology is based on Hall effect sensors (Figure 9.7.22) [56].

Figure 9.7.22: Three-dimensional magnetic field mapper based on Hall sensors.

Hall devices present important benefits such as easy to build, easy to use, low cost, low noise, fast, large range of sensitivity, sensitive to field direction and highly reproducible. Behind these prospects, Hall devices have also some limitations: offset, nonlinearity temperature sensitivity, drift, crosstalk between axes and orthogonality for 3D Hall's. Most of these impediments can only be solved by periodical recalibration of the device in a reference field system, e.g. using a thermally stabilized magnet certified with more accurate methods, e.g. NMR spectroscopy.

Scanning of the volume of interest can be done with a customized mechanical arm which precisely positions the measurement sensor or alternatively moving the magnet itself against a fixed sensor. The second one is preferred for rotational measurements, e.g. rotor magnets. Usually, to measure the position of the sensor, an independent set of linear or angular optical encoders are used.

Field mapping is also successfully implemented in a more sensitive area, e.g. Fluxgate mappers (CERN) and NMR mappers (medicine tomographs) and help for complete diagnosis of system homogeneity [57].

References

[1]   Jiles DC, Atherton DL. Theory of ferromagnetic hysteresis. J Appl Phys, 1984, 55, 2115.
[2]   Parker RJ. Advances in Permanent Magnetism, John Wiley & Sons, New York, 1990.
       ISBN 0-471-82293-0.
[3]   O'Handley RC. Modern Magnetic Materials: Principles and Applications, Wiley, New York,
       1999.
[4]   Cremer R, Pfeiffer I. Permanent-magnet properties of Cr-Co-Fe alloys. Physica, 1975, 80B, 164.
[5]   Radeloff C, Pfeiffer I. Study of the origin of coercive force in semihard magnetic Co-Fe-Ni-
       alloys with additions of Al and Ti. IEEE Trans Magn, 1975, 11, 1417–9.
[6]   Went JJ, Rathenau GW, Gorter EW, van Oosterhout GW. Ferroxdure, A class of new permanent
       magnet materials. Phillips Techn Rev, 1952, 13, 194–208.
[7]   Kojima H. Fundamental properties of hexagonal ferrites with magnetoplumbite structure.
       Handbook Ferromag Mater, 1982, 3, 305–91.
[8]   Hilzinger R, Rodewald W. Magnetic Materials Fundamentals, Products, Properties,
       Applications, Publicis, Erlangen, 2013.
[9]   Coey JMD. Magnetism and Magnetic Materials, Cambridge University Press, Cambridge,
       2009.
[10]   Cullity BD, Graham CD. Introduction to Magnetic Materials, 2nd edition, John Wiley & Sons,
       Inc., Hoboken, New Jersey, 2009.
[11]   Hoffer G, Strnat K. Magnetocrystalline anisotropy of YCo_5 and Y_2Co_{17}. IEEE Trans Magn, 1966,
       2, 487–9.
[12]   Sagawa M, Fujimura S, Togawa N, Yamamoto H, Matsuura Y. J Appl Phys, 1984, 55, 2083.
[13]   Michalowsky L. Neue Keramische Werkstoffe, Wiley-VCH, Weinheim, 1994. ISBN 978-3-527-
       30938-2.
[14]   Hadjipanayis GC (Ed). Bonded Magnets, Proceedings of the NATO Advanced Research
       Workshop on Science and Technology of Bonded Magnets, U.S.A, Newark, 22–5.
       August 2002. DOI: 10.1007/978-94-007-1090-0.
[15]   Warlimont H. (Ed). Magnetwerkstoffe und Magnetsysteme, DGM Informations-gesellschaft-
       Verlag, Oberursel, 9, 1991. ISBN 3-88355-167-8.
[16]   Strnat KJ, Strnat RMW. Rare earth-cobalt permanent magnets. J Magn Magn Mater, 1991, 100,
       38–56.
[17]   Croat JJ, Herbst JF, Lee RW, Pinkerton FE. High-energy product Nd-Fe-B permanent magnets.
       Appl Phys Lett, 1984, 44, 148–9.
[18]   Givord D, Li HS, Moreau JM. Magnetic properties and crystal structure of $Nd_2Fe_{14}B$. Solid
       State Commun, 1984, 50, 497–9.
[19]   Herbst JF, Yelon WB. Crystal and magnetic structure of $Pr_2Fe_{14}B$ and $Dy_2Fe_{14}B$. J Appl Phys,
       1985, 57, 2343.
[20]   Wang Y. Method of making Nd-Fe-B sintered magnets with Dy or Tb, US8480815B2.
[21]   Constantinides S. The important role of dysprosium in modern permanent magnets, Arnold
       Magnetic Technologies, Rev 150903a, 1–10. https://www.arnoldmagnetics.com, accessed
       on October 6th, 2021.
[22]   Dent PC. Rare earth elements and permanent magnets. J Appl Phys, 2012, 111, 07A721.
[23]   Huber C, Abert C, Bruckner F, Groenefeld M, Muthsam O, Schuschnigg S, Sirak K, Thanhoffer
       R, Teliban I, Vogler C, Windl R, Suess D. 3D print of polymer bonded rare-earth magnets, and
       3D magnetic field scanning with an end-user 3D printer. Appl Phys Lett, 2016, 109, 162401.
[24]   Huber C, Abert C, Bruckner F, Groenefeld M, Schuschnigg S, Teliban I, Vogler C, Wautischer
       G, Windl R, Suess D. 3D printing of polymer bonded rare-earth magnets with a variable
       magnetic compound density for a predefined stray field. Sci Rep, 2017, 7, 9419.

[25] Hubert C, Mitteramskogler G, Goertler M, Teliban I, Groenefeld M, Suess D. Additive manufactured polymer-bonded isotropic NdFeB magnets by stereolithography and their comparison to fused filament fabricated and selective laser sintered magnets. Materials, 2020, 13, 1916.

[26] Sonnleitner K, Huber C, Teliban I, Kobe S, Saje B, Kagerbauer D, Reissner M, Lengauer C, Groenefeld M, Suess D. 3D printing of polymer-bonded anisotropic magnets in an external magnetic field and by a modified production process. Appl Phys Lett, 2020, 116, 092403.

[27] The Effects of Temperature on Permanent Magnets – Applications Brief, Magnetfabrik, Bonn, 2008. www.magnetfabrik.de

[28] Mitchell IV. (Ed). Nd-Fe Permanent Magnets, Their Present and Future Applications, Elsevier Applied Science Publishers, London and New York, 1985, 241. ISBN 0-85334-405-1.

[29] Campbell P. Permanent Magnet Materials and Their Application, Cambridge University Press, Cambridge, 1996.

[30] McCallum RW, Lewis LH, Skomski R, Kramer MJ, Anderson IE. Practical aspects of modern and future permanent magnets. Ann Rev Mater Res, 2014, 44, 451–77.

[31] Koch J. Dauermagnetische Werkstoffe und ihre Anwendungen, Wiley, Weinheim, 1975.

[32] Coey JMD. Hard magnetic materials: A perspective. IEEE Trans Magn, 2011, 47, 4671–81.

[33] Grieb B. Magnetwerkstoffe Seminar 18.-19, February 2020, Magnequench GmbH/Haus der Technik e.V., Essen Magnettechnik, 2020.

[34] Seltene Erden in Permanentmagneten nach Öko-Institut e.V., https://www.umweltbunde samt.de/, accessed on October 6th, 2021.

[35] Pedu JC. Hysteresis brakes and clutches, US5238095A.

[36] Jang S-M, Lee S-H. Comparison of three types of permanent magnet linear eddy-current brakes according to magnetization pattern. IEEE Trans Magn, 2003, 39, 3004–6.

[37] Parker R. Magnetic latch, US3790197A.

[38] Gieras JF. Permanent Magnet Motor Technology: Design and Applications, CRC Press, Boca Raton, 2010.

[39] Alternator, https://en.wikipedia.org/wiki/Alternator, accessed on October 6th, 2021.

[40] Wiederick HD, Gauthier N, Campbell DA, Rochan P. Magnetic braking: Simple theory and experiment. Am J Phys, 1987, 55, 500–3.

[41] Hahn KD, Johnson EM, Brokken A, Baldwin S. Eddy current damping of a magnet moving through a pipe. Am J Phys, 1998, 66, 1066–76.

[42] Sodano H, Bae J-S. Eddy current damping in structures. Shock Vibr Dig, 2004, 36, 469–78.

[43] Doppelbauer M. Die Erfindung des Elektromotors 1800–1854, www.eti.kit.edu/1376.php, accessed on October 6th, 2021.

[44] Linear motor, https://en.wikipedia.org/wiki/Linear_motor, accessed on October 6th, 2021.

[45] Jamerson FE, Benjamin E. Worldwide electric powered two wheel market. World Elect Vehicle J, 2012, 5, 269–75.

[46] Electric Bikes Global Market Reports, www.businesswire.com, accessed on October 6th, 2021.

[47] Kellogg EW. Production of Sound, US1983377A.

[48] Angle sensor magnets– Applications Brief, Magnetfabrik, Bonn, 2013. www.magnetfabrik.de

[49] New sensor magnets for multichannel sensors – Applications Brief, Magnetfabrik, Bonn, 2014. www.magnetfabrik.de

[50] Halbach K. Nucl Instr Methods, 1980, Vol. 169, 1–10.

[51] Michalowsky L, Heinecke U, Schneider J, Wich H. Magnettechnik – Grundlagen und Anwendungen, Fachbuchverlag Leipzig GmbH, Leipzig, 1993.

[52] Cassing W, Kuntze K, Ross G. Dauermagnete: Mess- und Magnetisiertechnik, expert Verlag, Essen, 2014.

[53] Schüler K, Brinkmann K. Dauermagnete Werkstoffe und Anwendungen, Springer, New York, 1970.

[54] https://www.magnet-physik.de/messtechnik/hysteresegraphen.php (accessed on october 5th 2021).

[55] Weber W. Über die Anwendung der magnetischen Induction auf Messung der Inclination mit dem Magnetometer. Ann Phys, 1853, 2, 209–47.

[56] Keller P. Technologies for Precision Magnetic Field Mapping (Upload 2015/07), Metrolab Instruments, Geneva, Switzerland, 2015.

[57] Bottura L, Henrichsen KN. Field Measurements, CERN, Geneva, Switzerland, 2002.

9.8 Data storage materials

Günter Reiss

Data storage is one of the key technologies for our modern and information-driven society. While a single device could handle only two digital photos in 1956, this increased to 12 million photos in 2020, and the worldwide data volume that is stored grew from about 4.5 zettabytes in 2018 to 45 zettabytes in 2020 (zetta: 10^{21}) and is expected to maintain this growth rate in the next decade. Storing is digital, i.e. each information is encoded to a digital number such as 101110011 (=371 decimal), and each of the 0/1 bits is stored and possibly processed in one cell. This means that a storage cell must have two stable states associated with a digital 0 or 1 that can be written and read in a write- and read-time, t_W and t_R, respectively, and "stable" means that the stored value is maintained for a certain time t_S. We thus have three timescales associated with data storage. These scales are very different depending on the application, and the same is true for the amount of data that needs to be stored. This leads to the hierarchy of storage devices.

9.8.1 Data storage devices

The hierarchy of data storage shown in Figure 9.8.1 is caused by the very different requirements of the subunits of information technology. While processing units must be very fast, the main memory must be cheap, relatively fast and be able to take at least some gigabyte. Mass storage, however, must be cheap and non-volatile such that t_S is larger than about 10 years. Thus, the storage devices are also very different: In CPUs and GPUs, data are stored in relatively expensive but very fast devices (static random access memory, SRAM [1, 2]) that keep the information as long as power is on. The computer main memory consists of DRAM (dynamic RAM [3]), which can take moderate amounts of data, is moderately fast and cheap and is volatile in the sense that it needs a refresh of the information every few ms, even if power is on. In these two memory types, the information is either stored in a stable state of a transistor circuit (SRAM) or as charge in a capacitor accessed by a transistor (DRAM), and some of the related material issues will be addressed in Section 9.8.2.1. In contrast with these memory types, the mass storage requires non-volatility, meaning that the information is stable for at least 10 years even if power is off. In addition, the capacity must be very large and the price per bit very low. For these storage devices, very often magnetic materials are the most appropriate choices, because their magnetic state has the required stability and data can be packed densely.

One basic issue that needs to be addressed is the stability time t_S, also called data retention. If our storage device has two stable states 0 and 1, we have an energy

https://doi.org/10.1515/9783110733471-008

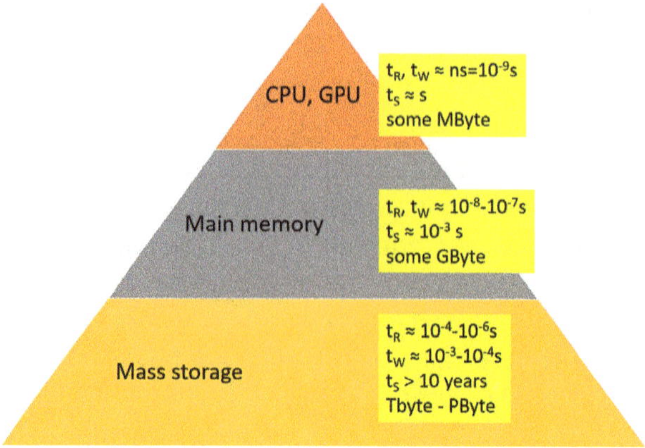

Figure 9.8.1: Data storage hierarchy with typical timescales and capacities for processing units: CPU and GPU, computer main memory and mass storage.

barrier ΔE that must be overcome in order to change the state from, e.g., 0 to 1 (Figure 9.8.2).

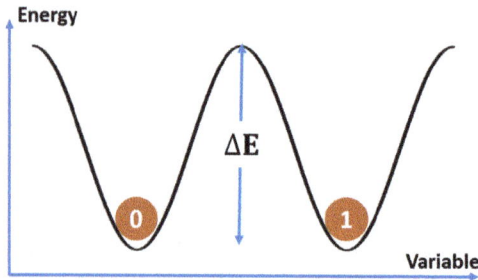

Figure 9.8.2: The energy as a function of the memory variable such as the angle of the magnetization or the charge on the gate capacitance of a MOSFET.

In different devices, the energy barrier is due to device-specific properties: in a MOSFET transistor [4], ΔE is the charging energy of the gate capacitor, while in a magnetic material of volume V and a magnetic anisotropy energy density K, ΔE is the energy for switching the magnetization between two easy directions. In the latter case, one has $\Delta E = K \cdot V$, and the variable in Figure 9.8.2 would be the angle between the magnetization and an easy axis, where the magnetization has lowest energy.

The retention time t_S is closely related to the energy barrier. At finite temperature T, the thermal energy $k_B T$ can lead to a switching of the state. The mean time τ the system of Figure 9.8.2 would stay in one of the states is given by $\tau = \tau_0 e^{\Delta E / k_B T}$ with k_B being the Boltzmann constant, and τ_0 the so-called attempt time (about 1 ns for typical magnetic materials). Thus, for the requirements put on non-volatile memories (t_S > 10 years), ΔE must be larger than about $40 \cdot k_B T$, which amounts to around 1 eV at room temperature.

Keeping these basic requirements for memories in mind, we will now address some of the materials that are keys for device performance.

9.8.2 Materials for memories

If we consider material issues in memory devices, we have to follow the hierarchy shown in Figure 9.8.1. Because there are very different memory types, the performance of these memories is determined by different materials and material properties. In, e.g., the electronic memories of CPUs, GPUs and main memories, the performance is defined by the packing density on the semiconductor chips, i.e. the so-called minimum feature size, which is the smallest feature of the micro- or nano-electronic layout, mostly the length of the conducting channel in MOSFETs, today between 5 and 10 nm. In addition, the charging energy of the gate capacitance and the delay times of the operations is determined by the size and the material of the gate oxide, the resistance of the MOSFET channels and the interconnects, and the parasitic capacitance between, e.g., the electric interconnects.

In contrast to the electronic memories, the mass storage requires high stability (long τ) and a very low price, which leads to the needs for a high energy barrier ΔE, relatively cheap materials and a high packing density of the bits.

In addition, the end of Moore's law [5], i.e. the unlimited downscaling, brings material issues related with memories even more in the focus both of research as well as of applications. The reason is, that, given downscaling reaches its limits, the device performance can be enhanced only by using materials that have better properties such as high dielectric constant or high magnetic anisotropy. Some of these issues will be discussed in the next sections.

9.8.2.1 Materials for electronic memories

Besides the semiconductors that have been already discussed in Section 9.1, insulating and metallic materials are key ingredients for the electronic memories such as SRAM or DRAM. The metallic materials are used for the wiring of the chips with interconnects; the initial material of choice was $Al_{1-x}Si_x$ (x = 0.01–0.05) and –since about 20 years– it is Cu. Cu interconnects are considered commonly as the best solution providing low resistivity and good hardness against electromigration [6]; hence, materials for interconnects will not be discussed further. Other issues related with material's properties are the conductance of the MOSFET channels and the delay times of the electronic memory's circuits. In order to increase the channel conductance, materials with high dielectric constant ε_R are required, while a decrease of the delay times needs a low ε_R.

9.8.2.1.1 High-*k* materials

The material that was and is used as dielectric to separate the gate metal from the conducting channel in MOSFETs is SiO_2 [7]. In general, the charge accumulated in the conducting channel and the mobility of the charge carriers determine the channel's conductance and thus an increase of the charge boosts the MOSFET performance. The charge in turn is proportional to the area of the gate contact, the inverse of the dielectric's thickness d_D separating the gate metal from the semiconductor, and the dielectric constant ε_R. The thickness d_D can be not lower than about 3 nm to avoid electron tunneling. Thus, the capacitance of the gate cannot be increased any more by lowering d_D, and consequently the search for dielectric materials with high ε_R is a major research direction for increasing MOSFET performance. High ε_R (often also denoted as κ, where the name "high-*k*" comes from) means that the material is strongly polarized by an electric field at the operation frequency of the MOSFET of usually some GHz. At this frequency, only the electronic and ionic contribution to the polarization is fast enough to follow the electric field [7]. In SiO_2, the ionic contribution is almost negligible, and its ε_R is around 4. Research therefore concentrates on finding materials that have a large ionic polarization at some GHz frequency. They, however, must satisfy some criteria in order to be useful, and the most important are: (i) a bandgap of at least 5 eV, (ii) a high enough ε_R to justify R&D costs and (iii) thermodynamic stability in contact with Si up to around 400 °C process temperature. A general trend for the search can be derived from the law of Clausius-Mossotti. The larger the ratio α/V_m (α is the polarizability and V_m is the volume of the unit cell) becomes, the larger will be the dielectric constant.

Among the various candidate materials, nitrides and oxides are promising. Nitrides can have a larger ε_R, such as Si_3N_4 (ε_R around 7.5), but they often have a bandgap only slightly larger or even lower than 5 eV. Thus, the main research is concentrating on oxides, and in particular materials derived from hafnia (HfO_2). Already the base material HfO_2, however, can solidify in seven different crystalline forms (such as c-, t- and m- HfO_2 [8]) already at a temperature around 450 °C, with ε_R of the oxide (*k*) ranging from about 2 to more than 30. Thus, the *k*-value on one side strongly depends on the preparation of the materials. On the other side, dielectric stability requires amorphous materials to avoid charge or material migration along grain boundaries. Thus, pure HfO_2 is a borderline case from the perspective of manufacturability.

In order to improve these material properties, there have been numerous attempts such as:

- Incorporation of Al_2O_3 in HfO_2 to form $Hf_xAl_yO_z$ solid solution. These materials can still have a large k of around 15 and they remain amorphous up to 900 °C. Some reports even show that the amorphous phase is stable up to 1000 °C during rapid thermal annealing ($Hf_{0.93}Al_{0.07}O_x$ deposited by MOCVD, [9]).
- Combining HfAlO alloys with SiO_xN_y in a multilayer film was successfully integrated in a CMOS process with promising properties of the MOSFET performance.
- Integration of silicates to form $(HfO_2)_x(SiO_2)_{1-x}$ has been shown to maintain an amorphous structure with good dielectric properties, but a slightly lower k-value (≈ 11).

For further miniaturization, the search for materials with better properties is still an important issue. Among those, oxides such as Pr_2O_3 in the hexagonal phase, La_2O_3, $LiNbO_3$, Sm_2O_3, Ce_2O_3 and Gd_2O_3 are investigated and all do have some positive and some negative properties with respect to manufacturability. For all these materials, alloying with aluminates or silicates and/or the incorporation of N can improve the properties, and research with different deposition processes (MOCVD, ALD, PVD) and conditions is ongoing.

9.8.2.1.2 Low-k materials

While a high-k is required for a large gate capacitance, low-k materials are needed for dielectric insulation on microelectronic chips in order to reduce parasitic capacitances. As already pointed out in Section 9.8.2.1.1, the ratio α/V_m determines the k-value and thus materials with large unit cell and low polarizability are required. At the 250 nm node of miniaturization, the interconnects and their parasitic capacitances became a bottleneck to further downscaling. Again, starting from SiO_2, which was the material used at feature sizes larger than about 200 nm, fluorinated silicon glass (SiO_xF_y) was introduced because the dielectric constant can be reduced from about 4.1 to 3.6 [10].

Many low-k dielectrics have been found already, but most of them cannot be integrated with CMOS technology. One of the first successful materials was from the organosilicate class SiCOH that was deposited by PECVD and showed an ε_R of 3 [11]. Further research mainly concentrates on making these materials and pure SiO_x porous, in order to obtain a lower dielectric constant, and values as low as around 2 have been achieved. Polymers such as PTFE (C_2F_4) have similarly low k-values, but their mechanical and thermal stability is not high enough for CMOS integration. Thus, similar to the high-k materials, research for lower k-values and manufacturability is ongoing.

9.8.2.2 Materials for magnetic mass storage memories

For the mass storage of data, ferromagnets are the most widely used materials even in times of FLASH memories. The reason is that for each SSD-bit sold on the market, more than one bit on hard disk drive capacity is installed for backup purposes. For long-term backup, tapes also still have a considerable market.

Hard disks started in the 1950s with a capacity of about 3.7 Mbyte, a weight of almost 1000 kg and a prize of almost US $10.000 per MByte. As in 2021, this improved to about 20 TByte (20.000.000 Mbyte), 0.06 kg and US $0.000.024 per Mbyte. This tremendous increase of the performance has been achieved by both technical breakthroughs as well as by improved materials and material combinations.

In general, a hard disk consists of a flat disk (called platter) that spins at speeds of about 4.000–15.000 rpm and that is covered by a thin-film system including ferromagnetic layers. The top film is usually a lubricant material such as amorphous carbon (a:CH). Information is written and read by a head that flies at a distance in the 10 nm range above the lubricant as shown in Figure 9.8.3.

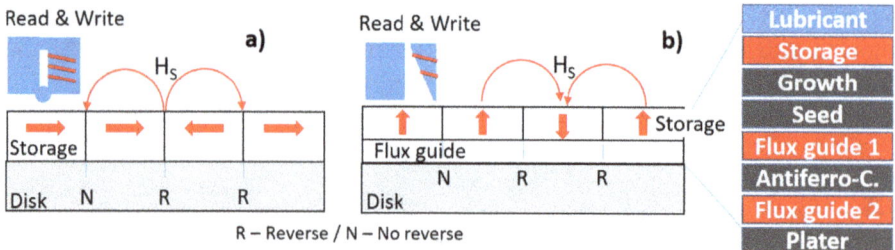

Figure 9.8.3: The principle of a magnetic hard disk for (a) in-plane recording and (b) out-of-plane recording with a typical sequence of the perpendicular recording thin-film system. The magnetization is shown as red arrows.

The read-and-write-head has both a magnetic field source that can change the magnetization direction of the storage layer as well as a read element that can detect the magnetic stray field H_S that emerges, if the magnetization in the storage layer is reversed. In real applications, the storage films are polycrystalline with a grain size that must be smaller than the bit size sketched in Figure 9.8.3. Today, this requires a mean grain size smaller than about 5 nm.

The magnetic films have been in the beginning optimized for in-plane recording (Figure 9.8.3 a), i.e. they had a large coercive field H_C and a large saturation magnetization M_S. For these purposes, materials such as electroplated Ni-Co-P alloys with $H_C >$ 70 kA/m (>900 Oe) have been used [12]. Later on, obliquely evaporated alloyed films of Co, Fe and Ni have been deposited as storage layers and due to the preferred direction, H_C was increased to around 1000 Oe. The addition of heavy metals such as Pt or W in particular to alloys from Co and Ni increased this further to ≈2000 Oe. Fe-Co-Cr films produced by e-beam evaporation and protected from the lubricant by Ru reached similar H_C's with a bit size of about 1 µm. In parallel, reactive sputtering was developed and enabled the use of oxides such as magemite ($\gamma\text{-}Fe_2O_3$) or, later on, magnetite (Fe_3O_4), which can have similar coercive fields after doping with, e.g., Os.

The need for higher packing density and reduced bit size drove the research in the direction of perpendicular recording, as shown in Figure 9.8.3b, which was shipped to customers first in 2005. As storage layer, sputter-deposited Co-Cr alloys have been used. After an annealing process these alloys adopt a special microstructure, where c-axis-textured and small Co grains with a perpendicular magnetic anisotropy are magnetically separated by a Co-Cr alloy at the grain boundaries. This enabled a higher packing density with a bit size well below 0.5 µm. The addition of Pt to these alloys allowed to reduce the grain size while maintaining the high perpendicular anisotropy. For these magnetically perpendicular materials, however, a flux guide layer is needed that has a high permeability μ. Often, permalloy is used for this purpose, because this material class based on Ni_xFe_y with $x \approx 80$, $y \approx 20$ has high μ, low magnetostriction and relatively high M_S, and its properties can be tailored by addition of dopands such as Mo.

Modern hard disk drives have packing densities of some TBits/inch^2, i.e., the mean bit size is in the range of some 10 nm. For these media, a layer system as sketched in Figure 9.8.3b is used with Ru as the antiferro-coupler, permalloy-based materials as flux guides and Co-Cr-O alloys as storage layers. Alternatively, magnetic multilayers such as $[Fe/Pt]_x$ or $[Co/Pt]_x$ are possible, where the perpendicular magnetic anisotropy stems from the orbital interaction at the interfaces between the ferromagnet and the heavy metal.

In research, also so-called particular media are considered, where either magnetic nanoparticles [13] are used, or where large area patterning with, e.g., laser interference lithography produces patterns of separated magnetic dots. In either case, one bit would be stored in one nanoparticle or dot. Such approaches promise to reduce the bit size further, but still have issues with manufacturability.

Magnetic tapes that are used for long-term backup are produced often with magnetic particles from γ-Fe_2O_3, CrO_2 or ferrites such as $SrFe_{12}O_{19}$ embedded in a polymer matrix. Such tapes reach densities of some 10 GBit/inch^2. Since data safety is an issue of ever-increasing importance, magnetic tapes are expected to maintain or even widen their market [14].

9.8.2.3 Materials for emerging memories

Emerging memories is an umbrella term for devices that can store information and that are either still not yet shipped or that have only niche markets so far. For research, emerging memories are very important drivers because new materials as well as new effects are investigated to realize memories. From the industrial aspect, the mix of memories that exists at present (see Figure 9.8.1) is nasty, and the limits of downscaling hinder the further improvement of the performance of data processing systems. One general goal is to find a universal non-volatile memory, where universal means that it can replace most of the existing memory types discussed in the foregoing sections. The preconditions for being a hot candidate are low costs, fast operation both in reading and writing, low energy consumption, endurance of 10^{15} read/write cycles, data retention of more than 10 years and – last, but not least – compatibility with existing CMOS processes.

This long wish list can be fulfilled at present only to some extent by the candidate devices that are discussed in the following.

The ferroelectric RAM (FeRAM (Figure 9.8.4) exists already since about 1975 and has been shipped for niche applications for many years. Therefore, the dielectric of the DRAM capacitor or the gate dielectric of a MOSFET is replaced by a ferroelectric material. Ferroelectric materials can be polarized by an electric field and they maintain their electric dipole moment if the field is reduced to zero. Thus, a capacitor keeps its charge and the information is maintained after the voltage (V_G in Figure 9.8.4) is removed. There are many of such ferroelectric materials, such as complex perovskites or layered perovskites, and the most common ones used for FeRAM are $Pb(Zr_xTi_{1-x})O_3$ (PZT), (Pb, La)TiO$_3$ (PLT) and $SrBi_2Ta_2O_9$ (SBT). Variants that have also been investigated are (Pb, La)(Zr,Ti)O$_3$ (PLZT), BaTiO$_3$-PbZrO$_3$ (BPZT). Alternative materials that have been considered in recent years are ferroelectric fluorite-structure oxides such as hafnia (HfO$_2$) and zirconia (ZrO$_2$) [15]. In either case, the reliable integration of the ferroelectric material with a typical thickness of 100 nm is relatively expensive, and aging effects due to high temperatures are the main reasons for failures.

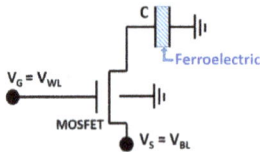

Figure 9.8.4: The principle of a ferroelectric random access memory.
The ferroelectric material is in this example integrated as dielectric in the capacitor C accessed by a MOSFET with the bitline voltage V_{BL} and the wordline voltage V_{WL}.
Other realizations exist, where the ferroelectric is in the gate capacitance.

Because the FeRAM is considered to be probably the suboptimal solution for a universal memory, other types are at present topics of research in electronics, solid-state physics and chemistry, and some of them are already shipped. Figure 9.8.5 shows four devices that are considered as possible solutions for the quest of universal memory.

A common feature of these devices is their memristive-like property. They change their electric resistance in dependence of the amount of charge Q that has flown through the device ($Q = \int I(t)dt$) or the so-called flux W ($W = \int V(t)dt$), with $I(t)$ being the time-dependent current, and $V(t)$ the voltage, respectively. Thus, their functional part is effectively a resistor with a value that can be programmed by the applied voltage or current in a reversible manner.

In the phase change random access memory (Figure 9.8.5a [16]), materials that can change their microstructure between crystalline and amorphous as a consequence of

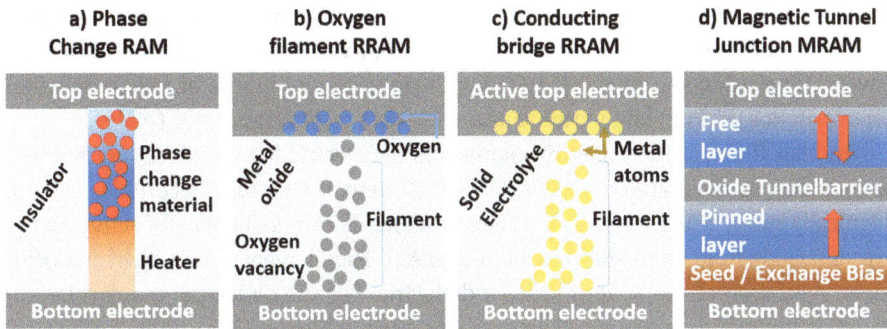

| a) Phase Change RAM | b) Oxygen filament RRAM | c) Conducting bridge RRAM | d) Magnetic Tunnel Junction MRAM |

Figure 9.8.5: The principles of emerging memories that are based on programmable resistors.

different heat treatments are used. In an SET-current or -laser pulse, the temperature is driven above the crystallization temperature but remains below the melting point. After a cooldown, the crystalline phase is set. To change the structure, a RESET pulse heats the material above the melting temperature, and a following fast cooldown quenches the microstructure into the amorphous state. The memristive property is in this case the large difference between the electrical resistivity ρ in the crystalline (low ρ) and the amorphous state (high ρ), which can be up to 3–4 orders of magnitude.

The most frequently used material class for the PCRAM are chalcogenides. In the strong glass-forming material $Si_{12}Te_{48}As_{30}Ge_{10}$ (STAG), the effect of reversible switching was observed first. In the last years, research concentrates more on materials that are situated along the $GeTe–Sb_2Te_3$ line in the Ge-Te-Sb phase diagram, because they show much faster recrystallization. A frequently used material is $Ge_2Sb_2Te_5$. There is, however, a vast amount of materials that do have phase change properties. In the phase diagram Ge-Te-Sb, Ge can be, e.g., fully or partly replaced by In, Ag or Sn, and Sb by Bi, Au or As. Although PCRAM is already in use for EEPROM replacement, there are still issues that need to be resolved with respect to a universal memory, such as the limited data retention (10^{12}–10^{14}) and the relatively long SET time (ca. 100 ns).

The resistive RAMs shown in Figure 9.8.5b and c work with conducting filaments in a dielectric that can be formed and removed by voltage pulses with different polarity. In the case of the oxygen filament RRAM [17], the electric field drives oxygen into the top electrode and the remaining filament has a high concentration of oxygen vacancies and thus a much smaller resistivity ρ than the surrounding dielectric. Similar to the PCRAM, ρ can decrease by 3–4 orders of magnitude after filament formation. The most common materials used for this device are metal oxides such as hafnium oxide (HfO_x), titanium oxide (TiO_x), tantalum oxide (TaO_x), nickel oxide (NiO), zinc oxide (ZnO), manganese oxide (MnO_x), magnesium oxide (MgO) and aluminum oxide (AlO_x), as well as zinc titanate (Zn_2TiO_4), and zirconium dioxide (ZrO_2). The thin films are often deposited by atomic layer deposition (ALD) due to the conformal growth and precise thickness control of this technique. In the last years, it turned out that the mechanism of filament formation is quite common in many material classes and also occurs in block copolymers such as poly(9-(4-vinylphenyl)carbazole)-b-poly(2-vinylpyridine) (denoted PVPCz-b-P2VP) with two different blocks of carbazole as an electron donor (PVPCz) and pyridine as a very weak electron acceptor (P2VP).

Another filament formation mechanism is the electric field-induced migration of metal atoms, as illustrated in Figure 9.8.5c [18]. Here, the bottom electrode is electrochemically inert, while the top electrode or anode is electrochemically active and is typically made from an oxidized metal (Cu or Ag). The filament is formed by the field-driven migration of metal atoms into a dielectric. For the dielectric, many different materials have been reported: chalcogenides such as $GeSe_x$, GeS_2, $GeTe$, Cu_2S, Ag_2S and others. The other class is oxide-based materials such as Ta_2O_5, SiO_2, ZrO_2, or GeO_x. Also amorphous Si and Si_3N_4 can form the dielectric. Some publications reported improved properties with bilayers such as $Cu-Te/GdO_x$, $Cu-Te/SiO_x$, MoO_x/GdO_x, $TiO_x/TaSiO_y$, $GeSe_x/TaO_x$, Ti/TaO_x, $Cu-Te/Al_2O_3$, TiW/Al_2O_3 and $CuTe-C/Al_2O_3$.

In general, these RRAMs are very promising candidates for a universal memory. The formation or removal of the filaments produces large resistance changes, that can be easily detected, and the switching is relatively fast and consumes little power. There are, however, still issues with the data retention (about 10 years) and endurance ($<10^{10}$) which are currently topics in solid-state research.

The fourth candidate device of Figure 9.8.5d is the so-called magnetic random access memory (MRAM, nowadays STT-MRAM [19, 20]) that effectively constitutes a tunnel junction with a metal oxide barrier (thickness 0.7–2 nm) and two ferromagnetic electrodes. One of these electrodes has a fixed magnetization (pinned layer), while the other one has a magnetization that can be set by a current pulse through the device, either in parallel or antiparallel direction, by spin-transfer torque (STT, free layer). Due to the electronic band structures of the ferromagnets and the barrier, the tunneling resistance R changes between the parallel (R_P) and the antiparallel state (R_{AP}). Usually, R_{AP} is larger than R_P, with record values reaching a factor of 20 at low temperature. Due to the hysteresis of the magnetization of the free layer, the device remains in the state that has been set by the current pulses after power is off.

The main research directions for this type of device target at an increase of the tunneling magnetoresistance TMR = $(R_{AP}-R_P)/R_P$ and a large stability against thermal agitation and external magnetic fields. It turned out that for an increase of the TMR, two main strategies are promising: the first one is an increase of the spin polarization P at the Fermi level E_F of the ferromagnets, i.e., the surplus of one spin direction over the other. Ideal materials have $P = 1$, which means that at E_F, only one spin type is present, and the TMR could be infinitely large at low temperature. While the normal ferromagnets have P around 0.5, some materials are very promising: metal oxides such as magnetite, Fe_3O_4, or CrO_2 do have this property in the bulk. It is, however, very challenging to form high-quality tunnel junctions with such materials as electrodes because they tend to form a thermodynamically more stable phase at the interfaces such as magemite, Fe_2O_3, or the semiconducting Cr_2O_3. Another important material class that can have $P \approx 1$ are the so-called full Heusler compounds [21] of the type A_2BC with four interpenetrating fcc-substructures. Among them, the Co-based Heusler compounds (such as Co_2MnAl, Co_2MnSi and Co_2FeSi) did show a high P or record values of TMR. One promising property of this material class is their rigid electronic band structure, meaning that E_F can be tuned by adding or removing charge carriers and thus shifting E_F to a value with large P.

Because it turned out that the preparation of MTJs with Heusler compounds is challenging, the spin filter effect of the barrier material was also investigated: therefore, the crystalline barrier has an electronic band structure that allows efficient tunneling only for highly symmetric electronic states of the electrode material. For the case of Fe/MgO/Fe, it was theoretically predicted [22] that the fully spin-polarized Δ_1-state of the ferromagnet tunnels much more efficiently than all other states, such that very high TMR values can be found. Experimental work subsequently showed that a TMR well above 200% is possible with the combinations Fe/MgO/Fe [23] and CoFe/MgO/CoFe [24]. Later on, the combination $Co_{x1}Fe_{y1}B_{z1}$/MgO/$Co_{x2}Fe_{y2}B_{z2}$ ($x_{1/2} = 2, \ldots, 6$, $y_{1/2} \approx 8-x_{1/2}$, $z_{1/2} \approx 2$) turned out to be easier to prepare while maintaining the high TMR values of today up to 600% at room temperature. Therefore, the ferromagnets are amorphous after preparation and adjust to the crystal structure of the MgO during annealing such that a coherent tunneling of charge carriers is possible. With the concentrations $x_{1/2}$ and $y_{1/2}$ (the boron concentration is around 20%), further properties of the MTJs can be tailored such as the magnetic anisotropy, independently for the pinned layer and the free layer (see Figure 9.8.5d). By the choice of the thicknesses and the interfaces in the layer stack, it is also possible to prepare both layers independently either magnetically in- or out-of-plane. Due to its flexibility, this system is at present the working horse for the efforts on the emerging memory STT-MRAM. Further candidates, such as the three-terminal device SOT-MRAM, are ahead.

Thus, although the resistance change of the MRAM is smaller than those of the other candidates for a universal memory, its potential is considered to be very high, due to unlimited endurance and a data retention larger than 20 years. Issues that need to be resolved are a further increase of the TMR, the downscaling in the range below about 20 nm and the stability of magnetically perpendicular electrodes at this size scale.

References

[1] Santhiya V, Mathan N. Review on performance of Static Random Access Memory (SRAM). Int J Adv Res Comp Commun Eng, 2015, 4, 403–6.
[2] Sebastian A, Le Gallo M, Khaddam-Aljameh R, Eleftheriou E. Memory devices and applications for in-memory computing. Nat Nanotechnol, 2020, 15, 529–44.
[3] Spessot A, Oh H. 1T-1C dynamic random access memory status, challenges, and prospects. IEEE Trans Electron Dev, 2020, 67, 1382–93.

[4]   Jaeger RC, Blalock TN. Microelectronic Circuit Design, 4th edition, McGraw Hill, New York, 2010. ISBN 978-0-07-338045-2.

[5]   Waldrop MM. Nature, 2016, 530, 145–7.

[6]   Pierce DG, Brusius PG. Electromigration: A review. Microelectron Reliab, 1997, 37, 1053–72.

[7]   Hall S, Buiu O, Mitrovic IZ, Lu Y, Davey WM. Review and perspective of high-k dielectrics on silicon. J Telecommun Inform Technol, 2007, 2, 33–43.

[8]   Rignanese G-M. Dielectric properties of crystalline and amorphous transition metal oxides and silicates as potential high-k candidates: The contribution of density-functional theory. J Phys: Condens Matter, 2005, 17, R357–79.

[9]   Kwa KSK, Chattopadhyay S, Jankovic ND, Olsen SH, Driscoll LS, O'Neill AG. A model for capacitance reconstruction from measured lossy MOS capacitance-voltage characteristics. Semicond Sci Technol, 2003, 18, 82–7.

[10]   Leobandung E, Barth E, Sherony M, Lo S-H, Schulz R, Chu W, Khare M, Sadana D, Schepis D, Boiam R, Sleight I, White F, Assaderaghi F, Moy D, Biery G, Goldblat R, Chen T-C, Davari B, Shahidi G. High performance 0.18 μm SOI CMOS Technology. Tech Dig Int Electron Dev Meet, 1999, 679.

[11]   Grill A, Perraud L, Patel V, Jahnes C, Cohen S. Low dielectric constant SiCOH films as potential candidates for interconnect dielectrics. Mater Res Soc Symp Proc, 1999, 565, 107.

[12]   Sato I. Thin film media for hard disks. IEEE Transl J Mangn Jpn, 1987, TJMJ-2.

[13]   Reiss G, Hütten A. Magnetic nanoparticles – applications beyond data storage. Nat Mater, 2005, 4, 725–6.

[14]   Lantz M. Why the future of data storage is (still) magnetic tape, IEEE spectrum, 28 Aug 2018 | 15:00 GMT, https://spectrum.ieee.org/computing/hardware/why-the-future-of-data-storage-is-still-magnetic-tape

[15]   Park MH, Lee YH, Mikolajick T, Schroeder U, Hwang CS. Review and perspective on ferroelectric HfO$_2$-based thin films for memory applications. MRS Commun, 2018, 8, 795–808.

[16]   Gallo ML, Sebastian A. An overview of phase-change memory device physics. J Phys D: Appl Phys, 2020, 53, 213002.

[17]   Zahoor F, Zulkifli TZA, Khanday FA. Resistive Random Access Memory (RRAM): An overview of materials, switching mechanism, performance, multilevel cell (MLC) storage, modeling, and applications. Nanoscale Res Lett, 2020, 15, 90.

[18]   Jana D, Roy S, Panja R, Dutta M, Rahaman SZ, Mahapatra R, Maikap S. Conductive-bridging random access memory: Challenges and opportunity for 3D architecture. Nanoscale Res Lett, 2015, 10, 188.

[19]   Na T, Kang SH, Jung S-O. STT-MRAM sensing: A review. IEEE Trans Circ Syst –II, 2021, 68, 12–8.

[20]   Reiss G, Schmalhorst J, Thomas A, Hütten A, Yuasa S. Magnetic tunnel junctions, in Springer tracts in modern physics, vol 227. Springer, Berlin, Heidelberg. DOI: https://doi.org/10.1007/978-3-540-73462-8_6.

[21]   Wollmann L, Nayak AK, Parkin SSP, Felser C. Heusler 4.0: Tunable materials. Ann Rev Mater Res, 2017, 47, 247–70.

[22]   Butler WH, Zhang X-G, Schulthess TC, Maclaren JM. Spin-dependent tunneling conductance of Fe/MgO/Fe sandwiches. Phys Rev B, 2001, 63, 054416.

[23]   Parkin SSP, Kaiser C, Panchula A, Rice PM, Hughes B, Samant M, Yang S. Giant tunnelling magnetoresistance at room temperature with MgO (100) tunnel barriers. Nat Mater, 2004, 3, 862–7.

[24]   Yuasa S, Nagahama T, Fukushima A, Suzuki Y, Ando K. Giant room-temperature magnetoresistance in single-crystal Fe/MgO/Fe magnetic tunnel junctions. Nat Mater, 2004, 3, 868–71.

9.9 Piezo- and pyroelectric materials

Adam Slabon

9.9.1 Introducing remarks

The capacitance C_0 between two plates aligned in parallel and separated by vacuum is defined as follows:

$$C_0 = \frac{\varepsilon_0 \cdot A}{d}$$

where ε_0 is the vacuum permittivity equal to 8.854×10^{12} F/m, A is the surface area of the plates and d is the distance between the plates. C_0 depends consequently just on the dimensions and distance of the plates. The application of an electric potential U to the plates results in stored electrical charge Q_0, which depends linearly on the value of U:

$$Q_0 = C_0 U$$

Figure 9.9.1 depicts a situation where a dielectric material, i.e. material with zero direct current electrical conductivity, is placed between the plates. Both the stored charge and capacitance increase reversibly to Q_1 and C_1, respectively [1]. The capacitance ratio before and after insertion of the dielectric material can be expressed as the dielectric constant ε':

$$\varepsilon' = \frac{C_1}{C_0}$$

The value ε' is strongly dependent on the polarizability of a material. For inorganic solids where the chemical bonding is mainly of ionic nature, the value of ε' can reach up to 10, whereas the ferroelectric materials exhibit values up to 10,000. An applied field E_0 can create an induced dipole, resulting in the polarization

$$p = qr$$

where the charges q are separated by the distance r. As such, the applied field E_0 can be altered by the material's internal field to yield the local field E_{local}. The sum of i dipole moments in the material leads to total polarization P according to

$$P = \sum_i p_i = \sum_i \alpha_i E_{\text{local}, i}$$

where α_i is the polarizability of the material.

The interaction of dielectric materials with the abovementioned external field can be described in terms of polarizability, permittivity and dielectric susceptibility.

https://doi.org/10.1515/9783110733471-009

Figure 9.9.1: Schematic illustration of a dielectric material placed between two plates with an applied external potential.

The latter is equal to the permittivity upon excluding the contribution from the vacuum permittivity:

$$\varepsilon_0 \chi = \varepsilon' \varepsilon_0 - \varepsilon_0$$

In opposite to ionic compounds with ion-conductive properties, e.g. $ZrO_2:Y^{3+}$, dielectric materials are characterized by lack of motion except the displacements of charges. The requirement for pyroelectric, ferroelectric and piezoelectric materials is that dielectric compounds crystallize in a non-centrosymmetric structure. The dipoles of *piezoelectric materials* can be modulated by external mechanical stress, leading to agglomeration of positive and negative charges on the opposite end of the corresponding crystal. Complementary, *pyroelectric materials* develop variations in the polarization depending on the temperature. From the structural point of view, 21 of the 32 crystallographic point groups are non-centrosymmetric. When subtracting the point group (432) that does not induce piezoelectricity, 20 piezoelectric point groups remain. The half of these point groups represents the polar dielectrics that exhibit pyroelectric properties. A subclass of pyroelectric materials that show spontaneous polarization, which can be reversed by application of an external electric field, is described as *ferroelectrics*. Figure 9.9.2 depicts the relationship between ferro-, piezo- and pyroelectric materials. Ferroelectric materials are simultaneously pyro- and piezoelectric materials, whereas not every piezoelectric material is necessarily a pyroelectric material.

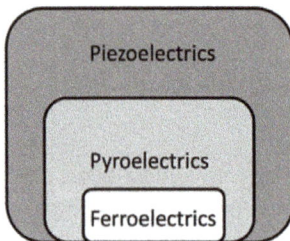

Figure 9.9.2: Relationship between ferroelectrics, pyroelectrics and piezoelectrics.

9.9.2 Piezoelectric materials

The piezoelectric effect allows to harvest mechanical energy and to successively convert it into other forms of energy as a result of dipole deformation and the creation of charge, which can be extracted to empower external devices. Piezoelectric materials exhibit generally one polar axis, symbolized as 3, being the direction of the applied mechanical stress and which has an effect on the amount of converted energy. Directions at right angles (90°) to 3 are described as 1. Accordingly, the 33-mode relates to the direction of applied stress that is parallel to the direction 3, whereas the 31-mode describes the situation where the applied stress is at a right angle to 3. For the fabrication of piezoelectric devices, the piezoelectric material is sandwiched between elastomers to ensure structural stability by minimizing the exposure to humidity, before the device is electrically wired from two sides. The resulting open-circuit potential of the piezoelectric material is defined as follows:

$$V = \frac{d_{ij}}{\varepsilon' \varepsilon_0} \sigma_{ij} g_e$$

where d_{ij} is the piezoelectric coefficient, ε' is the dielectric constant, ε_0 is the vacuum permittivity, σ_{ij} is the applied mechanical stress and g_e is the distance between the electrodes [2]. In 33-mode, the applied stress and acquired voltage are in the identical direction, while in 31-mode the voltage is acquired in perpendicular direction and stress is applied in axial direction (Figure 9.9.3).

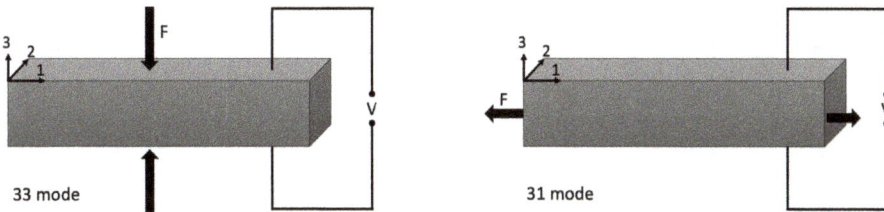

Figure 9.9.3: Operation modes of piezoelectric materials.

Representative inorganic compounds of piezoelectric materials are quartz, the Rochelle salt sodium potassium tartrate tetrahydrate ($NaKC_4H_4O_6 \times 4H_2O$), $BiFeO_3$, $BaTiO_3$, tourmaline, AlN, $K_{0.5}Na_{0.5}NbO_3$, $Bi_{0.5}Na_{0.5}NbO_3$, ZnO, hexagonal BN, $PbTiO_3$ and Pb $[Zr_xTi_{1-x}]O_3$ [3]. The lead zirconate-titanate system, commonly abbreviated as PZT, with the chemical composition $Pb[Zr_{0.5}Ti_{0.5}]O_3$ is together with quartz, one of the most used piezoelectric materials, and accounts for over 90% of all piezoelectric applications these days. However, the application of the toxic lead in PZT-based devices is inevitably connected to environmental issues after the end of the materials' lifetime.

Piezoelectric materials have a wide range of application based on mechanical energy conversion or waste energy harvesting. For the automotive sector, modern piezoinjectors in highly compact form for common-rail systems enable high-precision fuel injection into the cylinder to ensure economical combustion at different operating conditions. The small size of the injectors offers the opportunity to vary the given injection characteristics. Despite the technological advances in lithium-based batteries for electric vehicles, it is unlikely that the piezoinjectors will become obsolete in the near future. On the other side, piezoelectric motors can be precisely controlled by the applied voltage that determines the degree of expansion and subsequent contraction, even within the nanometer regime. One important advantage of piezoelectric motors in comparison to conventional motors is the possible application in harsh environments such as strong magnetic fields or very low temperatures. Transportation is inevitably connected to mechanical energy in the road, which can be harvested with piezoelectric generators. Industrial solutions that harvest the mechanical energy from asphalt to empower external devices, e.g. traffic lights, by application of piezoelectrics have already entered the market. One of the most used and commonly known applications of piezoelectrics is igniters, i.e. piezoelectric lighters (Figure 9.9.4), where a push of a button eventually releases a small hammer that hits the piezoelectric ceramic. The mechanical force generates a piezovoltage sufficient to create a spark that ignites fuel.

Figure 9.9.4: Electric arc generated by a piezoigniter element of a commercial gas lighter (photo by Thomas Fickenscher).

Piezoelectrics are also used today for a plethora of sensing application. Piezoelectric pressure sensors have a superior acquired sensitivity and more reliable results for dynamic pressure changes due to the high-frequency response and signal conversion. Similarly, piezoelectric sensors are suitable for sonar equipment because piezoelectrics have an exceptional response for ultrasonic sounds in the 50–200 kHz range. Piezoelectric transducers, i.e. electronic elements that convert signals, are important components for ultrasound imaging. Piezoelectric speakers are inexpensive and as such used in different electronic equipment, ranging from cell phones, ear buds and buzzers.

Piezoelectric materials also play a role in the manufacturing of electrified musical instruments. In microphones, sound vibrations can be converted to an electrical signal using piezoelectrics. Another example is the pickup in stringed instruments, e.g. electric guitars, where a piezoelectric material, placed between the body of the instrument and the string-supporting frame, converts the string vibration to an electric signal.

9.9.3 Pyroelectric materials

The pyroelectric effect is the change of the polarization P due to sequential change of temperature T. In opposite to the spatial temperature change $(dT/dx) \neq 0$ of the thermoelectric effect, the pyroelectric effect is driven by the temperature variation $(dT/dt) \neq 0$. This leads to the generation of an electric potential at the material boundaries under open-circuit conditions. Closing the circuit leads to a flow of electrical current, being described according to

$$i = \frac{dQ}{dt} = pA\frac{dT}{dt}$$

where Q is the pyroelectric charge, A is the surface area, T is the temperature and t is the time [4]. The pyroelectric coefficient p under constant stress and constant electric field is defined as follows:

$$p = \frac{dP}{dT}$$

The pyroelectric effect can be also expressed as

$$p = \frac{i}{A\left(\frac{dT}{dt}\right)}$$

The strong temperature-dependent polarization and creation of voltage enables highly precise sensing applications [5]. Passive infrared (PIR) motion sensors based on pyroelectric materials, e.g. ZnO, allow to monitor whether humans have entered the sensor range because the device is sensitive to the change in acquired IR radiation. Pyroelectric materials consisting of heterostructures, e.g. interfaces between n-ZnO and p-Si, are used for detection of photons from the IR and near-infrared (NIR) region. During irradiation with NIR light, the induced pyro-polarization at the material's interface allows to acquire a converted electric signal. Bolometers are devices that measure the incident electromagnetic radiation as a function of the material heating in order to establish a linear relationship between the temperature and the electric resistance. Such devices consisting of single crystals of the pyroelectric compound $LiNbO_3$ can detect a minimum temperature variation of 15 µK and use single-layer graphene as a corresponding transducer. One promising future application is the pyroelectric nanogenerator for harvesting thermal energy. This type of energy conversion could be used in self-powered wearable electronic devices and could be integrated with piezoelectric nanogenerators for simultaneous harvesting of biomechanical energy. However, this requires mechanically stable materials to ensure high performance and additionally the application of lead-free materials to circumvent toxicity.

9.9.4 Piezo- and pyroelectric materials for catalysis, photocatalysis and photoelectrochemistry

Catalysis (see also Chapters 7.1 and 7.2) is the key for industrial chemical technologies because it allows to enhance the reaction rate of chemical transformations. In opposite to homogeneous catalysis, where the substrate and catalyst are in the same phase, heterogeneous catalysis facilitates the separation of the catalyst from the reaction media. Besides heterogeneous catalysts that operate at elevated temperatures, e.g. iron-based catalysts for the production of ammonia, a tailored material's design enables to drive chemical reactions using the electric potential (*electrocatalysis*), light (*photocatalysis*) or both the electric potential and light (*photoelectrocatalysis*). A promising technology for solar energy conversion is photoelectrochemical water splitting to provide chemical energy in the form of hydrogen. The operating principle is based on the simultaneously occurring oxygen evolution reaction and hydrogen evolution reaction on two electrodes, whereas at least one of them must be a semiconducting material and is defined as a photoelectrode due to its role for light harvesting. If the latter is a p-type semiconductor, e.g. CdS, such photoelectrode has the function of a photocathode that reduces water. In case of an n-type semiconductor, e.g. ZnO, the thin-film photoelectrode is considered as a photoanode that oxidizes water to oxygen. This electrode configuration is depicted in the schematic illustration in Figure 9.9.5. Upon illumination of the thin-film photoanode with photons that have a higher energy than the bandgap of the semiconductor, electrons are being excited from the valence band to the conduction band and migrate further through the conductive substrate over the electric wiring to the metallic cathode, where they eventually reduce water. The simultaneously generated holes in the valence band of the photoanode migrate to the electrode-electrolyte interface, i.e. the semiconductor-water interface, where they oxidize water in a four-electron process.

Thin-film photoelectrodes in the form of nanostructures have to be synthesized on an electronically conductive substrate (Figure 9.9.6a) in order to enable efficient electron transport from the working electrode to the counter electrode. This can be achieved by hydrothermal growth directly on the conductive substrates or by spin-coating. The performance of materials for photoelectrochemical, electrocatalytic and/or photoelectrocatalytic applications is measured in terms of current density,

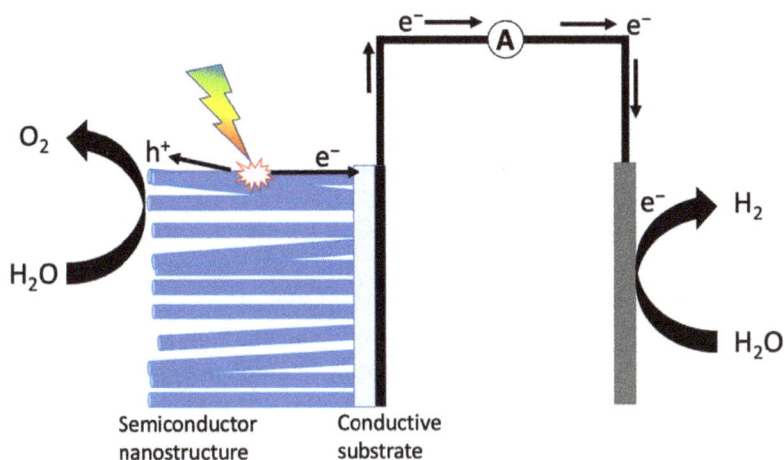

Figure 9.9.5: Operating principle of a photoelectrochemical cell for solar-driven water splitting. The n-type semiconductor nanostructures (blue) form a thin-film photoanode and are electrically wired to a metallic counter electrode for the complementary reduction of water. The cell is immersed in an aqueous electrolyte.

i.e. the recorded current value per surface area of the immersed electrode. For light-dependent application, photocatalyst performance is studied by using an external light source, such as a solar simulator (Figure 9.9.6b) or a light-emitting diode (LED) at one given wavelength. In case of heterogeneous catalysts that are also dielectric materials and their structure can be classified within the 20 piezoelectric point groups (*vide supra*), the catalytic activity can be ameliorated by either piezo-, pyro- or ferrocatalytic effects. In this chapter, we shall focus on the two former cases.

Figure 9.9.6: (a) Semiconducting thin-film photoelectrode on a fluorine-doped tin oxide (FTO) substrate. (b) Experimental for photoelectrochemical water splitting using solar simulator with an AM1.5 G filter (power output of 1 sun) (pictures were taken by Adam Slabon).

The *piezocatalytic effect* is based on the conversion of mechanical energy into chemical energy and can thus enhance the reaction rates due to the upsurge of the piezopotential by means of an external force. Piezocatalysis that is performed without illumination is based on the conversion of mechanical energy into chemical energy using low-frequency vibrations. The induced polarization leads to free charge carriers that can drive chemical reactions [6]. Several prospective piezocatalysts are known today, e.g. $BaTiO_3$, ZnO, $NaNbO_3$, $Pb[Zr_xTi_{1-x}]O_3$, $KNbO_3$, $BiFeO_3$, $ZnSnO_3$, $K_{0.5}Na_{0.5}NbO_3$, Bi_4NbO_8X (X = Cl, Br), C_3N_4, MoS_2 and CdS. Fabricating these compounds into anisotropic nanostructures can facilitate the mechanical excitation and consecutively improve the catalytic activity due to high specific surface area. For instance, the photocurrent density of $NaNbO_3$ nanorod photoanodes can be amplified by app. 30% to 1.02 mA/cm^2 under application of ultrasounds [7] . For freestanding materials in solution in the form of dispersed powder, piezocatalytic water-splitting on ZnO and $BaTiO_3$ nanofibers, without the application of a light source, occurs under vibration with ultrasonic waves due to strain-induced electric charge development on the materials' surfaces. Piezocatalysis is not solely constrained to catalytic reaction involving small inorganic molecules but can be also applied using mechanochemistry for organic transformation of small organic molecules [8].

The operating principle of pyrocatalysis is the spontaneous polarization due to a temperature gradient, which yields heightened adsorption and desorption rates of reactants on the catalyst surface. This enables to harvest waste heat from industrial processes and to use the thermal energy for catalytic reactions, such as oxidation of organic pollutants or water-splitting for hydrogen generation. The thin-film $NaNbO_3$ photoanode combines both pyrocatalytic properties and electronic requirements for an n-type semiconductor that is suitable for photoelectrochemical water oxidation. This synergistic effect of thermal excitation and illumination results in an augmented photocurrent density during water oxidation. Besides the prominent example of $BaTiO_3$, which combines ferro-, piezo- and pyrocatalytic properties, several related materials are considered as efficient pyrocatalysts, e.g. ZnO, $LiNbO_3$, $BiFeO_3$, two-dimensional phosphorene and $Pb[Zr_xTi_{1-x}]O_3$.

References

[1] West AR. Solid State Chemistry and its Applications, 2nd edition, Wiley, Weinheim, 2014. Student Edition, ISBN: 978-1-119-94294-8.
[2] Anton SR, Sodano HA. Smart Mater Struct, 2007, 16, R1-21.
[3] Sezer N, Koç M. Nano Energy, 2021, 80, 105567.
[4] Zhang D, Wu H, Bowen CR, Yang Y. Small, 2021, 2103960.
[5] Ryu H, Kim SW. Small, 2021, 17, 1903469.
[6] Li S, Zhao Z, Zhao J, Zhang Z, Li X, Zhang J. ACS Appl Nano Mater, 2020, 3, 1063–79.
[7] Li H, Yu YH, Starr MB, Li ZD, Wang XD. J Phys Chem Lett, 2015, 6, 3410–6.
[8] Kubota K, Pang Y, Miura A, Ito H. Science, 2019, 366, 1500–4.

10 Technical ceramics and hard materials

10.1 Sanitary ceramics, tableware, porcelain

Marcel Engels, Christoph Piribauer

Sanitary ceramics, tableware, porcelain as well as ceramic vitrified tiles, specified as fine ceramics, belong to the group of traditional silicate ceramic products, which also includes coarser ceramic products like bricks, roofing tiles and refractory products (Chapter 10.5). All products have in common that their composition is based on the use of a wide variety of naturally occurring raw clay materials, mainly clays and feldspars. These raw materials exhibit a wide range of characteristics, which influence their preparation as raw materials, their processing and the characteristics of the final product.

Fine ceramics are defined after Hausner [1, 2] as being based on compositions with a component particle size of the microstructure <0.1 mm and a light-colored body. Especially sanitary ware, tableware and porcelain are characterized by a bright body color after firing (in contrary to the often dark gray and red coloration of the coarser building ceramics) with, in the case of porcelain, even translucency. The properties of these often complex-shaped products are, next to their composition, strongly influenced by their firing temperature, which is needed to generate the required product characteristics, which include a combination of a highly aesthetic and impermeable surface with high chemical and mechanical resistance. Further classification of these products is based on the porosity, measured as the water absorption of the ceramic body. For sanitary ceramics, specified as vitreous china, stoneware and bone china tableware as well as vitreous stoneware and soft porcelain, kiln temperatures of 1200 to 1300 °C apply to generate a dense and water-impermeable body with porosity well below 2%. Sanitary ceramics and stoneware tableware have a white-yellowish body with a low water absorption below 0.5 wt%, but no translucency. Porcelain is fired at 1300 to 1450 °C and can be distinguished from stoneware tableware due to its defined translucency at a low porosity below 0.05 wt% [1].

10.1.1 Production of fine ceramics

The general production route for fine ceramic materials can be divided into different steps [4–6] (Figure 10.1.1). Starting with the raw materials' selection by mining, preparation, eventually beneficiation and initial storage in the quarry, the material is transported to the production site, where a ceramic mix is prepared from several different raw materials.

The mix is prepared either by the wet or the plastic route. In the wet route, hard materials like feldspars are ground by wet milling, dispersion of the clay materials

https://doi.org/10.1515/9783110733471-010

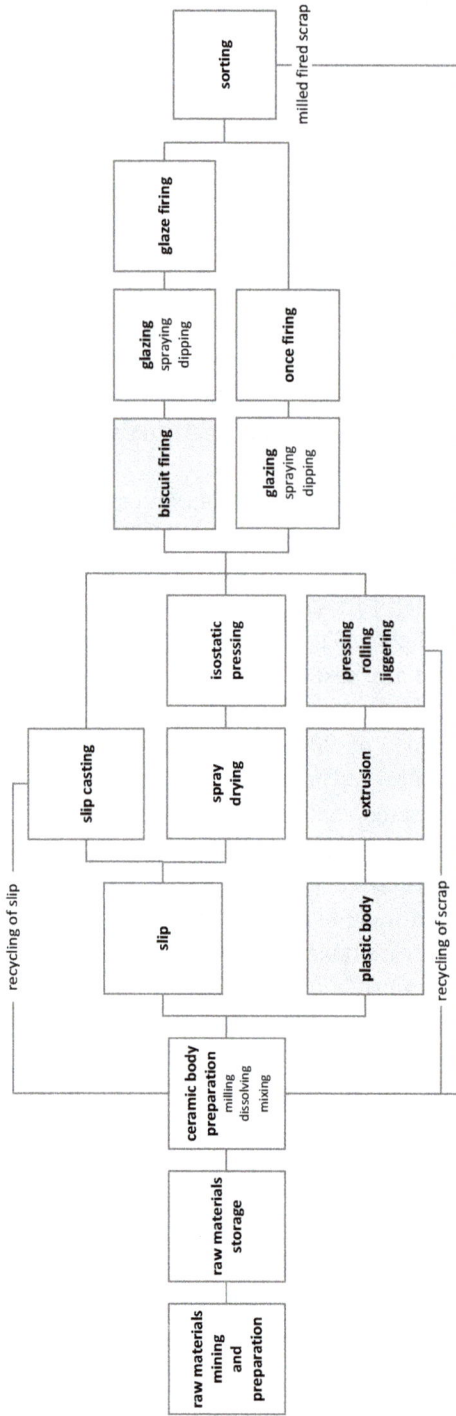

Figure 10.1.1: Schematic representation of the production of fine ceramics.

in water, mixing both up to a solid content of 50–60 wt% and sieving and eventually storing the composition for homogenization. The composition can then be spray dried before pressing or cast as slip under atmospheric or elevated pressure in gypsum or resin molds. The plastic route involves the mixing, aging and homogenizing of the material using up to 10 wt% water to generate a plastic body, which then can be extruded before shaping by pressing, for some products after additional drying, rolling or jiggering in specific molds. Subsequently, the so-called green body product is dried. During this process, the first shrinkage, the so-called drying shrinkage (up to 3–5%), occurs. Depending on the product, the firing temperature of the body can be well above 1200 up to 1400 °C after glazing, If the body is pre-fired before glazing, this so-called biscuit firing can be substantially lower, from 900 to 1100 °C, before the glazed body is fired in the so-called glost firing at 1200–1400 °C. During firing, the materials densify by phase transitions of the raw materials and liquid phase sintering, which strengthens the product, accompanied with a shrinkage between 5% and 10% and a significant reduction of porosity.

Not only due to environmental restrictions but also to economic reasons, the recycling of in-process waste materials has become increasing importance. Where for the high-quality porcelain this has its limits, for sanitary ware and stoneware tableware higher recycling rates are possible. The addition of in-process or external waste material into the ceramic process influences the properties of the ceramic mix. A general recycling of the ceramic slip is a common feature during the slip casting of hollow products like sanitary ware, where slip is removed after the required cake thickness is reached. This requires a significant storage volume for the slip before shaping and implies long-term stability of the material. Unfired material is re-introduced in the body preparation step, and fired material scrap must be processed further before re-use (e.g. by grinding). The latter is used as a filler to control the shrinkage and increase the strength of the material during firing, but the possible amount is limited due to the risk of contamination.

The selection of clay materials in these different production routes and for different products is mainly governed by their chemical and mineralogical composition [7], which exerts a strong influence on their processing characteristics [8, 9]. The basis for the use of these materials is their unique interaction with water, giving them a specific intrinsic shaping characteristic, influencing colloidal behavior and viscosity, plasticity, drying behavior, dry bending strength and shrinkage. They also define the phase transformations and liquid phase sintering in the clay-based body composition in the temperature range of 1000–1400 °C [2, 3] to enhance the creation of resilient and durable products, achieving the required mechanical, chemical and thermal resistance for sanitary, tableware and porcelain applications, the latter ranging from tableware to electrical porcelain insulators. To understand these product characteristics, the clay mineralogy and its influences on processing and product characteristics must be addressed.

10.1.2 Clay mineralogy

Significant for the processing of clay raw materials is their mineralogical structure as phyllosilicates with a layered structure, built from SiO_4-tetrahedra and Al/Mg/Fe(O, OH)$_6$-octahedron layers, bound by bridging oxygen ions, with negatively charged surfaces (Figure 10.1.2). The degree of substitution of the silicon and aluminum atoms in this structure by potassium, sodium, calcium, magnesium, titanium, manganese or iron, as well as the replacement of these cations with dissolved cations, specified by the so-called cation exchange capacity as a measurable characteristic defines their colloidal properties, their interaction with water as a main processing medium [10]. For minerals with more than two layers, the bonding of water between these layers and on the surface leads to the so-called intracrystalline swelling up to a manyfold of the original size, which significantly influences their viscosity during the preparation as a slip for casting or spay drying or their plasticity for extrusion, which are the main production routes for silicate ceramic products up to the current day.

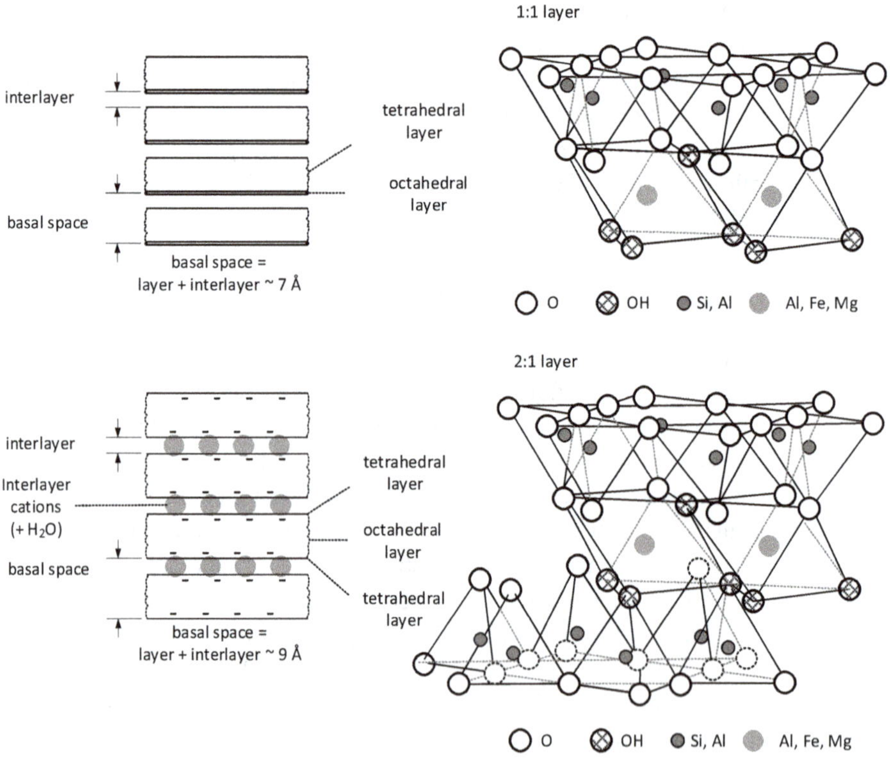

Figure 10.1.2: The layered structure of phyllosilicate materials (after [2, 7]).

Table 10.1.1 lists the main groups of phyllosilicate clay minerals, which are used in the production of sanitary ware, tableware and porcelain [2, 7–9]. A large variation in the layering of the silicates, their particle sizes as well as their degree of substitution by cations in the structure is available, due to their origin and their genesis by sedimentation, weathering, hydrothermal alteration, diagenesis by interactions with water, microbial activity, and compaction after deposition, and metamorphism, the change of minerals due to heat and pressure and the interaction with naturally occurring chemically active fluids. This implies that to characterize the clay materials for raw materials or process control, a more complex analytical approach must be taken [11–13]. Here, mineralogical characterization by X-ray diffraction, which is complex and under continuous development, electron microscopy, thermogravimetry and differential thermal analysis and particle size measurement techniques are applied.

Table 10.1.1: Clay mineral groups.

Mineral group	Lattice	Representatives	Typical formula	Interlayer cation/ swelling capacity	Specific surface BET in m^2/g	Cation exchange capacity in meq/100 g
Kaolinite	Two-layer*	Kaolinite	$Al_2Si_2O_5(OH)_4$	Non-swelling	1–10	2–10
		Fire clay	Various	Non-swelling	10–30	10–15
		Halloysite/ dickite/ nacrite	$Al_2Si_2O_5(OH)_4$	Non-swelling	5–30	10–50
Clay minerals resembling mica	Three-layer*	Muscovite	$KAl_2(Si_3Al)O_{10}$ $(F,OH)_2$	K^+, Na^+ non-hydrated, non-swelling	<3	<5
		Illite	$KAl_2(Si_3Al)O_{10}$ $(OH)_2$	K^+, H_3O^+ (Na^+, Ca^{2+}) non-hydrated, non-swelling	50–90	20–50
Smectite*	Three-layer*	Ca-/Na- montmorillonite/ bentonite	$(Ca,Na)_{0.35-0.7}$ $(Al,Mg,Fe)_2$ $(Si,Al)_4$ $O_{10}(OH)_2 \cdot$ nH_2O	Na +, Ca^{2+}, Mg^{2+}, Li^+ (K^+, Al^{3+}, $Fe^{2+/3+}$, H_3O^+) hydrated, swelling	70–120	60–150
		Beidellite/ nontronite / saponite/ hectorite	Different possibilities		30–70	

Table 10.1.1 (continued)

Mineral group	Lattice	Representatives	Typical formula	Interlayer cation/ swelling capacity	Specific surface BET in m^2/g	Cation exchange capacity in meq/100 g
Vermiculite[*]	Three-layer[*]	Vermiculite	$(Mg,Fe,Al)_3$ $(SiAlO_5)_2 \cdot 4H_2O$	Ca^{2+}, Mg^{2+} hydrated, swelling	50–80	110–180
Chlorite	Four-layer[*]	Chlorite	$(Mg,Fe,Al)_6(Si,$ $Al)_4\,O_{10}(OH)_8$	Non-swelling	5–20	10–40
Mixed layer	Regular/ irregular	Illite-smectite mixed layer	Various	Swelling	Various	Various
		Chlorite and smectite	Various	Swelling	Various	

[*]"Two-layer" consists of one octahedron + one tetrahedron layer, "three-layer" consists of one octahedron + two tetrahedron layers, "four-layer" consists of two octahedra + two tetrahedron layers [2, 3, 8, 9].

A typical aspect of the clay mineral particles is that they have specific shapes, from platelike particles with different shaped edges, from pseudohexagonal platelets with round well-defined edges for kaolinite, arranged in stacks or bulk shapes of various sizes, to flaky and fibrous illite shapes, ragged edge smectite and fibrous halloysite (Figure 10.1.3).

Generally, the fraction of the clay raw material with a particle size below 2 µm (based upon sedigraphic measurements) is regarded as the mineral component significant for its characteristics. The higher grain sizes consist of accompanying mineral components and quartz particles [12]. For a relevant evaluation of the particle size distribution, also the specific morphology of the clay particles must be considered. The historically available sedimentation techniques, based on the assumption of gravitational sedimentation of spherical particles in water, underestimate the amount of small platelike particles. The laser scattering techniques based upon the determination of particle perimeters in random orientation in the measurement solution, consequently, give different results. As a rule of thumb, the fraction below 2 µm using sedigraphy is comparable to the fraction <8 µm using laser scattering techniques [14]. The state-of-the-art possibilities of modern electron microscopy [15] help to evaluate the actual particle size and morphology (Figure 10.1.3).

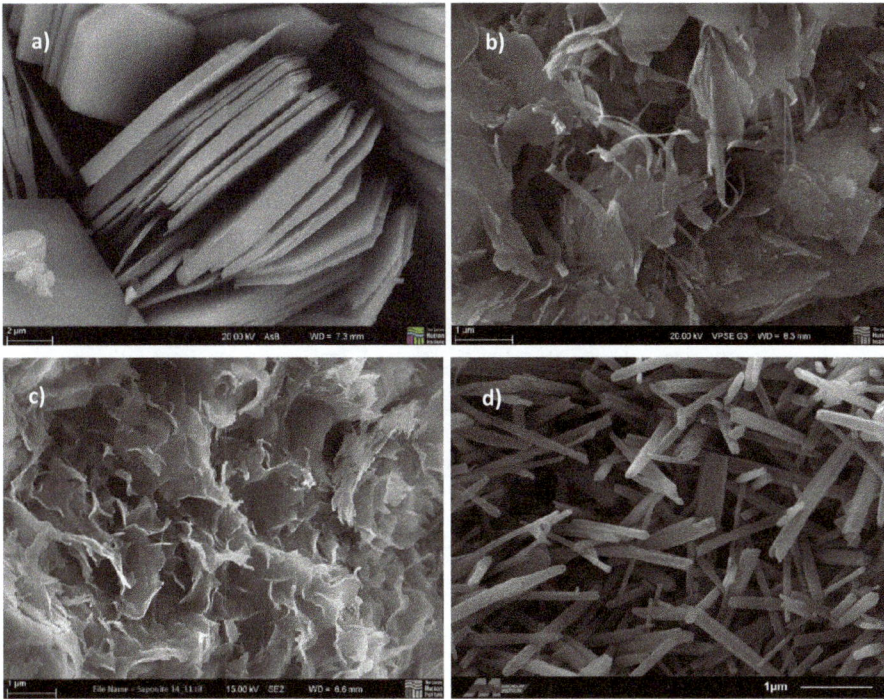

Figure 10.1.3: (a) kaolinite, (b) illite, (c) smectite, (d) tubular halloysite (images reproduced with courtesy of the Clay Minerals Society and the Clay Minerals Group of the Mineralogical Society; all images of Clay Gallery are available at https://www.minersoc.org/images-of-clay.html) [13].

10.1.3 Clay raw materials

A clear distinction must be made between clays and clay minerals: clays used in the production of whitewares, tableware and porcelain are not the singular minerals but naturally occurring mixtures (Table 10.1.2) [16]. A current definition of clay is that it is a naturally occurring material, composed primarily of fine-grained minerals, which behaves plastically at appropriate water contents and hardens when dried or fired. Clays can be used in the ceramic industry either in bulk form without any beneficiation or after processing, which can be done by physical or chemical treatment like wet milling and magnetic separation by sieving, flocculation, flotation and treatment with inorganic or organic compounds, thus removing impurities and increasing the actual clay mineral content.

The mostly used kaolin and ball clays are kaolinitic sedimentary clays that consist in a wide variety of compositions in the ranges of 20–80 wt% of kaolinite, 10–25 wt% mica and 6–65 wt% quartz. They contain lower amounts of accessory minerals and carbonaceous materials such as lignite. Due to the low amount of iron and organic components, they generate white-colored products when fired. Kaolin or China clay consists mainly of kaolinite with almost no iron content. Due to their regular particle packing, these clays exhibit a controlled rheology, stable and high casting rates, good mechanical properties of the dry shaped product. Their high fusion temperatures (above 1400 °C) assure high mechanical resistance and stability during and after firing. A special kaolinite component is halloysite, with a distinctive hollow tubular structure (Figure 10.1.3), which is especially suited for the manufacturing of products with a high degree of whiteness and translucency, like bone and fine china, thin porcelain and hotel tableware. Ball clays contain a more varied mixture of clay minerals, generating the required processing properties like plasticity and rheological stability in interaction with water during the shaping process [17].

Table 10.1.2: Examples of clay raw materials (adapted from [16]).

Composition	Sanitary kaolin	Porcelain kaolin	Sanitary ball clays	Stoneware tile clay	Porcelain clay
Chemical composition (wt%)					
SiO_2	45–55	45–55	45–55	56–70	55–60
Al_2O_3	30–37	33–37	30–35	20–28	25–30
Fe_2O_3	<1.0	<0.9	0.5–2.0	<1.0	<1.5
TiO_2	0.2–0.8	0.2–0.8	<1.3	<1.0	<1.6
CaO	<0.8	<0.8	0.2–0.5	0.2–0.7	0.2–0.5
MgO	<1.3	<1.3	<0.5	0.4–0.8	0.2–0.5
K_2O	<2.0	<1.6	0.7–2.5	<3.0	<2.5
Na_2O	<0.1	<0.1	0.2–0.6	0.2–0.6	0.2–0.4
l.o.i	<14	<14	8–15	7–10	8–10
Mineralogical composition (wt%)					
Kaolinite	>70	80–88		7–70	
Illite/Muscovite	ca. 20	ca. 10		11–32	
Montmorillonite		<1.0			
Mixed layer	0–10			0–10	
Feldspar	8	4–11		0–10	
Quartz	7–20	ca. 1.5		6–61	
Cristobalite					
Fraction <2 μm (wt%)	>40	>70		22–96	

Other significant additions to the ceramic body compositions are the feldspars [2, 18], tetrahedral aluminum tectosilicate minerals with a structural replacement of Si^{4+} by Al^{3+} ions, shifting the Al/Si ratio and compensation of the valence differences by Na^+, K^+, Ca^{2+} or Ba^{2+} ions. Due to the mixed compositions, feldspars exhibit a good fluxing behavior during firing: they support the formation of melting places at a lower temperature than the clay material and enhance the densification of the material during firing. The most common members of the feldspar group are the potassium-based orthoclase types, $K[AlSi_3O_8]$, the sodium-based albites $Na[AlSi_3O_8]$ and the calcium-based anorthites $Ca[Al_2Si_3O_8]$, as well as mixed sodium-calcium feldspars (plagioclase) and potassium-sodium (alkali) feldspars. Potassium feldspar is characterized by incongruent melting at 1150 °C with a high melt viscosity and a long melting range, whereas sodium feldspar melts congruently at 1118 °C, resulting in a less viscous melt with a shorter melting range. Main suppliers for the European industry are located in Germany, Poland, the Czech Republic and Ukraine. An often-used representative of the alkali feldspar group is the Scandinavian or South African granite-like nepheline syenite feldspar, with no quartz and a low iron content, so making a good fluxing agent for whiteware sanitary ceramics with low porosity and for tableware translucency. Pegmatite [2, 3], a rock component, consisting of a mix of mostly potassium feldspar, quartz and kaolin with low concentrations of coloring oxides like titanium and iron is used for white tableware. Other raw materials are quartz [2, 3, 19], introduced as fine-grained sand, which supports the strength of the material as a filler material and as a temperature-stable component, with a high melting viscosity, thus reducing the weakening and so-called pyroclastic deformation of the product during firing. Wollastonite [2, 3], $CaSiO_3$, is used to contribute CaO as a necessary reaction component during firing (see Section 10.1.5) with a low loss of ignition below 2% and low iron content (below 0.5%) as an alternative to calcium carbonates with high CO_2 degasification. Due to its cost, it is mainly used in sanitary ware to improve the mechanical properties of the green body and as addition to engobes and glazes to reduce cracking and crazing defects, improve surface durability and surface appearance for these products. Main suppliers are located in China and Finland.

10.1.4 Processing characteristic of clay minerals

To assess the processing and product characteristics of the fine ceramic products, the primary characteristics of the silicate-based clays as a primary component, the mineralogical, chemical, colloidal and particle characteristics, must be differentiated (Figure 10.1.4). The secondary characteristics, defining the processing of the raw materials and basis for the raw material control [8, 9], indicate the complexity of measurements to be performed to assure a relevant process control, exceeding particle size measurement and chemical analysis. Although the Al_2O_3/SiO_2 ratio, the iron content, the amount of soluble salts and loss of ignition, indicating the organic content, and particle size are used for a first indication, the mineralogical composition and its specifics, like the actual amount of swellable materials and the cation exchange capacity, largely influence the actual processing conditions [8, 9, 12, 13, 20]. In this case, the use of functional tests like rheology, plasticity measurements and drying and firing tests, addressing the actual processing conditions, must be considered. Especially in the ceramic body composition, where several clay raw materials, feldspars and other components are mixed, evaluating these properties requires a multiscale insight in the material characteristics, especially as there are time-dependent factors like the biological reactions of the organic matter in the clays [21] or the delayed release of soluble salts, especially important in the case of slip storage and recycling of excess casting slip.

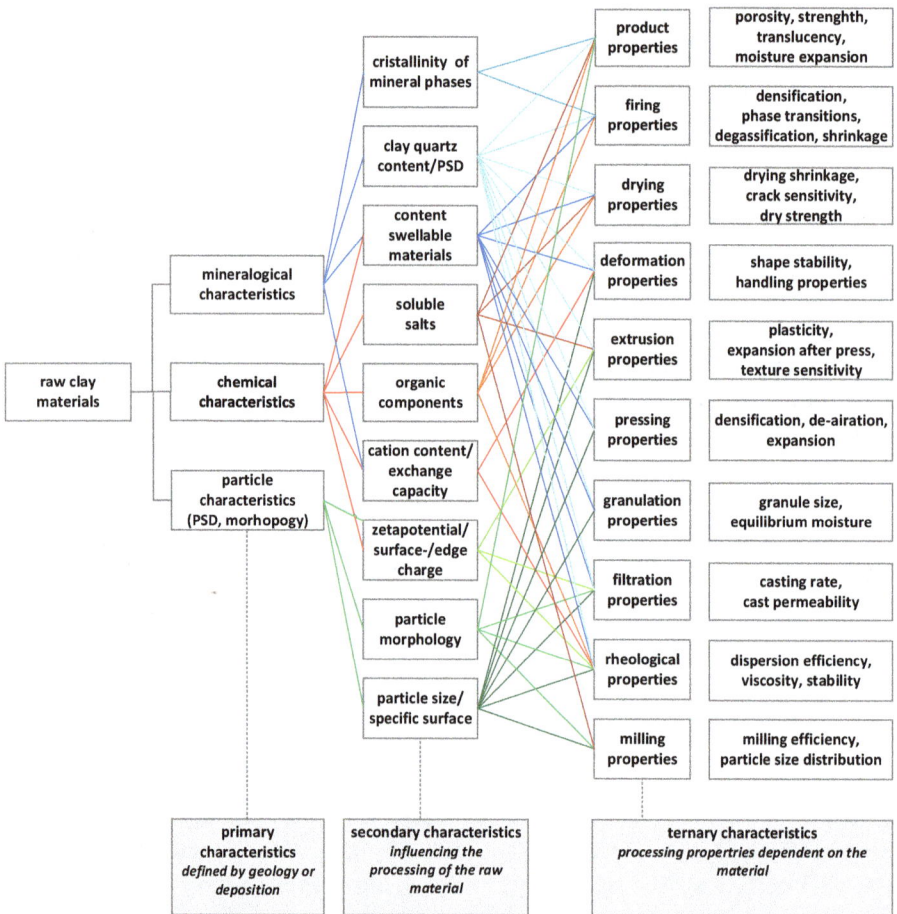

Figure 10.1.4: Raw clay material characteristics and their processing and product influences material quality.

Important characteristics, which are significant for fine ceramics production, are influenced by interactions of the body composition with water and ionic components therein, from additives or contaminations to the pH value. They define the plastic behavior and flow performance of a clay: in two-layer minerals, Ca^{2+} ions can, due to the double charge, generate side-to-side and side-to-surface bonding of the clay particles; a "house of cards" structure is possible, immobilizing water and supporting plastic behavior (leading to yield stresses in dispersions; Figure 10.1.5). Na^+ ions shield the effective negative charging of the clay particle, leading to a compact layer-to-layer contact: this results in dense structures, supporting liquid behavior (more compact cake in slip filtration).

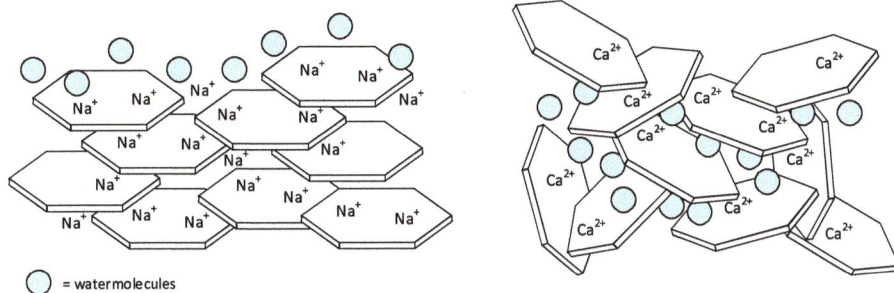

= water molecules

Figure 10.1.5: Schematic representation of the interaction of Ca^{2+} and Na^+ ions in aqueous kaolinite dispersions (adapted from [2]).

In three-layer minerals (like montmorillonite), Ca^{2+} ions can be included, supporting increased plasticity, as the double charge enhances water intake between the particle layers, causing intercrystalline swelling. In contrast to kaolinite and illite, an occupation by Na^+ ions results in a more plastic body due to the intercrystalline swelling processes. These effects also influence the equilibrium amounts of water a clay material surface can incorporate or interact with. Above and below these amounts, the sensitivity of the material to the moisture or water content can change significantly. Increasing the solid content can significantly alter viscosity or plasticity. For kaolin suspensions, a solid content above 20 wt%, viscosity builds up in time [13], the so-called thixotropy effect, indicating clay-water interaction; above 30 wt% the yield stress, the minimal stress needed to bring the solution in movement, increases strongly due to the clay-particle interaction; and above 65 wt% the viscosity strongly increases, indicating that the clay-particle interaction is no longer counteracted by water, leading to high interparticle friction in movement.

Special attention must be given to soluble salts [2, 17, 22]: these are the source for soluble ions, which in an aqueous clay solution are in equilibrium with the solid content as elementary components of the clay- and non-clay minerals, like Na^+, K^+, Mg^{2+}, Ca^{2+}, Fe^{2+}, Cl^- or SO_4^{2-} and complexing ions like Fe^{3+}, Al^{3+}, SiO_4^{2-} or PO_4^{3-}, which are the basis for oxo or hydroxo complexes or oligomers in short chains, fragments or clusters. The delayed release of these ions from natural impurities of processing aids from the beneficiation of the clay can cause instable solutions with over time rapidly decreasing or increasing in viscosity and slip stability.

10.1.5 Drying and firing characteristics of fine ceramics

The drying and firing of fine ceramics is a complex process, in which several stages and transitions occur which in effect define the product characteristics and the product quality [2, 3, 22, 23, 24]. These are identified using thermal analysis like dilatometric measurements to identify the shrinkage, water absorption measurements to estimate the densification and thermogravimetry and differential thermal analysis to evaluate the phase transitions and relevant degasifications during firing. For upscaling of these results, the actual heating rate must be considered, as higher heating rates might lead to residual phases of not completed phase transitions, as well as compromises between densification of the product with rapidly decreasing porosity and degasification or necessary oxidation.

The significant stages in drying and firing of fine ceramic bodies are according to Salmang et al. [2] and McColm [3]: from approx. 40 °C, free pore water evaporates from the porous, wet-shaped product, causing shrinkage of the product. Up to 100 °C, physically adsorbed water evaporates from the pores. In these stages, the ceramic structure is weakened by evaporation. Especially in the case of hydration, swelling of three-layered minerals like montmorillonite and mixed layer minerals leads to a high crack sensitivity during drying. From approx. 240 °C, organic components are oxidized to CO_2 and CO, which escape through the open porosity in the product. The reaction of iron with CO can lead to a gray/black colorization of the body, especially in the core. This decomposition can take place up to 800 °C, dependent on the porosity and the type of organic component. For whitewares, therefore, low iron and organic contents are preferred. At these temperatures, the interlayer water of the clay minerals is desorbed.

From 500 °C, for trilayered montmorillonites, the chemically bonded crystalline water is separated, leading to their structural collapse. From approx. 550 °C, kaolinite $Al_2O_3 \cdot 2SiO_2 \cdot 2H_2O$ reacts with metakaolinite $Al_2O_3 \cdot 2SiO_2$ with separation of chemically bonded water, leaving an amorphous residue of silica. This reaction is endothermal. At 573 °C, quartz transitions from the α- to β-modification, combined with volume expansion, mostly accommodated by the plasticity of the green body structure. At approximately 700 °C, the dissociation of calcite ($CaCO_3$, if present) occurs up to 895 °C, generating CaO and CO_2. From 900 °C, first, amounts of high-viscosity eutectic melting phases from sodium/sodium-potassium feldspar are formed, causing shrinkage and a reduction of the porosity of the product. This melt causes the dissolution of quartz and Fe_2O_3. From 950 °C, the metakaolinite transgresses into a metastable spinel phase in the amorphous silica matrix. In the presence of CaO, the spinel reacts with gehlenite ($Ca_2Al(AlSiO_7)$). The viscosity of the melting phase decreases, thus increasing the diffusion of quartz and supporting the generation of primary mullite. The firing shrinkage reaches its maximum, and the porosity is strongly decreased and closed.

From approx. 1000 °C, primary mullite ($3Al_2O_3 \cdot 2SiO_2$) is formed from the melting phase in an exothermal reaction. The residual kaolinite transforms to the spinel phase. From the present gehlenite phase, anorthite phase is formed subsequently, strengthening the product. At these temperatures, the amount of eutectic melting phases starts decreasing the product stability (sagging, pyroplastic deformation) which is countered by the dissolution of quartz in the melt, increasing its viscosity. Due to the long firing times, partial conversion of quartz into cristobalite can take place.

From approx. 1100 °C, the transition of the spinel into needle-like (secondary) mullite from the melt phase into the amorphous silica matrix further strengthens the product. From approx. 1200 °C, primary mullite and quartz dissolve in the melt, causing higher viscosity (supported by higher quartz and sodium-feldspar amount). The secondary mullite as well as quartz support a decrease in pyroplasticity. For porcelain, firing up to 1400 °C enhances the vitrification to increase its mechanical stability, and translucency based upon its glassy character.

The following cooling freezes the formed composition of phases. After a more rapid cooling rate, care must be taken at 573 °C, where the quartz transition is reversed from the β- to α-modification (combined with volume shrinkage). This can cause cracking if transgressed too quickly.

To optimize the surface characteristics, glaze is applied, which must also be fired. The glaze is a fired thin silicate-based glassy covering of the product, which can be transparent or colorized to smoothen and close the surface, as well as to improve its functionality regarding resistance. It is intended to interlock and react with the underlying body. For some products, the materials are first fired (biscuit firing) after which the glaze is applied and fired (glost firing). The significance for the aspect of mechanical resistance and durability is the so-called glaze-fit: the thermal expansion of the glaze must be matched to that of the body to generate compressive stresses, to prevent cracking and crazing. Especially in the case of glaze on biscuit bodies, due to a diminished reactivity of the body with the glaze, the "glaze-fit", the interlocking and differences in thermal expansion coefficients have to be well defined.

10.1.6 Sanitary ware

To generate the required quality of products, vitreous china sanitary ware bodies are composed of 25–35 wt% kaolinite clays, 20–28 wt% ball clays, 15–23 wt% feldspar or feldspathoid fluxes like nepheline-syenite, 17–31 wt% quartz and a small amount (3–9 wt%) of recycled fired scrap. The amount of ball clay in these compositions improves the whiteness of the bodies and its resistance to deformation at high temperatures, whereas the kaolin components provide the necessary processing properties. The feldspars used are mostly a combination of the sodium and potassium types accompanied by secondary materials such as anorthite and quartz. The potassium feldspar is preferred due to its high melting viscosity providing high product stability during firing and its long melting range, accommodating any temperature gradients in the industrial kiln (Figure 10.1.6). To produce sanitary ware, the recycling of milled fired scraps is permissible without interfering with the required product quality, as it increases the casting rate in the slip casting and reduces the risk associated with the volume change of the β- to α-quartz transformation during the cooling phase, which can cause cracking.

Figure 10.1.6: The firing of sanitary ware in a roller kiln with defined temperature zones, seen from the exit of the kiln (photo by FGK).

As substitution for the kaolin and ball clays, fire clay sanitary ware bodies contain about 40 wt% fired and ground refractory clays, and a lower percentage of quartz (10–16 wt%). This material is used for less complex products which require high firing stability. The products usually have higher porosity.

For sanitary ware and fireclay glazes, raw or frit-based (in which the components are pre-melted and subsequently quenched and milled) glazes are applied, mainly consisting of quartz, feldspar or feldspathoid, kaolin, alkaline earth metal carbonate and zinc oxide. Zirconium silicate is added as an opacifying agent. Due to the reduced availability of the latter, alternative raw materials are being investigated in the ongoing research.

10.1.7 Porcelain and tableware

Porcelain can be distinguished from stoneware tableware due to its high degree of glass phase during the firing fired at 1300–1450 °C, in which mullite and quartz is embedded, which lead to a defined translucency at a low porosity below 0.05%.

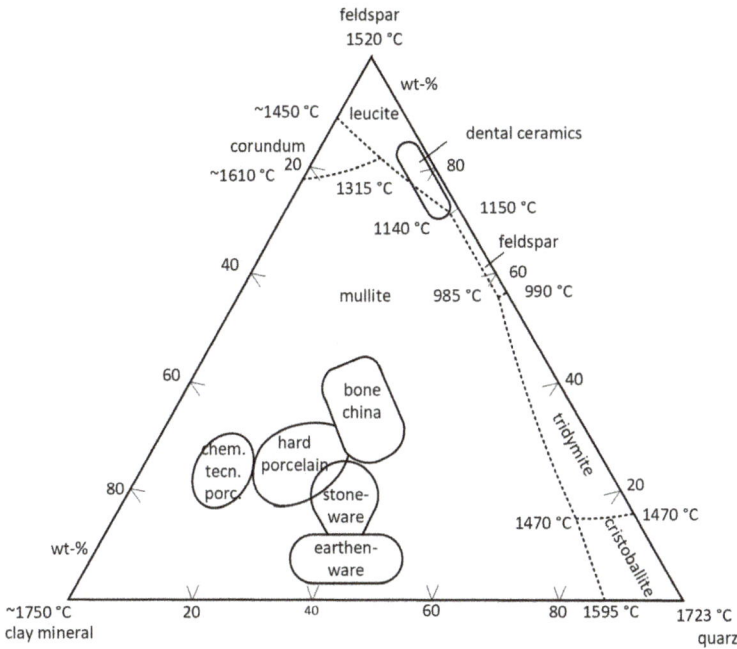

Figure 10.1.7: Different porcelain types according to their composition based upon clay minerals, feldspar and quartz (adapted from [2]).

This criterion is, for instance, one of the product criteria used in customs to control its adequate labeling. Where porcelain is based on simultaneous firing of the body and a feldspathic glaze, stoneware, tableware and the so-called vitreous china tableware is fired in two steps, where after the so-called biscuit firing above 1200 °C the glaze or glost firing is separately performed at a lower temperature (1000–1200 °C).

In Figure 10.1.7, the different porcelain types are specified according to their composition based upon clay minerals, feldspar and quartz. For tableware, the most traditional material is bone china, in which the clay component, mostly kaolin, is enriched with calcium-phosphate containing bone meal, generating a lighter shard with a defined translucency of the material. Although used for its aesthetic properties, the hardness of the material is lower than that of the feldspathic and hard porcelain, and therefore, in most cases limiting its use in restaurants and hotels. It is not to be confused with the traditional earthenware ceramics like Dutch porcelain with the so-called delfts blue underglaze decoration.

Hard porcelain has a high kaolin content (>50 wt%), quartz and feldspar content, and is covered by a sodium-based feldspathic glaze, which is simultaneously fired with the green body to generate mullite to harden the surface, which is required for gastronomical applications. The quartz supports the hardness and the impact resistance of the material, so the requirements for gastronomical use like chemical resistance, dishwasher safety and the resistance to cutlery abrasion are met. For this purpose, also alumina and zircon are added to the composition to increase the mechanical resistance.

For tableware and porcelain products intended for contact with food, especially in the case of colored glazes or decorations, special requirements apply regarding the leaching of heavy metal components, like lead and cadmium. This requires severe testing of the products by leaching emission tests after established standard tests before it is allowed on the market.

References

[1]   Hennicke HW. Zum Begriff Keramik und zur Einteilung keramischer Werkstoffe. Ber Dtsch Keram Ges, 1967, 44, 209–11.

[2]   Salmang H, Scholze H, Telle R (Ed). Keramik 7, Vollständig neubearbeitete und erweiterte Auflage, Springer-Verlag, Berlin, Heidelberg, 2007.

[3]   McColm IJ. Ceramic Science for Materials Technologists, Leonard Hill, Glasgow, 1983. ISBN 0249441632.

[4]   Kingery WD. Introduction to Ceramics, Wiley, New York, 1960. DLC 60053448.

[5]   Terpstra RA, Pex PPAC, de Vries AH. Ceramic Processing, Chapman & Hall, London, 1992.

[6]   Kingery WD. Ceramic Fabrication Processes, John Wiley & Sons Inc, New York, 1975.

[7]   Jasmund K, Lagaly G. Tonminerale und Tone: Struktur, Eigenschaften, Anwendungen und Einsatz in Industrie und Umwelt, Steinkopff, Darmstadt, Germany, 1993, 490.

[8]   Vogt S, Vogt R. Relationship between minerals and the industrial manufacturing properties of natural clay deposits and the clay bodies produced from them for the heavy clay industry, Part 1: Basic principles of clay mineralogy and primary data on clay minerals, ZI Jahrbuch 2003, Bauverlag, Gütersloh, 2003, 114–26.

[9]   Vogt S, Vogt R. Relationship between minerals and the industrial manufacturing properties of natural clay deposits and the clay bodies produced from them for the heavy clay industry Part 2: Manufacturing properties, utilization estimates and clay body optimization, ZI Jahrbuch 2004, Bauverlag, Gütersloh, 2004, 73–103.

[10] Meier LP, Kahr G. Determination of the cation exchange capacity (CEC) of clay minerals using the complexes of copper (II) ion with triethylenetetramine and tetraethylenepentamine. Clays Clay Miner, 1999, 47, 386–8.

[11] Emmerich K, Weidler P, Steudel A, Engels M, Schuhmann R. Charakterisierung von Mineraloberflächen in keramischen Tonen, Ceramic forum international (Berichte der dt. Keram Ges, 2012, 89, D1–D6.

[12] Diedel R, Latief O. Anforderungen an die Rohstoffanalytik vor dem Hintergrund moderner Prozesstechnologien, Ceramic forum international (Berichte der dt. Keram Ges, 2011, 88, D21–D26.

[13] Christidis GE. Industrial Clays, advances in the characterization of industrial minerals Chapter 9, European Mineralogical Union notes in mineralogy. Mineral Soc Great Britain Northern Ireland, 2011, 9, 341–414.

[14] Konert M, Vandenberge J. Comparison of laser grain size analysis with pipette and sieve analysis: A solution for the underestimation of the clay fraction. Sedimentology, 1997, 44, 523–35.

[15]   Becker C, Dohrmann R. Environmental scanning electron microscopy (ESEM) – a new method in clay science. In: Rammlmair D, Mederer J, Oberthür T, Heimann RB, Pentinghaus H (Eds). Applied Mineralogy, Balkema, Rotterdam, 2000.

[16]   Lorenz W, Gwosdz W. Bewertungskriterien Für Industrieminerale, Steine und Erden, Teil 1: Tone, Geol Jahrbuch, Reihe H, Heft 2, Schweizerbart Science Publishers, Hannover, 1997.

[17]   Lagaly G. Colloid clay science, Chapter 5. In: Bergaya Theng BKG, Lagaly G (Eds). Handbook of Clay Science, Elsevier, Amsterdam, 2006, 141–245.

[18]   Lorenz W, Gwosdz W. Bewertungskriterien für Industrieminerale, Steine Und Erden, Teil 7: Feldspäte und Andere Flussmittel, Geol JahrbucH, Reihe H, Heft 10, Schweizerbart Science Publishers, Hannover, 2003.

[19]   Lorenz W, Gwosdz W. Bewertungskriterien für Industrieminerale, Steine und Erden, Teil 3: Quarzrohstoffe, Geol JahrbucH, Reihe H, Heft 6, Schweizerbart Science Publishers, Hannover, 1999.

[20]   Petrick K, Diedel R, Peuker M, Dieterle M, Kuch P, Kaden R, Krolla-Sidenstein P, Schuhmann R, Emmerich K. Character and amount of i-s mixed-mixed layer minerals and physical/chemical parameters of two ceramic clays from Westerwald, Germany: Implications for processing properties. Clays Clay Miner, 2011, 59, 58–74.

[21]   Emmerich K, Schuhmann R, Petrick K, Menger-Krug E, Kaden R, Obst U, Dieterle M, Kuch P, Diedel R, Peuker M, Huber S, Fischer H, Beyer D, Zehnsdorf A, Krolla-Sidenstein P. Umfassende Materialcharakterisierung von tonmineralischen Rohstoffen zur Entwicklung mikrobiologischer Aufbereitungstechnologien. CFI-Ceram Forum Int, 2009, 86, 29–34.

[22]   Penner D, Lagaly G. Influence of organic and inorganic salts on the aggregation of montmorillonite dispersions. Clays Clay Miner, 2000, 48, 246–55.

[23]   Schulle W, Schmidt G. Single-firing of porcelain: Prospects and problems. Ceram Forum Int, 2000, 77, 31–5.

[24]   Zanelli C, Raimondo M, Dondi M, Guarini G, Tenorio Cavalcante PM. Sintering mechanisms of porcelain stoneware tiles, Proceedings of the Qualicer 2004: VIII world congress on ceramic tile quality, official chamber of commerce. Ind Navig, 2004, GI247–59.

10.2 Separation membranes

Ingolf Voigt

10.2.1 Introduction

Filtration membranes are thin, semipermeable layers that are used for the selective separation of substances. The membrane technology follows the filtration (particle filtration) with pore sizes <1 µm. In the pore size range of 1 µm to 50 nm, one speaks of macropores. The associated area of membrane filtration is microfiltration, which is used for sterile filtration and to separate turbid substances. This is followed by the area of the mesopores up to a pore size of 2 nm, which is called ultrafiltration in membrane technology. Pores smaller than 2 nm are called micropores. Water can flow through pores down to a pore size of about 1 nm. This membrane area is called nanofiltration. Even smaller pores can no longer be penetrated. Transport can only take place through diffusion. Such membranes are used in reverse osmosis, pervaporation and gas separation.

Inorganic membranes are distinguished from polymer membranes by their high chemical resistance, mechanical strength and thermal stability. They are preferably used in corrosive media and at higher temperatures. In the interests of high flow, the membranes must be very thin. Depending on the membrane material and manufacturing process, membrane thicknesses of 10 µm to 10 nm are used, which cannot be handled as freestanding membrane. The use of coarse-pored ceramic supports enables these membranes to be used at high transmembrane pressures of more than 30 bar without the pores being compressed. Fine-pored, very thin membrane layers are not deposited directly on the support but on intermediate layers that gradually adapt the pore size to the membrane. This creates a gradient in the pore size, which in membrane technology is called an asymmetrical membrane (Figure 10.2.1).

Membranes are preferably operated in cross-flow filtration. This means that the starting mixture (feed) flows over the membrane tangentially at a speed of up to 5 m/s. This suppresses the deposition of particles and the formation of the top layer and ensures a high flow through the membrane. Components that are smaller than the pore diameter can permeate through the membrane and form what is known as permeate. At the same time, the more permeable components accumulate and form the retentate.

https://doi.org/10.1515/9783110733471-011

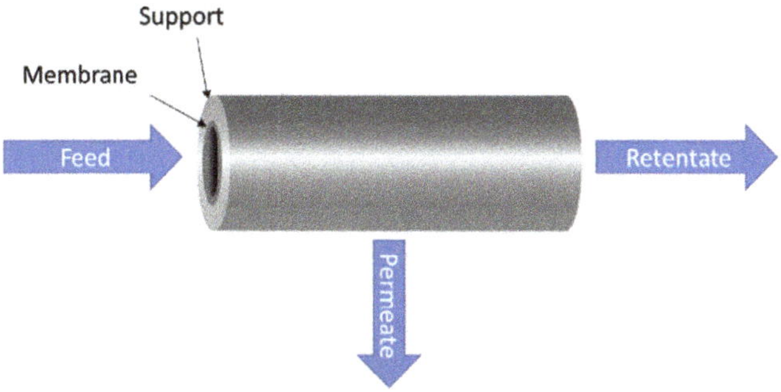

Figure 10.2.1: Scheme of cross-flow filtration using an asymmetrical, tubular membrane.

10.2.2 Inorganic membrane materials

10.2.2.1 Oxidic membrane materials

Regarding their chemical stability, the oxidic materials Al_2O_3, TiO_2 and ZrO_2 are mainly used as membrane materials. The purity of the oxides and the selection of the thermodynamically stable phase play an important role.

In the case of aluminum oxide, fused corundum is used, which is melted in an electric arc and crystallizes below 2050 °C to form α-Al_2O_3. Powder fractions with grain sizes between 15 µm and 500 nm are produced by grinding and sieving. These powders can be used to produce coarse-pored supports and macroporous microfiltration membranes. The finest corundum powders are produced using high-energy grinding and have particle sizes of around 200 nm. A bottom-up synthesis of fine Al_2O_3 powders through hydroxide precipitation and calcination leads to nanodisperse transition alumina, which are referred to as γ-Al_2O_3. Mesoporous membranes can be produced from this, but they have a limited chemical stability in the range of pH 1–10. The conversion to α-Al_2O_3 takes place at 1200 °C, which is associated with strong particle growth and the appearance of macropores:

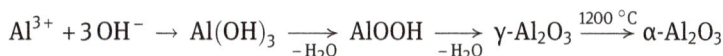

$$Al^{3+} + 3\,OH^- \rightarrow Al(OH)_3 \xrightarrow[-H_2O]{} AlOOH \xrightarrow[-H_2O]{} \gamma\text{-}Al_2O_3 \xrightarrow{1200\ °C} \alpha\text{-}Al_2O_3$$

There are two chemically stable modifications of titanium oxide that are used for membranes: rutile and anatase. The extraction from titanium iron ore (ilmenite, $FeTiO_3$) takes place through a sulfuric acid digestion, precipitation of the titanium oxide hydrate and calcination. First, anatase is formed, which converts to rutile at temperatures above 800 °C:

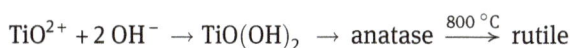

$$TiO^{2+} + 2\,OH^- \rightarrow TiO(OH)_2 \rightarrow anatase \xrightarrow{800\ °C} rutile$$

With the help of grinding and sieving processes, grain fractions in a range similar to that of corundum are obtained. Very pure, nanodisperse anatase powders with particle sizes of around 50 nm are produced by hydrolysis of $TiCl_4$ in the gas phase. When processing ceramics, the anatase-rutile phase transition must be considered, which leads to a weakening of the structure. Anatase powder is therefore only suitable for the preparation of mesoporous membranes.

For nanofiltration membranes, amorphous TiO_2 layers are used, which, like anatase, have a very good pH resistance of pH 0–14.

Zirconium oxide is considerably more expensive than alumina and titania and therefore does not play a role as a support material. However, it is very easily accessible via sol-gel technology and is used as a mesoporous membrane. In the case of zirconium oxide, a distinction is made between a cubic, a tetragonal and a monoclinic phase. The cubic and the tetragonal phases are very similar. The conversion to the thermodynamically stable monoclinic phase takes place at 1100 °C:

$$\text{cubic } ZrO_2 \rightarrow \text{tetragonal } ZrO_2 \xrightarrow{1100\,°C} \text{monoclinic } ZrO_2$$

The thermal fixation of the mesoporous membranes usually takes place at temperatures <600 °C so that there is no risk of phase transformation and membrane damage.

10.2.2.2 Nonoxidic membrane materials

From the group of nonoxidic materials, especially silicon carbide is used for membrane applications. Silicon carbide is produced by reducing SiO_2 with carbon using the Acheson process:

$$SiO_2 + 3\,C \rightarrow SiC + 2\,CO$$

The resulting SiC is grinded and fractionated by size by sieving. The minimum particle size here is also around 500 nm. Silicon carbide does not sinter in the classical sense. Ceramic components can only be manufactured by using sintering aids or by evaporation/recrystallization at very high temperatures of >2000 °C under inert conditions. The recrystallized SiC is of particular interest for membrane applications because it is highly porous and it shows the best chemical stability, which is noticeably reduced by sintering aids (e.g. SiO_2). SiC membranes have so far been limited to macroporous membranes. There are attempts to produce SiC from silicon-organic precursors via a chemical route. This requires a nonoxidic carrier, since the amorphous SiC is oxidized during the thermal treatment of the layers.

10.2.2.3 Metallic membrane materials

In the case of metals, there are filters made of porous sintered metal, which are not used as a membrane, but as a particle filter or for ventilation of bioreactors. However, it is interesting to use thin layers of palladium or palladium alloys as hydrogen-selective membranes. Such palladium layers are deposited on porous supports, mainly ceramic supports, by means of chemical vapor deposition (CVD) or electroless plating. Membrane thicknesses are on the order of a few micrometers. With rising raw material prices, especially for precious metals, a lot of effort has been made in recent years to reduce the layer thicknesses and thus the material consumption. This places high demands on the quality of the supports.

10.2.2.4 Carbon membrane materials

Carbon is known as a chemically inert material with special wetting and adsorption properties. For this reason, carbon membranes are particularly suitable for gas separation. Various carbon modifications come into consideration here. In the turbostratic carbon, one finds nanoscopic ordered areas that are statistically distributed in an amorphous matrix. The diffusion in the ordered areas along the network planes at a distance of $\ll 1$ nm determines the selective transport properties. A second interesting modification is carbon nanotubes (CNT), which can be obtained from ethylene by CVD. The prerequisite for this is suitable metallic catalyst particles that are previously arranged on the substrate. In the CVD process, carbon is formed, which forms at the three-phase contact between ceramic substrate and metal particle and ethylene and pushes the metal particle upward. A tube is formed, and the diameter of which is largely determined by the size of the particle. The metal particles are removed by an etching process. What remains is a dense lawn of CNTs. A third carbon modification that is currently attracting great interest among membrane researchers is graphene. A permeance that is orders of magnitude higher is expected as a 2D material. For the formation of the pores, the graphene network must be deliberately disturbed by high-energy radiation.

10.2.3 Membrane preparation

10.2.3.1 Porous ceramic supports

The support gives the membrane its external shape and mechanical strength. The pores are formed by the voids between the particles. For this purpose, powders with a narrow particle size distribution are selected, which are almost densely packed in the molding process due to the action of high compression forces. In the subsequent firing process, the particles sinter at the contact points. Cavities with a diameter of about 1/3, the particle size remains in the interstices. In the ideal case of the closest packing of spheres, an open porosity of 26% is obtained. In the real case, the porosity of the ceramic substrates is 30–35%. The shaping during extrusion into tubular substrates, in the simplest case, is designed as single-channel pipes. In the interest of a high membrane area per filter element, multichannel tube substrates are predominantly used technically (Figure 10.2.2).

Tape casting is used to produce flat, disk-shaped or plate-shaped substrates. For this purpose, powders are suspended in a suitable solvent and stabilized by adding dispersants. By adding binders and plasticizers, a viscous slip is obtained, which is casted and shaped with a doctor blade into a uniform film with a thickness of about 1 mm. After drying, a flexible film is obtained that can be cut or punched. A ceramic plate remains after the firing process.

Figure 10.2.2: Tubular ceramic substrates made from alumina (white samples) and titania (gray samples).

10.2.3.2 Membranes for liquid filtration

10.2.3.2.1 Macroporous membranes

Macroporous ceramic membranes are made from ceramic powders, similar to supports. The powders are dispersed in an aqueous-alcoholic solution, mixed with suitable stabilizers and binders and homogenized in a drum mill. The grinding balls are separated and the slip is deaerated before further use. The coating is carried out by dip-coating. In the case of the inner coating of pipes, the process can be simplified by filling the slip into the tubes from below and allowing the excess slip to run out again after a short dwell time. Layer formation is achieved through the suction effect of the porous substrate. The solids content of the slip is matched to the open porosity of the substrate in order to obtain a closed, as thin as possible, layer. This layer is dried and then sintered at temperatures between 1200 and 1400 °C. The slip technique described can only be used to the extent that commercial powders are available. With α-Al_2O_3, the smallest pore sizes that can be produced are around 70 nm (Figure 10.2.3), with TiO_2 30 nm.

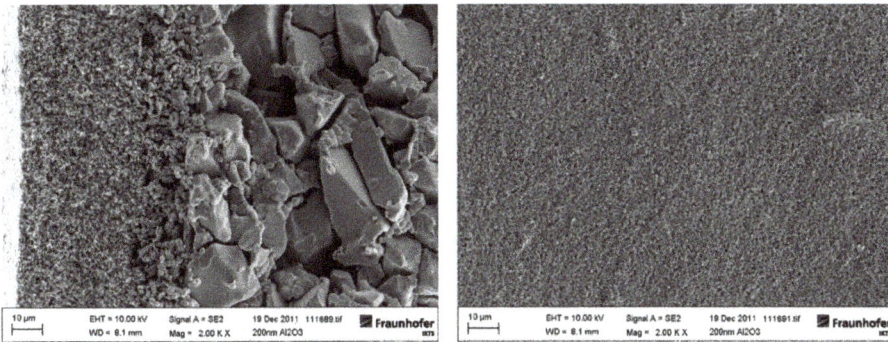

Figure 10.2.3: Cross section (left) and surface (right) of a 200 nm α-Al$_2$O$_3$ membrane.

10.2.3.2.2 Mesoporous membranes

Sol-gel technology can be used as a way out of the lack of availability of nanodisperse powders. The precipitation of the metal hydroxides leads to colloidal particles with a diameter of a few nanometers and, with further processing, membranes with a pore size between 3 and 30 nm. For reasons of purity and the controllability of the precipitation, one likes to start with metal alcoholates. In the case of aluminum, titanium and zirconium, these are propanolates or butanolates. The alcoholates can be purified by distillation before they are used. Precipitation takes place by dripping the alcoholic metal alcoholate solution into water (Figure 10.2.4). Every droplet gets contact to an infinite excess of water so that complete hydrolysis and hydroxide precipitation takes place. The surface of the hydroxide particles is positively charged by adding a small amount of nitric acid. The associated repulsive forces combined with the small size and weight of the particles lead to a stable suspension of the colloids, the so-called sol. Solid contents in these sols are very low and are between 2 and 5 wt%, depending on the oxide and alcoholate. If the concentration is increased, the repulsive forces come into play, the viscosity rises sharply and a gel is formed. The solids content of the sol depends on the oxide used and is 5–10 wt%. This sol-gel transition gave the method its name. The further processing of the membrane layers is very similar to the slip coating in the previous chapter. The sol is used for coating. The sol-gel transition takes place on the substrate surface. The drying of the gels is associated with high shrinkage, which builds up tensions in the material that lead to cracks when sintered. Crack-free conversion into a solid, stable, defect-free ceramic layer is only possible with very thin layers of <1 μm. This is in the sense of the membrane application but requires a very good surface and defect-free substrate (Figure 10.2.5):

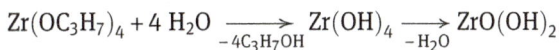

$$\text{Zr}(\text{OC}_3\text{H}_7)_4 + 4\ \text{H}_2\text{O} \xrightarrow[-\,4\text{C}_3\text{H}_7\text{OH}]{} \text{Zr}(\text{OH})_4 \xrightarrow[-\,\text{H}_2\text{O}]{} \text{ZrO}(\text{OH})_2$$

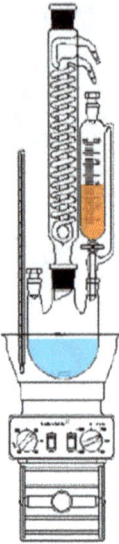

Figure 10.2.4: Scheme of the apparatus for sol-preparation (orange: diluted alcoholate, blue: water).

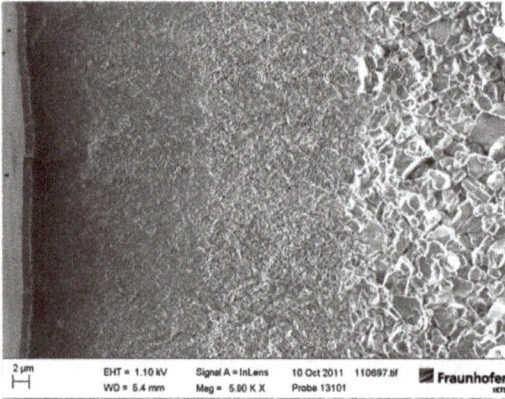

Figure 10.2.5: Cross section of a mesoporous TiO_2 membrane with a pore size of 5 nm.

10.2.3.2.3 Microporous membranes

Colloidal sol-gel chemistry is limited to about 5 nm in terms of small particle size. The sol-gel technique can also be carried out in an organic solvent and very little water. For this purpose, the alcoholate, diluted in the corresponding alcohol, is placed in advance and water, also strongly diluted in the alcohol, is added dropwise while stirring. In this way, only some of the alcoholate groups are hydrolyzed. The hydroxide groups formed can easily condense with the formation of a metal-oxygen-metal bond. The water formed in this way is available again for hydrolysis. Polymer chains are formed, which is why this method is also known as the polymer sol-gel technique:

$$nTi(OC_3H_7)_4 + 2nH_2O \xrightarrow[-2nC_3H_7OH]{} nTi(OC_3H_7)_2(OH)_2 \xrightarrow[-H_2O]{}$$

$$\left[-Ti(OC_3H_7)_2 - O - Ti(OC_3H_7)_2 - \right]_{\frac{n}{2}}$$

The solution of the polymers shows a similar viscosity behavior and a sol-gel transition as the aqueous sols and can be used directly for coating. Due to the high vapor pressure of the alcohol, drying can be reduced to a few minutes. In the case of larger membrane elements, one must consider the explosion protection here. The layers are burned in at temperatures of around 400 °C. This leads to amorphous oxide layers with a thickness of 50 nm and a pore size of about 1 nm (Figure 10.2.6).

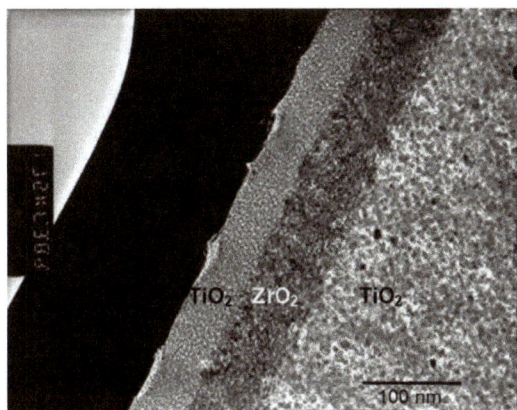

Figure 10.2.6: Cross section of a microporous TiO_2 membrane with a pore size of 0.9 nm on top of a 3 nm ZrO_2 membrane and a 5 nm TiO_2 membrane, both prepared by colloidal sol-gel technique.

10.2.3.3 Membranes for pervaporation and vapor permeation

If the pore size is further reduced, the dimension of single molecules is achieved with pore sizes smaller than 1 nm. In this case, we no longer speak of liquids or gases flowing through the membrane but of the diffusion of molecules. The driving force is no longer the differential pressure between feed and permeate but the concentration gradient of the component on the membrane surface of the feed and permeate side. Another prerequisite is the mobility of the component in the membrane, which can be described in a simplified manner by a diffusion coefficient. It should be noted that the diffusion coefficient is not constant but dependent on the concentration.

The use of porous membranes for molecular separation requires very defined pore sizes in the range from 0.2 to 0.8 nm without a pore size distribution. This is only possible with materials that have structural pores. In the case of inorganic materials, these are the zeolites, a great variety of which can be synthetically produced.

Zeolite membranes (Figure 10.2.7) are manufactured in two steps. First, the porous ceramic support is coated with a slip made of nano-zeolite particles. These zeolite particles cannot sinter to form dense zeolite layers by applying high temperatures. But they can act as seeds for a second crystal growth step to close the intergrain pores. To do this, the seeded supports are placed vertically in an autoclave and filled with an alkaline synthesis solution that contains Si, Al and Na (K, Ca) in a suitable ratio. The autoclave is closed and heated to the temperature at which the desired zeolite type shows the best crystal growth behavior. Depending on the type of zeolite, temperatures between 120 and 180 °C and a growing time of several hours are applied. After the autoclave has cooled, the membrane elements are removed and washed thoroughly. Loose zeolite crystals are rinsed off, and remnants of the alkaline synthesis solution are removed.

Figure 10.2.7: NaA-zeolite membrane (left: membrane surface, right: cross section).

10.2.3.4 Membranes for gas separation

10.2.3.4.1 Mixed ion electron conducting membranes

Ceramic membranes with a mixed conductivity of oxygen ions and electrons can advantageously be used for the separation of oxygen from air. Perovskites with an $ABO_{3-\delta}$ structure are suitable for this. The oxygen ions are transported via defects in the oxygen substructure. Electron transport is made possible by a change in valence at the cations (Figure 10.2.8). The conductivity for the oxygen ions has a comparatively high activation energy. Therefore, the membranes can only be used at high temperatures of 800–900 °C. If the membranes are defect-free (pore-free), they have an infinite selectivity. Pure oxygen is obtained in the permeate. To maintain the driving force, a reduced pressure of <200 mbar is used on the permeate side. The composition of the perovskites was optimized with regard to the phase stability in the specified temperature and oxygen partial pressure range and with regard to the concentration of oxygen vacancies.

One of the favored compositions is $Ba_{0.5}Sr_{0.5}Co_{0.8}Fe_{0.2}O_{3-\delta}$. The alkaline earth ions occupy the A position, and cobalt and iron share the B position (Figure 10.2.9). The oxygen stoichiometry varies between 2.9 (Co^{IV}, Fe^{III}) and 2.4 (Co^{III}, Fe^{II}) without the occurrence of a phase change. This gradient in composition along the membrane thickness enables the diffusion of oxygen ions.

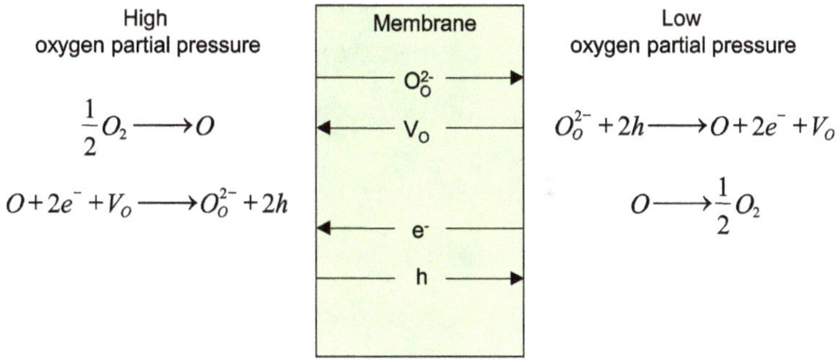

Figure 10.2.8: Principle of mixed oxygen ion electron conductivity.

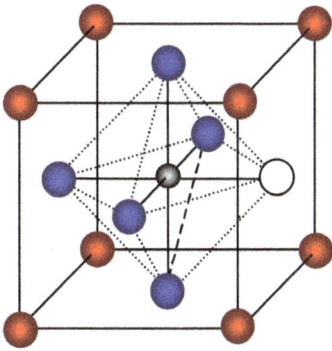

Figure 10.2.9: Unit cell of $Ba_{0.5}Sr_{0.5}Co_{0.8}Fe_{0.2}O_{3-\delta}$ (Ba, Sr: red, Co, Fe: gray, O: blue).

10.2.3.4.2 Carbon membranes

Microporous carbon is well known as an adsorbent. It is therefore not surprising that attempts are made in different ways to use carbon layers as membranes for gas separation. If the adsorption conditions are set on the feed side and the desorption conditions on the permeate side, a quasi-continuous pressure-swing adsorption can be operated.

A promising way of producing carbon membranes uses the pyrolysis of polymeric precursors. For this purpose, a porous ceramic carrier is first coated with a polymer that is cross-linked to form a thermoset. This is important in order to avoid softening and infiltration of the precursor into the porous support during heating throughout pyrolysis. Good results are achieved, for example, by using unsaturated polyester resins that are cross-linked with styrene. In an inert atmosphere, the polymer layers decompose to carbon in the temperature range of 700–800 °C (Figure 10.2.10). The structure of this carbon formed in this way is called turbostratic. These are disordered, nanocrystalline areas that are arranged in an amorphous matrix. The spacing of the lattice planes in the nanocrystalline areas is 0.4 nm, only slightly above the spacing of the lattice planes in graphite (0.335 nm). This distance determines the separability of the membranes.

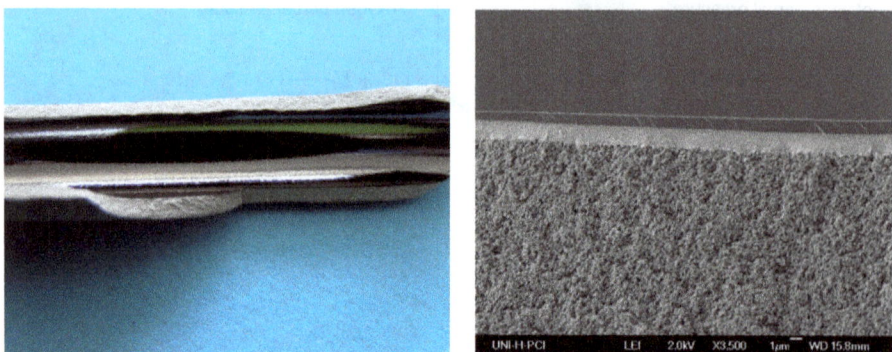

Figure 10.2.10: Carbon membrane on top of porous ceramic substrate.

10.2.3.5 Membrane application

10.2.3.5.1 Removal of turbidity and disinfection with ceramic MF membranes

Ceramic microfiltration membranes have found widespread use in food technology. They are used for the clarification and sterile filtration of juices and wine. A pore size of 200 nm is used, which ensures safe retention of microorganisms. Ceramic membranes have found widespread use in the production of fresh milk with a longer shelf life (extended shelf life milk, ESL milk). To do this, the milk is first broken down into cream and skimmed milk using a centrifuge. The cream is pasteurized at 75 °C, and the skimmed milk is filtered. Then both are put back together to make ESL milk. The advantage over long-life milk is that it tastes like fresh milk.

In principle, polymer membranes could also be used for these applications. With ceramic membranes, however, the systems can be cleaned with superheated steam, which is important for reasons of hygiene in food technology.

10.2.3.5.2 Recycling of hot water in textile finishing with ceramic NF membranes

The cleaning of colored wastewater from textile finishing was one of the first successful applications of ceramic NF membranes. A washing machine is divided into three main parts, whereas each part is fed with freshwater (Figure 10.2.11). The wastewater of the first part contains most of the dyes and can be treated with an NF membrane. A first plant for this treatment with a membrane area of 25 m² (feed and bleed, transmembrane pressure: 15 bar, cross-flow velocity: 5 m/s, permeate flux: 5 m³/h, concentration factor: 10–20) was installed in 2002 and was running continuously for 10 years before first membrane replacement in 2012 (Figure 10.2.12). Because of the three stages of the washing process, the permeate can be fed back in the first stage resulting in water and heat savings.

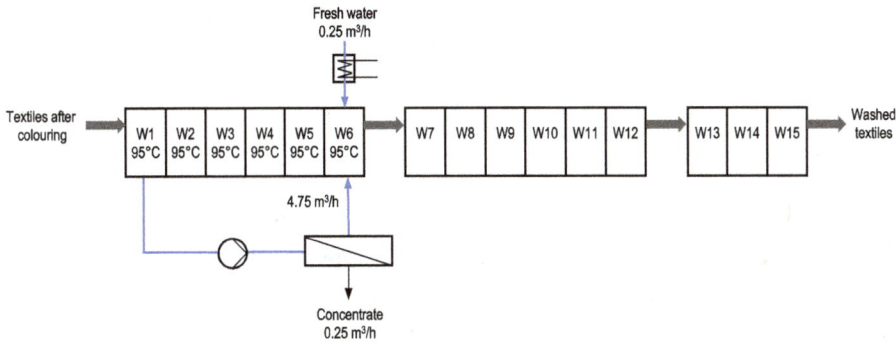

Figure 10.2.11: Scheme of implementing ceramic nanofiltration to recycle hot water from the first high-temperature part of a washing machine.

Figure 10.2.12: First NF plant with ceramic membranes built by the company Andreas Junghans.

10.2.3.5.3 Drying of natural gas

Natural gas has to be dried after exploitation or storage in underground gas storages in order to avoid hydrate formation and condensation in pipeline systems. Classically, this drying takes place by an absorption process using triethylene glycol (TEG). After absorption, the TEG has to be regenerated by distillation at temperatures of 190 to 205 °C consuming huge amount of energy, reducing lifetime of TEG and producing a vent gas, which has to be incinerated.

NaA zeolite membranes enable gentle, energy-efficient drying of the TEG at temperatures of 110–140 °C (Figure 10.2.13). To demonstrate the efficiency of the membranes, a pilot plant with a membrane area of 20 m^2 was put into operation at a natural gas storage facility in Stassfurt in 2018. In one winter period, 4 tons of water was removed from the TEG and 8 million tons of natural gas was dried with it. It has been shown that the operating costs of natural gas drying can be reduced by 30% in this way and CO_2 emissions can be reduced by 9.2 tons.

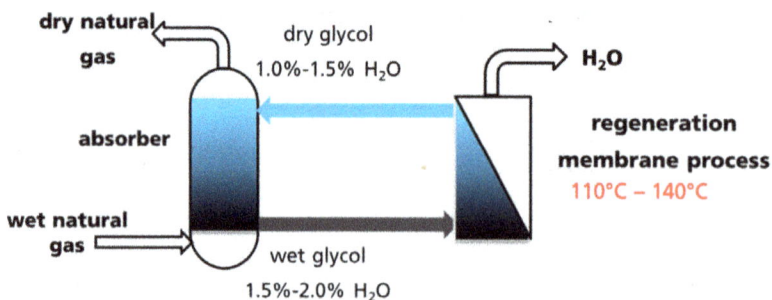

Figure 10.2.13: Scheme of TEG drying by using NaA zeolite membranes.

10.2.3.5.4 O_2 separation from air

Thin membrane layers on a porous support bear a high risk of defects. In gas separation even small defects lead to a significant decrease in selectivity. That is why thin-walled capillaries are preferred for perovskite membranes. These are closed on one side and inserted into a metal plate with a silicone seal on the open side. Since the silicone seal cannot withstand the high temperatures, a ceramic perforated stone is used, which thermally insulates the hot area from the seal area. In a perforated brick, 157 capillaries are combined to form a standard membrane unit (SMU) with an active membrane area of 0.75 m^2 (Figure 10.2.14). In the application of the membranes, it is important that the heat of the oxygen-depleted air and that of the oxygen is transferred to the fresh air. With regenerative heat exchangers in the form of ceramic honeycombs, heat recovery >90% is possible. In total, this leads to an energy demand of 0.7 kWh/m^3 for the separation of oxygen from the air. This is only a little higher than cryogenic air separation with the advantage of generating oxygen in a decentralized and modular manner, as required.

10.2.3.5.5 H_2/C_3H_8 separation

Propene is produced by dehydrogenation of propane at 600–700 °C over platinum catalysts. A conversion of about 50% is achieved in equilibrium.

Figure 10.2.14: Standard membrane units with 157 capillaries and a total membrane area of 0.75 m^2.

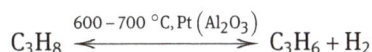

$$C_3H_8 \xleftarrow{\quad 600-700 \text{ °C, Pt}\left(Al_2O_3\right) \quad} C_3H_6 + H_2$$

Although the conversion can be increased by increasing the temperature, there is increased coke formation and deactivation of the catalyst. Alternatively, one tries to shift the equilibrium by separating off the hydrogen formed. Turbostratic carbon membranes on ceramic supports are suitable for this. After pyrolysis, these membranes show a molecular sieve separating behavior with ideal permselectivities of 10,000 and H$_2$ permeances of 5 m^3/(m^2 · h · bar), which are confirmed in the mixed gas measurement (Figure 10.2.15). By tempering the membranes in air at temperatures between 300 and 400 °C, the spacing between the lattice planes can be increased and the separation properties can be optimized for the process.

10.2.4 Summary

Inorganic membranes have found their way into many applications due to their robustness. The production of very fine-pored membranes requires a deeper understanding of phase formation and layer preparation. The membrane composition ranges from simple oxides, zeolites and perovskites to non-oxides, carbon and metals. They play a key role in many applications in process engineering, food technology and climate and environmental protection. With the transformation of chemistry (crude oil → natural gas → CO$_2$) and energy supply (fossil fuels → regenerative energy sources), the importance of inorganic membranes will continue to increase.

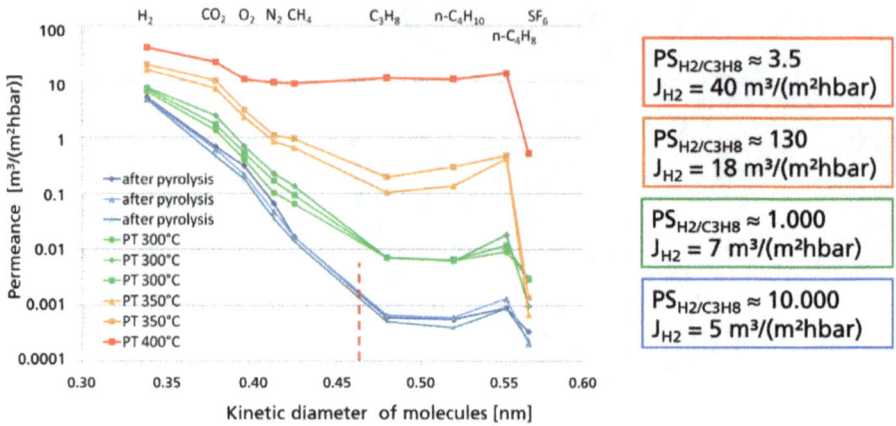

Figure 10.2.15: Single gas permeance and ideal permselectivity of carbon membranes after pyrolysis and after post treatment (PT) in air at temperatures between 300 and 400 °C.

References

[1]   Voigt I. Keramische Membranen. In: Ohlrogge K, Ebert K (Eds). Wiley-VCH, Weinheim, 2006, 103–129. ISBN 978-3-527-30979-9.

[2]   Voigt I, Adler J, Weyd M, Kriegel R. Ceramic filters and membranes. In: Riedel R, Chen IW (Eds). Ceramic Science and Technology, Volume 4: Applications, Wiley-VCH, Weinheim, 2013, 117–167.

[3]   Melin T, Rautenbach R. Membranverfahren, 3rd Updated and Expanded Edition, Springer-Verlag, Berlin, Heidelberg, 2007. ISBN 978-3-540-34327-1.

10.3 Functional ceramics

Jan Werner

10.3.1 Preliminary note

Although ceramics belong to the oldest man-made materials, they are still very important and do not include ceramic construction materials in civil engineering (tiles, bricks, pipes) and household products (sanitary ware, tableware, porcelain, stoneware, earthenware, pottery). For this topic, see Chapter 10.1.

Ceramics also play a key role as extremely durable components with high thermal and chemical resistance and mechanical stability for various mainly technical and industrial applications or in environments that demand high mechanical strength and toughness and long term stability even under harsh chemical and thermal conditions e.g. in metallurgy and concrete fabrication, the chemical, pharmaceutical and food industry, as well as in plant and mechanical engineering, transportation and automotive industry. For this topic we refer to Chapter 10.4.

Also of high importance are the heat-resisting ceramic materials and products of the refractory industry sector. Without them, neither the production of steel and light metals nor the manufacture of cement and concrete for the construction industry would be conceivable. For this topic, please refer to Chapter 10.5.

In this Chapter the group of so-called functional ceramics is presented. They have special optical, electrical and/or magnetic properties and are indispensable in a broad variety of technical products of our daily life. They fulfill their functions in capacitors, resistors, actuators, sensors, radiation converters, optical components and many more in nearly all kind of electronic devices, computers, telephones, consumer electronic products, light sources, in transportation and so on. Some are excellent chemical catalysts, and some are even well proven since decades as successful bio-compatible materials in modern surgery and dentistry.

Some of the properties of the ceramic material systems presented here are also characteristic for other material systems, too, especially for single crystals, glasses and powders or powder-containing composites of the same or similar chemical composition. These are described in detail in the respective chapters of this book, insofar as not all properties are only typical exclusively for ceramics. As on the one hand the ceramic production process allows an easier, faster, less energy consuming and/or cheaper fabrication and on the other hand, when beneficial effects arise from the polycrystalline ceramic microstructure, causing the typical ceramic properties such as high thermal, mechanical and chemical resistance, they will be presented here. For more general information on all kind of ceramic materials, their properties and manufacturing processes, please refer to the relevant textbooks and technical literature.

https://doi.org/10.1515/9783110733471-012

Basically, the specific properties of ceramic materials depend on their precise chemical composition, including mineral phases, dopants, defects/vacancies as well as on the macro, micro and nanostructure, grain size distribution, grain morphology and orientation, porosity and surface characteristics, including inner interfaces and outer surfaces. This means that the properties may be modified and optimized by the proper choice of materials and ceramic fabrication method.

10.3.2 Brief introduction into ceramics and their fabrication

10.3.2.1 Definition and explanation of terms

The term ceramics is used in this chapter solely for inorganic, non-metallic, solid materials with a highly densified polycrystalline structure, produced by sintering (high-temperature treatment of compacted powders). In other contexts, semicrystalline (glass ceramic) and amorphous (glass) materials as well as set inorganic binders (lime, cement, gypsum and liquid glass) are also referred to as ceramic materials. Thus, ceramics differ from metals and organic solids (e.g. polymeric materials) by their inorganic non-metallic character (i.e. chemical composition and type of chemical bonding), from setting binders by the sequence of thermal processing and shaping and from single crystals by their polycrystalline structure with grain sizes ranging from a few nanometers to a few millimeters, typically mostly in the micron or submicron range. Ceramics can generally be divided into three main material categories:

- silicate ceramics, SiO_2-containing oxide ceramics: alkali aluminum silicates (porcelains), magnesium silicates (steatites), magnesium aluminum silicates (cordierites), aluminum silicon oxide (mullites) and other ceramics, produced with naturally occurring raw materials, mainly clay, kaolin, quartz, feldspar and lime (earthenware, stoneware, porcelain, bone china),
- oxides (except silicates): alumina = Al_2O_3, magnesia = MgO, spinel = $MgAl_2O_4$, yttria = Y_2O_3, yttria alumina garnet (YAG) = $Y_3Al_5O_{12}$, zirconia = ZrO_2,
- non-oxides: carbides – silicon carbide (SiC), tungsten carbide (WC), borides – titanium diboride (TiB_2), lanthanum hexaboride (LaB_6), nitrides – aluminum nitride (AlN), silicon nitride (Si_3N_4) and
- composite materials: particulate reinforced composites – zirconia-toughened alumina (ZTA), alumina-toughened zirconia (ATZ), fiber-reinforced ceramic matrix composites (CMCs), combinations of different oxides or non-oxides (e.g. oxycarbides, oxynitrides and oxysulfides, as SiOC, AlON, SiAlON or RE_2O_2S, with RE = rare earth element, e.g. La, Gd, Y).

It is also useful to distinguish ceramic products by different applications, which allows dividing them into four main types of usage:

- white wares, including tableware, cook and kitchenware, wall tiles, pottery products and sanitary ware,
- structural clay products, including bricks, pipes, floor and roof tiles,
- refractories, such as kiln linings, gas fire radiants, steel and glass making crucibles,
- technical ceramics, also known as engineering, advanced, special or fine ceramics, including gas burner nozzles, ballistic protection, vehicle armor, nuclear fuel pellets, coatings of jet engine turbine blades, ceramic disk brakes, space shuttle heat protection tiles, ball bearings, missile nose cones, biomedical implants (bone grafts, artificial joints), dental ceramics and a large variety of functional ceramics like piezoelectric, magnetic, ion-conducting, superconductive, high resistivity, radiation-converting ceramics, biosensors and many others.

10.3.2.2 General ceramic fabrication process

Besides recently strongly arising various techniques of additive manufacturing (AM) and special thermal processing techniques like spark plasma sintering (SPS) or field-assisted sintering technology (FAST), sometimes also called current-activated pressure-assisted densification (CAPAD), the classical ceramic fabrication processes typically include the following steps:

- Powder processing and forming via
  a) fabrication of liquefied, good moldable slurries for slip casting by mixing, grinding and homogenization of raw material powders, often referred to as ceramic powders, with water or other solvents, utilizing surface active additives, e.g. deflocculants, dispersants, plasticizers, lubricants, defoaming agents and binders, or
  b) fabrication of plasticized, good moldable, pasty masses for extrusion forming with a higher solid content as said molding slurries, or
  c) fabrication of dried, good flowing and easy compactable granulates by spray drying or other granulation techniques, e.g. pelletizing, or special mixing techniques for different press forming techniques, as e.g. axial pressing and/or cold isostatic pressing, followed by
- optional intermitted machining of thus compacted so-called green ceramics by cutting, punching, drilling, structuring,
- thermal processing, sometimes subdivided into several subsequent heating processes, like binder burnout, sintering (in electrically driven furnaces with resistance-heating elements, microwave generators, induction heating coils or gas-fired burners) and finally, optional post treatment like pressure-assisted sintering alias hot pressing (HP) or hot isostatic pressing (HIP),

–   finishing of the so-called as-fired ceramics, meaning optional machining of the parts or some of their functional surfaces – where necessary – by e.g. grinding and polishing, lapping etc. to a desired/needed surface quality.

The properties and performance of functional ceramics depend mainly on at least these three main important aspects: Chemistry, crystallography and microstructure of the ceramics:

a) Chemistry – fundamental chemical compositions and stoichiometry, trace elements, sintering aids, dopants.

b) Crystallography – fundamental crystalline structures/phase composition (polymorphism, polymorphs, coexisting phases, transitional phases, secondary phases, amorphous proportions), mineralogy.

c) Microstructure – fundamental physical characteristics: crystallinity, amorphous phases, density/porosity, open/closed porosity, pore size distribution, pore morphology, pore form anisotropy, pore localization (intergranular/intragranular porosity), grain size distribution, grain boundary composition/defects/segregation, crystal/grain orientation, homogeneity, inhomogeneities, anisotropic grain orientation, domain formation, etc.

Especially in comparison to glasses, glass ceramics and single crystals with comparable or identical chemical composition, crystallographic and microstructural aspects and differences become particularly relevant. In these cases, the grain structure and orientation as well as the grain boundary properties are of great importance. Special shaping processes with an influence on the orientation of anisotropic shaped powder particles, such as tape casting or extrusion, can be used to produce special properties, or also the influence of applied electric or magnetic fields, e.g. during the shaping process, as in the case of the production of ferrite magnets.

Figure 10.3.1 illustrates the entire ceramic production chain and demonstrates the influence of the processing steps on the characteristics of the ceramic microstructure, which determine the ceramic properties.

10.3.3 Functional ceramics

Functional ceramics are generally ceramics that not only exhibit the typical properties of structural ceramics but are also designed for applications that require additional special properties such as optical, electrical or magnetic properties. Somewhat more broadly, bioceramics and ceramic catalysts can also be regarded as ceramics with special functionalities and thus as functional ceramics. Functional ceramics have in common that, unlike structural ceramics or conventional ceramics, they usually play an active role in any process. A systematic overview is given in Table 10.3.1.

This book chapter is primarily intended to provide a general overview of the large group of functional ceramics. The inclusion of the sometimes-non-trivial physical

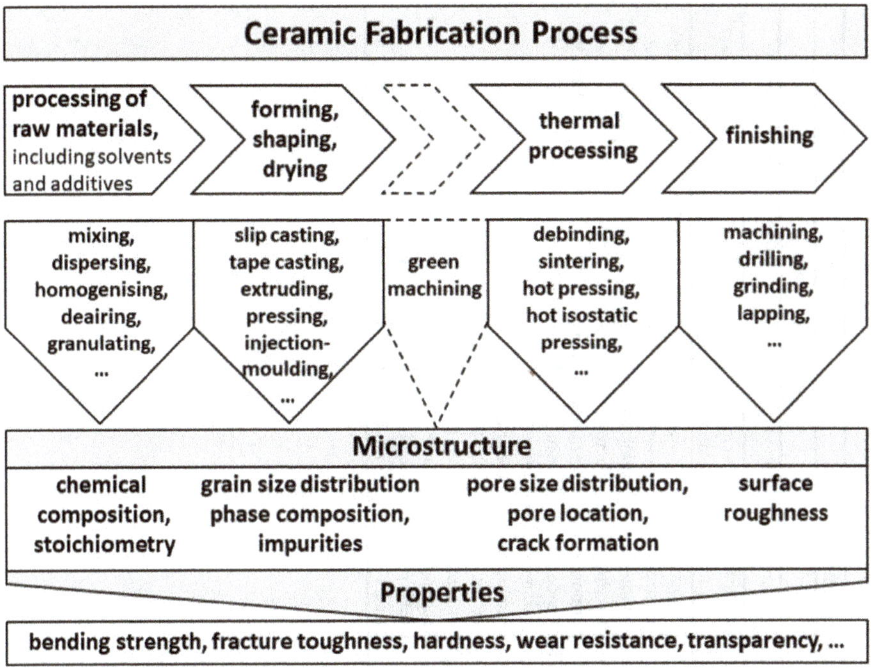

Ceramic Fabrication Process

processing of raw materials, including solvents and additives	forming, shaping, drying		thermal processing	finishing

mixing, dispersing, homogenising, deairing, granulating, ...	slip casting, tape casting, extruding, pressing, injection-moulding, ...	green machining	debinding, sintering, hot pressing, hot isostatic pressing, ...	machining, drilling, grinding, lapping, ...

Microstructure

chemical composition, stoichiometry	grain size distribution phase composition, impurities	pore size distribution, pore location, crack formation	surface roughness

Properties

bending strength, fracture toughness, hardness, wear resistance, transparency, ...

Figure 10.3.1: Flow chart of a conventional ceramic production chain, demonstrating the influence on the microstructural characteristics and on the properties of the produced ceramic components.

backgrounds of optics, electricity, magnetism, biology and medicine as well as chemical catalysis, would go far beyond the scope of this book and this chapter. They are mentioned where necessary, but not explained in detail. For a more in-depth understanding of these backgrounds, a supplementary study of the literature on the fundamentals of the respective topics is recommended.

Furthermore, the most important and established material systems and compositions are presented in the respective subchapters, primarily. An exhaustive presentation of nearly all scientifically investigated material systems would be almost impossible, as a multitude of composition variants, mostly with systematic substitution of different components, have been and are being produced and investigated for their properties, some of which also have the potential for multiple uses, so this is avoided for the sake of clarity and no claim is made to completeness. Therefore, it is recommended to study the references given and the secondary literature cited therein.

10.3.3.1 Optical ceramics

Translucent and transparent ceramics have gained considerable interest during the last decades. Light-transmitting ceramics have increasingly been exploited successfully

Table 10.3.1: Systematic overview of different ceramic types and applications.

Traditional ceramics		Advanced ceramics								
		Structural ceramics				Functional ceramics				
		wear resistant, load bearing, thermally stable, insulating, non-corrosive				ceramics with special optical, electrical, magnetic, catalytical, or biological functionalities				
Whitewares	Structural clay products	Abrasives	Refractories	Wear parts	Electrical insulators	Optical ceramics	Electrical ceramics	Magnetic ceramics	Ceramic catalysts	Biofunctional ceramics
pottery, earthenware, stoneware, porcelain, fine China, table ware, sanitary ware	bricks, pipes, tiles	grinding wheels	high temperature resistant products	tribological ceramics in technical & medical application	power line insulators, high voltage insulators	optical lenses, prisms and windows, transparent armor, electro-optical ceramics; luminescent ceramics, LED-phosphors, scintillators, solid state lasers	insulators, dielectrics, semiconductors, piezo-electric ceramics, electric conductors, ionic conductors, super-conductors	hard magnetic ceramics, soft magnetic ceramics, semi-hard magnets, motors, communication, MW technology	photo-catalysts, CO-oxidizing catalysts	ceramics in aesthetical dentistry and special medical diagnostics and therapies

as transparent armor and for several high-tech applications, e.g. in electrical light sources and optical technologies. They can be utilized as highly refractive lenses, electro-optical lenses or beam deflectors, high-power solid-state laser media, inorganic scintillators in radiation detectors e.g. special camera optics, diagnostic medical imaging and luminescent compounds in highly efficient light-emitting diodes, to mention just a few prominent examples.

High purity ceramics, mainly oxides with cubic crystal structure and thus, isotropic optical properties, can achieve transparencies comparable to that of optical glasses and single crystals. Additionally, in most cases the optical properties are accompanied by typical ceramic characteristics, such as high mechanical strength, surface hardness, wear resistance, thermal shock resistance and chemical stability. Furthermore, the characteristics of the ceramic fabrication route can provide certain advantages. Therefore, the number of examples for the successful implementation of transparent ceramic components in diverse applications has continuously increased over the last years and is expected to even expand further.

Transparent and translucent ceramics have gained increasing interest during the past decades, beginning with the development and invention of translucent polycrystalline alumina (PCA) by Coble [1] at General Electrics. Due to the optical anisotropy of the hexagonal crystal structure of alumina in combination with the relatively large mean grain size of these ceramics, initially only translucency could be achieved. However, the resulting translucency has proven to be sufficient for the use of PCA in arc tubes of high-pressure sodium discharge lamps. Far more important is the fact that this was the starting point of the development of the high-quality transparent ceramics of today.

For a ceramic with a high chemical purity in the range of 99.99 mass% or more and with a mineral phase purity at least below the detection limits of modern X-ray analysis, the main objective in the production of transparent ceramics is to achieve the highest possible density by avoiding any residual porosity in the final ceramic body, as this would cause scattering and thus transmission losses.

Therefore, it is of utmost importance to generate a green body with an optimal particle packing and green density. Krell et al. [2] and other researchers showed that the chosen processing route has substantial impact on the required sintering temperature and the necessary dwell time, correlated to the homogeneity and packing density of fine-grained ceramic powders. Therefore, wet shaping techniques like gel casting (GC), pressure casting (PC) and electrophoretic densification (EPD) are advantageous compared to axial pressing (AP) or cold isostatic pressing (CIP). For small, complex-shaped parts that are being produced in large or medium-sized quantities, powder injection molding (PIM) (high-pressure injection molding – HPIM as well as medium and low-pressure injection molding – MPIM and LPIM) are attractive particularly with regard of fabrication costs. Recent own studies, yet unpublished, also show the great potential of using high-quality powders in additive manufacturing technologies.

10.3.3.1.1 Passive optical ceramics

With the aim of a more efficient (i.e. lighter) transparent armor, bulletproof ceramics made of magnesium aluminum spinel ($MgAl_2O_4$) [3, 4] and aluminum oxynitride (ALON) with the chemical formula $(AlN)_x(Al_2O_3)_{1-x}$, $0.30 \leq x \leq 0.37$ (also with cubic spinel structure) [5] are successfully used today, especially for transparent mobile vehicle protection. The focus of the functionality is on high mechanical strength in the sense of optimum protection of people and material in the event of being fired upon by armor-piercing projectiles, even in the case of a multiple hit scenario, with simultaneously the best optical properties in the sense of a high total forward transmission (TFT) with minimal scattering, corresponding to a high real inline transmission (RIT).

Optically highly refractive materials were developed by Murata company in Japan. They are based on $Ba(Mg,Ta)O_3$, partially substituted with tetravalent ions of Ti, Zr, Hf or Sn on the B-site of the perovskite structure (e.g. $Ba(Mg,Zr,Ta)O_3$, BMZT, allowing to change the original hexagonal structure to a cubic system. Furthermore, LaAlO-Sr(Al,Ta)O$_3$, LAO-SAT and $La_2Zr_2O_7$, LZO with a cubic pyrochlore structure were investigated and found to be applicable [6–8]. In terms of applications as a lens, the focal distance can be shortened by a high refractive index for enabling a compact design for the optical system. By integrating such LUMICERA® lenses into the design of zoom lenses developed by Casio®, their size has been reduced by almost 20%.

Another highly refractive material with high transmission and low scattering losses at nearly zero porosity is fully stabilized cubic zirconia (ZrO_2) [9]. It allows ceramic fabrication of high-quality passive optical compounds like prisms and lenses and shows also high potential for ultraslim and ultralight optical devices, such as multi-lens array microscopes or multi-channel cameras. Using near-net shape forming techniques, materials consumption of such valuable powders can be minimized as well as expenses for further mechanical processing steps and surface finishing.

Weight and size reduction of optical instruments, apparatus and devices are also important aspects in today's aerospace, drone or model flight reconnaissance or ground investigations, as well as in the miniaturization of microscopes and micro-optical components in general.

Miniaturized optical components such as microlens arrays and microprisms can be produced, by using specific powder preparation and processing methods and by using near-net-shape thermoplastic molding [10, 11].

Also, the so-called sesquioxides of the REE yttrium, scandium and lutetium, yttria (Y_2O_3), scandia (Sc_2O_3) and lutetia (Lu_2O_3) all exhibit refractive indices with n > 1.9 [12], which is remarkably high compared to optical glasses that are limited to lower values in this respect. They all have a cubic (C-type) structure, which makes them interesting as transparent optical ceramics. Due to their low thermal expansion, high thermal resistance and mechanical strength, they are potential candidates for various applications like transparent windows in industrial applications under severe conditions, IR-transparent missile domes, etc. However, because of their very high melting points of >2400 °C, reaching highest density and optical quality by sintering is a challenge.

10.3.3.1.2 Active optical ceramics

Active optical ceramics are materials that change their properties (e.g. refractive index) under the influence of electromagnetic fields, such as electro-optic ceramics, or those that can convert radiant energy, typically in the ultraviolet (UV) and visible (VIS) spectral regions, into radiation at another desired wavelength, typically in the visible or infrared (IR) spectral range. Such ceramics are used in numerous applications, for example in luminescent LED converters, polycrystalline solid-state lasers or scintillators for medical imaging. In addition to the structure-forming properties of rare earths in various compounds described in the previous sections, there are also several applications that open up when specific effects can be achieved in existing structures (which may also contain rare earth ions, such as cubic sesquioxides, pyrochlores or garnets), also by specific ion exchange in these structures by rare earth ions known as doping. Such dopants, when excited by electromagnetic radiation, can emit part of the absorbed radiation energy in the form of light in the wavelength range UV-VIS-IR. Such luminescent ceramics, which are designed for use in lasers, scintillators and light converters, some of which have already been introduced successfully are presented in the following sections.

Electro-optical ceramics

Electro-optical ceramics belong to the class of active optical functional ceramics. Electric and magnetic fields induce changes in materials optical and dielectric coefficients and refractive indices. Thus, their refractive indices can be controlled by applying an external electrical field. Electro-optical materials are suitable for converting electrical to optical information, and vice versa. Frequently, those materials are applied in the form of single crystals (e.g. $LiNbO_3$, $LiTaO_3$) or liquid crystals. Ceramic materials are also suitable for applications like electrically controlled panels, color filters or light modulators. The core properties of such ceramics are optical transparency, high electro-optical coefficient, rapid-response time and low energy consumption. The predominant feature is light modulation. When using an electro-optic device in the shape of a prism or lens a beam can be deflected or focused by varying the refractive index.

In comparison to single crystals, ceramics offer the potential of a more cost-efficient production, especially in case of large-scale devices. Well-known examples of transparent oxide electro-optic ceramic materials are $Pb_{1-x}La_x(Zr_yTi_{1-y})_{1-x/4}O_3$ (PLZT), $Pb(Mg_{1/3}Nb_{2/3})O_3$-$PbTiO_3$ (PMN-PT) and $Pb(Zn_{1/3}Nb_{2/3})O_3$-$PbTiO_3$ (PZN-PT). All these materials have high-quality electro-optical effects. Alternative lead-free materials like $LiNbO_3$ or $RbTiOPO_4$ require relatively high operating voltages of several thousands of volts for electro-optical modulators. High operating voltages complicate electrical controlling and miniaturization [13]. Alternative materials with a high electro-optical effect can be potassium tantalate niobate ($KTa_{1-x}Nb_xO_3$; KTN) or barium strontium titanate (($Ba,Sr)TiO_3$; BST). First studies deal with the development of transparent KTN-based ceramics [14–16].

Lead lanthanum zirconate titanates with variable Zr/Ti ratios and with general compositions $Pb_{1-x}La_x(Zr_yTi_{1-y})_{1-0.25x}V^B_{0.25x}O_3$, for example $Pb_{0.91}La_{0.09}(Zr_{0.65}Ti_{0.35})O_3$ (PLZT 9/65/35) crystallize in a cubic ABO_3 perovskite structure and exhibit ferroelectric and piezoelectric properties due to their crystallographic structure. PLZT materials have been developed based on ferroelectrics known earlier, such as barium titanate ($BaTiO_3$) or lead zirconate titanate $Pb(Zr,Ti)O_3$ (PZT). The partial substitution of the divalent Pb^{2+} ion by the trivalent La^{3+} ion improves the piezoelectric and electro-optical properties on the one hand and favors the densification during sintering to high density and thus transparent ceramics on the other hand. Well-densified polycrystalline ceramics with high optical transparency can be produced by sintering in oxygen atmosphere, comparable to the fabrication of dense PZT ceramics [17] or by hot pressing [18].

When exposed to an electric field, a polarization of the unit cell and the material occurs due to a shift of the Zr^{4+}/Ti^{4+} ions located in the central BO_6 octahedron. The polarization can cause changes in the refractive index as well as in the macroscopic shape, which can be observed and used as piezoelectric effects [19, 20]. Thus, a large number of useful electro-optical components for various technical applications can be produced from PLZT, such as very fast operating electro-optical switches, electro-optical filters and polarizers [21], microactuators, wireless actuators in micro-electromechanical systems (MEMS) and micro-opto-electromechanical systems (MOEMS), which advantageously combine electrical, optical and mechanical properties [22] and electro-optical devices in modern fiber-optic communication [23].

Compositional changes within this ferroelectric system can significantly alter the materials properties and behavior under applied electric fields or temperature variations. This allows such a system to be tailored to a wide range of converter applications. For instance, PLZT ceramics have been suggested for use in optical devices [24–27] because of their good transparency from the visible to the near-infrared, and their high-refractive index ($n \approx 2.5$), which is advantageous in light wave guiding applications [28, 29]. Using a CO_2-IR laser, the use of a PLZT ceramic as a thermal lens could also be successfully demonstrated for focusing a red HeNe-laser beam [30]. Ferroelectrics will also play a major role in the next generation of optical devices. They can make a valuable contribution to wireless laser communication systems or in ground-based fiber optic networks to transmit large data volumes at very high speed. The miniaturization of practical devices and advances in nanotechnology have recently made it possible to produce ferroelectric materials with actuator effects in the nanometer range. Beyond their industrial and technological use, ferroelectric materials may also play an important role in consumer goods in the future [30].

LED converters

Luminescent ceramics play an important role in so-called phosphor-converted light-emitting diodes (pcLEDs) as they can effectively be used for the conversion of ultraviolet or blue primary radiation into visible light, ~400–700 nm.

Compared with luminescent inorganic powders (also referred to as "phosphors") embedded in polymeric matrices, overall light emission with customized color temperature and high-power output can be achieved by combination of a suitable excitation source with appropriate ceramic light converter materials. On the one hand, luminescent ceramics offer higher efficiency and long-term stability. On the other hand, they permit operation at temperatures much higher than polymer based luminescent materials, which are severely limited due to their thermal instability at temperatures already below 200 °C. Therefore, ceramic-based pcLEDs, based on Ce:YAG enable the generation of high-performance white light sources. If combined with other luminescent ceramic materials, high-quality white-emitting LEDs with high color-rendering indices, tailored color temperature and high efficiency can be obtained.

As mentioned earlier, luminescent inorganic powders play an important role in modern solid-state lighting (SSL). In modern white light LEDs, the white color impression is created by mixing several primaries. This is often achieved by blue or ultraviolet light from a semiconductor chip made of indium-gallium-nitride (In,Ga)N, which partially excites one or more emitter materials, which in turn emit in the yellow, green or red spectral range. In such so-called phosphor-converted light-emitting diodes (pcLEDs) a certain proportion of the exciting blue light is transmitted through the emitter materials, so that in the optimum case a complete spectrum is generated in the visible range. A large number of luminescent materials – many of them based on REE-doped inorganic compounds – have been developed and successfully proven in recent decades as stable, durable and efficient light converters, thus helping modern LED lighting to achieve the breakthrough. And, the search for technologically and commercially even better materials continues unabated.

With increasing demand for improved performance and efficiency, powder-based LED converter solutions are reaching their limits. In such white light LEDs, the powders are usually embedded in a polymer matrix, of silicone or epoxy resin, so that inhomogeneity, strong scattering losses and poor heat dissipation can cause problems. At high-power levels, operating temperatures are reached which can lead to damage to the polymer matrix and associated negative optical effects such as brown discoloration.

Compared to such luminescent powders, embedded in polymeric matrices, overall light emission with customized color temperature and high-power output can be achieved by combination of a suitable excitation source, i.e. a blue to ultraviolet emitting LED or a laser diode LD with an appropriate ceramic converter, as tiny ceramic plate of approx. 1×1 mm^2 and only a few hundreds of microns thick. This platelet is directly attached on top of the LD/LED chip or alternatively at some distance from the LD/LED chip as so-called remote phosphor approach. Thus, on the one hand, luminescent ceramics offer higher efficiency and long-term stability. On the other hand, they permit operation at temperatures much higher than polymer luminescent materials composites, which are severely limited due to their thermal instability at temperatures already below 200 °C.

The ceramic fabrication techniques are the same as for other optical and luminescent ceramics, described in the chapters before (in detail depending on the respective materials, especially regarding the sintering parameters). The ceramic converters are most often produced by pressing, sintering, polishing and finally slicing to the demanded size, while tape-casting techniques are discussed as alternatives to pressing.

Whenever new luminescent materials are developed and found suitable, it is necessary to determine the best production parameters for the individual processing steps.

The most common system by far to date is a combination of cerium-doped YAG [31] with a blue LED. In this simple but effective binary concept, the relatively broadband emission of Ce:YAG with an emission intensity maximum in the yellow region of the spectrum already produces a cold white light. By choosing the optimum LED in combination with a converter with adjusted chemical composition and the right thickness of the ceramic plate and an appropriate amount of scattering (e.g. by tailoring the porosity of the ceramic in the range of up to 3 vol.%), satisfactory results can be achieved [32–34].

A green shift of the emitted light can be realized by changing from Ce:YAG to Ce:LuAG, or by using varieties with adjusted yttria and lutetia proportions, i.e. Ce:(Y,Lu)AG, or cerium-doped yttrium aluminum gallium garnet Ce:YAGG [35], respectively.

With the desire for better color rendering and a warmer white light, concepts with additional orange to red components were developed. Particular reference is made here to the europium doped nitride compounds of the type (Ca,Sr)AlSiN$_3$:Eu^{2+} and (Ca,Sr)$_2$Si$_5$N$_8$:Eu^{2+} [36, 37] and Mn^{4+} activated K$_2$SiF$_6$:Mn^{4+} emitters [38]. Due to the challenging and costly production and limited chemical stability, there is still a sustained demand for advanced red emitters.

An alternative concept is based on the use of UV-emitting LEDs whose light is generated entirely by a combination of red, green and blue emitters (instead of the blue LEDs discussed so far in combination with a red and a green converter). A suitable blue phosphor is Eu^{2+}:BaMgAl$_{10}$O$_{17}$ (Eu^{2+}:BAM), which can also be produced in the form of monolithic translucent ceramics [39].

While it is relatively easy to create mixtures when using powders, the combination of several ceramics with each other is more elaborate and can be achieved by a layer-wise assembly of several individual ceramics [40], which may have to be separated from each other by protective layers to prevent reactions during the bonding process at high temperatures, or by coating one phosphor with another [41]. In an alternative conceptual approach, a composite was produced from a low-melting matrix of an europium-doped $Li_3Ba_2La_3(MoO_4)_8:Eu^{3+}$ red phosphor [42, 43] sinterable at low temperatures, in which a stable phosphor $Ce^{3+}:(Y,Lu)_3Al_5O_{12}$ was embedded [44]. Combined with an adequate blue-emitting LED, color rendering and efficiency could be further improved compared to conventional warm white LEDs.

Polycrystalline solid-state lasers

The excitation of transition metal ions (e.g. Cr^{3+}) or RE metal ions (e.g. Nd^{3+}, Yb^{3+}) in solids allows light amplification by stimulated emission of radiation (laser). Therefore, besides glasses and single crystals, nowadays solid-state lasers based on transparent ceramics also provide new possibilities.

Since the pioneering works on Nd:YAG ceramic laser host materials with laser output in the mW range by Ikesue and co-workers [45], impressive technical progress has been achieved, resulting in ceramic laser systems which nowadays can gain more than 100 kW output power [46]. Such high-power lasers are of interest for various present and future high-performance applications.

In the system RE_2O_3-Al_2O_3, namely Y_2O_3-Al_2O_3, there are three phases of yttrium oxide-aluminum oxide compounds, a monoclinic $Y_4Al_2O_9$ phase (YAM), an orthorhombic $YAlO_3$ perovskite phase (YAP) and a cubic $Y_3Al_5O_{12}$ phase with garnet structure (YAG).

However, the reaction of a mixture of Y_2O_3 and Al_2O_3 with stoichiometric YAG composition does not take place directly to the target compound but proceeds in several temperature-dependent steps [47]. First, the YAM phase forms in the temperature range between 900–1100 °C:

$$2\,Y_2O_3 + Al_2O_3 \rightarrow Y_4Al_2O_9\,(\text{YAM})\ [900-1100\,°C]$$

Together with still unreacted alumina, YAM forms the YAP phase in the temperature range between 1100 and 1300 °C:

$$Y_4Al_2O_9 + Al_2O_3 \rightarrow 4\,YAlO_3\,(\text{YAP})\ [1100-1300\,°C]$$

Between 1400 and 1600 °C, the YAP phase finally reacts with previously unreacted alumina to form the desired YAG phase:

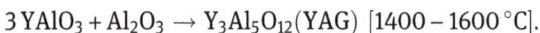

$$3\,YAlO_3 + Al_2O_3 \rightarrow Y_3Al_5O_{12}\,(\text{YAG})\ [1400-1600\,°C].$$

Similar processes can also be found when using other more reactive starting compounds or via synthesis routes with wet chemical precipitation (co-precipitation of poorly soluble hydroxides) and their calcinates. The starting materials are often precalcined between 900 and 1000 °C to prevent sintering of loose agglomerates of the nano or submicron-scale particles.

Such powders are frequently mixtures of still unreacted Al_2O_3 and some of the yttria-alumina phases, YAM, YAP and YAG. However, these can be further processed by ceramic technology, as the complete conversion into YAG usually takes place during the further sintering process. This form of reactive sintering generally also allows the fabrication of defect free and therefore highly transparent ceramics.

In the Y_2O_3-Sc_2O_3-Al_2O_3 system exists a wide range of garnet-structured solid solutions. In some laser applications e.g. instead of YAG the corresponding $Y_3ScAl_4O_{12}$ (yttria scandia alumina garnet, YSAG) is utilized. Also, a manifold of other garnets like gadolinium gallium garnet (GGG) is well known. However, these garnets are rarely used non-doped, but almost exclusively as host materials for laser or LED converter materials.

Also, the so-called sesquioxides of the *RE* element yttrium, scandium and lutetium, yttria (Y_2O_3), scandia (Sc_2O_3) and lutetia (Lu_2O_3) could be used as high-power solid-state laser hosts, as they can be doped with other laser active trivalent RE^{3+} ions. However, because of their very high melting points of $T_{melt} > 2400$ °C, reaching highest density and optical quality by sintering is a challenge.

A laser is a source of coherent, quasi-monochromatic and sharply focused radiation in the visible and adjacent regions of the electromagnetic spectrum (far-infrared, infrared, ultraviolet and X-ray). In principle, every laser consists of three components: first, an active laser medium, which largely determines the properties of the laser, e.g. a gas, a solid material (solid-state laser, SSL) or a diode; second, a pumping mechanism that supplies energy to the laser medium, e.g. a flash lamp or an electrically driven gas discharge emitting at a wavelength shorter than the laser wavelength; and third, a laser resonator, a system of transparent and semitransparent mirrors and other optical elements that provides optical back-coupling and thus induced emission of the radiation. Depending on the specific design and the choice of components, there are several different types of lasers, which differ mainly in the attainable power (between a few microwatts and many kilowatts) and frequency characteristics.

Since the construction of the first functional prototype in 1960 at Hughes Research Laboratories [48] (a ruby, i.e. chromium-doped aluminum oxide ($Cr:Al_2O_3$) laser with a red emission line at 694.3 nm), the laser has become widespread in science, technology, industry, medicine, entertainment and even a divers consumer products.

Nowadays, many *RE* element are utilized as dopants in solid-state laser materials like neodymium (Nd), ytterbium (Yb), holmium (Ho), thulium (Tm) and erbium (Er). Neodymium (Nd) is a common dopant in various solid-state laser crystals, including yttrium orthovanadate ($Nd:YVO_4$) and yttrium aluminum garnet (Nd:YAG). These lasers can generate high-power infrared emission at 1064 nm that is used for cutting, welding and marking metals and other materials, but also in spectroscopy, medical surgeries and many other applications.

In the beginning, there were only two types of SSL gain media: single crystals and glasses doped with laser-active ions. Compared to most crystalline materials, ion-doped glasses usually exhibit much broader amplification bandwidths, allowing for large wavelength tuning ranges and the generation of ultrashort pulses. Drawbacks are inferior thermal properties (limiting the achievable output powers) and lower laser cross sections, leading to a higher threshold pump power. The fabrication of single crystals (e.g. via Czochralski growth from a melt) is time consuming and expensive, and size and geometries with which they can be produced are limited.

The first lasing in a $Dy^{2+}:CaF_2$ ceramic was demonstrated in 1966 [49]. A further breakthrough was the fabrication of a Nd:YAG ceramic in laser quality in 1995 by Ikesue et al. [50]. In 2002 a next milestone was achieved by demonstrating an output power of >1 kW using a Nd:YAG ceramic [51]. Great improvements in powder synthesis, forming and sintering have led to further remarkable achievements. For example, 105 kW output power was realized from a Nd:YAG ceramic laser system in 2009 [52].

Meanwhile, ceramic laser materials with different composition based on lutetia alumina garnet (e.g. Nd:LuAG) and yttria-scandia alumina garnet, e.g. Yb:YSAG are under investigation. Compared to glass and single-crystal laser technologies, advanced ceramic lasers are anticipated to be highly attractive alternatives in the future.

Although the first ceramic laser gain medium was based on polycrystalline cubic $Dy:CaF_2$, further development mainly concentrated on doped oxide ceramic lasers. With regard to some technical advantages such as high transparency in a wide wavelength region from the VUV to the IR, a decreased probability of non-radiative transition probability (i.e. higher efficiency), and a lower melting point compared to most oxide-based lasers. During the last years interest in these materials increased and successful preparation use of RE element-doped CaF_2 ceramics was reported on $Yb:CaF_2$ [53, 54], $Er:CaF_2$ [55] and $Nd,Y:CaF_2$ [56].

Further detailed information on laser ceramics with respect to the various innovation steps can be found in several reviews and the literature cited therein [57–59].

Scintillators

Since the discovery of ionizing radiation (X-rays) more than 100 years ago, it has been used in a wide variety of technical and medical applications. To make the radiation – invisible and harmful to the human eye – usable, materials are needed that can convert the received radiation into visible radiation. Such materials are called scintillators. Sometimes, the excited state is metastable, so the relaxation back down from the excited state to lower states is delayed. Scintillating materials are currently widely used in many detection systems addressing different fields of application, such as medical imaging (CT scanners and gamma cameras in medical diagnostics), civil protection, industrial control and oil-drilling exploration. Among the desired technical properties of scintillators are their fast operation speed, high density, radiation hardness and durability of operational parameters. Especially because of the latter, stable inorganic scintillators are of great interest and benefit. In addition to a multitude of monocrystalline scintillators, there is an interest in polycrystalline ceramic alternatives – comparable to the development of the laser materials described above. Although the possibility of the use of ceramics was not considered at the beginning of the 2000s [60], several ceramic scintillators have been investigated to date. These are often also *RE* element-doped materials of the pyrochlore structure type like Eu^{3+}:$(La,Gd)_2Hf_2O_7$ [61] or of the garnet structure type like cerium doped lutetia alumina garnet (Ce:LuAG) [62], cerium doped gadolinium-yttrium gallium-aluminum garnet Ce:$(Gd,Y)_3(Ga,Al)_5O_{12}$ (Ce:GYGAG) [63], cerium doped yttria alumina garnet (Ce:YAG) [64] or cerium doped gadolinium gallium aluminum garnet Ce:$Gd_3Ga_3Al_2O_{12}$ (Ce:GGAG) [65] which have in principle already been described in the previous sections. They will therefore not be treated in detail again here. Instead, reference is made to a comprehensive current overview [66].

10.3.3.2 Electrical ceramics

Contrary to the common opinion that ceramics are generally electrical insulators, they cover a very wide range that includes not only highly insulating ceramics, as used in insulators and as dielectrics in capacitors, but also electrical resistors with very versatile properties, as well as semiconducting materials and even materials with good electrical conductivity [67]. These include those with electrical conduction through electron flow, but also ionic conductors. In addition, ceramics belong to the extraordinary group of high-temperature superconductors. The group of piezoelectric ceramics, which serve in a very extensive field of applications, must also be emphasized.

10.3.3.2.1 Insulators and dielectrics

Insulators

Ceramics with very low-electrical conductivity and high-dielectric breakdown strength have been used successfully as electrical insulators for over one hundred years. Other important property characteristics for this application – as in their use as substrate materials (ceramic circuit boards) – are high thermal conductivities, low AC voltage losses and low dielectric constants. In the case of electronic housings, impermeability and corrosion resistance are further important characteristics.

For small insulators, electro-porcelain (e.g. for spark plugs of combustion engines, still in use today), aluminum oxide (Al_2O_3), steatite ceramic mainly based on natural soapstone magnesium silicate $Mg(Si_4O_{10})(OH)_2$ and cordierite $2\,MgO \cdot 2\,Al_2O_3 \cdot SiO_2$ are used. Quartz porcelain or high-alumina porcelain have been used for over a century and to this day for large-format high-voltage insulators [68–73].

For ceramic printed circuit boards, mainly high-quality aluminum oxide (Al_2O_3) and in special cases beryllium oxide (BeO) [74] or aluminum nitride (AlN) [75–77] are used. Depending on the application, the individual overall property range is decisive for material selection and economic aspects are considered.

Dielectrics, capacitor ceramics, dielectrics for energy storage, microwave resonators

The dielectric properties of a ceramic material determine its functionality. These properties include the relative permittivity or dielectric constant, the dielectric loss, its inverse value, the quality factor and the temperature coefficient of the resonant frequency [78].

In most capacitors, the ceramic material acts as a dielectric between one or more alternating layers of ceramic and a metal layer acting as electrodes. Most ceramic capacitors are constructed as multilayer ceramic capacitors (MLCCs) whose electrical behavior and thus the application are determined by the composition of the ceramic material and the layer structure. Ceramic capacitors are divided into two application classes:

- Class 1 ceramic capacitors are used in resonant circuit applications and are characterized by their high stability and low losses. In the past porcelain, steatite (talc), mica and other silicates were used. Today, most Class I dielectrics are based on simple oxides, e.g. TiO_2 (rutile) and perovskite titanates, e.g. $CaTiO_3$ and modified $(Ca,Sr)(Zr,Ti)O_3$ [79]. Typically, Class 1 ceramic capacitors show low loss accompanied by low capacitance density.
- Class 2 ceramic capacitors are used for buffer, by-pass and coupling applications and are characterized by their high volumetric efficiency. For class 2 capacitors ferroelectric materials based on barium titanate $BaTiO_3$ are utilized, using additives like aluminum and magnesium silicate or aluminum oxide. They may also be doped with rare earth elements with smaller rare earth ion radius, such as Dy_2O_3, Ho_2O_3, Er_2O_3 or Y_2O_3. Despite the discussion about environmental aspects, some lead-containing materials such as $Pb(Mg_{1/3}Nb_{2/3})O_3$, $Pb(Fe_{1/3}Nb_{2/3})O_3$ and $Pb(Fe_{1/3}W_{2/3})O_3$ are used in niches due to their special dielectric properties [80]. Typically, Class 2 ceramic capacitors show high loss accompanied by high capacitance density.

Former Class 3 ceramic capacitors are barrier layer capacitors. They are no longer standardized [81].

The main driving force for the continuous development of the MLCC technology is the large-scale application of these passive components in consumer electronics [82]. However, in the last few years there has been an increasing demand for ceramic capacitors with higher energy density for power electronic devices and pulse power applications, including hybrid and electrical vehicles. While ceramic capacitors have a high-power density (fast charge/discharge performance), the energy density is relatively low and extensive studies have been undertaken to develop new dielectric materials with enhanced breakdown field and improved storage efficiency [83–85]. $BaTiO_3$-$BiScO_3$ dielectrics exhibited superior energy storage capability at the same applied electric field, relative to commercial MLCCs [86, 87].

Supercapacitors

Compared to ordinary capacitors, supercapacitors have significantly higher capacitance at lower voltages. Their properties lie between those of accumulators and electrolytic capacitors. Their energy storage capability exceeds that of electrolytic capacitors by a magnitude or two and the charging and power supply of the battery is much faster, tolerating significantly more charge-discharge cycles than rechargeable batteries. Supercapacitors are mainly used in applications requiring many rapid charge-discharge cycles in transportation applications, elevators and cranes, where they are used for regenerative braking, short-term energy storage or burst-mode power delivery. Smaller units are used as power backup for static random-access memory (SRAM) [88]. In contrast to conventional capacitors, supercapacitors use instead of conventional solid dielectrics, electrostatic double-layer capacitance and electrochemical pseudocapacitance, which both contribute to the overall capacitance of the capacitor [89]. Various materials such as carbon-based materials, oxides, hydroxides, sulfides, carbides, nitrides, metal-organic frameworks (MOFs) and two-dimensional transition metal carbides and nitrides (MXene) have been developed for supercapacitor electrodes. Although ceramic materials are often overlooked due to their low specific surface area and relatively low surface activity, their corrosion resistance, high-temperature resistance, radiation resistance and thermal shock resistance still make them attractive for the development of supercapacitor electrodes. Ceramic materials – often nanostructured – of different kind have been investigated, from single oxides like Al_2O_3 and SnO_2 to binary oxide ceramics like $NiFe_2O_4$ and $Co_xZn_{0.04-x}Fe_2O_4$ spinel ferrites or $ZnMn_2O_4$ and $NiCo_2O_4$ spinels, as well as $BaTiO_3$ ceramics or even complex composite multi-element ceramics like $CaCu_3Ti_4O_{12}$ and $Li_3V_2(PO_4)_3$, $Li_{1.3}Al_{0.3}Ti_{1.7}P_3O_{12}$ and even hierarchically intermixed phases like $CuO/La_{0.7}Sr_{0.3}CoO_{3-\delta}$ (with CuO nanoparticles embedded in a matrix of a porous $La_{0.7}Sr_{0.3}CoO_{3-\delta}$) as ceramic electrode materials have been investigated. A comprehensive review article concisely highlights the recent research progress made in the field of ceramics for supercapacitors [90].

In recent years, research is going on to introduce new types of electrode materials to satisfy the increasing demands for efficient energy storage devices. Among various energy storage devices, the supercapacitors gained a lot of attention owing to their long cyclic life, environment-friendly nature and high-power density. In

this view, global research has been reported to this rapid development of fundamental and applied aspects of supercapacitors [91].

Microwave dielectrics

Microwave dielectric ceramics are used for a variety of applications ranging from
wireless communication (including mobile communications, ultra-high-speed local
area networks, intelligent transport system, satellite communications) to electronic
consumer products. Wireless communication applications include wireless FAX,
cellular phones, global position satellite (GPS), military radar systems, intelligent
transport system (ITS) and direct broadcast satellites. The range of materials is
rather broad and includes materials as Mg_2SiO_4 and Zn_2SiO_4 silicates, $Mg_2Al_4Si_5O_{18}$,
representatives of the olivine group in which the endmembers forsterite (Mg_2SiO_4),
fayalite (Fe_2SiO_4) and larnite (Ca_2SiO_4) or feldspar solid solutions between albite-
anorthite ($NaAlSi_3O_8$-$CaAl_2Si_2O_8$) and many more [92], but also non-silicate-based
ceramics like $(Ca,Co,Mn,Ni,Zn)Nb_2O_6$ niobates, $(Mg,Ca)TiO_3$, $(Zr,Sn)TiO_4$, $BiNbO_4$,
TiO_2, $BaTiO_3$, Bi_2O_3 and ceramics of the tungsten bronze type system, as e.g. BaO-
RE_2O_3-TiO_2 with RE = rare earth element like Sm, Nd, Gd, Pr or La [92, 93].

Resistors, NTC thermistors, PTC thermistors, varistors VDR/MOV, chemical gas and humidity sensors

Resistors are electronic components that provide specific amount of resistance to
an electrical current in electronic circuits. Ceramic resistors use ceramics to control
the resistor's resistive value. Ceramic resistors are made of a combination of finely
powdered carbon and ceramic material. These two powders combine in specific ratios to determine the final value of the resistor. The higher the ratio of carbon in the
mix, the lower resistive value the ceramic resistor will have. A higher ratio of ceramic material, on the other hand, will mean a higher resistive value of the resistor.
Once the proper ratios are established, the mixture is compressed to create its
shape and then kiln fired to set the ceramic. It is common for these types of resistors
to have an external shell of pure ceramic material to serve as an external insulator.
Ceramic resistors are widely used in many different types of electronic circuits and
devices. The materials used include $ZrSiO_4$ and Al_2O_3. In high-voltage applications,
ceramic resistors are used as elements in surge arresters. For example, ZnO [94] and
SiC [95] are used here.

A special kind of resistors with temperature-dependent resistance behavior are
used in the form of ceramics with a negative temperature coefficient (NTC ceramics), e.g. for electrical resistance thermometers, and in the form of ceramics
with a positive temperature coefficient (PTC ceramics), e.g. self-regulating heating
elements for fan heaters or in sensor technology as level indicators. Such semiconducting ceramics are summarized under the term thermistors.

Ceramic negative temperature coefficient resistors (NTCR) serve as temperature sensors in domestic, laboratory, medical and industrial applications with high sensitivity and accuracy. Most compositions can be described as more or less complex AB_2O_4 spinels, such as $NiMn_2O_4$ or $(Ni,Mn)_3O_4$, $NiCo_2O_4$ or $(Ni,Co)_3O_4$ or $NiFe_2O_4$ or $(Ni,Fe)_3O_4$, as well as $(Ni,Mn,Co)_3O_4$ and $(Ni,Mn,Co,Fe)_3O_4$. But also ABO_3 perovskites such as $LaCo_{1-x}(Al,Ti)_xO_3$, $La_{0.7}Sr_{0.3}Mn_{1-x}(Al,Ti)_xO_3$, $Y(Mn,Ni)O_3$ and $A_2B_2O_7$ pyrochlores such as $La_2Zr_2O_7$ [96] have been investigated recently [97].

Positive temperature coefficient of resistivity (PTCR) materials are widely used in the electronics industry for applications including temperature sensors, time delay circuits and current limiters for overvoltage or overcurrent protection, overheat protection, current stabilizers and others. The main groups of ceramic PTCR materials are $BaTiO_3$ based or quasi-$BaTiO_3$ based ternary perovskite compounds. $BaTiO_3$ is an insulator at room temperature, but donor doped $BaTiO_3$ with ions such as La^{3+}, Y^{3+}, Sb^{3+} and Nb^{5+} presents semiconducting behavior at room temperature and an anomalous increase in resistivity near the ferroelectric-paraelectric Curie transition temperature T_C. Divalent ions such as Pb^{2+}, Ca^{2+} and Sr^{2+} are used as additives to substitute the Ba^{2+} ions in the $BaTiO_3$ structure and tetravalent ions such as Zr^{4+}, Hf^{4+} and Sn^{4+} are used as substitutes for Ti^{4+} ions in the $BaTiO_3$ structure. Both can be used with the intention of adjustment of specific electrical properties [98].

A variable resistor (varistor) is a voltage-dependent resistor (VDR). Ceramic metal oxide varistors are often abbreviated as MOVs. Above a certain threshold voltage, which is typical for the respective varistor, the differential resistance decreases abruptly. Specifically, no current or only a very small current flows at low voltages (leakage current range), while above a certain voltage (breakdown range) the current flow increases strongly (high current range). This effect can be utilized by means of mainly ZnO-based ceramic varistors as overvoltage protection both of sensitive semiconductor assemblies and to protect substations, e.g. against lightning strikes. The varistor effect is not only a material property of ZnO but is based on the different conductivity of the grains (electrically conductive) and the grain boundaries (electrically insulating). Additives such as Bi_2O_3, Sb_2O_3, Mn_3O_4, Co_2O_3, Co_3O_4, Fe_2O_3, Pr_2O_3, MnO_2 and others can be used to tailor the microstructure and adjust the desired current-voltage behavior [99, 100].

Chemical sensors

Sensors are key elements in the rapidly evolving fields of measurements, instrumentations and automated systems. The recent progresses made in improving the reliability and lowering the cost of microprocessors and interface circuits has resulted in a higher demand for sensors, which convert physical or chemical quantities in various environments into an electrical signal. The identification and quantification of chemicals has become an important issue in a variety of industrial applications. Intelligent systems based on sensors and controls can be used for health and safety (e.g. medical diagnostics, air quality surveillance and detection of harmful or hazardous gases). There is a continuing need for the development of fast, sensitive, robust, reliable and low-cost sensors for applications under severe industrial conditions found in heat treating, metal processing and casting, glass, ceramic, pulp and paper, automotive, aerospace, utility and power, chemical and petrochemical and food-processing industries. The ability to detect gases is based on the change of the physico-chemical properties in dependence of the surrounding gas atmosphere. Ceramic gas sensors are distinguished either according to the sensor materials or according to the principle of changing physical-chemical properties [101, 102].

The change in electrical conductivity of a metal oxide in the presence of a gas is a relatively simple method of detection. Ceramic relative humidity sensors work utilizing the mechanism of the chemical and physical adsorption of water vapor on the materials surface in a quantity that is proportional to the relative humidity of the surrounding environment. The adsorption generates a consequent variation of the capacitance or impedance [103].

In particular, titanium dioxide (TiO_2), tin dioxide (SnO_2), zinc oxide (ZnO), nickel oxide (NiO), copper(II) oxide (CuO), copper(I) oxide (Cu_2O), tungsten oxide (WO_3), molybdenum oxide (MoO_3), gallium oxide (Ga_2O_3), indium oxide (In_2O_3), iron oxide (Fe_2O_3), chromium oxide (Cr_2O_3) as well as lanthanum ferrite ($LaFeO_3$) and mixed-valent manganese oxide (Mn_3O_4), cobalt oxide (Co_3O_4) and doped variants of these materials have shown high surface reactivity towards carbon monoxide (CO) and carbon dioxide (CO_2), oxygen (O_2), hydrogen (H_2), ammonia (NH_3), nitric oxides (NO_x), hydrocarbons (HCs) and volatile organic compounds (VOCs) and have a sufficient band gap, so that such semiconducting metal oxides (SMO) or metal oxide semiconductors (MOS) can be used as valuable inorganic chemical gas sensors both in industry and for measuring the alcohol content in exhaled breath [104–107].

10.3.3.2.2 Piezoelectric ceramics

Piezoceramics (see also Chapter 9.9), a significant group of piezoelectric materials, are ferroelectric materials with polycrystalline structures (perovskite, tetragonal/rhombo-hedral crystals). Above the Curie temperature (T_C), these crystals exhibit simple cubic symmetry in structure. There are no dipoles present in this state, as the positive and negative charge sites are coincident due to the centrosymmetric structure. However, this symmetry is no longer present below the Curie temperature, where the charge sites are no more coinciding. This results in the formation of electric dipoles due to localized charge separation. Neighboring dipoles with parallel orientation realign locally to form Weiss domains. These Weiss domains tend to be randomly oriented in a raw piezoelectric material, hence the material does not show any piezoelectric response. However, when the material is heated above its Curie temperature in the presence of a strong electric field, these dipoles will be oriented in the direction of the applied field. On cooling, the material tends to maintain the dipolar orientation achieved during heating. Piezoelectric materials comprise a polar structure and generate an electric potential in the direction of their polar direction when an external mechanical stress is applied. Inversely, they will undergo mechanical strain when an electric field is applied to the materials. The direction of the strain is dependent on the polarization of the material and the orientation of the potential difference. Using these phenomena, piezoelectric materials have been used in numerous applications such as electromechanical transducers, sensors, actuators, sonar transducers, for ultrasonic applications and energy harvesting [108].

Lead zirconate titanate (PZT), a solid solution and binary system of ferroelectric lead titanate ($PbTiO_3$) and anti-ferroelectric lead zirconate ($PbZrO_3$) with a large range of Zr:Ti ratio and the general formula $Pb(Zr_xTi_{1-x})O_3$ ($0 \leq x \leq 1$) [109] is the most widely used piezoelectric ceramic material due to its superior performance.

PZT belongs to the perovskite structure family with general formula ABO_3, where A is a divalent cation (Pb^{2+}) that occupies the corners of the cube, B is a tetravalent cation (Zr^{4+}/Ti^{4+}) that occupies the 1/2 1/2 1/2 position and O are the oxygen atoms at the faces of the cubic structure. When an electric field is applied to the PZT unit cell, the Ti^{4+} or Zr^{4+} ion is displaced along the direction of the applied field. A wide variety of cations can be substituted in the perovskite structure [110, 111].

However, the toxicity of lead has raised concerns over the use of PZT. Therefore, over the past two decades, a large number of lead-free piezoceramics were explored, among others mainly several perovskite materials, as barium titanate ($BaTiO_3$, BT), bismuth sodium titanate ($Bi_{0.5}Na_{0.5}TiO_3$, BNT), bismuth potassium titanate ($Bi_{0.5}K_{0.5}TiO_3$, BKT), potassium sodium niobate ($K_xNa_{1-x}NbO_3$ with $0 < x < 1$, KNN) potassium niobate ($KNbO_3$), lithium niobate ($LiNbO_3$), lithium tantalate ($LiTaO_3$) as well as sodium tungstate (Na_2WO_4) and zinc oxide (ZnO). Promising results were found, exploring the barium zirconate titanate – barium calcium titanate system ($xBa(Zr_{0.2}Ti_{0.8})O_3-(1-x)(Ba_{0.7}Ca_{0.3})TiO_3$, BZT–BCT) [112–116].

10.3.3.2.3 Ceramic conductors

Ceramic materials are known for their hardness, compressive strength, resistance against thermal and chemical attack and a pronounced electric resistivity. Therefore, ceramic materials such as porcelain, cordierite, steatite, alumina, etc. have traditionally been utilized as electric insulators. In contrast, some ceramics are excellent electrical conductors. Electric conductivity can be based on different effects, such as electron flow, diffusion of ions or vacancies of the material itself or of secondary phases, dopants, oxygen deficiencies, ion valence differences, charge transfer, polarization effects, etc. To present all this in detail would go far beyond the scope of this chapter and this book. Therefore, reference is made to the literature cited – here, review articles were selected as far as possible – and the literature quoted therein.

The broad range of electrical conducting properties and different types of conductivity makes ceramic materials very important and interesting for many applications, such as Ohmic resistors for electrical sensors (see above), ceramic heating elements, such as SiC, $MoSi_2$ and calcium doped $LaCr_2O_4$ [117–122], ionic conductors, such as $Na_3Zr_2Si_2PO_{12}$ (NASICON-type), $Li_4(Ge,Si,V)O_4$ (γ-Li_3PO_4-type) or $Li_7La_3Zr_2O_{12}$ (LLZO), perovskite-type or garnet-type [123–125], electrolytes and electrodes in batteries [126, 127] and fuel cells (solid oxide fuel cells, SOFCs), such as yttria-doped zirconia Y-ZrO_2, an oxygen-ion-conducting ceramic electrolyte membrane [128–133] and even ceramic superconductors [134–139], i.e. materials with no electrical resistance below the Curie temperature T_C. Conductive ceramics also allow special fabrication techniques, using electrical current for electrical discharge machining (EDM) [140–143]. Tin-doped indium oxide, better known as indium tin oxide (ITO), typically mixtures with an approximate composition of In_2O_3: $SnO_2 = 9$:1, do not only exhibit good electrical conductivity but also high transparency, when applied in thin layers. Therefore, ITO is used in a wide variety of applications e.g. solar cells, liquid crystal displays and thin-film resistors in integrated electronic circuits, among others [144, 145].

Some ceramic materials, such as $Ca_3Co_4O_9$ and $Ca_{3-x}Tb_xBi_yCo_4O_{9+\delta}$ are recently under investigation due to their promising thermoelectric behavior [146–148].

10.3.3.3 Magnetic ceramics

The magnetism of solid matter (see also Chapters 9.6 and 9.7) is based on the sum of all magnetic moments of all elementary particles with a magnetic moment of which the solid is composed and their interaction with each other. A significant contribution to the magnetic moment of ceramic materials is made by the magnetic moments of the ions forming them and their arrangement in the crystal structures of the oxide phases. Oppositely oriented magnetic moments equalize each other, so that in particular the spin orientation and spatial arrangement of unpaired electrons in the crystal structure, their cooperative orientation and their environment play the essential role for the magnetic behavior. The magnetic moments of the ions of a structure or substructure are each oriented in parallel over larger spatial regions, the Weiss domains, separated by the Bloch walls.

Magnetism of solids is a cooperative phenomenon. The macroscopic magnetization is composed additively of the individual contributions of the building blocks (atoms, ions, quasi-free electrons) of the material. In the case of many materials, the individual building blocks already have no magnetic moment. However, even of materials whose building blocks carry non-vanishing magnetic moments, only a few exhibit macroscopic magnetization. The reason for this is that the individual building blocks (magnetic substructures) usually arrange themselves in such a way that their contributions equalize each other.

The behavior of materials in a magnetic field is divided into five categories: diamagnetic, paramagnetic, ferromagnetic, anti-ferromagnetic and ferrimagnetic. A more detailed explanation of these special physical properties can be found in the relevant literature and in physics textbooks.

In ceramic ferrimagnets, there are two substructures with opposite magnetization but different absolute magnitude, resulting in an overall magnetic moment. Typical compounds are the oxides, fluorides, etc. of the 3d metals manganese, chromium, iron, cobalt and nickel.

$$MnO \cdot Fe_2O_3, \ FeO \cdot Fe_2O_3, \ NiO \cdot Fe_2O_3, \ CuO \cdot Fe_2O_3 \ \text{or} \ MgO \cdot Fe_2O_3.$$

When an external magnetic field is applied, an inductive alignment of the magnetic moments in the material in the field direction occurs due to displacement of the Bloch walls between the Weiss domains until saturation magnetization occurs. When the external magnetic field is completely removed, a certain magnetization, the remanence, remains, which can only be removed again by applying a magnetic field with the opposite orientation up to the coercitivity (demagnetization). Multiple magnetizations in different field directions thus result in a typical hysteresis curve for magnetic materials. A distinction is made between materials with a broad hysteresis loop, i.e. with high remanence and high coercivity (hard ferrites) and materials with a narrow hysteresis loop due to low remanence and low coercivity (soft ferrites).

Compounds of iron oxide with other metal oxides form the main group of magnetoceramic materials as ferrites of the general composition $Me_xFe_yO_x$. The magnetism of ferrite ceramics is based on a ferrimagnetic behavior, i.e. the antiparallel coupling of magnetic moments of the cations of the respective 3d electrons of the metal ions on different Wyckoff sites via the oxygen ions within a suitable crystal structures. This includes ferrites with spinel-structure, such as $MeFe_2O_4$ with Me = Mn, Ni, Cu, Co, Fe, Zn, Mg, ferrites with magnetoplumbite structure, such as $MeFe_{12}O_{19}$ with Me = Ba, Sr as well as ferrites with garnet-structure, in particular $Y_3Fe_5O_{12}$, yttrium-iron-garnet (YIG) [149]. Magnetic oxides, their composites and nanoparticles are uniquely suited for a wide variety of applications in new technologies, including device miniaturization, power efficiency improvement and health sector innovations. The interest in these materials is due to such properties as high resistivity, low dielectric and magnetic losses, good corrosion resistance and favorable mechanical characteristics [150].

Ferrites are often classified in terms of their magnetic properties as soft (low coercivity, low remanence, easy to demagnetize), hard (high coercivity, high remanence, difficult to demagnetize) or semi-hard magnetic materials.

10.3.3.3.1 Hard-magnetic ceramics

Hard ferrites are used for the production of ceramic permanent magnets and are used in a variety of household and technical products. Hexagonal ferrites, also known as hexaferrites or simply hard ferrites, crystallize in a hexagonal magnetoplumbite structure. These M-type called hexaferrites have the general formula $MFe_{12}O_{19}$, M = Ba^{2+} or

Sr^{2+}, and are typically abbreviated as BaM and SrM. A strong uniaxial anisotropy causes the M-type ferrites to be very robust against demagnetization (magnetically hard) and therefore also very useful as permanent magnets for all kinds of industrial and household products and applications, as well as small electric motors, microwave devices, magneto-optic media, telecommunication, electronic industry, loudspeaker magnets and as magnetic recording media, e.g. on magnetic strip cards. Like for any other functional material, the magnetic properties of M-type hexagonal ferrites do not only depend on their structure and chemical composition, or the temperature of operation. They are also highly influenced by the method used for the synthesis of the magnetic powders, as well as by the strategies followed to consolidate the powders, and very importantly, by the magnetic alignment or orientation of the densified piece [151].

10.3.3.3.2 Soft-magnetic ceramics

Mainly non-permanent magnets based on MnZn ferrites and NiZn ferrites are used for electrical transformers, inductors and antennas. $Y_3Fe_5O_{12}$, yttrium iron garnet (YIG), is a soft ferrite and mainly used as core electromagnetic frequency transformer, radio antennas, etc. Industrial electronic components made of soft ferritic materials are primarily used in the application fields of high-frequency technology. The system of manganese-zinc ferrites (MnZn ferrites) as carriers of soft magnetic properties has played a leading role since the beginning of ferrite development. The high-magnetic moment of this material group shows a strong application potential. However, this requires further efforts in research and development to improve and optimize this material system. The possible applications of MnZn ferrites in the field of high-frequency technology are manifold [152–154].

10.3.3.3.3 Semi-hard magnetic ceramics

The properties of cobalt ferrite with the chemical formula $CoFe_2O_4$ ($CoO - Fe_2O_3$) lie between those of soft and hard magnetic materials. It is therefore usually often classified as a semi-hard magnetic material [155]. Due to its high saturation magnetostriction, it is mainly used for magnetostrictive applications such as actuators and sensors. Magnetic field-assisted densification, magnetic annealing or uniaxial densification during spark plasma sintering allow the adjustment of the magnetostrictive properties by inducing uniaxial magnetic anisotropy [156].

Spinel ferrites are synthesized by various techniques such as sol-gel, hydrothermal, ultrasonically assisted hydrothermal processes, reverse micelle synthesis, citrate precursor techniques, electrochemical synthesis, combustion methods, solid-state reaction and mechanical alloying, among them, also the co-precipitation shows a great promise as an effective synthetic approach [157].

10.3.3.4 Ceramic catalysts

In the author's opinion, catalytically active materials should also be considered as "functional". And even though the catalytic processes of chemical oxidation and reduction considered here are caused by electron transfer reactions, ceramic catalysts should not be regarded as electroceramics, but from the special point of view of their specific catalytic effect mode of operation. Therefore, a small separate chapter is devoted to this group of materials.

When ceramics are referred to in connection with catalysis, they are mostly addressed in the form of temperature and chemical-resistant substrates with suitable mechanical strength, often cordierite $Mg_2Al_3[AlSi_5O_{18}]/Mg_2Al_4Si_5O_{18}$ (or also, depending on the requirements, e.g. of aluminum oxide (Al_2O_3) [158]) finds application. Typical forms are granules, open-cell foams or extruded blocks with parallel channels in a honeycomb structure. The actual solid catalysts may either be intermixed with the catalyst carrier phase or applied as coatings, also called wash coats. These can be inorganic, non-metallic, i.e. ceramic powders, immobilized on the substrate or also precious metals (e.g. palladium and/or platinum). More rarely, the ceramic itself is the catalyst, most often it is applied also as a coating on inert ceramic supports, e.g. in catalytic reactors or on ceramic products with catalytic surfaces, such as bricks, tiles, etc.

10.3.3.4.1 Photocatalytic ceramics

Heterogeneous photocatalysis based on semiconducting inorganic materials has attracted considerable interest in recent decades and is used in a variety of applications for the purification of water, air and surfaces, as well as for solar-assisted energy generation by water splitting (see below). Basically, photocatalysis is caused by the excitation of an electron by a photon. The electron is raised to an energy level in the conduction band, leaving a hole in the valence band. These generated charge carriers enable redox reactions at the catalyst surface, which can be used for the degradation of pollutants or for antiviral, antibacterial or antifungal purposes [159]. Titanium dioxide (TiO_2) is a semiconductor whose band gap, especially in the anatase modification, causes the generation of electron-hole pairs when irradiated with UV radiation and is superior compared to other photocatalysts such as ZnO, ZrO_2, SnO_2, WO_3, CeO_2, ZnS and Fe_2O_3 [160]. As described in the introduction to this chapter the catalysts are typically immobilized on support structures or product surfaces [161]. In the presence of water, reactive hydroxyl radicals can thus be formed, which on the one hand lead to the oxidative degradation of pollutants. Thus, organic pollutants can be degraded in wastewater [162], and harmful gases as NO_x and volatile organic pollutants (VOCs) in the air [163]. In combination with another effect, the superhydrophilicity of such surfaces, which also occurs with the process, surfaces can be created that are easy to clean. In addition, anti-microbial, antiviral and antifungal properties can be obtained. Through suitable doping of the TiO_2, it is possible to modify the band gap in such a way that the desired effects can also be stimulated in the visible spectral range. Supported TiO_2 catalysts in ceramic materials were extensively reviewed recently for the photocatalytic degradation of contaminants in liquid effluents [164].

10.3.3.4.2 Photoelectrochemical ceramics

Functional ceramics are also of fundamental and technological interest in the context to energy application. Some ceramics are excellent candidates in view of their optical, mechanical, thermal, electrical and corrosion-resistant properties as photocatalytic material for water splitting under the presence of photon and electrical energy, with respect to water redox levels. Photoelectrochemical (PEC) water splitting enables the direct use of solar energy to generate hydrogen and oxygen and is thus considered a promising contribution to the future energy supply. Preferred long-term stable materials under oxidizing conditions in aqueous solution are semiconductor oxides such as TiO_2 [165, 166], ZnO [167], WO_3 [168] and $BiVO_4$ [169], as photoelectrode materials [170].

10.3.3.4.3 Catalytic ceramics for oxidative gas purification

Some ceramic materials have been proven very efficient for gas purification application, such as oxides of Mn, Co, Fe and Cu, which are employed for the catalytic oxidative removal of volatile organic compounds (VOC) [171], and the catalytic oxidation of toxic carbon monoxide (CO) and nitrogen monoxide (NO). A very efficient oxidative catalyst system is Hopcalite, mainly a CuO/MnO mixture. One of its main applications is the oxidation of carbon monoxide to carbon dioxide. However, it can also be used for the oxidative degradation of VOCs. Hopcalite mixtures are therefore also used in air filter systems for cleaning the breathing air of divers or in filter components of breathing devices in mining or firefighting. When the gas flows over the surface of the catalyst, carbon monoxide (CO) and oxygen (O_2) molecules attach to the surface of the catalyst particles, where they are oxidized to form carbon dioxide. Also, perovskite catalysts, such as a $LaCuO_3$, $LaMnO_3$ and $LaCoO_3$ [172] and double perovskites, such as La_2CuMnO_6 and La_2CoMnO_6 [173] and even complex compositions like a mixed oxide $Ce_{0.12}Zr_{0.40}Mn_{0.48}O_2$ catalyst on a cordierite support [174] are under investigation in the field of ceramic-based catalytic combustion in recent years.

10.3.3.5 Bioceramics

The most general definition of functional ceramics includes ceramic materials with special optical, electrical or magnetic properties, distinguishing them, on the one hand, from the classical products of silicate ceramics (typically based on the raw material clay), and, on the other hand, from the materials of the so-called structural ceramics, which are used in various applications, always when their high mechanical, thermal and/or chemical resistance is required. In case of conventional implant materials based on alumina (Al_2O_3), zirconia (ZrO_2) or composites thereof (alumina-toughened zirconia – ATZ and zirconia-toughened alumina – ZTA) that are well established in modern surgery and dentistry, bioceramics are often attributed to structural ceramics, due to their high strength, fracture toughness and wear resistance, which makes them very useful as long-lasting materials in joint replacement and as dental root implants. These materials are assessed as highly biocompatible and inert even decades after implantation [175–178].

10.3.3.5.1 Bioinert ceramics with special medical functionality

In modern medicine 8 mol% Y_2O_3-stabilized cubic zirconia (YSZ) ceramics are also under investigation for potential use as transparent windows implanted in the skull for ultrasonic and IR-laser-brain therapy [179, 180].

In the field of aesthetic dental restorations, the highly vitreous glass-ceramic materials (feldspar ceramics, lithium disilicate ceramics and recently developed variations thereof) should be mentioned on the one hand, and on the other hand the still increasing efforts to further enhance the translucency of mechanically high load-bearing zirconia ceramics by optimizing the composition and microstructure in the sintered state [181–183]. In addition to the described use of glass-ceramic and ceramic materials in the replacement of natural teeth or tooth structure, highly translucent ceramic brackets have also been successfully implemented in modern dentistry for many years for the aesthetic and barely visible correction of tooth misalignments [184].

10.3.3.5.2 Bioactive functionality, bioactivity

Finally, in contrast to the above mentioned nearly inert materials for joint replacement, some materials also serve as biomaterials and implants in the human body as so-called bone replacements or bone grafts. Serving to fill voids in defect or lost bone (e.g. after the removal of tumors or cysts) and as bioactive coating on metal-based implants for improved ingrowth. Such ceramics are typically based on bioactive hydroxyapatite ($Ca_5(PO_4)_3OH$, HOAp), resorbable ß-tricalciumphosphate $Ca_3(PO_4)_2$ and biphasic calcium phosphate (BCP) mixtures, and therefore might also be attributed to the group of functional ceramics. Here only a small insight can be given into this fascinating world of biofunctional ceramics, with a reference to the literature [185–191].

10.3.4 Conclusions and outlook

The aim of this chapter was to show how versatile functional ceramics are and what a great influence they have on almost all areas of our daily private and professional lives, from technology in industry and commerce, to household, transport, in the field of data storage and telecommunications, to various areas of modern medicine.

Functional ceramic materials offer many fascinating optical, electrical, magnetic, biological and catalytic effects and thus enable a wide range of applications without which our modern life would be almost inconceivable. Further developments both in the field of materials research (new materials and microstructure design) and in the manufacturing and processing procedures will lead to further exciting developments in the future.

In the author's opinion, it can be assumed that, in addition to the ceramic materials themselves, the development of complex composite materials and functionalization of non-functional materials will continue and that on that base more and more multi-functional materials will be developed in the future.

In combination with other trends such as the miniaturization of components and the use of additive manufacturing of components – also with the possibilities of modern multi-material 3D printing technologies – many exciting developments can be expected.

References

[1] Coble RL. Transparent Alumina and Method of Preparation. 3026210, A US Patent, 20 March 1962.

[2] Krell A, Klimke J. J Am Ceram Soc, 2006, 89, 1985–92.

[3] Tsukuma K. J Ceram Soc Jpn, 2006, 114, 802.

[4] Parish MV, Pascucci MR, Gannon JJ, Harris DC. Window Dome Technol Mater Proc XV, 2017, 89, 101790G.

[5] Ramisetty M, Sastri S, Kashalikar U, Goldman LM, Nag N. Am Ceram Soc Bull, 2013, 92, 20.

[6] Kintaka Y, Kuretake S, Tanaka N, Kageyama K, Takagi H. J Am Ceram Soc, 2010, 93, 1114–9.

[7] Kintaka Y, Kuretake S, Hayashi T, Tanaka N, Ando A, Takagi H. J Am Ceram Soc, 2011, 94, 4399–403.

[8] Kintaka Y, Hayashi T, Honda A, Yoshimura M, Kuretake S, Tanaka N, Ando A, Takagi H. J Am Ceram Soc, 2012, 95, 2899–905.

[9] Tsukuma K, Yamashita I, Kusunose T. J Am Ceram Soc, 2008, 91, 3.

[10] Zwick M, Werner J, Kratz N, Vetter S. DE Patent Application 10 2017 104 168 A1, 2018.

[11] Zwick M, Werner J. DE Patent Application 10 2017 104 166 A1, 2018.

[12] Lide DR (Ed). CRC Handbook of Chemistry and Physics, 90th edition, CRC Press/Taylor and Francis, Boca Raton, Index of Refraction of Inorganic Crystals, 10-245-10-248, 2010.

[13] Haertling GH. J Am Ceram Soc, 1999, 82, 797.

[14] Debely PE, Gunter P, Arend H. J Am Ceram Soc, Bull, 1979, 58, 606.

[15] Dubernet P, Ravez J, Pigram A. Phys Stat Sol, 1995, 152, 555.

[16] Yoshikawa K, Asaka T, Higuchi M, Azuma Y, Katayama K. Ceram Int, 2008, 34, 609.

[17] Murray T, Dungan R. Ceram Ind, 1964, 82, 74.

[18] Haertling G, Land C. J Am Ceram Soc, 1971, 54, 1.

[19] Haertling G. J Am Ceram Soc, 1971, 54, 303.

[20] Haertling G. Ferroelectrics, 1987, Vol. 75, 25.

[21] Haertling G. J Am Ceram Soc, 1999, 82, 797.

[22] Wang X, Huang J, Tang Y. Adv Mech Eng, 2015, 7, 1.

[23] Jiang H, Zou Y, Chen Q, Li K, Zhang R, Wang Y, Ming H, Zheng Z. Proc SPIE, 2005, 5644, 598.

[24] Glebov A, Smirnov V, Le M, Glebov L, Sugama A, Aoki S, Rotar V. IEEE Photon Technol Lett, 2007, 19, 701.

[25] Liberts G, Bulanovs A, Ivanovs G. Ferroelectrics, 2006, 333, 81.

[26] Wei F, Sun Y, Chen D, Xin G, Ye Q, Ai H, Qu R. IEEE Photon Technol Lett, 2011, 23, 296.

[27] Ye Q, Dong Z, Fang Z, Qu R. Opt Expr, 2007, 15, 16933.

[28] Kawaguchi T, Adachi H, Setsune K, Yamazaki O, Wasa K. Appl Opt, 1984, 23, 2187.

[29] Thapliya R, Okano Y, Nakamura S. J Lightwave Technol, 2003, 21, 1820.

[30] Sabat R, Rochon P. Ferroelectrics, 2009, 386, 105.

[31] Blasse G, Bril A. Appl Phys Lett, 1967, 11, 53.

[32] Nakamura S, Fasol G. The Blue Laser Diode: GaN Based Light Emitters and Lasers, Springer, Berlin, 1997.

[33] Chen L, Lin CC, Yeh C-W, Liu R-S. Materials, 2010, 3, 2172.

[34] Raukas M, Kelso J, Zheng Y, Bergenek K, Eisert D, Linkov A, Jermann F. ECS J Solid State Sci Technol, 2013, 2, R3168.

[35] Hua H, Feng S, Ouyang Z, Shao H, Qin H, Ding H, Du Q, Zhang Z, Jiang J, Jiang H. J Adv Ceram, 2019, 8, 389.

[36] Müller-Mach R, Müller G, Krames M, Shchekin O, Schmidt P, Bechtel H, Chen C-H, Steigelmann O. Phys Stat Solid RRL, 2009, 3, 215.

[37] Schnick W. Phys Stat Solid RRL, 2009, 3, A113.

[38] Sijbom H, Verstraete R, Joos J, Poelman D, Smet P. Opt Mater Exp, 2017, 7, 3332.

[39] Cozzan C, Brady M, O´Dea N, Levin E, Nakamura S, DenBaars S, Seshadri R. AIP Adv, 2016, 6, 105005.

[40] Pricha I, Rossner W, Moos R. J Am Ceram Soc, 2015, 99, 211.

[41] Schricker A, Mai K, Basin G, Mackens U, Vogels J, Wejers A, Zijtfeld K. US Patent 10,205,067 B2, 2016.

[42] Katelnikovas A, Plewa J, Sakirzanovas S, Dutczak D, Enseling D, Baur F, Winkler H, Kareiva A, Jüstel T. J Mater Chem, 2012, 22, 22126.

[43] Böhnisch D, Baur F, Jüstel T. Dalton Trans, 2018, 47, 1520.

[44] van de Haar M, Werner J, Kratz N, Hilgerink T, Tachikirt M, Honold J, Krames M. Appl Phys Lett, 2018, 112, 132101.

[45] Ikesue A, Furusato I, Kamata J. J Am Ceram Soc, 1995, 78, 225.

[46] Sanhgera J, Kim W, Villalobos G, Shaw B, Baker C, Frantz J, Sadowski B, Aggarwal I. Ceram Laser Mater, 2012, 5, 258–77.

[47] Fabrichnaya O, Seiffert HJ, Ludwig T, Aldinger F, Navrotsky A. Scand J Metall, 2001, 30, 175.

[48] Maiman T. Nature, 1960, 187, 493.

[49] Carnall E, Hatch E, Parsons W. Mater Sci Res, 1966, 3, 165.

[50] Ikesue A, Kinoshita T, Kamata K, Yoshida K. J Am Ceram Soc, 1995, 78, 1033.

[51] Lu J, Ueda K, Yagi H, Yanagitani T, Akiyama Y, Kaminskii AA. J Alloys Compd, 2002, 341, 220.

[52] Bishop B. Northrop Grumman scales new hights in electric laser power, achieves 100 kW from a solid-state laser (March 18th, 2009), https://news.northropgrumman.com/news/re leases/photo-release-northrop-grumman-scales-new-heights-in-electric-laser-power-achieves-100-kilowatts-from-a-solid-state-laser; last accessed on February 24th, 2022.

[53] Aubry P, Bensalah A, Gredin P, Patriarche G, Vivien D, Mortier M. Opt Mater, 2009, 31, 750.

[54] Aballea P, Suaganuma A, Druon F, Hostalrich J, Georges P, Gredin P, Mortier M. Optica, 2015, 2, 288.

[55] Zhou W, Cai F, Zhi G, Mei BN. Mater Sci Pol, 2014, 32, 358.

[56] Sun Z, Mei B, Li W, Liu X, Su L. Opt Mater, 2017, 71, 35.

[57] Ikesue A, Aung Y, Taira T, Kamimura T, Yoshida K, Messing G. Annu Ref Mater Res, 2006, 36, 397.

[58] Lupei V. Opt Mater, 2009, 31, 701–6.
[59] Sanghera J, Kim W, Villalobos G, Shaw B, Baker C, Frantz J, Sadowski B, Aggarwal I. Opt Mater, 2013, 35, 693.
[60] Derenzo S, Weber M, Bourret-Courchesne E, Klintenberg MK. Nucl Instrum Meth, 2003, 505, 111.
[61] Wang Z, Zhou G, Zhang J, Qin X, Wang S. Opt Mater, 2017, 71, 5.
[62] Kuntz J, Roberts J, Hough M, Cherepy N. Scr Mater, 2007, 57, 960.
[63] Cherepy N, Seeley Z, Payne S, Beck P, Swanberg E, Hunter S, Ahle L, Fisher S, Melcher C, Wei H, Stefanik T, Chung Y-S, Kindem J. Proc SPIE, 2017, 9213, 921302.
[64] Osipov V, Ishchenko A, Shitov V, Maksimov R, Lukyashin K, Platonov V, Orlov A, Osipov S, Yagodin V, Viktorov L, Shulgin BV. Opt Mater, 2017, 71, 98.
[65] Ye Y, Liu P, Yan D, Xiu X, Zhang J. Opt Mater, 2017, 71.
[66] Dujardin C, Auffray E, Bourret-Courchesne E, Dorenbos P, Lecoq P, Nikl M, Vasilev A, Yoshikawa A, Zhu R-Y. IEEE Trans Nucl Sci, 2018, 65, 1977.
[67] Buchanan RC (Ed). Ceramic Materials for Electronics, 3rd edition, CRC Press, Boca Raton, 2004, 692.
[68] Rowland D. Electr Eng, 1936, 55, 1142.
[69] Rigterink MD. J Am Ceram Soc, 1958, 41, 501.
[70] Hawley R. Vacuum, 1968, 18, 383.
[71] Nanni P, Viviani M, Buscaglia V. In: Nalwa HS (Ed). Handbook of Low and High Dielectric Constant Materials and Their Applications, Academic Press, San Diego, 1999.
[72] Penn S, Alford N. In: Nalwa HS (Ed). Handbook of Low and High Dielectric Constant Materials and Their Applications, Academic Press, San Diego, 1999.
[73] Roman OV, Shmuradko VT, Panteleenko FI. Refract Ind Ceram, 2021, 61, 499.
[74] Akishin GP, Turnaev SK, Vaispapir VY, Gorbunova MA, Makurin YN, Kiiko VS, Ivanovskii AL. Refract Ind Ceram, 2009, 50, 465.
[75] Slack GA, Tanzilli RA, Pohl RO, Vandersande JW. J Phys Chem Solids, 1987, 48, 641.
[76] Watari K, Shinde SL. MRS Bull, 2001, 26, 440.
[77] Kuramoto N, Takada K. Key Eng Mater, 2003, 247, 467.
[78] Narang SB, Bahel S. J Ceram Proc Res, 2010, 11, 316.
[79] Pan M-J, Randall CA. IEEE Electr Insul Magn, 2010, 26, 44.
[80] Kishi H, Mizuno Y, Chazono H. Jpn J Appl Phys, 2003, 42, 1.
[81] IEC 60384-1:2021, Fixed capacitors for use in electronic equipment – Part 1: Generic specification, International Standard 2021,163. https://webstore.iec.ch/publication/62499 (Accessed march 3rd 2022)
[82] Ho J, Jow TR, Boggs S. IEEE Electr Insul Mag, 2010, 26, 20.
[83] Yao ZH, Song Z, Hao H, Yu Z, Cao M, Zhang S, Lanagan MT, Liu H. Adv Mater, 2017, 29, 1601727.
[84] Palneedi H, Peddigari M, Hwang G-T, Jeong D-Y, Ryu J. Adv Funct Mater, 2018, 28, 1803665.
[85] Prateek, Thakur VK, Gupta RK. Chem Rev, 2016, 116, 4260.
[86] Ogihara H, Randall CA, Trolier-McKinstry S. J Am Ceram Soc, 2009, 92, 1719.
[87] Buscaglia V, Buscaglia MT, Canu G. In: Pomeroy M (Ed). Encyclopedia of Materials: Technical Ceramics and Glasses, Elsevier, Amsterdam, 2021, 311.
[88] Tehrani Z, Thomas D, Korochkina T, Phillips C, Lupo D, Lehtimäki S, O'Mahony J, Gethin D. Energy, 2017, 118, 1313.
[89] Bueno PR. J Power Sources, 2019, 414, 420.
[90] Zeng X, Song H, Shen Z-Y, Moskovits M. J Materiomics, 2021, 7, 1198.
[91] Kumar NS, Suvarna RP, Naidu KCBN, Boddula R. Frontiers Ceram Sci, 2020, 3, 12.
[92] Kamutzki F, Schneider S, Barowski J, Gurlo A, Hanaor DAH. J Eur Ceram Soc, 2021, 41, 3879.

[93]   Narang SB, Bahel S. J Ceram Proc Res, 2010, 11, 316.

[94]   Christodoulou CA, Vita V, Mladenov V, Ekonomou L. Energies, 2018, 11, 3046.

[95]   Hinrichsen V. Metal-oxide Surge Arresters in High-voltage Power Systems – Fundamentals, 3rd edition, Siemens, Erlangen, Germany, 2011. www.siemens.com/energy/arrester accessed march 3rd 2022.

[96]   Chen X, Li X, Gao B, Kong W, Zhao P, Chang A. J Eur Ceram Soc, 2022, 42, 2561.

[97]   Feteira A. J Am Ceram Soc, 2009, 92, 967.

[98]   Chen YL, Yang SF. Adv Appl Ceram, 2011, 110, 257.

[99]   He J. Introduction of varistor ceramics. In: He J (Ed). Metal Oxide Varistors, Wiley-VCH, Weinheim, 2019, 1–30.

[100] Meshkatoddini MR. Metal oxide ZnO-based varistor ceramics. In: Sikalidis C (Ed). Advances in Ceramics – Electric and Magnetic Ceramics, Bioceramics, Ceramics and Environment, InTech Open, Rijeka, Croatia, 2011, 329.

[101] Akbar S, Dutta P, Lee C. Int J Appl Ceram Techn, 2006, 3, 302.

[102] Gurlo A. Ceramic gas sensors. In: Riedel R, Chen I-W (Eds). Ceramics Science and Technology, Wiley-VCH, Weinheim, 2014.

[103] Pelino M, Cantalini C, Faccio M. Act Pass Electr Comp, 1994, 16, 69.

[104] Patil SJ, Patil AV, Dighavkar CG, Thakare KS, Borase RY, Nandre SJ, Deshpande NG, Ahire RR. Front Mater Sci, 2015, 9, 14.

[105] Jaouali I, Hamrouni H, Moussa N, Ncib MF, Centeno MA, Bonavita A, Neri G, Leonardi SG. Ceram Int, 2017, 44, 4183.

[106] Krishna Prasad NV, Prasad V, Ramesh S, Phanidhar SV, Venkata Ratnam K, Janardhan S, Manjunatha H, Sarma MSSRKN, Srinivas K. Front Mater, 2020, 7, 2020.

[107] Ji H, Zeng W, Li Y. Nanoscale, 2019, 11, 22664.

[108] Waqar S, Wang L, John S. Electr Text, 2015, 173–97.

[109] Jaffe H. J Am Ceram Soc, 1958, 41, 494.

[110] Gonnard P, Troccaz M. J Solid State Chem, 1978, 23, 321.

[111] Panda K, Sahoo B. Ferroelectrics, 2015, 474, 128.

[112] Maeder MD, Damjanovic D, Setter N. J Electroceram, 2004, 13, 385.

[113] Damjanovic D, Klein N, Li J, Porokhonskyy V. Function Mater Lett, 2010, 3, 5.

[114] Acosta M, Novak N, Rojas V, Patel S, Vaish R, Koruza J, Rossetti GA Jr, Rödel J. Appl Phys Rev, 2017, 4, 041305.

[115] Liu W, Cheng L, Li S. Curr Comput-Aided Drug Des, 2019, 9, 179.

[116] Yan X, Zheng M, Gao X, Zhu M, Hou Y. J Mater Chem C, 2020, 8, 13530.

[117] Gazda M, Mielewczyk-Gryń A. Crystals, 2019, 9, 173.

[118] Kim KJ, Lim K-Y, Kim Y-W. J Ceram Soc Jpn, 2014, 122, 963.

[119] Wötting G, Martin W. Ceram Appl, 2014, 2, 1.

[120] Zheng J, Lu K. J Am Ceram Soc, 2021, 104, 2460.

[121] VandeGoor G, Stigesser P, Berroth K. Solid State Ionics, 1997, 101–103, 1163.

[122] Van der Biest O, Vleugels J. J Eur Ceram Soc, 2007, 27, 1247.

[123] Honma T, Okamoto M, Togashi T, Ito N, Shinozaki K, Komatsu T. Solid State Ionics, 2015, 269, 19–23.

[124] Zhao G, Suzuki K, Seki T, Sun X, Hirayama M, Kanno R. J Solid State Chem, 2020, 292, 121651.

[125] Porz L, Knez D, Scherer M, Ganschow S, Kothleitner G, Rettenwander D. Sci Rep, 2021, 11, 8949.

[126] Fergus JW. J Power Sources, 2010, 195, 4554.

[127] Chen A, Qu C, Shi Y, Shi F. Front Energy Res, 2020, 8, 571440.

[128] Minh Nguyen Q. J Am Ceram Soc, 1993, 76, 563.

[129] Heuer AH, Hobbs LW (Eds). Advances in Ceramics, Vol 3, Am Ceram Soc, Columbus, OH, USA, 1981.

[130] Baumard JF, Papet P, Abelard P. Adv Cer, 1986, 24b, 779.

[131] Cheikh A, Madani A, Touati A, Boussetta H, Monty C. J Eur Ceram Soc, 2001, 21, 1837.

[132] Breeze P. The solid oxide fuel cell. In: Breeze P (Ed). Fuel Cells, Academic Press, New York, 2017, 63–73.

[133] Cassir M, Jones D, Ringuedé A, Lair V. Handbook of Membrane Reactors, Elsevier, Amsterdam, Vol. 2, 2013, 553–606. DOI: doi.org/10.1533/9780857097347.3.553.

[134] Takada K. Encyclopedia of Electrochemical Power Sources, Elsevier, Amsterdam, 2009, 328–36.

[135] Müller-Buschbaum H. Angew Chem Int Ed, 1989, 28, 1472.

[136] Bussmann-Holder A, Keller H. Z Naturforsch, 2019, 75b, 3.

[137] Cava RJ, Batlogg B, van Dover RB, Murphy DW, Sunshine S, Siegrist T, Remeika JP, Rietman EA, Zahurak S, Espinosa GP. Phys Rev Lett, 1987, 58, 1676.

[138] Anderson PW. J Phys Conf Ser, 2013, 449, 012001.

[139] Hadi MA. J Phys Chem Sol, 2020, 138, 109275.

[140] Olivier M, Heß R, Gommeringer A, Kern F, Herrig T, Bergs T ESAFORM, 24th Int Conf Mater Form Proc Liège, Belgium, 2021.

[141] Gommeringer A, Schmitt-Radloff U, Ninz P, Kern F, Klocke F, Schneider S, Holsten M, Klink A. Procedia CIRP, 2018, 68, 22.

[142] Gommeringer A, Kern F, Gadow R. Ceramics, 2018, 1, 4.

[143] Malek O, Lauwers B, Perez Y, De Baet P, Vleugels J. J Eur Ceram Soc, 2009, 29, 3371.

[144] Udawatte CP, Yanagisawa K. J Am Ceram Soc, 2021, 84, 251.

[145] Medvedovski E, Alvarez NA, Szepesi CJ, Yankov O, Lippens P. Funct Bioceram, 2013, 112, 243.

[146] Noudem JG, Lemonnier S, Prevel M, Reddy ES, Guilmeau E, Goupil C. J Eur Ceram Soc, 2008, 28, 41.

[147] Koumoto K, Funahashi R, Guilmeau E, Miyazaki Y, Weidenkaff A, Wang Y, Wan C. J Am Ceram Soc, 2013, 96, 1.

[148] Romo-De-La-Cruz C-O, Chen Y, Liang L, Williams M, Song X. Chem Mater, 2020, 32, 9730.

[149] Modern Ferrite Technology, Springer, Boston, MA, USA, 2006.

[150] Jotania RB, Mahmood SH (Eds). Magnetic Oxides and Composites, Materials Research Forum LLC, Millersville, PA, USA, 2018.

[151] Granados-Miralles C, Jenuš P. J Phys D Appl Phys, 2021, 54, 303001.

[152] Lucke R, Esguerra M, Wrba J. CFI/Ber DKG, 2004, 81, 32.

[153] Lucke R. Adv Res Workshop Mod Transform, 2004, 322.

[154] Ott G, Wrba J, Lucke R. J Magn Magn Mater, 2003, 254–255, 535.

[155] Hosni N. J Alloys Compd, 2016, 694, 1295.

[156] Olabi A. Mater Des, 2008, 29, 483.

[157] Hosni N, Zehani K, Bartoli T, Bessais L, Maghraoui-Meherzi H. J Alloys Compd, 2017, 694, 1295.

[158] Visconti CG. Trans Ind Ceram Soc, 2012, 71, 123.

[159] Qian R, Zong H, Schneider J, Zhou G, Zhao T, Li Y, Yang J, Bahnemann DW, Pan JH. Catal Today, 2019, 335, 78.

[160] Mo J, Zhang Y, Xu Q, Lamson JJ, Zhao R. Atmos Environ, 2009, 43, 2229.

[161] Carneiro JO, Teixeira V, Azevedo S, Fernandes F, Neves J. J Nano Res, 2012, 18–19, 165.

[162] McCullagh C, Robertson JMC, Bahnemann DW, Robertson PKJ. Res Chem Intermed, 2007, 33, 359.

[163] Cao J, Huang Y, Zhang Q. Catalysts, 2021, 11, 1276.

[164] Danfá S, Martins RC, Quina MJ, Gomes J. Molecules, 2021, 26, 5363.

[165] Fujishima A, Honda K. Nature, 1972, 238, 37–8.
[166] Masai H, Sakurai H, Koreeda A, Fujii Y, Ohkubo T, Miyazaki T, Akai T. Sci Rep, 2020, 10, 11615.
[167] Zhang B, Wang ZY, Huang BB, Zhang XY, Qin XY, Li HL, Dai Y, Li YJ. Chem Mater, 2016, 28, 6613.
[168] Liu XE, Wang FY, Wang Q. Phys Chem Chem Phys, 2012, 4, 7894.
[169] Qiu YC, Liu W, Chen W, Chen W, Zhou GM, Hsu PC, Zhang RF, Liang Z, Fan SS, Zhang YG. Sci Adv, 2016, 2, e1501764.
[170] Zheng L, Wang M, Li Y, Ma F, Li J, Jiang W, Liu M, Cheng H, Wang Z, Zheng Z, Wang P, Liu Y, Dai Y, Huang B. Nanomater, 2021, 11, 2404.
[171] Lei Z, Hao S, Zhang L, Yang J, Yusu W. Environ Sci Pollut Res, 2020, 27, 23695.
[172] Zang M, Zhao C, Wang Y, Chen S. J Saudi Chem Soc, 2019, 23, 645.
[173] Pan KL, Pan GT, Chong S, Chang MB. J Environ Sci, 2018, 69, 205.
[174] Azalim S, Brahmi R, Agunaou M, Beaurain A, Giraudon J-M, Lamonier J-F. Chem Eng J, 2013, 223, 536.
[175] Heimann RB. CMU J, 2002, 1, 23.
[176] Thamaraiselvi TV, Rajeswari S. Trends Biomater Artif Organs, 2004, 18, 9.
[177] Bal BS, Garino J, Ries M, Rahaman MN. Hip Int, 2007, 17, 21.
[178] Scholes SC, Joyce TJ. In: Pignatello R (Ed). Advances in Biomaterials Science and Biomedical Applications, IntechOpen, London, 2013.
[179] Damestani Y, Reynolds CL, Szu J, Hsu MS, Kodera Y, Binder DK, Park BH, Garay JE, Rao MP, Aguilar G. Nanomed: Nanotech, Biol Med, 2013, 9, 1135.
[180] Gutierrez MI, Penilla EH, Lorenzo VA, Garay JE, Aguilar G. Adv Healthcare Mater, 2017, 6, 1700214.
[181] Carrabba M, Keeling AJ, Aziz A, Vichi A, Fonzar RF, Wood D, Ferrari M. J Dent, 2017, 60, 70.
[182] Fathy SM, Al-Zordk W, Grawish ME, Swain MV. Dent Mater, 2021, 37, 711.
[183] Pekkan G, Pekkan K, Bayindir BC, Özcan M, Karasu B. Dent Mater J, 2020, 39, 1–8.
[184] Elekdag-Türk S. In: Aslan BI, Uzuner FD (Eds). Current Approaches in Orthodontics, IntechOpen, London, 2018.
[185] Heise U, Osborn JF, Duwe F. Int Orthop, 1990, 14, 329.
[186] LeGeros RZ, Lin S, Rohanizadeh R, Mijares D, Legeros JP. J Mater Sci: Mater Med, 2003, 14, 201.
[187] Bouler JM, Pilet P, Gauthier O, Verron E. Acta Biomater, 2017, 53, 1.
[188] Eliaz N, Metoki N. Materials, 2017, 10, 334–437.
[189] Dorozhkin SV. J Ceram Sci Technol, 2018, 9, 353.
[190] Brunello G, Panda S, Schiavon L, Sivolella S, Biasetto L, Del Fabbro M. Materials, 2020, 13, 1500.
[191] Hench LL. J Mater Sci: Mater Med, 2015, 26, 86–9.

10.4 Structural ceramics

Nadja Kratz

10.4.1 Definition

The term structural ceramics is a classification of ceramics that have particularly high mechanical strengths, e.g. flexural strengths up to 1 GPa, toughness, as well as corrosion resistance, thermal shock resistance and reliability – even under application conditions such as temperatures >1000 °C. Advanced ceramics are defined in DIN V ENV 12212 as "highly developed, high-performance ceramic material that is predominantly non-metallic and inorganic and possesses certain useful properties". This group of ceramics is also referred to as engineering ceramics or technical ceramics, which, like structural ceramics, are not standardized terms. Unlike traditional ceramics, which are made from naturally occurring raw materials, structural ceramics are made primarily from synthetically produced raw materials. Traditional ceramic products include tiles, bricks, white goods (sanitary ware and tableware) (Chapter 10.1 Sanitary ceramics, tableware, porcelain) and refractory products (Chapter 10.5 Refractories).

10.4.2 Fields of applications

Structural ceramics provide components for the application markets of aerospace, defence, automotive, medical technology, electronics, telecommunications, scientific equipment, paper and textile industry, forming technology, chemical, energy and environmental technology as well as for semiconductor processing. Examples of structural ceramics applications include bearings, seals, armor, linings, nozzles and cutting tools (Chapter 10.7 Abrasives). Figure 10.4.1 shows an overview with examples of products.

The application fields of structural ceramics can be divided into the subgroups of construction ceramics, ballistic protection ceramics, space ceramics, cutting ceramics and grinding ceramics [1]. The so-called construction ceramics are used in machine, apparatus and engine construction. Dominant materials here are silicon carbide, aluminum oxide and dispersion ceramics of aluminum oxide and zirconium dioxide [2–4]. Ballistic protection ceramics are used as armor to protect people, vehicles and objects. Such ceramics have high hardness, high strength and high modulus of elasticity, and the lowest possible density. Relevant materials used here are boron carbide [5], aluminum oxide and for viewing windows transparent spinel ceramics [6]. Another subset of structural ceramics is now considered to be ceramics for space travel. For the re-entry of spacecraft into the atmosphere, high

https://doi.org/10.1515/9783110733471-013

Figure 10.4.1: Examples of products made of advanced ceramics (source/copyright: https://www.ceramtec-industrial.com/de/industrien/uebersicht-a-z).

thermal and thermomechanical stresses (thermal shock) occur, which could only be resisted by fiber reinforcement of the ceramics [1, 7].

Cutting ceramics are dense indexable inserts with defined cutting edges that are used for machining. Turning, milling and grinding tools with carbide cutting edges are widely used in machining technology. With ceramics, higher cutting speeds and thus higher material removal rates can be achieved here since ceramics have a higher (hot) hardness, wear resistance, thermal shock resistance and edge strength than metal tools. Dispersion ceramics of aluminum oxide and zirconium oxide are used as materials, as are mixed ceramics of aluminium oxide and hard materials such as titanium carbide (TiC) and/or titanium nitride (TiN). Compared with oxide ceramics, these materials further increase hardness, hot hardness and wear resistance and significantly improve thermal shock resistance and edge strength. Silicon nitride and boron nitride are also used [8]. For further details, refer to Chapter 10.7. The first grinding tools made of ceramics were produced at the time of the Industrial Revolution by Peter Fuchs in the "Kannenbäckerland". Besides the Westerwald binding clay, emery from the Greek island of Naxos was used as the main ingredient [9]. Emery consists of the mineral corundum (aluminium oxide) with admixtures of other minerals such as magnetite (iron oxide) and spinel (magnesium aluminium oxide). The abrasive ceramic was thus the first ceramic to be used because of its high hardness. It belongs to the group of ceramics used for their special mechanical properties, which are nowadays summarized as structural ceramics [1]. In addition to the classification of ceramics according to application-specific aspects, the classification can also be made according to material-specific aspects: oxide ceramics, non-oxide ceramics (Chapter 10.6) and silicate ceramics (Chapter 10.1) [1, 10].

10.4.3 Production and properties

The relevant materials assigned to structural ceramics are metals or semimetals that form bonds with the elements oxygen, nitrogen, carbon or boron. The respective chemical, electrical, thermal and mechanical properties result from the bond character, which is determined by the proportion of ionic or covalent bonding. Oxide ceramics (such as aluminum oxide and zirconium dioxide) are characterized by a predominant proportion of ionic bonding. In (poly-)crystalline solids, the type of bond also determines the mobility of the atoms and ions in the crystal structure and thus the diffusion properties. Therefore, in general, the melting temperatures of oxides are lower than those of non-oxides. The diffusion properties are decisive for the sintering behaviour of the materials. In the ceramic manufacturing process, material diffusion is used to consolidate the powdered raw materials. The composition of the raw materials and their properties, such as purity, grain size and specific surface area, determines the sintering activity as well as the respective sintering mechanisms that take place. Mechanisms such as solid, liquid or gas phase sintering can also occur in interaction. Due to their properties, oxide ceramics are usually compacted by pressureless sintering processes, i.e. thermal treatment under atmospheric pressure, at temperatures in the range of 1100–1800 °C. Non-oxides (carbides, borides, silicides, nitrides) usually have a high proportion of covalent bonding; therefore, it is very difficult to compress them in their single-phase forms in a pressureless sintering process under oxygen exclusion (inert gas) ($\gg$1800 °C). Hence, sintering aids and/or pressure-assisted sintering processes are often used. Typical shrinkage values are in the range of approx. 18–22%. Exceptions are the SiC variants silicon-infiltrated SiC (SISIC) and recrystallized SiC (RSIC) or certain reaction-bonded ceramics, which show no significant shrinkage. This makes it possible to produce very complex or large components. Non-oxide ceramics are generally characterized by their high hardness, high chemical and thermal resistance, and low coefficient of thermal expansion (see also Chapter 10.6).

Another type of consolidation of ceramic materials is reaction bonding (also related to reaction sintering, reaction forming or self-bonding) [11, 12]. While in the sintering process the material transport takes place via atomic place change processes (one-component reactions), real chemical reactions (two component reactions) cause the consolidation in reaction-bonding. Today, several different consolidation measures are grouped under the term reaction bonding [13]. Reaction bonding can have advantages over sintering. For example, in some processes, consolidation can be achieved without shrinkage being present, which would be unthinkable in the case of sintering. Low process or process initiation temperatures are present in many reaction bonding processes (the latter only in autocatalytic solid-state reactions). Some processes lead to products with a particularly high degree of purity. Due to low process temperatures, even the fibers, which are usually not very thermally stable, can be incorporated, so that high-performance fiber-reinforced ceramics, also known as ceramic matrix composites – CMCs, are possible by reaction bonding. During the material synthesis process, reactions occur between the aggregate states: solid/gaseous, solid/liquid or liquid/gaseous [1, 14, 15]. Table 10.4.1 provides a comparative overview of relevant properties of typical structural ceramics. The respective materials are discussed in more detail in the following sections.

10.4.4 Materials

The materials relevant for structural ceramics are divided here according to the material-specific aspects: oxide ceramics and non-oxide ceramics. In each case, the dominant materials are listed and considered in terms of their most important properties.

Oxide ceramics can be single oxides (such as Al_2O_3, ZrO_2, SiO_2 and MgO) or mixed oxides (such as $BaTiO_3$, $Pb(Zr,Ti)O_3$ and Al_2TiO_5). In this section, the oxide materials dominating structural ceramics, namely alumina (Al_2O_3) and zirconia (ZrO_2), are given in more detail. Initially, the focus is on the single-phase materials in each case. In the further course of the chapter, examples of mixed oxide and dispersion ceramics are explained. Compared to oxide ceramics, non-oxide ceramics are characterized by higher covalent and lower ionic bond contents and represent a material based on the compounds of boron, carbon, nitrogen and silicon. Carbon products from amorphous graphite are not counted among them. The strong bonding energies in the material result in high chemical and thermal resistance, hardness and strength. Carbides possess electrical conductivity. Due to their high covalent bonding content, nitrides are also insulating up to the high temperature range. The main representatives for structural ceramic applications are silicon carbide (SiC), boron carbide (B_4C), boron nitride (BN), silicon nitride (Si_3N_4) and SiAlON ceramics. Carbides, borides and silicides are also discussed in Chapter 10.6.

10.4.4.1 Alumina

Alumina is obtained from the raw material bauxite, which is fabricated by the Bayer process. Depending on further purification and processing, the alumina qualities differ considerably in the content of impurities, mainly SiO_2, Fe_2O_3 and Na_2O. There are over 25 different aluminum hydroxide modifications, which is why calcination plays a decisive role here. At 1020 °C, *alpha*-Al_2O_3 formation starts, so that at higher temperatures, only corundum is stable. The calcination of the alumina also influences the powder characteristics, such as specific surface area, grain size and crystal size distribution, which in turn also influences the processing and sintering properties. Alumina is sintered at temperatures in the range of 1500–1700 °C, depending on the purity or the sintering aids used. Na_2O impurities, caused by leaching in the Bayer process, play a special role. They can accumulate on the grain surfaces and

Table 10.4.1: Comparison of properties of typical structural ceramics.

	Alumina Al$_2$O$_3$ 99.99% [16]	Zirconia ZrO$_2$ Y-TZP [16]	Alumina-toughened zirconia ATZ 20/80 [16]	Zirconia-toughened alumina ZTA 88/12 [16]	Silicon carbide SiC (SSiC) [16]	Silicon nitride Si$_3$N$_4$ (SSN) [16]	Cubic boron nitride 90–95% CBN + 5–10% Co alloy [17, 18]	Boron carbide B$_4$C [19, 20]
Density (g/cm^3)	>3.9	>6.0	>5.4	>4.1	3.1	3.22	3.1	2.52
Flexural strength (MPa)	340	1100	1000	600	420	750	–	–
Young's modulus (GPa)	380	210	220	360	410	300	680	450–470
Fracture toughness (MPam$^{1/2}$)	4.3	10.5	>6.5	4.5	3.2	7	10	2.9–3.7
Coefficient of expansion 20 . . . 1000 °C (10^{-6} K^{-1})	8.5	10.5	9	9	4.1	3.3	3.2	5
Thermal shock resistance (K)	190	300	300	250	350	450	–	–
Max. application temperature (°C) – oxidizing/inert	1650/1650	950/950	1200/1200	1000/1000	1500/1800	1100/1300	900/2000	1000/2200

be present in high concentrations after sintering, correspondingly at the grain boundaries of the ceramic. The alkali content of the alumina therefore largely determines the application properties and thus the quality of a powder. With increasing purity, flexural strength, Young's modulus and thermal conductivity increase. For translucent alumina, starting powders up to a purity >99.95 wt% are used, which are therefore often produced by decomposition of ammonium aluminium sulphate or via other aluminium salts [14, 21]. The production of alumina via the oxidation process was developed by Claussen, Le and Wu in 1989 [22]. The microstructure formed in situ also makes reaction-bonded alumina an excellent matrix material for metal-ceramic composites. Fibers or metallic ligaments can be introduced into the ceramic matrix since the reaction temperatures are relatively low at 400–1000 °C, and thus, the stability of the reinforcing components is not threatened [23]. Al_2O_3 has good strength and dimensional stability, high hardness and wear resistance, and high resistance to acids and alkalis. Examples of applications are tools for forming technology, high-performance substrates and resistance carriers for the electrical industry, tiles in the field of wear protection and ballistics, thread guides in textile technology, sealing and control discs for water taps and valves, heat sinks for lighting technology, electronic substrates, grinding tools as well as protective tubes in thermal processes or catalyst carriers for the chemical industry. It is also used as a corrosion-resistant and inert material in the food industry and for the manufacture of medical implants [11, 21, 24, 25].

10.4.4.2 Zirconia

Naturally occurring minerals are baddeleyite (ZrO_2) and zircon ($ZrSiO_4$). Both minerals are contaminated by the chemically very similar hafnium oxide as well as by other accompanying substances, especially radioactive oxides. Usually, the technically pure ZrO_2 is obtained from $ZrSiO_4$. There are various synthesis approaches for this. One variant to produce particularly fine powder is the thermal decomposition of $ZrSiO_4$ at temperatures between 800 and 1200 °C, whereupon after a reaction with carbon and chlorine, zirconium tetrachloride ($ZrCl_4$) is present as an intermediate product, which is then converted by hydrolysis into zirconium oxychloride and finally calcined [26, 27]. Any radioactive oxide impurities that may remain must be critically controlled, especially when ZrO_2 is used in medical applications.

Zirconia can occur in three modifications: Unstabilized, pure ZrO_2 is monoclinic (*m*) at room temperature, transformation to the tetragonal (*t*) phase occurs at approx. 1170 °C and the cubic (*c*) modification forms at approx. 2370 °C. The transformation to the high temperature structural modifications is due to thermally induced vacancies on oxygen sites. The high mobility of oxygen anions also accounts for the special property of ZrO_2, enabling its use as an oxygen sensor and oxygen ion conductor in functional applications [11]. A well-known example here is the oxygen ceramics made from partially pressure probe (lambda sensor) used in automotive applications.

Zirconia offers high chemical and corrosive resistance at high temperatures up to 2400 °C – well above the melting point of alumina. In its pure form, due to changes in crystal structure, it is only suitable in certain ways for mechanical or thermal applications. However, the microstructure and resulting properties of ZrO_2-based ceramics can be varied over a wide range. Claussen [28] and Rösler [29] give good overviews of this. Ceramics made from partially stabilized zirconia (PSZ) and tetragonal zirconia polycrystals (TZP) have achieved technological relevance. Stabilized zirconia with calcium, magnesium or yttrium oxide additives can be used to meet the highest requirements in terms of strength, hardness or even toughness. The material has low thermal conductivity (20% that of aluminum oxide) and is ionic conductive above 600 °C, which is very advantageous for applications such as fuel cells. Other typical applications include precision ball valves (balls and seats), high-density grinders, thread guides, cutting blades, medical prostheses, pump seals, valves and drive gears, high-frequency heating elements and components for measuring instruments [24]. Due to its relatively high coefficient of thermal expansion ($\alpha = 11 \cdot 10^{-6}$ K^{-1}) [30], ZrO_2 is ideally suited for material composites with steels. Figure 10.4.2 shows a typical polycrystalline structure of a densely sintered yttrium-stabilized ZrO_2 ceramic.

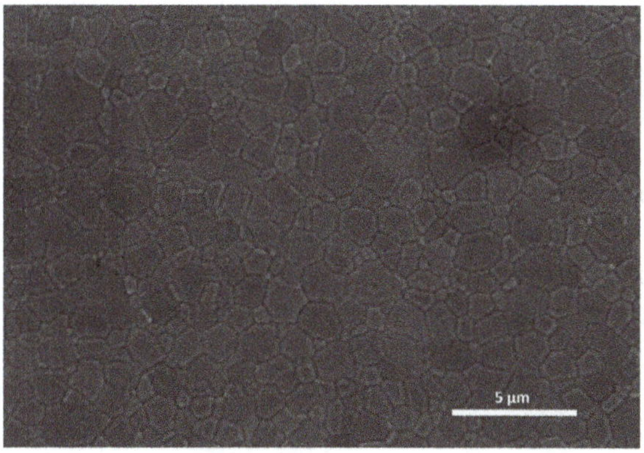

Figure 10.4.2: Polycrystalline microstructure of a densely sintered yttrium-stabilized ZrO_2 ceramic (copyright: FGK, Forschungsinstitut für Glas/Keramik GmbH).

10.4.4.3 Mixed oxide ceramics: dispersion ceramics

As already mentioned in the case of zirconia, pure materials often do not have sufficient mechanical properties. Thus, only partially stabilized zirconia (PSZ) and tetragonal zirconia polycrystals (TZP) ceramics have achieved technical relevance. So-called mixed oxide ceramics or dispersion ceramics are mixtures of individual oxides that have been specifically developed to optimize the required properties. Another example is zirconia toughened alumina ceramics (ZTA). These ceramic composites have not only the hardness and wear resistance of aluminum oxide but also the fracture toughness of zirconium dioxide. Thus, the properties of the mixed ceramics are between alumina and zirconia. The best properties here are shown by AZ15, a mixed ceramic of Al_2O_3 with 15–16% ZrO_2 by volume [11]. This makes them more interesting not only for use in structural ceramics but also for the field of bioceramics. Both alumina and zirconia-based ceramic oxides have long been used as biomaterials (dental ceramics or orthopedic joint replacements) because of their high wear resistance and excellent biomedical compatibility. The functional ceramics barium titanate and lead zirconate titanate (PZT) are also well-known examples of mixed oxides (see also Chapter 10.3). Just as pure alumina is used in a variety of ways, it has also assumed a dominant role in various dispersion ceramics. In some mixed ceramics, non-oxides are also used as a second phase. One example is the hard material Al_2O_3-TiC, which is used as a cutting material for metal cutting and has particularly high book resistance values and bending strength [11]. Another example is a dispersion ceramic of zirconia (Y-TZP) and tungsten carbide (WC), where the bimodal WC reinforcement (40%vol) improved the performance of the composite against wear [31].

10.4.4.4 Silicon carbide

Along with silicon nitride, silicon carbide is one of the most important representatives of non-oxide structural ceramics. SiC is produced on a large industrial scale via the Acheson process from quartz sand and petroleum coke ($SiO_2 + 3\,C \rightarrow SiC + 2\,CO$) at temperatures of 2000 °C. High purity SiC is produced via gas phase reactions from silicon halides and hydrocarbons or by decomposition of methyl or methylchlorosilanes or other organometallic compounds. There is a wide range of materials made of silicon carbide, which differ according to their bonding types (foreign or species-bonded) and manufacturing processes. Dense materials include silicon-infiltrated SiC (SiSiC), sintered SiC (SSiC), liquid phase-sintered SiC (LPSSiC) and hot isostatically pressed SiC (HPSiC, HIPSiC). Materials with open porosity include silicate-bonded SiC, recrystallized SiC (RSiC) and nitride or oxynitride-bonded SiC (NSiC). Most SiC grades are usually sintered at temperatures above 2000 °C in an inert gas atmosphere and have high strength up to very high application temperatures. Due to its very good corrosion resistance, SiC is also used as a hot gas filter, known as a material for diesel particulate filters. Also, the very good thermal shock resistance, low thermal expansion and very high thermal conductivity make silicon carbide a predestined material for high temperature applications, such as high load-bearing kiln furniture (beams, rolls, plates, etc.). A prominent example is brake pads made of carbon-fiber-reinforced silicon carbide (C/SiC), which withstand the special wear and extreme thermal cycling. As a semiconductor, SiC also has good prospects in the field of power electronics. Semiconductors made of silicon carbide could take power electronics in batteries and sensors to a new level, providing significant support for the breakthrough of electromobility and digitalization in industry [32, 33]. Further information of silicon carbide can be found in Telle [11] and Kriegesmann [34], among others.

10.4.4.5 Boron carbide

After diamond, cubic boron nitride and boron oxide, boron carbide are the hardest known materials. Boron carbide powders and pastes are used for lapping and surface finishing. The non-oxide ceramic is characterized by hardness and wear resistance (even at low operating temperatures) and is characterized by exceptionally long service life at low cost. B_4C components are used in extreme wear applications such as linings for mortars, ball mills and as sandblasting nozzles. Boron carbide offers excellent ballistic protection here with reduced weight at the same time and is therefore used for light armor for personal and object protection. Customized ceramic panels made of boron carbide are used here, for example, as are hot-pressed boron carbide panels for pilot seats in helicopters. At temperatures above 1100–1400 °C in a non-oxidizing environment, it is the hardest known material ever. Boron carbide is chemically inert (resistant to hot nitric acid and hydrogen fluoride) and almost insoluble in water. It is not noticeably attacked by chlorine or oxygen until temperatures exceed 1000 °C. In addition, B_4C is electrically conductive and has good thermal conductivity and excellent heat resistance [11, 35].

10.4.4.6 Boron nitride

α-Boron nitride is structurally comparable to graphite (therefore, also called white graphite), and consists of a planar, hexagonal crystal structure in layers. In powder form or as a surface coating, hexagonal boron nitride α-BN is often used as a release agent and lubricant in foundry technology due to its poor wettability with molten metals. In its cubic modification β-BN, it is used as cutting ceramic due to its high hardness, thermal conductivity and thermal stability [36]. The industrially used name CBN is derived from the initial letters in "Cubic Boron Nitride". CBN was first synthesized in 1957 and has been available on the European market since 1969. The ceramic bonds, which have been in use since 1982, made it possible for the first time to dress grinding tools economically. With a Knoop hardness of 45 GPa at room temperature, CBN is the second hardest known material after diamond (88 GPa). Due to the graphitization of diamond, which begins at 700 °C, it can only be used up to a maximum of 900 °C. CBN, on the other hand, can be used up to 1400 °C and is therefore the hardest known material above 900 °C. Since CBN does not contain carbon, it is suitable – unlike diamond – for machining hardened and unhardened steels containing carbon [35, 37].

10.4.4.7 Silicon nitride and SiAlON ceramics

Silicon nitride is another dominant material in the field of structural ceramic non-oxides. As a ceramic with high toughness, strength and excellent thermal shock resistance, it is predestined for machine components with very high dynamic stresses and reliability requirements. Like silicon carbide, the known Si_3N_4 ceramic variants differ in their bonding methods and manufacturing processes. Usually, silicon nitride is sintered at temperatures between 1750 and 1950 °C, as in the case of low pressure sintered silicon nitride (SSN) and gas pressure sintered silicon nitride (GPSSN). In the case of hot-pressed silicon nitride (HPSN) and hot isostatic-pressed silicon nitride (HIPSN), the material is fully densified, which further increases strength and wear and chemical corrosion resistance, and also involves higher process costs. The dense Si_3N_4 materials are used in metalworking as indexable inserts, as balls in rolling bearing technology, as rollers and rings in mechanical engineering and forming technology [38]. Or as valves, turbocharger rotors for internal combustion engines, turbine blades, glow plugs, liquid metal handling, thermocouple sheaths, welding fixtures and fasteners, and welding nozzles [24]. Silicon nitride balls are used as spindle bearings for cutting tools, turbomolecular pumps, dental drills and special bearings for instrumentation [39]. Low-pressure-sintered silicon nitride (SSN) is a relatively inexpensive material that has medium bending strengths and can be used to make large-volume components for metallurgy, for example. Components made of silicon nitride and SIALONe are also used in gas turbine bearings [40, 41]. Silicon aluminum oxynitrides (SIALON) are considered variants of silicon nitride ceramics. This involves enrichment with alumina with the aim of synthesizing defined solid solution forms that achieve similar good properties to silicon nitride at lower sintering temperatures. Here, too, there are many variants whose properties can be varied via the composition. In some cases, SIALON even has a higher fracture toughness, which is why SIALON is used for cutting tools. Due to the low wettability by aluminium or non-ferrous metal melts, SIALON is often used for thermocouple protection tubes in smelting operations [38, 42].

10.4.4.8 Composites

Due to their brittleness, monolithic ceramics are susceptible to defects that induce stresses in the microstructure. Therefore, classical structural ceramic applications are limited to parts subjected to compressive loading or limited tensile or multi-axial loading. Similarly, impact and thermal stress are critical. Porous ceramics, typically used as refractory materials, and some dense monolithic ceramics based on Si_3N_4 and SiC, have good thermal shock properties. Dispersion or fiber reinforcement of ceramic materials significantly increases their tolerance to defects and dynamic or thermal loading by positively affecting crack propagation mechanisms. See also [29]. There are several multilayered approaches to this, which essentially differ in their basic features by the intercalation principle (particles or fibers) and by the matrix material (metal or ceramic). Two groups of composites can be distinguished:

metal-matrix composites (MMCs) and ceramic-matrix composites (CMCs). Systems in which materials with a ceramic matrix are synthesized by selective conversion of metallic or metal-containing powder mixtures occupy a special position within the last group [15, 43].

10.4.5 Development trends

Structural ceramics will continue to be developed further in the future and will have new outstanding properties. The focus will be on increasing service life, failure tolerance and the development of self-healing mechanisms through targeted microstructure design as well as the further development of ceramic manufacturing for microstructure applications. Additive technologies will also contribute to the development of new design methods for micro-components or multifunctional components in this area. Metal-ceramic composites or hard metal-ceramic composites are a predestined basis for the realization of multifunctional components because, in addition to the optimization of the mechanical properties, the electrically conductive second phases enable, for example, sensor technology or heating of the components. In this way, weight can be saved, and the system technology can become smarter. In addition, further approaches to near-net-shape processing will be pursued in the future to be able to manufacture structural ceramic components in an energy and resource-efficient manner. The careful use of resources also includes sustainable recycling concepts for the return of high-quality raw materials and the recovery of rare earths, which must be further developed [44]. Last but not the least, the digitalization of the ceramic manufacturing process will play an essential role as an innovation driver for the efficient and resource-saving production of economically strategic high-performance components [45].

References

[1]   Kriegesmann J, Kratz N. Definition, Systematik und Geschichte der Keramik. Einteilung der Keramik nach anwendungsorientierten Gesichtspunkten. Keram Zeitschr, 2015, 67, 276–87. DOI: 10.1007/BF03400383.
[2]   Ziegler G, Heinrich J, Wotting G. Relationships between processing, microstructure and properties of dense and reaction-bonded silicon nitride. J Mater Sci, 1987, 22, 3041–86. DOI: 10.1007/BF01161167.
[3]   Claussen N. Fracture toughness of Al_2O_3 with an unstabilized ZrO_2 dispersed phase. J Am Ceram Soc, 1976, 59, 49–51. DOI: 10.1111/j.1151-2916.1976.tb09386.x.
[4]   Garvie RC, Hannink RH, Pascoe RT. Ceramic steel? In: Sōmiya S, Moriyoshi Y (Hrsg). Sintering Key Papers, Springer, Dordrecht, Netherlands, 1990. 978-94-009-0741-6.
[5]   US Patent 3,729,372 A.

[6] Krell A, Strassburger E. Order of influences on the ballistic resistance of armor ceramics and single crystals. Mater Sci Eng A, 2014, 597, 422–30. DOI: 10.1016/j.msea.2013.12.101.

[7] Fitzer E, Gadow R. Fiber-reinforced silicon carbide. Am Ceram Soc Bull, 1986, 65, 326–35.

[8] Peters C. Herstellung und Einsatzverhalten von Keramik-Hartmetall-Verbundbohrwerkzeugen, Vulkan-Verlag, Essen, 2006. 9783802787317.

[9] 150 Jahre Fuchs, 1838–1988; Peter Fuchs, Schleifmittel GmbH, Fuchs'sche Tongruben GmbH & Co. KG, Peter Fuchs, Brennhilfsmittel GmbH & Co. KG, Ransbach-Baumbach, Hundsdorf; Geschichte der Firmen, Fuchs P (Hrsg.), Verlag Ransbach-Baumbach. ISBN 10: 3980082741.

[10] Kriegesmann J, Kratz N. Definition, Systematik und Geschichte der Keramik. Einteilung der Keramik nach werkstoffspezifischen Gesichtspunkten. Keram Zeitschrift, 2015, 67, 227–30. DOI: 10.1007/BF03400377.

[11] Telle R (Hrsg). Keramik, Springer, Berlin, Heidelberg, 2007. DOI: 10.1007/978-3-540-49469-0. 978-3-540-63273-3.

[12] Popper P, Davies DGS. The preparation and properties of self-bonded silicon carbide. Powder Metall, 1961, 4, 113–27. DOI: 10.1179/pom.1961.4.8.009.

[13] Haggerty JS, Chiang Y-M. Reaction-based processing methods for ceramics and composites. In: Wachtman JB (Hrsg). A Collection of Papers Presented at the 14th Annual Conference on Composites and Advanced Ceramic Materials, Part 1 of 2: Ceramic Engineering and Science Proceedings, Volume 11, Issue 7/8. Ceramic Engineering and Science Proceedings. Hoboken, John Wiley & Sons, Inc, NJ, USA, 1990, 757–81. DOI: 10.1002/9780470313008.ch19. 9780470313008.

[14] Salmang H, Scholze H, Telle R. Keramik (German Edition), Springer, Dordrecht, 2007. 978–3540494690.

[15] Verbundwerkstoffe. In: Telle R (Hrsg). Keramik, Springer Berlin Heidelberg, 2007, 947–66. DOI: 10.1007/978-3-540-49469-0_11. 978-3-540-63273-3.

[16] BCE Special Ceramics GmbH: Vergleichstabelle Technischer Keramik, 2021. https://www.bce-special-ceramics.de/vergleich/bce-materialtabelle.htm, downloaded at 15. 09.2021

[17] Arunachalam R, Mannan M. Machinability of nickel-based high temperature alloys. Machining Sci Technol, 2000, 4, 127–68. DOI: 10.1080/10940340008945703.

[18] McKie AL. Mechanical properties of cBN-Al composite materials dependence on grain size of cBN and binder content. https://www.semanticscholar.org/paper/Mechanical-properties-of-cBN-Al-composite-materials-McKie/411cd37d3c06f837905cc19ee81c3d1b11ad10c3#extracted, downloaded at 16.09.2021. Master thesis,University of the Witwatersrand, Johannesburg.

[19] 3M Deutschland GmbH: Borcarbid; KERAMIKPROFI © 2021. https://www.technical-ceramics.com/werkstoffe/borcarbid/, downloaded at 20. 09.2021

[20] Lucideon: Boron Carbide (B$_4$C) – Properties and Information about Boron Carbide, 2001. https://www.azom.com/article.aspx?ArticleID=75, downloaded at 20. 09.2021

[21] Foliensatz Technische Keramik für Hochschulen, Informationszentrum Technische Keramik. http://www.keramverband.de/content/pdf/foliensatz75dpi.pdf. downloaded at: 27. 06.2021.

[22] Claussen N, Le T, Wu S. Low-shrinkage reaction-bonded alumina. J Eur Ceram Soc, 1989, 5, 29–35. DOI: 10.1016/0955-2219(89)90006-X.

[23] Claussen N, Janssen R, Holz D. Reaction Bonding of Aluminum Oxide (RBAO). J Ceram Soc Jpn, 1995, 103, 149–58.

[24] Thomas M. Keramische Werkstoffe in Konstruktion und Entwicklung, 2007. https://www.konstruktionspraxis.vogel.de/keramische-werkstoffe-in-konstruktion-und-entwicklung-a-114229/, (ID:228198), downloaded 25. 10.2021

[25] CeramTec GmbH: Aluminiumoxid. https://www.ceramtec-industrial.com/de/werkstoffe/aluminumoxide, downloaded at 27.06.2021

[26] Linsmeier K-D. Technische Keramik. Werkstoff für höchste Ansprüche. Die Bibliothek der Technik, Band 208, Verlag Moderne Industrie, Landsberg/Lech, 2000. 3478932386.

[27] Bengisu M. Engineering Ceramics. Engineering Materials, Springer, Berlin, Heidelberg, 2001. DOI: 10.1007/978-3-662-04350-9. 978-3-642-08719-6.

[28] Rühle M, Claussen N, Heuer AH. (Hrsg). Science and Technology of Zirconia II. Advances in Ceramics, Volume 12, American Ceramic Society, Columbus, Ohio, 1984. 0-916094-64-2.

[29] Rösler J, Bäker M, Harders H. Mechanical Behaviour of Engineering Materials. Metals, Ceramics, Polymers, and Composites, Springer, Berlin, Heidelberg, 2010. 978-3-642-09252-7.

[30] CeramTec GmbH: Zirkonoxid. https://www.ceramtec-industrial.com/de/werkstoffe/zirkonoxid, downloaded at 27.06.2021

[31] Ünal N, Kern F, Öveçoğlu ML, Gadow R. Influence of WC particles on the microstructural and mechanical properties of 3mol% Y_2O_3 stabilized ZrO_2 matrix composites produced by hot pressing. J Eur Ceram Soc, 2011, 31, 2267–75. DOI: 10.1016/j.jeurceramsoc.2011.05.032.

[32] SGL Carbon: Warum Siliziumkarbid-Halbleiter eine große Zukunft haben, 2021. https://www.sglcarbon.com/fuer-eine-smartere-welt/warum-siliziumkarbid-halbleiter-eine-grosse-zukunft-haben/, downloaded at 28. 06.2021

[33] Hain S, Bertelshofer T. Effiziente Auslegung von SiC-Leistungselektroniken für Elektroantriebe. ATZelektronik, 2020, 15, 58–62. DOI: 10.1007/s35658-020-0219-x.

[34] Kriegesmann J. Processing of silicon carbide-based ceramics. Compr Hard Mater, 2014, 2, 89–175. DOI: 10.1016/B978-0-08-096527-7.00023-4.

[35] 3M Deutschland GmbH: Bornitrid; KERAMIKPROFI © 2021. https://www.technical-ceramics.com/werkstoffe/bornitrid/, downloaded at 20. 09.2021

[36] Kreß J. Auswahl und Einsatz von polykristallinem kubischem Bornitrid beim Drehen, Fräsen und Reiben, Vulkan-Verlag GmbH, Essen, 2007. 3802787412.

[37] Kirchgatter M. Einsatzverhalten Genuteter CBN-Schleifscheiben mit keramischer Bindung beim Außenrund-Einstechschleifen, Dissertation, Technische Universität Berlin, Berlin, 2010.

[38] Informationszentrum Technische Keramik: Brevier Technische Keramik. Lauf: Fahner, Verlag Hans Carl, Nürnberg, 1999. 3-924158-36-3.

[39] Pascucci MR, Katz RN. Modern day applications of advanced ceramics. Interceram, 1993, 42, 71–8.

[40] Sibley LB, Zlotnick M. Considerations for tribological application of engineering ceramics. Mater Sci Eng, 1985, 71, 283–93. DOI: 10.1016/0025-5416(85)90238-1.

[41] Bennett A. Requirements for engineering ceramics in gas turbine engines. Mater Sci Technol, 2013, 2, 895–9. DOI: 10.1179/mst.1986.2.9.895.

[42] Izhevskiy VA, Genova LA, Bressiani JC, Aldinger F. Progress in SiAlON ceramics. J Eur Ceram Soc, 2000, 20, 2275–95. DOI: 10.1016/S0955-2219(00)00039-X.

[43] Nestler DJ. Beitrag zum Thema Verbundwerkstoffe – Werkstoffverbunde. Status quo und Forschungsansätze. Chemnitz, Technische Universität, Fakultät für Maschinenbau, Habilitationsschrift, 2013. Chemnitz, Münster: Universitäts-Verlag mv (Monsenstein und Vannerdat) 2014. 978-3-944640-12-9

[44] Deutsche Keramische Gesellschaft, Verband der Keramischen Industrie e.V. u. Deutsche Gesellschaft für Materialkunde e.V.: Expertenstudie – Zukunftspotentiale von Hochleitungskeramiken. http://www.expertenstudie-hlk.dkg.de/, downloaded at 27.06.2021

[45] Gemeinschaftsausschuss ‚Hochleistungskeramik' der Deutschen Keramischen Gesellschaft (DKG) und Deutschen Gesellschaft für Materialkunde: Digitalisierung der keramischen Fertigung – Herausforderungen und Chancen, 2021. https://dgm.de/fileadmin/DGM/Netzwerk/Ausschuesse/GA-Hochleistungskeramik/2021-DKG-DGM-Strategiepapier-Digitalisierung.pdf, downloaded at 27.06.2021

10.5 Refractories

Peter Quirmbach, Almuth Sax

10.5.1 Definition, classification, application

Refractory materials are a material group with essential meaning for widespread industrial high temperature processes, like metallurgy, special steel production, glass production, energy generation and waste incineration. In consequence, on the multiple specifications of the industrial processes, refractories have to fulfill numerous requirements:
– High temperature resistance
– High volume stability
– High chemical resistance/corrosion resistance
– High erosion resistance
– High mechanical stability

According to an international agreement, refractories are non-metallic, ceramic materials with a softening point higher than 1500 °C. For highly refractory materials, the softening point is higher than 1800 °C. The softening point, i.e. the temperature of significant deformation of a refractory body due to beginning melt formation, is traditionally determined by a defined, standardized heating of a so-called pyrometric cone according to ISO/R 836. The pyrometric cone is an approx. 5 cm high-sloped pyramid with triangular base, where one side is shorter than the others. The pyrometric cone is heated with a defined heating rate until the pyramid bends down to the base plate. The corresponding temperature is the fall point and defines the softening point of the material. Nowadays, other physical properties like high temperature compressive strength give more precise information about mechanical behavior and temperature resistance of refractory construction materials.

Due to the primary requirement of a high temperature resistance, refractory materials are based predominantly on high-melting oxides. Table 10.5.1 gives an overview of the commonly used oxides and non-oxides and their melting and decomposition temperature, respectively. Beside oxides, due to its ability to form an oxidic passivation layer, SiC and SiC-Si_3N_4-based refractory materials are also used. For applications with reducing atmosphere also carbon as a monolithic material and also as bonding phase has a broad spectrum of application.

Refractory materials are classified according to different criteria – besides classification, temperature and designated application, the following properties are used:
– Delivery form for the industrial application:
  – shaped products: bricks, disks, crucibles, cubes, etc.
  – unshaped products: concrete, cement, mortar-materials

https://doi.org/10.1515/9783110733471-014

Table 10.5.1: Melting temperature and melting/decomposition temperatures of important unary and binary oxides and non-oxides for use as refractory material.

Compound	Melting temperature [°C]	Compound	Melting/decomposition temperature [°C]
SiO_2	1710	$Al_6Si_2O_{13}$	Approx. 1850
Al_2O_3	2054	$ZrSiO_4$	1675
MgO	2800	$MgAl_2O_4$	2116
CaO	2580	$MgCr_2O_4$	2390
ZrO_2	2680	SiC	>2300
Cr_2O_3	2435	C	>3500

- Chemical behavior:
    - acid: silica, chamotte
    - amphoteric: bauxite, corundum, mullite/andalusite/sillimanite, zirconia, zircon
    - basic: magnesia, magnesite, dolomite, chrome magnesite
- Type of bond:
    - chemically bonded refractories, i.e. phosphate-bonded, silicate-bonded, carbon-bonded
    - hydraulically bonded products, calcium aluminate-cement-bonded
    - ceramically bonded products
    - fused cast products
- Total porosity:
    - dense: <45%vol porosity
    - insulating: >45%vol porosity

With respect to the various fields of application, the chemical resistance, i.e. the type of chemical behavior is the main criterion for selection of suitable refractory products. The iron and steel industry, which is the main consumer of refractories, uses primarily basic products, and thus, basic refractories with 40% are the main part of refractory production. About 30% are alumina-rich refractories, followed by chamotte materials with 14% and silica bricks with 8%.

10.5.2 Fabrication of refractories

In dependence on the requirements of final properties and the type of bond, three main routes of fabrication of refractory materials can be differed:
- The common coarse-ceramic fabrication of products with grain size of the coarse grain in the range from 6 up to 25 mm

- the fine-ceramic fabrication with fine-grained ceramic raw materials with less than 1 mm grain size and
- the fabrication via melt (fused cast products)

The dominating fabrication route for common refractories is the coarse-ceramic production of bricks and also monolithic products.

In the first step, raw materials are crushed by breaking and milling, followed by a fractionation in various grain fractions by sieving or air separation. The batch mixture then is produced by defined quantities of the several grain fractions according to a specific grain fraction receipt. The latter guaranties that a maximum package density of the future brick can be achieved. Usually, at least three grain fractions, coarse grain, medium grain and fine grain, are used. However, modern batch mixture for special products may have a multiple variety of grain fractions with fine grain portions up to 1 μm. Additionally, the grain shape – cubic, splintery, round – is integrated. Table 10.5.2 shows some examples for typical grain compositions of usual refractory products [1–3].

Table 10.5.2: Typical grain compositions of common refractory products [1].

Refractory product	Grain fraction [%]			Addition
	1–5 mm	0.1–1 mm	<0.1 mm	
Silica brick	30	40	30	2.5% chalk/2% water/1.5% lignosulfonate
Chamotte brick (dry-pressed)	35	25	40 (inclusive clay)	15% clay/5% water
Magnesia brick	45	20	35 (10% < 0.06 mm)	2% water/lignosulfonate
Resin-bonded magnesia-carbon brick	55	30 (inclusive 5–20% graphite)	15	3% resin
Conventional refractory concrete	35	30	35 (inclusive cement)	15% cement/9% water
Low-cement refractory concrete	35	25	40 (10% < 0.01 mm)	5% cement/5% ultra-fine oxide powder/5% water/0.1% liquefier

Mixing of the various grain fractions of the specific receipt is usually carried out in a compulsory mixer (intensive mixer), which is able to guarantee an optimal homogenization process of the batch mixture, consisting in different materials of diverse grain size and mineralogy, without segregation by irreversible grain separation [4].

The applied shaping process depends on the moldability (plasticity, water content), the required product properties and the complexity of the final product. The usual shaping techniques are [5, 6]:

- dry pressing by hydraulic uniaxial press of batch mixtures with up to 5% humidity and pressures of 40–200 N/mm^2
- isostatic pressing with pressure up to 1000 N/mm^2 of special products with complex geometry (casting tube, shroud, crucible)
- manually or mechanical ramming, used for complex geometry and small numbers of pieces
- extrusion of plastic batch mixtures and
- slip casting of special products made from mostly fine-grained ceramic batch mixtures with water content of 10–20%

The pressed preforms, especially large parts, have to be completely dried. To avoid drying cracks or other defects, the required time for drying may be some days up to weeks for large formats.

Conventional refractory bricks will be fired subsequent to the drying process at temperatures up to 1200 °C $\leq T \leq$ 1800 °C depending on the brick type. During firing, the characteristic microstructure of the refractory material is generated by solution and precipitation processes, secondary crystallization, phase transformation and liquid phase formation. Figure 10.5.1 illustrates in a schematic drawing the main constituents of the refractory microstructure. Essential parts of the typical microstructure of refractory materials are:

- the coarse grain, which remains mostly unchanged during firing
- the bonding matrix, which is formed just during firing. In most products, the binding matrix is more porous than the coarse grain skeleton
- the pores, their volume and size distribution influence all physico-mechanical properties of the final brick

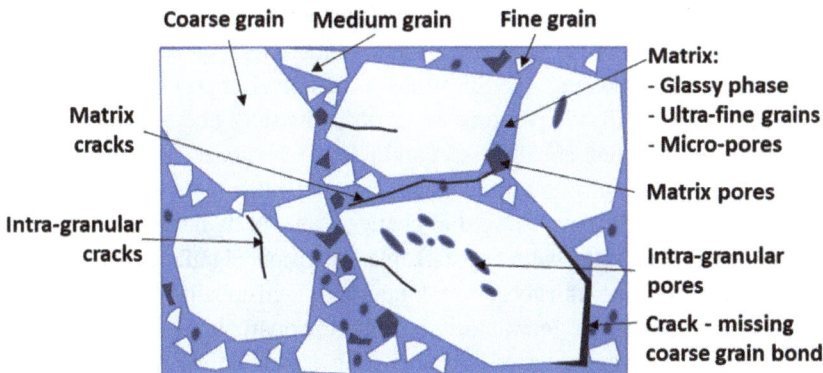

Figure 10.5.1: Schematic representation of the microstructure of refractories after firing.

Chemical-bonded bricks or bricks with organic bonds are used mostly un-fired. They are only tempered at temperatures above 150 °C, but clearly below 1000 °C for improving strength by initiating chemical reactions and to remove volatile components or hydration water.

After firing, for some applications with high requirements with respect to dimensional accuracy, a final treatment by cutting or grinding is necessary.

For transport and storage, common standard pallets are used. Only hydration-sensitive materials, for instance dolomite bricks, has carefully impermeable packaging by the use of polymer or metal foils.

10.5.3 Refractories in the system SiO_2-Al_2O_3

Refractories in the system SiO_2-Al_2O_3 include a large number of common refractory materials for widespread fields of application. In order of silica content, the following materials belong to this group:
– silica bricks with >93% SiO_2
– acid chamotte with 70–90% SiO_2
– chamotte with 55–70% SiO_2
– alumina-rich chamotte with 45–55% SiO_2
– sillimanite and andalusite bricks with 30–45% SiO_2
– mullite bricks with 25–30% SiO_2
– bauxite bricks with 15–25% SiO_2 and
– alumina bricks with ≤ 15% SiO_2

Figure 10.5.2 illustrates the position of the various materials and the typical raw materials in the system SiO_2-Al_2O_3.

Silica bricks have a chemical silica content of at least 93 wt%, and usually, it is between 95 and 97 wt% SiO_2. Typical raw materials are quartzite and pure quartz sand, which mainly consist in the low-temperature modification α-quartz. Considering the unitary system SiO_2, the initial α-quartz modification undergoes a number of structural phase transitions during heating. Modifications of technical importance are quartz with its low (α) and high (β) temperature form, tridymite (α, β and γ-form) and the high-temperature modification cristobalite (α and β-form). Figure 10.5.3 illustrates the different structural phase transitions of SiO_2 with their transition temperatures. During first heating of silica bricks, these structural changes have to be respected since the structural changes are connected with a remarkable expansion. Significant high expansion values occur during the transition from β-quartz to α-cristobalite (+ca. 17%) and from β-quartz to α-tridymite (ca. 14.5%). Additionally, the quartz inversion with ±0.8% vol change at 573 °C has to be respected during heating and cooling of silica bricks.

During firing of silica bricks, the initial α-quartz from quartzite is transferred to the high-temperature modification β-cristobalite. An extremely low heating rate and

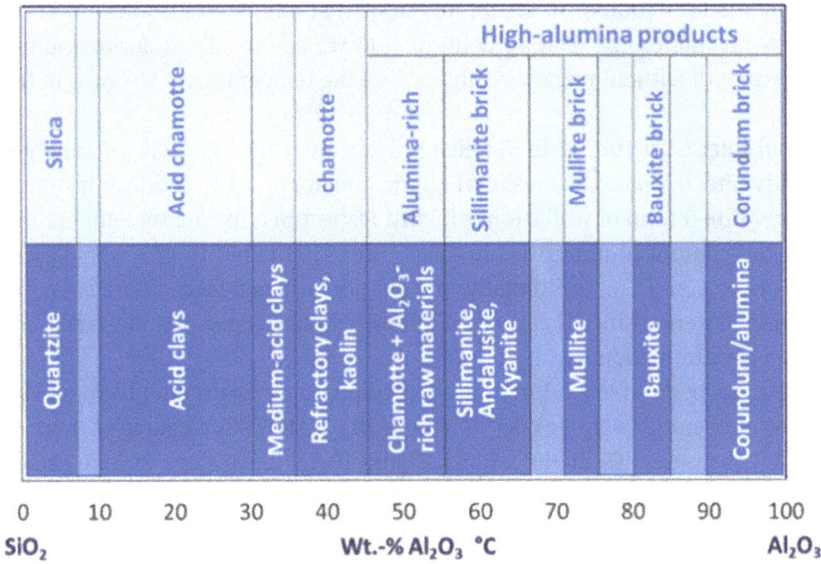

Figure 10.5.2: Refractory materials in the system SiO$_2$-Al$_2$O$_3$ – types of bricks and typical raw materials.

a total duration of the thermal treatment of up to 10 days guarantee, that defects due to volume expansion during change in modification are avoided. Since the β-quartz/α-tridymite-transformation and also the β-tridymite/α-cristobalite-transformation are irreversible, a reconversion in the finally fired product is excluded.

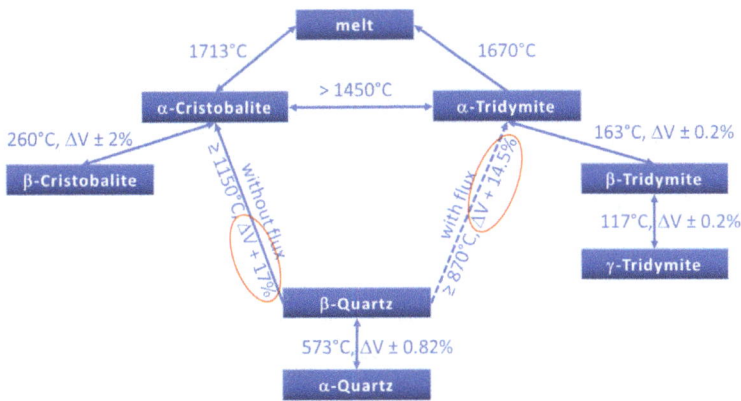

Figure 10.5.3: Structural phase transitions of SiO$_2$.

To the batch mixture of silica bricks, calcium hydroxide is usually added for achieving a final CaO content of 1–4 wt%. During firing, CaO forces the ceramic bond in the bricks by the formation of wollastonite ($CaSiO_3$) at temperatures above 600 °C. $CaSiO_3$ has a favorable high melting point of 1816 °C, so the silicate formation improves the strength without negative influence on the temperature resistance of the bricks.

The final microstructure of fired silica bricks consists of 35–45% cristobalite, 40–50% tridymite, 0.5–max. 3% residual quartz and about 5% pseudo-wollastonite (high-temperature β-form of wollastonite). With respect to a maximum temperature resistance, the content of impurities in silica bricks is limited to <2.5 wt% Fe_2O_3, 1.5% Al_2O_3 and <0.2% TiO_2. Additionally, the quantity of residual α-quartz is an important quality criterion since it causes an undesired after-expansion due to its late modification transformation [7].

A specific property of silica bricks is their comparably low thermal expansion, especially at temperatures above 600 °C. Figure 10.5.4 shows the course of thermal expansion for the pure SiO_2 modifications. Figure 10.5.5 illustrates the comparison of thermal expansion for a number of typical refractory products [8]. The low thermal expansion leads to an excellent thermal shock resistance of silica bricks in the temperature range above 600 °C in operating conditions, where rapid temperature changes or large temperature gradients along the cross-section of the lining occur. Only at lower temperatures T < 600 °C, temperature changes must pass carefully, i.e. slowly.

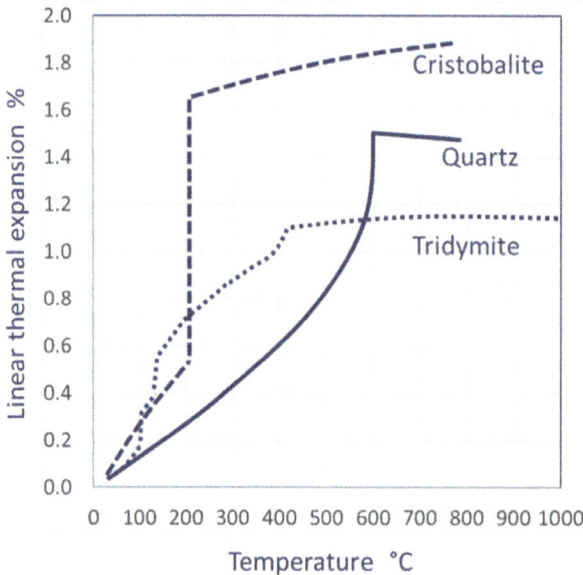

Figure 10.5.4: Thermal expansion of different SiO_2 modifications (according to [8]).

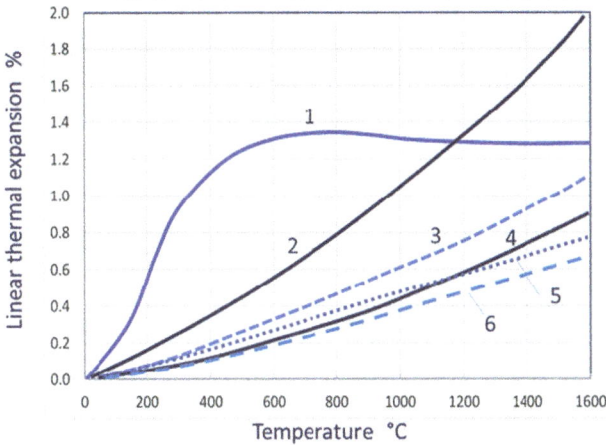

Figure 10.5.5: Thermal expansion of refractory bricks: (1) silica, (2) magnesia, (3) corundum (99% Al_2O_3), (4) corundum (90% Al_2O_3), (5) chamotte, (6) sillimanite (according to [9]).

The base raw materials for chamotte bricks are refractory clays and kaolin with a high clay mineral content. Major constituents with respect to chemical composition are SiO_2 and Al_2O_3 with small amounts (each 0.1–3.0%) of TiO_2, Fe_2O_3, CaO, MgO, K_2O and Na_2O as well as 0.8–2.0% organic substances. The clay is burned at first at temperatures >1200 °C for fabrication of the so-called raw chamotte. After cooling, the raw chamotte is milled, fractionated and then combined to the final batch mix. These batch mixes are formed plastically by extrusion or similar or by hydraulic pressing, dried and then fired at temperatures of 1250 °C $\leq T \leq$ 1500 °C.

During firing, the clay raw material undergoes a series of reactions, at which interlayer and crystal water are released by the formation of metakaolinit, followed by the transformation of metakaolinit into the stable phase mullite and SiO_2-rich amorphous-binding phase:

(1) ~150–180 °C release of adsorbed water and interstitial water
(2) ~450–550 °C dehydration of crystal water with formation of metakaolinite:

$$Al_2O_3 \cdot 2SiO_2 \cdot 2H_2O \rightarrow Al_2O_3 \cdot 2SiO_2 + 2H_2O \qquad (10.5.1)$$

(3) 700–900 °C: partial decomposition of metakaolinite with the formation of free SiO_2, which forms a silicate melt together with impurities, like alkalis, earth alkalis or iron oxides

$$Al_2O_3 \cdot 2SiO_2 \rightarrow Al_2O_3 + 2SiO_2 \qquad (10.5.2)$$

(4) >950 °C: transformation of metakaolinite into mullite and release of SiO_2

$$3Al_2O_3 \rightarrow 2SiO_2 \rightarrow 3Al_2O_3 \cdot 2SiO_2 + 4SiO_2 \qquad (10.5.3)$$

Finally, the microstructural components in chamotte bricks are mullite, cristobalite and a glassy phase, which contains other ions as impurities. In dependence on the raw material quality, a certain quantity of residual SiO_2 grains can also be observed. Figure 10.5.6 shows as an example the microstructure of a silica-rich chamotte with silica grains in a fine-grained mullite-glassy phase matrix and bright titanium oxide-rich impurities.

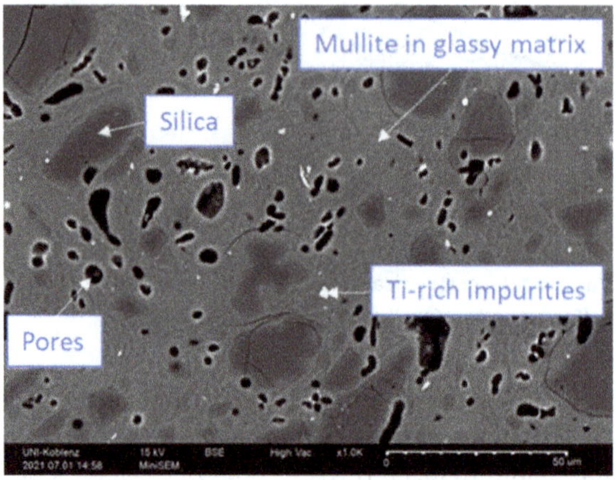

Figure 10.5.6: Microstructure of a silica-rich chamotte with silica-grains, surrounded by sharp cracks, in a fine-grained mullite-glassy phase matrix and bright titanium oxide-rich impurities.

The temperature resistance of chamotte bricks is mainly determined by their mullite content, i.e. by the initial Al_2O_3-content in the clay. The quantity relations are illustrated by the phase diagram SiO_2-Al_2O_3 in Figure 10.5.7. The relevant concentration range of chamotte bricks is between 55–70% SiO_2 and 30–45% Al_2O_3, respectively. In this range, the liquidus temperature increases with increasing Al_2O_3 concentration, and the equilibrium amount of mullite increases too. Deviations from this trend may be caused by coarse-grained quartz-fractions in the raw material since quartz coarse grain is low reactive and will remain mostly inert in the microstructure. The effective Al_2O_3:SiO_2 ratio is then shifted to higher alumina concentrations. Furthermore, impurities like alkali, earth alkaline and iron oxides lower the melting temperature so that SiO_2-rich melts can be formed at lower temperatures near 1000 °C. Hence, the final temperature resistance of chamotte bricks is determined by the Al_2O_3:SiO_2 ratio and the content of other oxide impurities. Thus, their concentration is limited to typical values of ≤3% Fe_2O_3, ≤3% TiO_2, <1% (CaO + MgO) and <3.5% (Na_2O + K_2O).

The group of alumina-rich refractories includes materials with an amount of more than 55% up to 100% Al_2O_3. Table 10.5.3 shows the different types of bricks in order of their Al_2O_3-content and the temperature resistance, characterized by the characteristic $T_{0.5}$-temperature, determined by the refractory-under-load (RUL)-test.

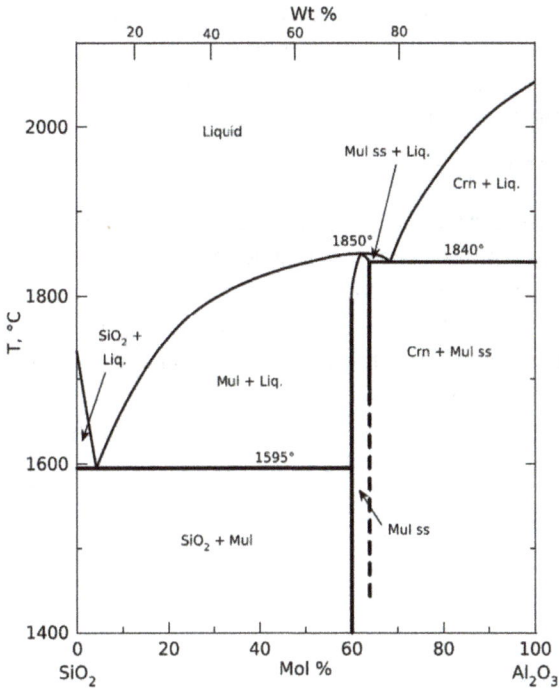

Figure 10.5.7: Phase diagram of the system SiO_2-Al_2O_3 according to [10].

The RUL-test determines the length change of a cylindric sample with 50 mm diameter and 50 mm height under a constant load in dependence on temperature. $T_{0.5}$ is the temperature, where 5% compression is observed and is used as a characteristic measure for the temperature resistance of refractories.

Table 10.5.3: Types of Al_2O_3-rich bricks, typical Al_2O_3-content and $T_{0.5}$-temperature [1].

Brick type	Al_2O_3 content [wt%]	DE $T_{0.5}$ [°C]
Corundum-containing bricks	50–65	1420–1500
Andalusite, sillimanite bricks	55–65	1590
Sintered mullite bricks	~72	1650
Fused mullite bricks	~75	>1700
Bauxite bricks	80–85	1480
Sintered corundum bricks	90–95	>1700
Pure fused corundum bricks	99	>1700

Andalusite, sillimanite and also kyanite are modifications of the compound Al_2SiO_5, which are used as high purity Al_2O_3-SiO_2 raw materials. During firing of the bricks, the compound is transformed according to eq. (10.5.4) into the thermodynamic stable compound mullite ($Al_6Si_2O_{13}$) by the release of 1 mol SiO_2 per formula unit Al_2SiO_5.

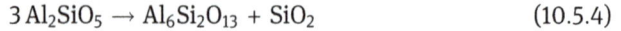

$$3\,Al_2SiO_5 \rightarrow Al_6Si_2O_{13} + SiO_2 \tag{10.5.4}$$

By the transformation, a characteristic structure of a mullite-framework with a SiO_2-rich amorphous phase is formed, which has a positive influence on the high temperature properties compared to the chamotte-type materials. Under load, the andalusite or sillimanite bricks have a $T_{0.5}$ of >1500 °C, while a chamotte brick achieves about 1200 °C $T_{0.5}$ temperature. For this reason, a preferable high Al_2O_3 content and a low impurity content is required for a maximum temperature resistance. In Figure 10.5.8, the temperature resistance of several refractory materials is compared by their change of height versus temperature graphs from the RUL-test.

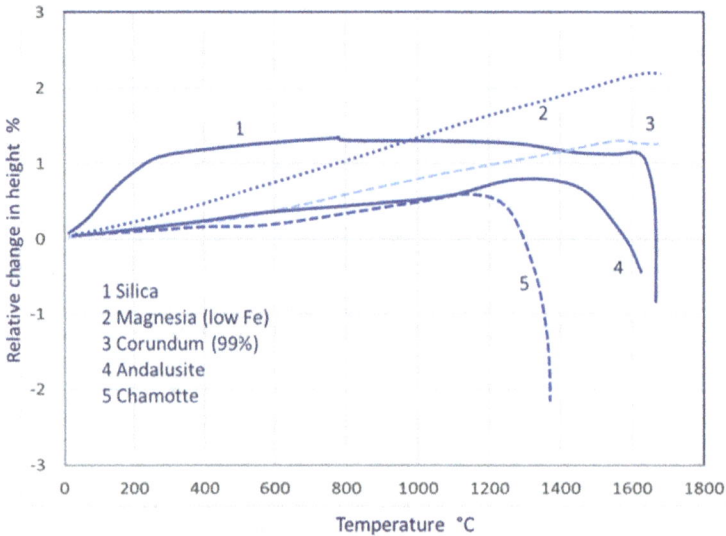

Figure 10.5.8: Relative change in height in dependence on temperature after RUL-test. Comparison of different types of refractories.

The direct use of mullite-raw materials instead of the conversion of andalusite or sillimanite can reduce the amount of SiO_2-rich amorphous phase in the brick and because of this increase the temperature resistance of alumina-rich Al_2O_3-SiO_2 bricks. For those bricks, only synthetical mullite raw materials are used, whereby sintered mullite is less expensive while fused mullite raw materials are used for high-grade mullite bricks with a high temperature resistance of about 1700 °C.

Bauxite bricks with an alumina content of 80–85% are fabricated on the base of natural bauxite raw materials. Refractory bauxites are usually diasporic or boehmitic bauxites with an Al_2O_3 content of 80–85% (water-free sample). In order to guarantee the required temperature resistance, the SiO_2 content is limited to 10–15%, and the maximum impurity concentrations are 2% Fe_2O_3 and 3% TiO_2. The production technology applied on bauxite bricks is in principle the same as for dry-pressed chamotte bricks.

For very high demands with respect to temperature and also corrosion resistance, pure corundum bricks with 90–99% Al_2O_3 are used. Raw material base is sintered or fused Al_2O_3 and, where appropriate, small amounts of kaolin, mullite or fine-grained, reactive alumina as binder. Shaping of theses bricks is carried out usually by dry or isostatic pressing or slip-casting. Due to the high purity raw material base, the firing temperature is usually above 1700 °C. After firing, the microstructure of corundum bricks contains an extraordinarily small amount of glassy phase. In mullite or kaolin-bonded bricks, the mullite crystals grow prismatic or columnar on the corundum grains and act as a bond or bridge between the alumina bulk grains.

10.5.4 Basic refractories

Materials with MgO or CaO as main constituents are referred to as basic refractories, derived from the fact that these oxides form hydroxides in contact with water. Basic refractories are used and preferred in the steel and cement production, whereby the predominant quantity of basic products is used in steel industry.

The main reason for the use of basic refractories is, beside the high temperature resistance, their excellent corrosion resistance in contact with steel slag or cement clinker. However, two significant disadvantages have to be considered: on the one hand, the strong hydration sensitivity of MgO and especially CaO (for instance during installation of a refractory lining), and on the other hand, the strong thermal expansion at the application temperature. While the hydration is connected with a strong destructive volume expansion (CaO: approximately 200%), the high thermal expansion coefficient of basics oxides leads to a poor thermal shock resistance.

As raw materials for magnesia refractory products, natural magnesite ($MgCO_3$) or seawater magnesia (base: $MgCl_2$) is used. The schematic representation in Figure 10.5.9 illustrates the manufacturing routes for MgO raw material from magnesite and seawater magnesia, respectively.

The fundamental quality criterion for MgO raw materials is their chemical purity, especially with respect to the CaO and SiO_2 content. With respect to the concentration of the two oxides, the mass ratio of CaO:SiO_2 is of special importance since this ratio significantly determines the temperature resistance of the later-produced brick. In dependence on the CaO:SiO_2 ratio, different calcium silicate phases with strongly different melting temperatures are formed during firing. Figure 10.5.10 illustrates the

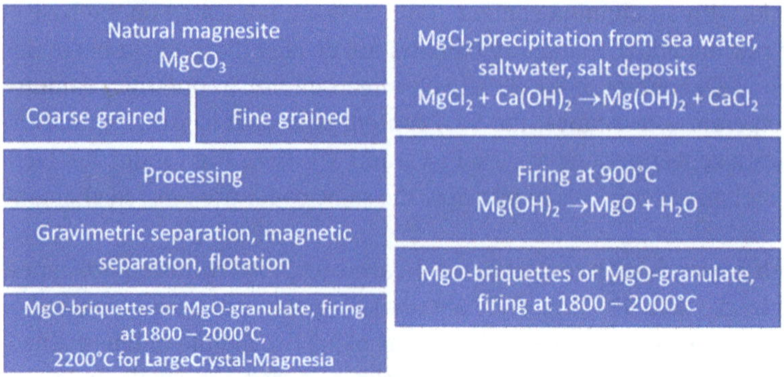
Natural magnesite
$MgCO_3$
Coarse grained
Fine grained
Processing
Gravimetric separation, magnetic separation, flotation
MgO-briquettes or MgO-granulate, firing at 1800 – 2000°C,
2200°C for LargeCrystal-Magnesia
$MgCl_2$-precipitation from sea water, saltwater, salt deposits
$MgCl_2 + Ca(OH)_2 \rightarrow Mg(OH)_2 + CaCl_2$
Firing at 900°C
$Mg(OH)_2 \rightarrow MgO + H_2O$
MgO-briquettes or MgO-granulate, firing at 1800 – 2000°C

Figure 10.5.9: Manufacturing routes for MgO raw material from magnesite and seawater magnesia.

relevant phase equilibria in the ternary system MgO-CaO-SiO_2. At a $CaO:SiO_2$ ratio close to 1, the dominating secondary phase is monticellite ($CaMgSiO_4$ [CMS]) with a rather low melting point of 1490 °C. In contrast, dicalcium silicate (Ca_2SiO_4 [C_2S]) is a high-melting binary silicate with a melting temperature of 2130 °C and is built at a $CaO:SiO_2$ ratio close to 2. Since the melting temperature of the calcium silicate secondary phases finally determines the temperature resistance of the hole brick, a CaO: SiO_2 ratio close to 2 is desired for high-quality magnesia raw materials.

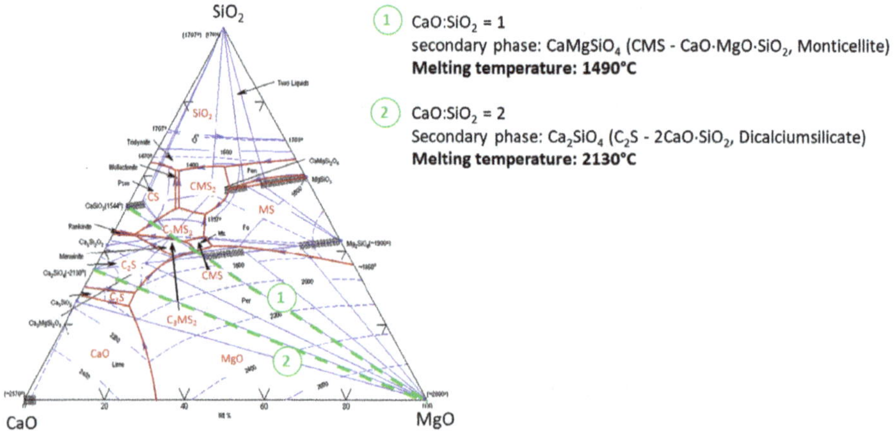
SiO_2
CaO
MgO
CS
CMS
MS
C_2S
C_3MS_2
1
2
1 $CaO:SiO_2 = 1$
secondary phase: $CaMgSiO_4$ (CMS - $CaO \cdot MgO \cdot SiO_2$, Monticellite)
Melting temperature: 1490°C
2 $CaO:SiO_2 = 2$
Secondary phase: Ca_2SiO_4 (C_2S - $2CaO \cdot SiO_2$, Dicalciumsilicate)
Melting temperature: 2130°C

Figure 10.5.10: Ternary system MgO-CaO-SiO_2 with reaction path of two different $CaO:SiO_2$ ratios.

A technological disadvantage of MgO is its high thermal expansion coefficient. It is connected with a significant length change during heating and cooling, respectively, which cause high thermal strain, as can be seen from eq. (10.5.5) for the thermal-induced strain in a refractory material.

$$\sigma_D = \alpha \cdot \Delta T \cdot E \qquad\qquad (10.5.5)$$

with: α: thermal expansion coefficient
ΔT: temperature difference
E: elastic modulus

Combined with the high brittleness of oxide ceramics in general, represented by a high elastic modulus, the high thermal expansion coefficient causes a high sensitivity of pure MgO-based materials against rapid temperature changes. For this reason, additional components for the increase of microstructural elasticity are used in MgO-based refractories.

For application in the steel industry, magnesia-carbon refractories are used. Here, carbon in the form of graphite or soot, together with synthetic resin as binder, increases the elasticity of the MgO-based brick. According to their carbon content, magnesia-carbon refractories are classified into three groups:
–   Fired carbon containing magnesia bricks with <2% C
–   Carbon-bonded magnesia bricks with <7% C
–   Carbon-bonded magnesia bricks with >7% C

The (partly) liquid or solid resin is added during mixing of the refractory batch mixture, which is shaped by common uniaxial pressing. During pre-firing of the MgO-C-bricks or during the first heating of MgO-C-lining, the resin is pyrolyzed and transferred in a predominately amorphous carbon structure. The carbon addition causes a significant improvement of microstructure flexibility due to its high thermal conductivity and its low thermal expansion. Additionally, the resistance in contact with corrosive steelmaking slags is improved in two ways: first, carbon shows a non-wetting behavior in contact with oxide steelmaking slags and by this acts as an effective infiltration barrier. Second, iron oxide, which is a liquifying component in steelmaking slags, is reduced by carbon into metallic iron, which causes an increase in slag viscosity and reduces the infiltration of the bricks too.

One disadvantage of the carbon bond in magnesia-carbon bricks is the poor oxidation resistance of carbon in air at application temperatures of $T > 1600$ °C, which leads to a loss of carbon in the surface-near microstructure of the bricks. A second disadvantage is the interaction of carbon and MgO by carbothermal reduction: at high temperatures, carbon reduces MgO by the formation of CO(g) and Mg(g)-vapor. Figure 10.5.11 illustrates the reaction path during carbothermal reduction in MgO-C-bricks. Mg(g)-vapor formation in the bulk of the brick increase the porosity and can weaken the brick structure. At the brick surface, at higher O_2 concentration, Mg(g)-vapor is oxidized with the formation of secondary MgO, which forms a dense layer at the brick surface. Carbothermal reduction of the secondary calcium silicate phases results in an increase of undesired low melting monticellite [CMS]-phase [1]. Especially vacuum

Reduction inside the brick:	Oxidation at the hot face:
$MgO + C \rightarrow CO(g) + Mg(g)$	

$Mg(g) + \frac{1}{2}O_2(g) \rightarrow MgO$ formation of secondary magnesia

$Ca_2SiO_4 + C \rightarrow CO(g) + SiO(g) + 2CaO$

$CaO + MgO + SiO(g) + \frac{1}{2}O_2(g) \rightarrow CaMgSiO_4$ formation of secondary C-M-S phases

$SiO_2 + C \rightarrow CO(g) + SiO(g)$
(SiO_2 from graphite ash)

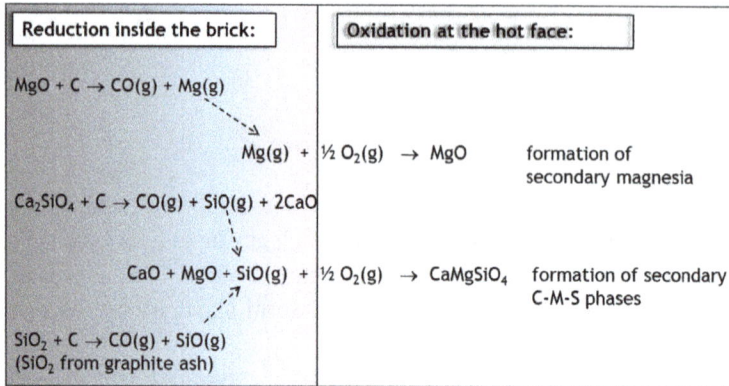

Figure 10.5.11: Reaction path during carbothermal reduction in MgO-C-bricks [1].

facilities, like vacuum oxygen decarburization, already at temperature above 1400 °C, a perceptible part of the microstructure is consumed by this phenomenon.

MgO-spinel refractory materials are used as lining in the hot zone of rotary kilns for cement fabrication. The oxidizing conditions in this application exclude the use of carbon for providing elasticity. The addition of spinel-type compounds or spinel-forming oxides results in a shrink coating of the MgO grains by spinel during cooling since MgO has a higher thermal expansion coefficient than spinel. Additionally, when spinel-forming oxides are used, the volume increase during spinel-formation during firing of the bricks occurs. Both phenomena lead to the formation of a network of fine microcracks, which provide the elasticity of the brick. Typical spinel-type compounds for use in MgO-based bricks are magnesium aluminate ($MgAl_2O_4$), magnesium chromite ($MgCr_2O_4$) and hercynite ($FeAl_2O_4$).

References

[1] Routschka G, Wuthnow H. Handbook of Refractory Materials, Vulkan Verlag, Essen, 2012.
[2] Harders F, Kienwo S. Feuerfestkunde, Springer Verlag, Berlin, 1980, 77–86.
[3] (a) Dinger D, Funk JE. Particle packing. Interceram, 1994, 43, 87–9; (b) idid 150–4; (c) ibid 350–3.
[4] Frey KJ, Löbe R, Nold P. Optimization of preparation plants for refractory bodies. Proceedings UNITECR2001, 945–67.
[5] Ramakrishnan V. Modern developments in the fabrication process of high-grade refractory bricks. Sprechsaal, 1987, 120, 880–5.
[6] Kremer R. Quality improvement of shaped refractories by vacuum pressing technique. Stahl und Eisen Special, 2002, 129–32.

[7]   Brunk F. Silica bricks. In: Routschka G, Wuthnow H (Eds). Refractories, Vulkan Verlag, Essen, 2012.
[8]   Salmang H, Scholze H, Telle R. (Eds). Keramik, Springer Verlag, Berlin, 2007.
[9]   Didier Feuerfest Technik: Feuerfeste Baustoffe und ihre Eigenschaften, Didier-Werke, Wiesbaden, 1974.
[10] Aramaki S, Roy R. Revised phase diagram for the system Al_2O_3 – SiO_2. J Am Ceram Soc, 1962, 45, 229–42.

10.6 Carbides, borides, silicides

Markus F. Zumdick

10.6.1 Carbides, borides and silicides as hard materials

The extraordinary hardness of hard materials is due to the strong bonding forces between the individual components of the atomic structure. This results, for example, in the very high melting points of approx. 1500 °C (β-TiSi$_2$) to almost 3900 °C (TaC) or very low thermal expansion coefficients. Therefore, they are used as wear resistant as well as high-temperature materials. Unfortunately, they are usually very brittle, which limits their applicability in pure form [1].

Hard materials can be divided into two groups [1, 2]:

- "metallic hard materials": mainly compounds of the transition metals of the Periodic Table groups IVb to VIb with the elements B, N, C and Si
- "non-metallic hard materials": compounds of the elements B, C, N and Si with each other and some oxides such as Al$_2$O$_3$ or ZrO$_2$.

This chapter deals with the metallic hard materials, i.e. carbides, borides and silicides. Historically, only the carbides, borides and nitrides were considered as metallic hard materials. Since, over the years, silicides have found an interest in the development of heat and scale resistant alloys and their hardness partly approaches that of molybdenum or titanium carbides, they were finally included in this group.

A large part of the metallic hard materials can be classified as so-called intercalation compounds with regards to their structural composition. This applies in particular to carbides and nitrides of transition metals. The term "intercalation structure" was defined by C. Hägg because, to a certain extent, small atoms are intercalated in the voids of a metallic host lattice. The stability of these structures is closely related to the radius ratio $0.41 \geq r_x/r_{Me} < 0.59$ (X = non-metallic, Me = metallic component). The host structures usually have the cubic or hexagonal dense or the hexagonal primitive structure. Above this radius ratio, more complex structures occur, just as in the borides, where the critical radius ratio of 0.59 is reached in almost all cases. Between the mostly simple structures of the carbides and the complex ones of the borides are those of the silicides, which occur mainly as disilicides, such as MoSi$_2$ or TiSi$_2$ [2]. Hard metals are sintered alloys of high-melting metallic carbides, borides or also silicides and low-melting metals of the iron group (above all cobalt) as a binder. The former components provide wear resistance and the latter a certain toughness and flexural strength of the materials [3].

https://doi.org/10.1515/9783110733471-015

10.6.2 Sintered hard metals

The following four types of sintered hard metals are distinguished [2, 4] – note: for hard metals, the terms hardmetals or cemented carbides are usual; in German, only the term *Hartmetall* exists [4]:

- WC-Co carbides: Although the carbides of this type mainly consist of only two components and likewise two phases, the WC and the binder cobalt, they have the greatest importance in terms of quantity. The reason for this is that, compared to the other types, they have higher strength values up to application temperatures of approx. 600 °C with simultaneously lower abrasive wear.
- WC-(Ti,Ta,Nb)C-Co carbides: Adding titanium carbide and/or tantalum (niobium) carbide to the WC-Co carbides results in three-phase materials with improved high-temperature properties. This applies in particular to the oxidation resistance, the hot hardness or hot strength as well as the diffusion resistance compared to iron-based alloys.
- Cermets: The main hard phase of cermets is a Ti-based cubic carbide, nitride or carbonitride. The main focus here is on efforts to maximize the hot hardness of the bonding phase by investigating solid solutions and precipitation hardenability.
- Special carbides: in these carbides, the binder phase cobalt is completely or partially replaced by nickel or nickel and chromium in order to increase the corrosion resistance of the material.

If we also take into account that the properties of components made of hard metals can be significantly influenced by coatings, we obtain a range of applications that cannot be achieved by other material groups. This is impressively illustrated for cutting materials, for example, in the following diagram (Figure 10.6.1).

10.6.2.1 The different carbides and their applications
The development of WC-Co carbides is dating back to the early 1920s, when the first patents on carbides were published. In 1927, the first carbide was marketed worldwide under the brand name WIDIA ("WIe DIAmant" = like a diamond) by the Friedrich Krupp Company. They were primarily used as materials for wire drawing dies and wear parts [4].

The properties of cemented carbides are given by the combination of vastly different constituents, e.g. soft and ductile Co-based binder with hard and wear-resistant WC or cubic carbides. By selecting appropiate raw materials, composition and suitable processing parameters, a wide range of mechanical properties can be achieved. In particular, their unique combination of hardness and toughness makes them attractive for many industrial applications (Figure 10.6.2). In WC-Co hard metals, the tough binder represents the minority component, usually with a mass fraction of 3–24%. The

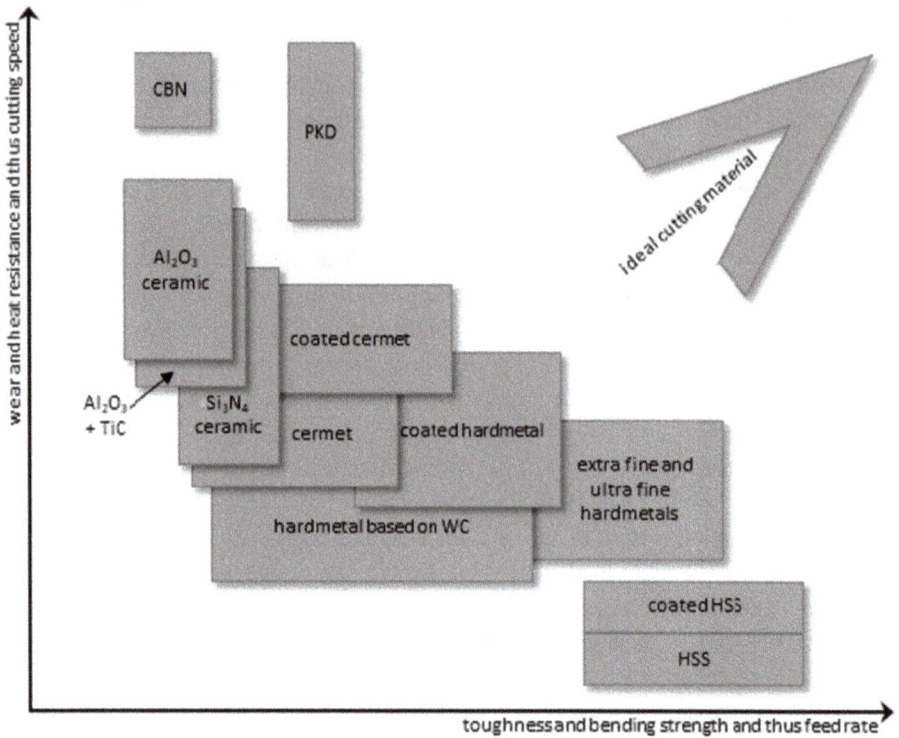

Figure 10.6.1: Due to the broad, adjustable range of properties, sintered carbides cover a much wider range of applications than other materials (such as cubic boron nitride and polycrystalline diamond) and material classes (ceramics and steels). Graphic according to [5, 6].

interaction between the binder and the carbide results in the microstructure, which is decisive for the component properties.

By reducing the carbide particle size, a significant improvement can be achieved in the combination of wear resistance/hardness and strength [4]. This is of special interest for inserts with sharp edges (for milling applications), drills (aerospace industry) or microdrills (electronic industry for composite machining, e.g. to drill holes into circuit boards), where high precision and dimensional control are necessary [4].

While, today, WC with a grain size of 100–200 nm is already readily available on an industrial scale (e.g. the WC DN powders from H.C. Starck Tungsten GmbH), it must be taken into account, however, that grain growth takes place during the necessary sintering process in carbide component manufacture at temperatures between 1300 and 1450 °C, i.e. the WC grain coarsens. The best-known method of suppressing this grain growth is the addition of so-called grain growth inhibitors. The most effective inhibitor is V, which also has a significant effect on the hardness of the carbide, followed by Mo, Cr and Ti, Ta or Nb – the latter ones with positive effects on the high-temperature properties, especially the high-temperature wear. The

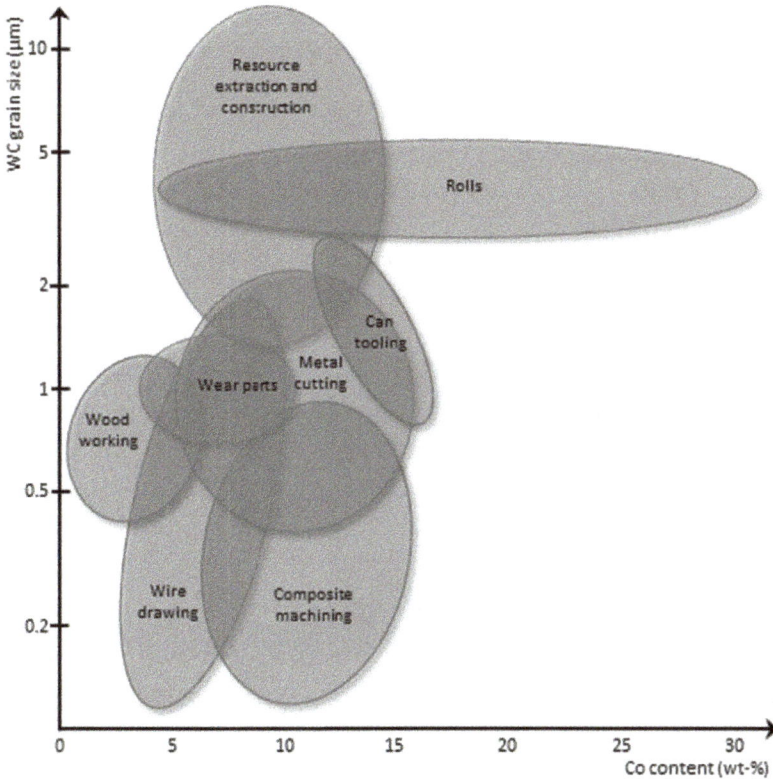

Figure 10.6.2: Applications of WC-Co carbides depending on the grain size and the cobalt content. Graphic according to [4].

total amount of inhibitors added is usually <2%, as otherwise precipitation would lead to embrittlement of the material. As a rule, grain growth inhibitors are added in the form of their carbides, i.e. as VC, MoC, Cr_3C_2, TiC, TaC and (Ta,Nb)C. The current understanding of the effect of grain growth inhibitor additions is the formation of thin cubic (Me,W)C (Me = V, Cr, Ti, etc.) films on the surfaces of WC grains during sintering of the hard metal component, thereby not only lowering the interfacial energy but also acting as a kinetic barrier. So far, the process of grain growth inhibition is not fully understood [4].

The main group of hard metals used for processing of short-chip ferrous materials are straight WC-Co hard metals with a Co mass fraction of 3–11%. The particle size of the WC phase is 0.5 to 5 µm, where materials with a WC grain size of <1 µm are mainly used for fine machining or woodworking.

Long-chip materials, like steels, are machined using hard metals based on WC-TiC-(Ta)C-Co (where Ta contains the element Nb in proportion up to 30%), with contents of TiC up to 35% and TaC up to 8% by weight.

Other hard metals of the aforementioned group, but with lower TiC contents (up to 5% by weight), are used for working with high-alloyed steels and non-ferrous metals which cannot be easily machined.

However, due to the superior performance of CVD and PVD-coated WC-Co grades, the use of WC-TiC-(Ta,Nb)C-Co hard metals is very limited nowadays. The coatings with a thickness of only 5 to 15 µm consist of TiC, TiN, Ti(C,N) or Al_2O_3. With PVD, coatings of (TI,Al)N, (Hf,Al)N or (Ti,Si)C are also possible [1].

Figure 10.6.3. shows typical components for turning and woodworking. These, as well as the pictures in Figure 10.6.4, were supplied by Boehlerit GmbH & Co. KG.

Figure 10.6.3: Components for woodworking (left) and turning (right); (courtesy of Boehlerit GmbH & Co. KG).

A large proportion of WC-Co carbides with a WC grain size of up to 5 µm are used for non-machining applications, where the service life of the component depends on its wear resistance. Typical examples are components such as rolls, forging dies (like dies for can tooling) or from the automotive or agricultural sectors (see Figure 10.6.4).

As already listed above, the main hard phase of cermets is a Ti-rich cubic carbide, nitride or carbonitride. This hard phase normally exhibits a core-rim structure with the hard phase itself as core with a rim formed by a (Ti,Me)(C,N) solid solution (Me = Mo, W, Ta, Nb). Binders are normally Ni and Co in different proportions. Cermets are mainly used as cutting tools in surface finishing due to their high chemical stability and good high-temperature hardness as compared to WC-based cemented carbides. The main drawback of cermets is their comparatively poor toughness, especially at low temperatures [4]. While the proportion of cermets in cutting materials in Japan is quoted as about 25%, the figures are much lower in Europe, but with a trend upwards [1].

10.6.2.2 Cast tungsten carbide (CTC)

Despite the energy transition and the growing share of regenerative energies, a significant part of global energy production is based on fossil fuels. The increasing

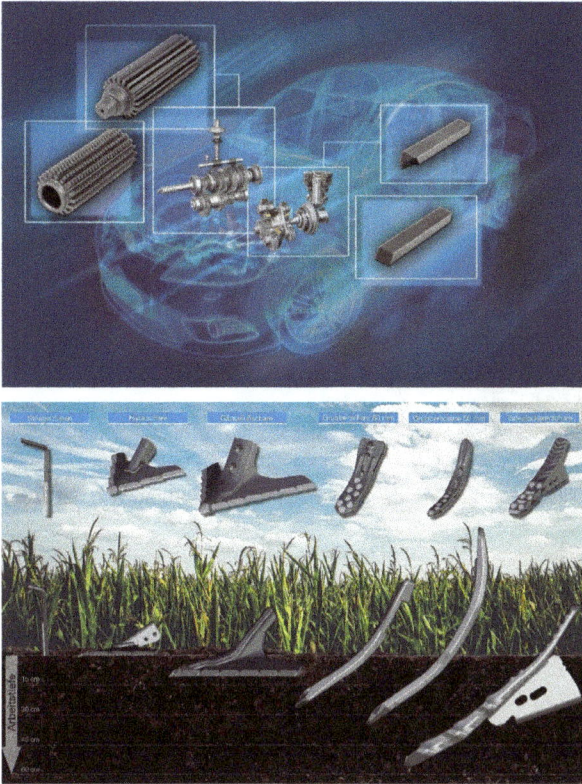

Figure 10.6.4: Components from the automotive sector (upper picture) and from the agricultural sector (lower picture, the light gray areas of the components are made of hard metal); courtesy of Boehlerit GmbH & Co. KG.

scarcity of fossil fuels means that both new technologies and new materials are needed to make more effective use of already tapped oil and natural gas deposits. One development are drill bits made of infiltrated, so-called matrix powders that can withstand the wear stresses that occur when working in rock layers of varying hardness in interaction with drilling fluid.

The main component of many matrix powders is cast tungsten carbide, which is an eutectic phase mixture of WC and W_2C. A characteristic feature is a lamellar structure of WC and W_2C, the "feather structure", as shown in Figure 10.6.5. When comparing CTC with WC, this structure is the reason for the significantly increased wear and abrasion resistance of CTC, which makes this material particularly suitable for the aforementioned complex wear stresses [7].

In contrast to WC, CTC is very quickly attacked by binders such as Co, Ni or Fe at high temperatures. This means that it cannot be used in conventionally produced hard metals. For drill bits, therefore, a fill of CTC is infiltrated with Cu-based

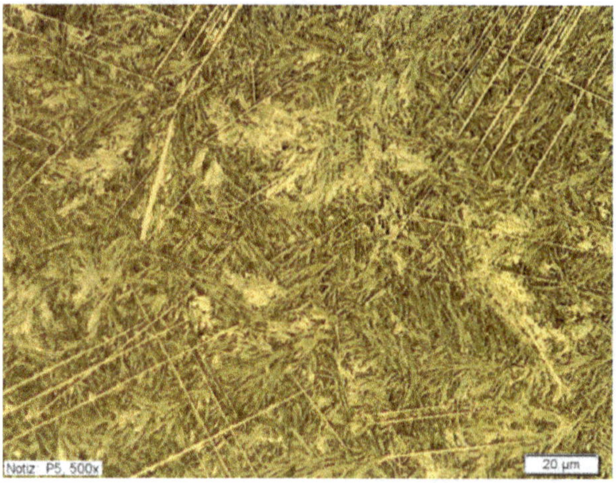

Figure 10.6.5: Lamellar structure of cast tungsten carbide. Etched sample (WC bright, W_2C dark).

binders at relatively low temperatures of 1200 °C. The CTC is then used in the drill bits. Inserts, e.g. of polycrystalline diamond (PDC), are subsequently brazed in.

10.6.3 Metal borides

Boron combines with a large number of metals and semimetals to form binary or higher solid compounds, the so-called borides. They are unique in the number of stoichiometries; compositions corresponding to at least 24 Me:B ratios (Me = metals and semimetals) between 5:1 and 1:66 are known. However, the most common are the monoborides (MeB), diborides (MeB_2), tetraborides (MeB_4), hexaborides (MeB_6), dodecaborides (MeB_{12}) and hectoborides (MeB_{66}). The most characteristic properties of the metal borides are their high melting points, extreme hardness and, in many cases, high electrical and thermal conductivities, fair corrosion resistance, good wear resistance and thermal shock resistance. The metal borides display resistance to oxidation in air at elevated temperatures and to attack by molten metals, basic slags and molten salts [8].

Although boride cermets that combine the extreme hardness of refractory borides with the high toughness and ductility of a metallic binder have recently been developed, they are far away from attaining large-scale application. Niche products are TiB_2-Fe cermets for machining aluminum alloys or ternary cermets based on Mo_2FeB_2, Mo_2NiB_2 and WCoB that are used as machine parts for injection molding, bearings, wire-drawing cones and cutting tools [8].

Metallized films or paper are widely used in the packaging industry. The metal layer improves the barrier properties of the films and thus protects the packaging contents from external influences such as moisture and oxygen. Candy wrappers and chip bags with an ultrathin aluminum layer are typical examples of applications, as are high-quality labels for glass bottles. In the electronics industry, the conductive layers are used, among other things, for the production of capacitors and flexible printed circuit boards [9].

Therefore, a large part of the annual TiB_2 production of about 250 ton is used for the production of evaporator boats; components that act as a receptacle for the aluminum to be evaporated. Widely used are hot-pressed crucibles made of both TiB_2-BN and TiB_2-BN-AlN. Because of the TiB_2, these composites show high corrosion resistance against molten aluminum and good electrical conductivity. By varying the BN content, the electrical conductivity of the boats can be controlled. So, the crucibles can be heated in direct current. AlN has very good thermal conductivity and is therefore particularly suitable for large evaporator boats so that they heat up homogeneously and quickly. In the production of metal layers, the continuously fed aluminum wire is heated under high vacuum so that it first melts and then evaporates. Due to special surface structuring and texture, a melt pool is formed that is optimally distributed over the boat, thus ensuring a homogeneous vaporization cloud. The metal vapor then precipitates onto a substrate located above the vapor cloud [8, 10].

As already mentioned, metal borides exhibit a large number of stoichiometries. This is resulting, among other things, in the fact that the electrical characteristics cover a wide spectrum: the MeB_{66}, MeB_6 and MeB_{12} phases of Be, Mg, Ca, Eu, Al and Si as semiconductors; TiB_2, ZrB_2 and the majority of transition-metal borides as metallic conductors; and, last but not least, NbB, YB_6 and ZrB_{12} as superconductors. In addition, LaB_6 and other lanthanide and actinide borides (YB_6, ThB_6, GdB_6, CeB_6, etc.) belong to the best-known high-temperature electron emitters. Most of these borides are intensively colored: for example, GdB_6 is blue and LaB_6 is purple [8].

LaB_6 and CeB_6 are vacuum stable and are characterized by an extremely low electron exit function. They are therefore used, among other things, in plasma technology and as an electron source (hot cathode) in some electron microscopes (Figure 10.6.6). The single-crystal cathodes consist of tiny tips sitting on a support. In a scanning electron microscope, the resulting luminance leads to better image resolution and a better signal-to-noise ratio compared to the tungsten cathode. In microanalytical applications (EDX), a smaller beam diameter and better counting statistics are achieved. The lifetime is higher with up to 3000 h compared to the tungsten cathode but requires a vacuum in the cathode chamber of 10^{-5} Pa or better.

If lanthanum hexaboride is added to a transparent polymer in the form of very fine particles in low concentrations, it can be mixed in as a laser absorber without significantly changing the visible optical properties of the polymer. Absorption of laser radiation of wavelength 1064 nm from the commonly used Nd:YAG laser is achieved. The material can thus be used for laser marking or laser welding of such materials [11–13].

Figure 10.6.6: Characteristic for many borides: their intense color, like here LaB_6, which is used e.g. as cathode for plasma-assisted coating processes; courtesy of Sindlhauser Materials.

10.6.4 Metal silicides

Among the metallic hard materials, silicides have the lowest melting temperatures and largely also the lowest hardness values. Since they are also very brittle, they are not suitable for use in hard metal alloys. The silicides have found technical significance only in metallurgical areas where high scale resistance and chemical resistance are important. For example, they are used as oxidation protection on refractory metallic surfaces [14].

Molybdenum disilicide ($MoSi_2$) has gained technical importance as a heating element in high-temperature furnaces. For a long time, however, its application was hindered due to its brittle nature at low temperatures, inadequate creep resistance at high temperatures, accelerated oxidation below 500 °C ("pesting") and its relatively high coefficient of thermal expansion compared to potential reinforcing fibers [15].

At higher temperatures, however, a passivating, dense, glassy SiO_2 top layer forms. In addition, the high-temperature creep resistance could be significantly improved by particle reinforcement, e.g. with SiC. Today, the $MoSi_2$-based heating elements (Figure 10.6.7) are used in high-temperature furnaces with operating temperatures up to 1800 °C. Additions of aluminum silicate further optimize the resistance of the heating element and, as SiO_2 former, also improve oxidation protection [16].

Molybdenum disilicide is also used in $MoSi_2$-Si_3N_4 composites as glow plug. They are used to help start engines by heating them, installed in the combustion chamber of the diesel engine. They can be heated at a higher heating rate, that is, the higher heat resistance than a metal glow plug with the result that the diesel engine can be started faster. Other applications for the $MoSi_2$-Si_3N_4 composite include igniters for gas and oil heaters, gas cookers, lambda probes and gas sensors, and electrodes for steel sheet pressure welding machines [17, 18].

Figure 10.6.7: Heating conductor prepared for installation (left) and in the furnace (right).

References

[1]   Schatt W, Wieters K-P. Hard materials and hard material compounds. In: Schatt W, Wieter K-P. (Eds). Powder Metallurgy. Processing and Materials. Shrewsbury SY1 1HU, UK, European Powder Metallurgy Association (EPMA), 1997, 442–476.
[2]   Kieffer R, Benesovsky F. Hartstoffe, Springer-Verlag, Vienna, Austria, 1963.
[3]   Holleman AF, Wiberg E, Wiberg N. Anorganische Chemie, 103rd edition, Walter De Gruyter GmbH, Berlin, Germany, 2017.
[4]   Garcia J, Collado Ciprés V, Blomqvist A, Kaplan B. Cemented carbide microstructures: A review. Int J Refract Met Hard Mater, 2019, 80, 40–68.
[5]   AFC Hartmetall, Unser Lexikon. Accessed April 30, 2021, at https://www.afcarbide.de/de/hart metall/lexikon.
[6]   Schneidstoff. Accessed April 30, 2021, at https://de.wikipedia.org/wiki/Schneidstoff.
[7]   Häslich FT. Flüssigphaseninfiltration von Matrixpulvern unter Normalatmosphäre. Bachelor-Thesis, Jena, Germany, Ernst-Abbe-Fachholschule Jena, 2012.
[8]   Greim J, Schweitz KA. Boron carbide, boron nitride, and metal borides. Ullmann´s Encycl Indust Chem, 2012, 6, 219–36.
[9]   3M-Verdampferschiffchen-3-0-fuer-robuste-Prozesse. Accessed May, 20, 2021 at https://www.pressebox.de/pressemitteilung/3m-deutschland-gmbh/3M-Verdampferschiffchen-3-0-fuer-robuste-Prozesse/boxid/1003073.
[10] verdampferschiffchen-aus-keramik. Accessed May 20, 2021, at https://www.technical-ceramics.com/produkte-im-einsatz/verdampferschiffchen-aus-keramik.
[11] Lanthanhexaborid. Accessed May, 21, 2021, at https://www.chemie-schule.de/KnowHow/Lanthanhexaborid.

[12] Elektronenkanone Accessed May 21, 2021, at https://de.wikipedia.org/wiki/
       Elektronenkanone.
[13] Lanthanhexaborid. Accessed May 21, 2021, at https://de.wikipedia.org/wiki/
       Lanthanhexaborid.
[14] Büchner W, Schliebs R, Winter G, Büchel KH. Industrielle Anorganische Chemie, Verlag
       Chemie, Weinheim, Germany, 1984.
[15] Hebsur MG. MoSi$_2$-base composites. In: Bansal NP (Eds). Handbook of Ceramic Composites,
       Springer, Boston, MA, 2005. Accessed May 21, 2021, at https://doi.org/10.1007/0-387-
       23986-3_8.
[16] Kollenberg E. (Ed). Technische Keramik, Vulkan-Verlag, Essen, Germany, 2004.
[17] Yamada K, Kamiya N. High temperature mechanical properties of Si$_3$N$_4$-MoSi$_2$ and Si$_3$N$_4$-SiC
       composites with network structures of second phases. Mater Sci Eng A, 1999, 261, 270–7.
[18] Composites, Si$_3$N$_4$-MoSi$_2$, as the material for heating elements of new generation. Accessed
       May 21, 2021, at https://www.icimb.pl/centrala_en/science/technology-innovations/272-
       heating-elements.

10.7 Abrasives and hard transition metal oxides

Dominik Wilhelm

10.7.1 Abrasives in grinding industries

Abrasives are used since the beginning of mankind. The cavemen had to sharpen their knives in order to hunt or defend themselves against aggressors. The Egyptians had to cut and shape their stones in order to build their pyramids. Today's abrasives are mostly used in special grinding tools, having a huge variety of different profiles, like straight wheels, cups, cuboid-shaped blocks, abrasive papers and many more. The field of applications are very versatile like knife and tool industry, automotive industry, medicinal industry and steel, stone and glass industries, and many more. Construction and aftermarket represent important and versatile applications as well [1–3].

The abrasive can be applied in resin-bonded (like phenol resin, epoxy resin or polyurethanes), vitrified-bonded or metal-bonded grinding tools. Even loose abrasives are used in vibratory finishing. A bonded grinding tool represents an interplay of the bond combined with the appropriate abrasive grain, pores and other additives. Its specific composition depends on the grinding process, the work piece material, the required abrasion, the resulting surface and many more. So when choosing the proper grinding tool, a huge variety of factors have to be considered. On closer examination towards the abrasive grain, the grain size, its grain shape and grinding properties have to be incorporated. Usually, there are four different abrasive grain types, which can be classified into two groups. Silicon carbide (SiC) and α-Al_2O_3, also called alumina, are named conventional abrasives. Diamond and cubic boron nitride, abbreviated as c-BN, are named as superabrasives [1–3].

10.7.1.1 Conventional abrasives in the grinding industry

As already mentioned, SiC and alumina are named conventional abrasives. They are often used as abrasives in grinding industry. Their Knoop hardnesses are in the range of ~25 GPa for SiC and 13.5–22.2 GPa for alumina. The hardness of the specific abrasives depends on their purities. In case of alumina, the usage of dopants like ZrO_2 or Cr_2O_3 plays an additional role concerning its hardness [1–6].

10.7.1.2 Silicon carbide

The Acheson process represents the production process of SiC. It leads to silicon carbide, which differs in its purity and therefore in its color. Green SiC consists of >98% SiC, whereas the black variant exhibits an SiC content of 95–98%. Green SiC has a Knoop hardness of ~28 GPa, whereas black SiC has a hardness of 26.3 GPa. The higher the purity of SiC, the more brittle is the abrasive. So, green SiC cuts sharper

https://doi.org/10.1515/9783110733471-016

than the black SiC. Green SiC does sliver sharp-edged while grinding, whereas black SiC is less brittle and therefore more stable towards pressure and cracks. Pure SiC is colorless and is not used in grinding industries. Figure 10.7.1 presents green and black SiC [1–10].

Figure 10.7.1: Left: green SiC. Right: black SiC [11].

Although it has a higher Knoop hardness than alumina, SiC is thermally and chemically less stable than alumina. When grinding steel or metals, like iron or nickel, it has a lower wear resistance because of the metals affinity to carbon. Its application fields are not as manifold as those from alumina. They are focused on glass, ceramic or stone machining to the tooling of carbides and non-ferrous metals [1–3].

10.7.1.3 Alumina

The invention of the Higgins electric arc furnace at the beginning of the twentieth century revolutionized the production of grinding tools. Herewith, it was possible to produce fused alumina with constant purities. Prior to this, grinding tool manufacturers had to use naturally occurring aluminum oxide, which had no consistent purity and therefore a high degree of fluctuation regarding its grinding behavior [1–3, 6]. As already mentioned, alumina is applied in different purities. However, when adding dopants to the production process, the alumina with the highest purity is used.

Bauxite represents the starting material for all fused alumina abrasives. Depending on the purity, different colors appear. Brown fused alumina (BFA) contains up to 4% TiO_2 as well as significant amounts of Fe_2O_3, SiO_2 and alkalines. It has the lowest purity of all alumina abrasives. It is a tough and blocky-wearing abrasive with a Knoop hardness of ~20 GPa. Low titania brown fused alumina (LTBFA), in variants also known as semi-friable alumina, contains up to 2% titanium dioxide or other impurities. Its color differs from the color of BFA, making it slightly lighter. The overall color scheme of BFA will vary between dark and light brown with tints of yellow and green and can also feature different shades of blue. In general, an increase of the TiO_2 content leads to tougher abrasives. However, this reduces the hardness. Therefore, semi-friable alumina is less tough compared to BFA but has a higher friability and therefore enables a cooler grinding process [1–4, 7, 8, 10, 11].

White fused alumina (WFA) represents the material with the highest purity (>99% Al_2O_3), making it almost completely transparent. To achieve this clarity, the alumina has to be purified prior to fusion, using the Bayer process. WFA has a Knoop hardness of about 20.7 GPa. Regarding its grinding properties, it shows a higher friability than BFA and LTBFA. It wears splintery and hence represents the alumina abrasive with the lowest toughness. Because of these material properties, new grinding edges are generated continuously during the process [1–4, 7, 8, 10].

It is possible to vary the material and grinding properties by adding dopants to the production process. The addition of Cr_2O_3 to WFA during the production process leads to pink and ruby alumina. Pink alumina contains about 0.5% Cr_2O_3, whereas the content of Cr_2O_3 is about 2–3% for ruby alumina. The addition of chromium(III) oxide increases the friability of the abrasive. Ruby alumina is tougher than pink alumina and WFA but not as tough as LTBFA and BFA, respectively. Regarding the toughness of the abrasives, pink and ruby alumina close the gap between WFA on one side and LTBFA and BFA on the other side. Pink and ruby alumina are, among others, used for dry grinding applications of steel. Figure 10.7.2 shows grains of brown fused alumina, white fused alumina as well as pink and ruby alumina [1–3].

With ZrO_2, a further transition metal oxide is used as a dopant for the production of alumina abrasives. The production of alumina zirconia takes place in a Higgins furnace by adding zirconia to white fused alumina during the production process. After cooling, a fine structure of α-alumina can be obtained, containing a high amount of tetragonal zirconia. Most of alumina zirconia grains contain about 25% ZrO_2. Depending on the content of zirconia, the Knoop hardness of alumina zirconia ranges from 14.3 GPa (10% ZrO_2) to 19.2 GPa (40% ZrO_2). Alumina zirconia is a very tough abrasive and wears blocky. It is among others used in hot-pressed resin bonds for conditioning of rough steel or billets of titanium or nickel alloys. The requirements to the abrasive and the grinding wheel are very high because conditioning of billets is a rough process. It requires an enormous abrasion on hot work pieces without usage of coolants. Furthermore, it is used for railway track grinding [1–3, 6].

10.7.1.4 Sintered alumina

Sintered alumina, also known as ceramic alumina, was first developed in the 1990s and further improved in recent years. It has a Knoop hardness of ~13.5 GPa and thus the lowest measured hardness value of the described alumina types. However, it closes the gap between conventional abrasives and superabrasives because of its

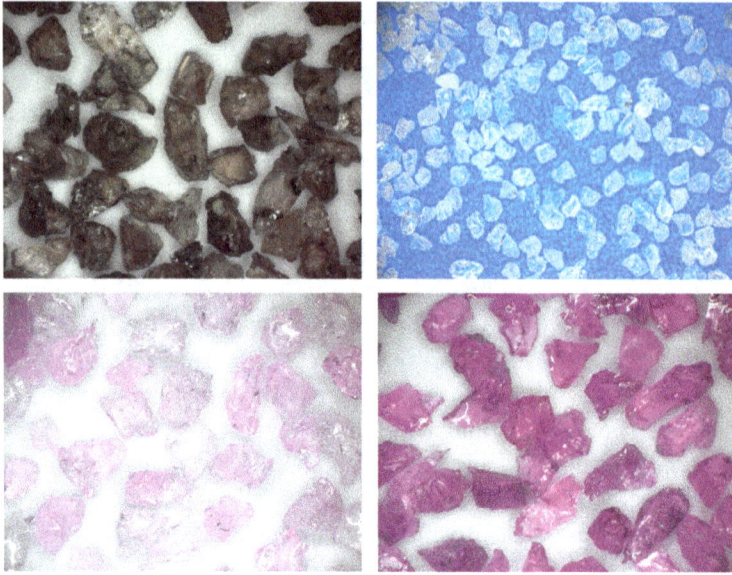

Figure 10.7.2: Top left: BFA. Top right: WFA. Bottom left: pink alumina. Bottom right: ruby alumina [12].

grinding properties. This is justified by the reduced crystallite size (a submicron crystallite structure) inside a grain, which makes the abrasive harder, tougher and self-sharpening compared to other alumina abrasives. This self-sharpening reduces the grinding temperature on the work piece and finally leads to longer life times of the grinding wheels. Sintered alumina is, among others, used for grinding high-speed steel [1–3, 6, 13–15].

It is mostly produced via a sol-gel route, using boehmite (γ-AlO(OH)), water and a suitable acid as starting materials. By adding additive agents, a homogeneous fine-crystalline microstructure can be achieved. After dehydration, the alumina is sintered. It is possible to create particular profiles like rods, triangles and many more. Therefore, the gel has to be treated *via* extrusion or pouring, before dehydrating [1–3, 6, 13–15]. Figure 10.7.3 shows the rod-shaped sintered alumina as well as triangle-shaped alumina.

10.7.1.5 Superabrasives

As already mentioned, diamond and c-BN are named superabrasives. Their Knoop hardness of 64 and 45 GPa, respectively, are the highest known. Although they are very expensive, they have to be used when machining very hard materials. Furthermore, they are used in order to decrease grinding costs because their hardness and grinding properties lead to increased wear resistance during the process [1–4, 6–10].

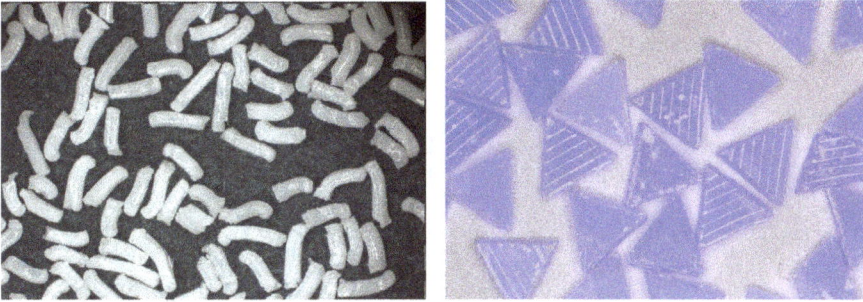

Figure 10.7.3: Left: rod-shaped alumina. Right: triangle-shaped alumina [12].

10.7.1.6 Diamond

Diamond represents the hardest known material. Furthermore, it has a high wear resistance. Therefore, it can be used to grind and cut the hardest materials. It is used to machine hard metals, stones and ceramics. However, being the high-pressure modification of carbon, it is unsuitable for machining steels or metals with a high affinity to carbon. In addition, diamond is used in dressing and conditioning tools in order to profile sharp or clean the grinding wheel before the grinding process and after every grinding cycle. Hence, even natural diamond is applied. Figure 10.7.4 shows two dressing tools, a single-point dressing tool and a dressing form roll [1–4, 6–10].

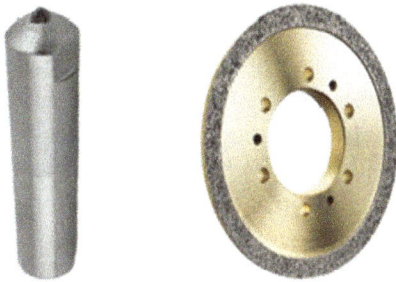

Figure 10.7.4: Left: single-point dressing tool. Right: dressing form roll [12].

Depending on the manufacturing process, in detail, the combination of catalyst, applied pressure and temperature, the grain shape ranges from splintery to blocky as well as the color of the diamond which can vary from colorless, greenish, yellowish-orange to gray. Figure 10.7.5 shows splintery and colorless diamond-abrasive grains as well as a yellowish and blocky diamond [1–3].

Its thermal conductivity is the highest known, leading to an excellent evacuation of the accruing heat out of the grinding zone. Thus, diamond finds application in the field of construction, in which cooled as well as uncooled processes appear. However, because of its extraordinary sensitivity towards thermal shock, it is important to avoid quenching of a red-hot diamond, which has to be considered when applied in

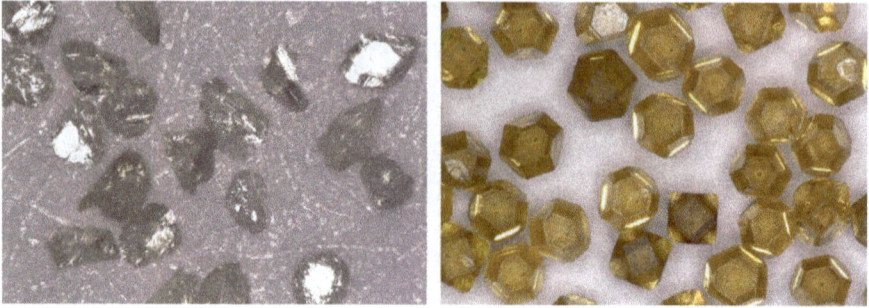

Figure 10.7.5: Left: splintery diamond grain. Right: blocky diamond grain [12].

uncooled processes. In the field of construction, diamond is especially used on blades for angle grinders, cutting, table, wire saws and many more [1–3, 6–10].

10.7.1.7 Cubic boron nitride

c-BN (β-BN), also called "inorganic diamond", is exclusively man-made. It represents the high-pressure modification of boron nitride and is synthesized the same way as diamond *via* high-pressure and high-temperature methods. α-BN (hexagonal modification) is used as starting material together with small amounts of catalysts like Li_3N [1–4, 6–10].

Equally to diamond, c-BN has a high wear resistance. It is many times higher than those of conventional abrasives. However, c-BN has the higher thermal and chemical resistance of the two superabrasives. For instance, this enables tooling of materials, which cannot be machined with diamonds, like iron, nickel, cobalt or molybdenum. With simultaneous consideration of these aspects, it does not surprise that c-BN is increasingly applied over other abrasives when grinding alloys or hardened steels like tool steels [1–3].

The usage of coolants represents an important aspect when grinding with c-BN. Due to the retransformation to α-BN at high temperatures, the hardness of c-BN decreases. This is why the arising temperatures have to be controlled precisely. However, its hardness is still higher than the hardness of conventional abrasives. When using water-based coolants, the formation of a passive layer of boron oxide (B_2O_3) has to be considered. Boron oxide dissolves in water, resulting in an increased wear of the grinding wheel. Therefore, CBN grinding applications will use mostly oil-based coolants; however, uncooled applications do exist. Equally to diamond, the grain morphology can be controlled during the production process. It ranges from blocky to splintery. Different shades of color do appear, depending on the type and the amount of the dopant. Figure 10.7.6 presents a splintery and a blocky c-BN [1–3, 10].

Figure 10.7.6: Left: splintery c-BN grain. Right: blocky c-BN grain [12].

10.7.2 Transition metal oxides in grinding industries

Polishing represents a final step of abrasive usage. A polishing step is applied to work pieces with extraordinary high requirements to the surface like optical glasses, artificial joints, elements for electronic industries and many more. In mechanical polishing, it has to be considered that the hardness of the polishing powder should not be higher than the hardness of the work piece. This avoids the formation of large grooves [1–3, 16].

In recent decades, the process of the so-called chemomechanical polishing (CMP) was developed. The hardness of the applied abrasives is in the same range or even softer than the work piece material. The main aspect using this technique is the chemical affinity between the polishing agent and the work piece material. Several transition metal oxides are applied for this polishing purpose, like cerium oxide (CeO_2), chromium oxide (Cr_2O_3), iron(III) oxide (Fe_2O_3) and iron(II,III) oxide (Fe_3O_4). The ability of an additional chemical reaction at the silica-CeO_2 interface makes cerium oxide the perfect fit for chemomechanical polishing of glasses. Furthermore, it is used as polish in order to generate mirror surfaces on silicon wafers. The hardness of chromium oxide is in the same range than the hardness of Si_3N_4. It was discovered that the function of Cr_2O_3 can be used in order to polish Si_3N_4 ceramics. Although the hardnesses of the two inorganic materials are almost equal, there is no mechanical but chemical abrasion because of the detected formation of Cr_2SiO_4 and chromium nitride (CrN). Fe_2O_3 and Fe_3O_4 were identified as suitable for chemomechanical polishing of Si_3N_4 as well [16–20].

10.7.3 Hard transition metal oxides

The oxides out of the elements of the titanium group (Ti, Zr, Hf) represent those oxides with the highest hardness. Their Knoop hardnesses range from 7 to 16 GPa. In the case of titanium dioxide (TiO_2), the hardness ranges from 7 to 11 GPa. Hafnium dioxide (HfO_2) has a hardness of 9.5 GPa, while zirconium dioxide (ZrO_2) represents the material with the highest hardness (14–16 GPa). Furthermore, they represent the highest melting and most stable transition metal oxides [4, 7–10, 21–23].

Titanium dioxide is mostly applied as a white pigment (see Chapter 1.4) because of its high refractive index and its lack of absorption of the visible light. Because of its refractoriness and its simplicity to sinter, TiO_2 is applied in ceramics as an opacifier. Furthermore, it is used as an oxygen sensor for automotive exhausts. This application results from the electrical resistance of titanium dioxide, which depends on the gaseous environment. Furthermore, it is used for preparation of catalysts in order to remove nitrous fumes from power plant exhausts. Ti_4O_7, resulting from rutile after reduction, is applied in conducting ceramics [4, 7–10].

Hafnium dioxide has only a very small abundance and therefore only minor industrial importance. It is exclusively obtained as a byproduct of the zirconia production, leading to a much higher price compared to ZrO_2. However, its physical properties make it more beneficial for particular applications. The phase transition from the monoclinic α-phase to the tetragonal β-phase appears at higher temperatures (1790 °C for HfO_2 and 1100 °C for ZrO_2). Furthermore, the volume change of this transition is smaller than that of ZrO_2 (3.4% in case of HfO_2 and 7.5% in the case of ZrO_2). This is an excellent example to favor HfO_2 as a refractory compared to zirconia [4, 7–10].

As already mentioned above, zirconia is used as a dopant during alumina synthesis, leading to alumina zirconia, a very tough and blocky wearing abrasive. In general, ZrO_2 is used in its cubic, high-temperature, γ-modification. However, this modification has to be stabilized with dopants like CaO, MgO, Y_2O_3 or Sc_2O_3 because it is only stable at temperatures above 2300 °C. By stabilizing zirconia, this modification is firm even at room temperature. Because of its extraordinary resistance towards molten metals, stabilized zirconia is a popular material for crucibles. Further applications of ZrO_2 are in the fields of refractories, ceramics, thermal coatings, solid electrolytes, glasses as well as container material in steel industries. Moreover, it is possible to synthesize stabilized zirconia crystals, having comparable optical properties to diamond. This makes zirconia interesting as synthetic gemstones [4, 7–10].

References

[1]   Rowe WB. Principles of Modern Grinding Technology, 2nd edition, Elsevier, Amsterdam, 2014. ISBN 978-0-323-24271-4.
[2]   Jackson MJ, Davim JP. Machining with Abrasives, 1st edition, Springer, Boston, MA, USA. ISBN 978-1-4419-7301.
[3]   Marinescu ID, Hitchiner MP, Uhlmann E, Rowe WB, Insaki I. Handbook of Machining with Grinding Wheels, 2nd edition, CRC Press, Boca Raton, 2016. ISBN: 978-1-4822-0668-5.

[4]   Holleman AF, Wiberg N. Anorganische Chemie, 102. Auflage, De Gruyter, Berlin, 2007. ISBN 978-3-11-017770-1.
[5]   Abderrazak H, Bel Hadj Hmida ES. Silicon carbide: Synthesis and properties. In: Gerhardt R. (Ed). Properties and Application of Silicon Carbide, IntechOpen, Rijeka, 2011. ISBN 978-953-307-201-2.
[6]   Coes L Jr. Abrasives. In: Applied Mineralogy, Springer-Verlag, Wien, 1971. ISBN 3-211-80968-6.
[7]   Bertau M, Müller A, Fröhlich P, Katzberg M. Industrielle Anorganische Chemie, 4. Auflage, Wiley-VCH, Weinheim, 2013. ISBN: 978-3-527-33019-5.
[8]   Brook RJ. Concise Encyclopedia of Advanced Ceramic Materials, Pergamon Press, Elsevier, Amsterdam, 1991. doi: 10.1016/C2009-1-28294-3. ISBN: 978-0-08-034720-2.
[9]   Riedel R. (Ed). Handbook of Ceramic Hard Materials Volume 1+2, Wiley-VCH, Weinheim, 2000. ISBN 3-527-29972-6.
[10]   Briehl H. Chemie der Werkstoffe, 3. Auflage, Springer Vieweg, Wiesbaden, 2014. ISBN: 978-3-658-06224-8. DOI: 10.1007/978-3-658-06225-5.
[11]   Passos ER, Rodrigues JA. Cerâmica, 2016, 62, 38–44.
[12]   Picture from Tyrolit Schleifmittelwerke Swarovski K.G., 2021.
[13]   Brunner G. Schleifen mit mikrokristallinem Aluminiumoxid, Fortschritt-Berichte VDI, Reihe 2, Fertigungstechnik, Nr. 464, Berichte Aus Dem Institut Für Fertigungstechnik Und Spanende Werkzeugmaschinen, Universität Hannover, VDI-Verlag, 1997. ISBN 3-18-346402-0.
[14]   König W, Ludewig T, Stuff D. Sol Gel-Korunde eröffnen neue Leistungspotentiale, Springer Verlag, Berlin, wt-Produktion und Management, 1995, Vol. 85, 22–9.
[15]   Hausberger P Dissertation, Leopold-Franzens-Universität Innsbruck, 1992.
[16]   Marinescu ID, Tonhoff HK, Inasaki I. Handbook of Ceramic Grinding and Polishing, 1st edition, Elsevier, Amsterdam, 1999. ISBN: 978-0-8155-1424-4.
[17]   Borra CR, Vlugt TJH, Yang Y, Offerman SE. Metals, 2018, 8, 801–16.
[18]   He Q. Appl Nanosci, 2018, 8, 163–71.
[19]   Kikuchi M, Takahashi Y, Suga T, Suzuki S, Bando Y. J Am Ceram Soc, 1992, 75, 189–94.
[20]   Bünzli J-CG, McGill I. Rare Earth Elements, in: Ullmann's Encyclopedia of Industrial Chemistry, 5th edition, Vol. Ass, Wiley-VCH, 2018.
[21]   Shackelford JF, Alexander W. Materials Science and Engineering Handbook, 3rd edition, CRC Press, Boca Raton, 2001. ISBN: 0-8493-2696-6.
[22]   Morscher GN, Pirouz P, Heuer AH. J Am Ceram Soc, 1991, 74, 491–500.
[23]   Tapily K, Jakes JE, Stone DS, Shrestha P, Gu D, Baumgart H, Elmustafa AA. J Electrochem Soc, 2008, 155, H545-51.

10.8 Surface hardening and hard coatings

Torben Buttler

Thin-film coatings are thin layers of a few nanometers to several micrometers. The surface properties can be optimally adjusted to the intended application. The core of the component (the substrate) can be made from a low-cost, ductile and easily machinable material and provided with perfectly matched surface properties. Properties such as corrosion resistance, hardness, ductility, electrical conductivity and color can be influenced. This results in a large spectrum of applications ranging from the semiconductor industry to the coating of hard metals to increase wear resistance [1]. The requirements for hard coatings applied to hard metals for the machining of metals are manifold. For example, the wear resistance should be increased and the adhesion and diffusion between tool and workpiece reduced. Abrasion and oxidation processes are also relevant. The substrate has to prevent the hard coating from breaking in, which is the so-called eggshell effect. The goal is, for the hard coating, to provide effective protection against all acting wear mechanisms [2].

In the following chapter, selected processes, coating adhesion, coating structure and the most common coatings for hard metals are presented. Although hard metals already have a very high wear resistance, the wear behavior can be further improved with a hard coating. Values for hardness or Young's modulus are not given, as these depend on many different parameters such as the deposition process, the process parameters and the chemical composition [3].

10.8.1 The coating process

Thin films are deposited by two processes: CVD (chemical vapor deposition) and PVD (physical vapor deposition). Both processes operate in an evacuated chamber, i.e. in a vacuum. The two processes differ significantly in the process temperatures. The CVD process takes place at approx. 800–1000 °C, while the PVD process is usually carried out at 400–500 °C. The thermal properties are correspondingly different as are the thermal stresses during the coating process for the substrate.

When coating by the CVD process, the coating material reacts with the component surface. To achieve this, at least two gases (reactants) are introduced into the chamber with the component, which react with each other. The introduced gas is thermally activated and dissociates. The energy required to activate this metastable state can be supplied in a wide variety of ways. Accordingly, many different process variants of the CVD process are available. Another possibility is the excitation by plasma. Process variants to be mentioned here are PACVD (plasma-assisted CVD) and PECVD (plasma-enhanced CVD), which work at lower temperatures of 400–600 °C. The layer is deposited on the component surface, and at least two further reaction

https://doi.org/10.1515/9783110733471-017

products are formed, which have to be removed. There is not a gaseous compound available for every desired layer, so the doping of the layers with different elements is limited. The high deposition temperatures are also a limiting factor. Changes in material properties (e.g. hardness) may be possible, but also warpage of the component may result from the high process temperatures. The advantage of the CVD process is a very good layer adhesion due to diffusion processes [4, 5].

All PVD processes can be roughly divided into three process steps, the generation of a vapor phase, the transport of the particles to the substrate and the layer formation on the mold surface. Essentially, a distinction is made between thermal evaporation (e.g. arc-PVD) and atomization (sputtering) [6, 7]. Arc-PVD is an established PVD process. Advantages include good process stability, very good coating adhesion due to a high degree of ionization of the particles transferred into the gas phase and low costs. When the arc strikes the target, larger atomic clusters that have not been transferred into the gas phase are ejected. These clusters accumulate incoherently in the coating and can break out under load. This results in defects on the surface, which are a major disadvantage of the process. Arc-PVD has been further developed and modified over time, allowing a large number of process parameters to be optimally adjusted [8]. Sputtering describes the physical (non-thermal) vaporization of atoms by impulse transmission. The inert process gas is ionized and the ions bombard the surface. Single atoms are dissolved out of the target and accumulate on the component surface as a homogeneous layer. Advantages of the process are a very homogeneous and imperfection-free surface and good mechanical properties. Disadvantages are a low-layer deposition rate and associated higher costs. Sputtering has also been further developed and modified in many ways. One of the most widely used modifications is magnetron sputtering (MS). In this process, a magnet is attached behind the target, which increases the ionization rate of the process gas and thus the coating deposition rate [9].

10.8.2 Layer adhesion and layer build-up

10.8.2.1 Layer adhesion
Metals show metallic bonds, leading to electrical conductivity and very good forming properties. Carbide-based materials are a composite of ceramic particles and a metallic matrix. The carbides have a high percentage of ionic and covalent bonds. This results in high hardness and good corrosion protection. But the disadvantage of this chemical inertness is that coatings do not adhere well to carbides [10].

The surface of the hard metals is differentiated. It can have a metallic as well as a ceramic character as indicated in Figure 10.8.1. The roughness also influences the coating growth since large carbide particles can lead to shielding effects. The coating grows heterogeneously, fewer atoms accumulate in the shadow of the carbides and the coating is weakened at this point. Cracks and breakouts can occur if a

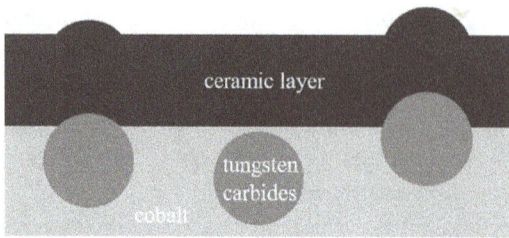

Figure 10.8.1: Schematic representation of different carbide positions in the coating of hard metals by PVD.

carbide is very close below the workpiece surface, as exemplified in the center of Figure 10.8.1. The grain boundaries at the transition between the carbide and the matrix should be understood as notches, which is why there is a particularly high intrinsic stress. In order to improve the adhesion of the layer, a monoelementary adhesion promoter layer is applied before the high-strength ceramic top layer is applied. This exhibits lower cohesion than ceramics, but ensures optimum layer adhesion due to its good adhesive properties [11, 12].

10.8.2.2 Layer structure

Pure monoelement diffusion barriers differ significantly in properties from binary, ternary or quaternary materials. These properties can be electrical conductivity, mechanical properties or optical properties. The advantages and disadvantages of the coatings can be combined by a smart choice of layers. Figure 10.8.2 shows the most important coating systems. A monolayer is shown on the left. Here, a coating is deposited without changing the process parameters. However, the nucleation of the layer is subject to defects, resulting in intrinsic stresses in the layer [12]. Under load, the layer can fail at these imperfections, resulting in cracks. These defects are subject to defect propagation; the intrinsic stress increases with increasing coating thickness. This can lead to cracks in the coating without external force. In order to be able to realize higher coating thicknesses, interlayers are introduced into the coating system [12]. The interlayer homogenizes the layer growth and interrupts the defect propagation. Layer thicknesses of 60 μm or more can be achieved in multilayer composites. The individual layers in such multilayer composites are usually 500–700 nm thick.

Another way to combine properties of the layers are graded layers. The transition between the layers is smooth. Figure 10.8.2 shows the example of a TiN/TiO coating in which the reactive gas flow was continuously adjusted during the coating process. Thus, no hard transitions of the layer are visible. Another build-up variant is the use of nanolayers since not only the chemical composition of the multilayer influences the properties but also the thickness of the layer. The individual layers in the nanolayer composite are only 10–75 nm thick. However, a coating is composed of up to

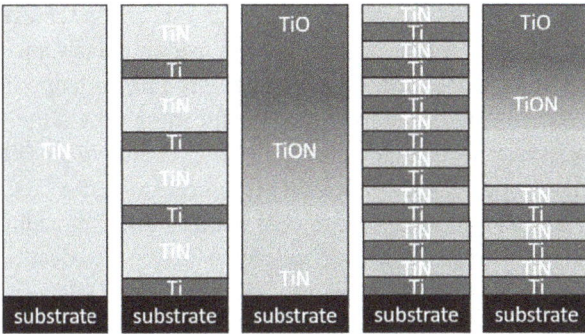

Figure 10.8.2: Schematic representation of different layer systems, from left to right: monolayer, multilayer, graded layer, nanolayer and combined layer.

several hundred layers [13]. A combination of the individual systems is possible as shown here on the right in Figure 10.8.2.

10.8.3 Chromium-based coating systems

Chromium is a metal with a body-centered cubic structure and a density of 7.14 g cm^{-3} [14]. Doping with nitrogen produces the phases chromium nitride (CrN) and CrN$_2$. These differ in their crystal structure: CrN has a cubic face-centered space structure; CrN$_2$ has hexagonal symmetry. The two phases also differ visually: gray CrN versus black CrN$_2$. The formation of the structure can be specifically influenced by process parameters. Due to the lower number of slip planes in the hexagonal structure, CrN$_2$ shows a higher hardness and also a significantly reduced ductility. Therefore, the cubic face-centered CrN variant is mainly used. By additional doping with aluminum, the ternary chromium aluminum nitride (CrAlN) can be produced. The aluminum atoms substitute chromium atoms, while nitrogen continues to interstitially intercalate. This allows the hardness, Young's modulus and corrosion resistance to be increased when compared to the CrN coating [15]. Chromium-based coatings are used especially for machining non-ferrous metals such as copper and titanium.

10.8.4 Titanium-based coating systems

Titanium is a hexagonally close-packed metal with a density of 4.5 g cm^{-3} [16]. As with CrN, doping with nitrogen results in a face-centered cubic structure in which the nitrogen atoms fill octahedral voids (rocksalt variant). Titanium nitride (TiN) is characterized by its high corrosion resistance, high surface hardness and adhesion resistance. This has made it one of the leading coatings since the 1980s [17]. In the titanium-based system, too, binary TiN was extended with aluminum to form ternary titanium aluminum nitride ($Ti_{1-x}Al_xN$). A significantly improved oxidation resistance is typical for TiAlN. Due to the substitution with aluminum, a very fine structure is formed. At a low aluminum content, the titanium atoms are substituted by aluminum. The structure shifts to a hexagonal wurtzite structure from a value of $x = 0.7$ onwards. For the coating of cutting materials, a hybrid structure of ductile face-centered cubic TiAlN domains and high hardness hexagonal wurtzite TiAlN domains is of particular interest. Various extensions of the TiAlN system are being tested with, for example silicon, to form quaternary TiAlSiN. This way, the oxidation resistance can be further increased [18]. Another common coating of hard metals is the titanium carbon nitride (TiCN) coating. This compound is characterized by its high hardness, good adhesion resistance, a low coefficient of friction and a blue-gray color. Thus, the wear resistance could be further reduced in contrast to the simple TiN coating [19]. TiAlN and AlTiN coatings are preferred for machining high hardness steels.

10.8.5 Carbon-based coating systems

Carbon is known as coal or diamond. Both substances exhibit different properties and yet consist of the same element. This is also used for carbon coatings. These are divided into crystalline, amorphous or plasma polymerization carbon coatings. Crystalline coatings are divided into graphite (sp^2 hybridization; π-bonding contributions) and diamond (sp^3 hybridization; solely σ-bonds). While there is a relatively large distance between the atomic layers of the sp^2 coatings, in the sp^3 formation, there is a strong three-dimensional covalent bonding. Slippage and thus plasticization of the atomic layers are almost impossible. Amorphous carbon coatings are also called diamond-like carbon coatings (DLC). These exhibit a mixture of sp^2 and sp^3-hybridized carbon, and the ratio of the bond types is crucial for the properties of the coating. In further experiments, the coating was doped with hydrogen to strengthen the hydrophobic properties [20]. Using multilayer technology, the coating systems presented here were also linked. For example, the combination of CrN and DLC coatings further increased the corrosion resistance [21]. The carbon coatings are used when ultra-high-strength steels or hard metals have to be machined.

10.8.6 Alternative coating systems

Multilayer systems are used to develop new coatings and adapt them to ever-increasing requirements. For example, the oxides of the elements aluminum (Al), zirconium (Zr) and hafnium (Hf) are integrated into the multilayer composite. As a monolayer, these coatings would not be life-enhancing due to their low ductility and poor adhesion to the substrate. Therefore, the coatings are used as a composite with already-known coatings such as the DLC coatings or TiAlN coatings.

References

[1] Bobzin K, Bagcivan N, Immich P, Pinero C, Goebbels N, Krämer A. PVD – Eine Erfolgsgeschichte mit Zukunft. Mat-wiss Werkstofftech, 2008, 39, 5–12.

[2] Klocke F. Schneidstoffe und Werkzeuge. In: Klocke F (Ed). VDI-Buch, Fertigungsverfahren 1: Zerspanung mit geometrisch bestimmter Schneide, 9th edition, Springer, Berlin, Heidelberg, 2018, 113–228. [Online]. Available: https://doi.org/10.1007/978-3-662-54207-1_4.

[3] Cunha L, Andritschky M, Pischow K, Wang Z. Microstructure of CrN coatings produced by PVD techniques. Thin Solid Films, 1999, 355-356, 465–71. DOI: doi.org/10.1016/S0040-6090(99)00552-0.

[4] Schwander M, Partes K. A review of diamond synthesis by CVD processes. Diamond Rel Mater, 2011, 20, 1287–301. DOI: doi.org/10.1016/j.diamond.2011.08.005.

[5] Hetzner H. Systematische Entwicklung amorpher Kohlenstoffschichten unter Berücksichtigung der Anforderungen der Blechmassivumformung. [Online]. Available: https://opus4.kobv.de/opus4-fau/frontdoor/index/index/docId/5004; access 07.07.2021

[6] Baptista A, Silva F, Porteiro J, Míguez J, Pinto G. Sputtering Physical Vapour Deposition (PVD) coatings: A critical review on process improvement and market trend demands. Coatings, 2018, 8, 402. DOI: 10.3390/coatings8110402.

[7] Bach F-W, Möhwald K, Laarmann A, Wenz T (Eds). Moderne Beschichtungsverfahren, 2nd edition, Wiley-VCH, Weinheim, 2006.

[8] Murrenhoff H. Umweltverträgliche Tribosysteme, Springer, Berlin, Heidelberg, 2010.

[9] Kelly PJ, Arnell RD. Magnetron sputtering: A review of recent developments and applications. Vacuum, 2000, 56, 159–72. DOI: 10.1016/S0042-207X(99)00189-X.

[10] Ibach H, Lüth H. Festkörperphysik, Springer, Berlin, Heidelberg, 2009.

[11] Inspektor A, Salvador PA. Architecture of PVD coatings for metalcutting applications: A review. Surf Coat Technol, 2014, 257, 138–53. DOI: 10.1016/j.surfcoat.2014.08.068.

[12] Bouzakis K-D, Makrimallakis S, Katirtzoglou G, Skordaris G, Gerardis S, Bouzakis E, Leyendecker T, Bolz S, Koelker W. Adaption of graded Cr/CrN-interlayer thickness to cemented carbide substrates' roughness for improving the adhesion of HPPMS PVD films and the cutting performance. Surf Coat Technol, 2010, 205, 1564–70. DOI: 10.1016/j.surfcoat.2010.09.010.

[13] Martínez E, Romero J, Lousa A, Esteve J. Wear behavior of nanometric CrN/Cr multilayers. Surf Coat Technol, 2003, 163–164, 571–7. DOI: 10.1016/S0257-8972(02)00664-3.

[14] Greenwood NN, Earnshaw A. Chemie der Elemente, 1st edition, VCH, Weinheim, Basel (Schweiz), Cambridge, New York, 1990.

[15] Bagcivan N, Bobzin K, Brögelmann T, Kalscheuer C. Development of (Cr,Al)ON coatings using middle frequency magnetron sputtering and investigations on tribological behavior against polymers. Surf Coat Technol, 2014, 260, 347–61. DOI: 10.1016/j.surfcoat.2014.09.016.

[16] Zwicker U. Titan und Titanlegierungen, Springer-Verlag, Berlin Heidelberg, 1974. DOI: 10.1007/978-3-642-80587-5.

[17] Cremer R, Witthaut M, von Richthofen A, Neuschütz D. Determination of the cubic to hexagonal structure transition in the metastable system TiN-Al. Fresenius J Anal Chem, 1998, 361, 642–5.

[18] Tillmann W, Dildrop M. Influence of bias voltage and sputter mode on the coating properties of TiAlSiN. Mat-wiss Werkstofftech, 2017, 48, 855–61. DOI: 10.1002/mawe.201600731.

[19] Siow PC, Ghani JA, Ghazali MJ, Jaafar TR, Selamat MA, Che Haron CH. Characterization of TiCN and TiCN/ZrN coatings for cutting tool application. Ceram Int, 2013, 39, 1293–8. DOI: 10.1016/j.ceramint.2012.07.061.

[20] Paul R, Das SN, Dalui S, Gayen RN, Roy RK, Bhar R, Palet AK. Synthesis of DLC films with different sp2/sp3 ratios and their hydrophobic behaviour. J Phys D Appl Phys, 2008, 41, 55309. DOI: 10.1088/0022-3727/41/5/055309.

[21] Decho H, Mehner A, Zoch H-W, Stock H-R. Optimization of chromium nitride (CrN$_x$) interlayers for hydrogenated amorphous carbon (a-C:H) film systems with respect to the corrosion protection properties by high power impulse magnetron sputtering (HiPIMS). Surf Coat Technol, 2016, 293, 35–41. DOI: 10.1016/j.surfcoat.2016.01.037.

11 Carbon- and sulfur-based materials

11.1 Carbon

Dogukan H. Apaydin, Bernhard C. Bayer, Jean-Charles Arnault, Dominik Eder

11.1.1 Introducing remarks

Carbon is the most versatile chemical element for designing molecules and materials. It exists in a variety of solid allotropes with a wide range of properties and resulting functionalities for different applications. Four electrons in its valence shell allow each carbon atom to covalently bind to up to four different partner atoms via predominantly two binding modes, known as sp^2 (graphite-like) and sp^3 (diamond-like) hybridization. These two basic binding motives, and their combinations, yield well-defined molecular and supra-molecular structures extending in all dimensionalities (zero to three-dimensional: 0D, 1D, 2D and 3D) and exhibiting distinct physical and chemical properties. Together with purposely engineered carbonaceous nanomaterials, such as glassy carbon, nanodiamonds, carbon dots, fibers and aerogels, carbon structures have evolved as key components in functional materials and devices capable of advancing such socioeconomic fields as structural engineering, electronics, energy conversion and storage, catalysis, sensors, medicine and photonics.

In this chapter, we will first discuss sp^2 allotropes, initially focusing on 3D bulk carbon materials (*graphite, glassy carbon*) and then on sp^2 low-dimensional carbon (2D *graphene* and 1D *carbon nanotubes*). This is followed by a discussion of sp^3 carbon structures (*diamond, lonsdaleite*) in both bulk and nanoparticle form as well as a brief coverage of 0D *fullerenes* and 1D sp^1-bonded *carbyne*.

11.1.2 Graphite

11.1.2.1 Structural characteristics

Graphite is the thermodynamically most stable form of carbon (at standard conditions) and has been utilized in various applications since the Iron Age. It ideally consists of infinite sheets of sp^2-hybridized carbon (termed graphene layers, see below) with the in-plane atoms arranged in a honey-comb lattice and with the individual sheets stacked parallel to each other (Figure 11.1.1). Graphite has an intraplanar C–C bond length of 1.42 Å and forms layers with a spacing of 3.354 Å [1], which are held together by van der Waals forces. These weak interactions between the layers render graphite a suitable material in solid lubricant applications. Furthermore, it enables the facile exfoliation of these layers to yield individual graphene sheets (see below). Graphite materials are usually characterized by the dimensions

https://doi.org/10.1515/9783110733471-018

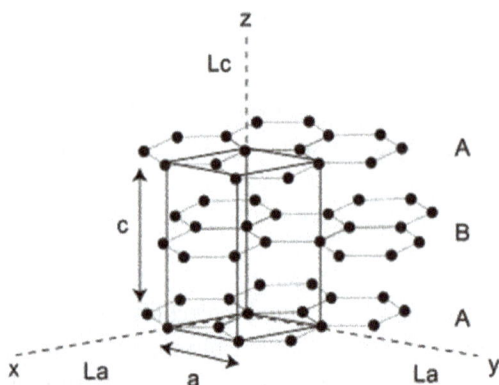

Figure 11.1.1: Graphite crystal structure with unit cell dimensions of 2.46 Å (*a* axis) and 6.71 Å (*c* axis). Reproduced with permission from ref. [4].

of the crystallites. L_a for the size of in-plane crystallites and L_c for the perpendicular to the graphene planes (Figure 11.1.1) [2]. The existence of bonding (π) and anti-bonding (π^*) orbitals at its valence band classifies graphite as a semi-metal [3], which renders graphite a promising candidate for electrochemical applications as unobstructed flow of electrons is essential in an electrode.

11.1.2.2 Synthesis

Synthetic graphite is a material consisting of graphitic carbon structures, which can be obtained by either graphitizing non-graphitic carbon materials at high temperatures, chemical vapor deposition (CVD) from hydrocarbons at temperatures above 2200 °C, decomposition of thermally unstable carbides or via crystallization from metal melts supersaturated with carbon [5]. In the mid-1890s, Edward Goodrich Acheson was trying to fabricate carborundum (silicon carbide; SiC) using silica and amorphous carbon in an electric furnace. He studied the effects of high temperature on carborundum and discovered that overheating carborundum above 4150 °C causes silicon to vaporize leaving the carbon behind as graphitic carbon. Acheson's technique for producing silicon carbide and graphite is now named the Acheson process [6]. In principle, all materials consisting of carbon can be used as starting material for graphite synthesis (with the boundary condition that enough carbon remains after thermal degradation). The reaction for the synthesis of graphite generally consists of the following steps that all impact the quality of the produced graphite [7]: bond cleavage leading to free radical formation, molecular rearrangement, thermal polymerization/polycondensation, aromatization, aliphatic side-chain elimination and dehydrogenation.

11.1.2.3 Properties

Graphite exhibits a very high temperature stability and chemical inertness (particularly under non-oxidizing conditions) and is thus the material choice for refractory materials like crucibles, oven inlays and electrodes. It has a thermal coefficient of expansion of $1.2–8.2 \times 10^{-6}$ K^{-1}. In addition, graphite is electrically conductive with an electrical resistivity of $5–30 \times 10^{-6}$ Ωm, making it a robust electrode material for industrial applications. With a Young's modulus between 9 and 11 GPa and a hardness of 1–2 Mohs, graphite can be mechanically shaped easily [8].

11.1.2.4 Key applications

Graphite is used in quite a wide range of industrial applications. Graphite electrodes are used in arc furnaces which constitute the majority of steel processing furnaces as well as in aluminum smelting [9]. Furthermore, graphite finds usage as neutron moderator in nuclear reactors and in carbon fiber-reinforced plastics. Mostly limited for research purposes is a special type of graphite called *highly oriented pyrolytic graphite* (HOPG).

HOPG is a special type of carbon analogous to single crystals that is produced by exposing pyrolytic graphite (formed by high-temperature decomposition of gaseous hydrocarbons) to high temperature and pressure. HOPG is made of graphene layers stacked parallel to each other in a lamella-like fashion with an angular spread of less than 1° [10]. The structural anisotropy of HOPG is pronounced, thus affecting its physical properties (thermal, electrical and optical properties) in different directions (Table 11.1.1).

Table 11.1.1: Physical properties of HOPG at 300 K [11].

Property	Along graphene layers	Perpendicular to graphene layers
Electrical resistivity (Ωcm)	$3.5–5.0 \times 10^{-5}$	0.15–0.25
Thermal conductivity ($Wm^{-1} K^{-1}$)	1700 ± 100	8 ± 1
Thermal expansion (K^{-1})	-1×10^{-6}	20×10^{-6}

Moreover, because of its atomically flat surface and low defect concentration, HOPG is used as a calibration standard, e.g. for scanning probe microscopy, as a nanoelectrode in electrochemical and biological sensing and as a (electro)catalyst with heterogeneous metal deposition and carbon surface functionalization [12, 13].

11.1.3 Glassy carbon

11.1.3.1 Structural characteristics

Glassy carbon (GC) combines glassy characteristics with ceramic properties of graphite (Figure 11.1.2). This leads to both a high thermal stability and extreme resilience towards chemical attack. The rate of oxidation in case of glassy carbon under oxygen, carbon dioxide or water is lower than for any other carbon form [14].

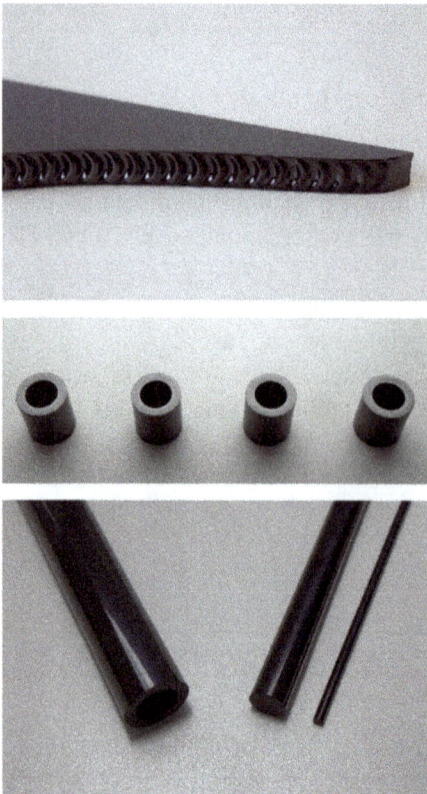

Figure 11.1.2: Glassy carbon components: (top) a plate with residual stress break, (middle) crucibles and (bottom) tube and cylinders. The components were kindly provided by HTW Hochtemperatur-Werkstoffe GmbH. Photos by Thomas Fickenscher.

The microstructure of GC has remained uncertain for many decades due to the complex internal organization of carbon atoms at the atomic scale (Figure 11.1.3). In the late 1940s, R. E. Franklin at Laboratoire Central des Services Chimiques de l'Etat in Paris, investigated the microstructure of GC with X-ray diffraction (XRD) and noticed that the ratio of graphitic and non-graphitic carbon depends not only on the temperature but also on the starting material. Eventually, she produced the first glassy carbon by thermally annealing a mixture of Saran (a polymer of polyvinylidene chloride) together with a small amount of polyvinyl chloride [15].

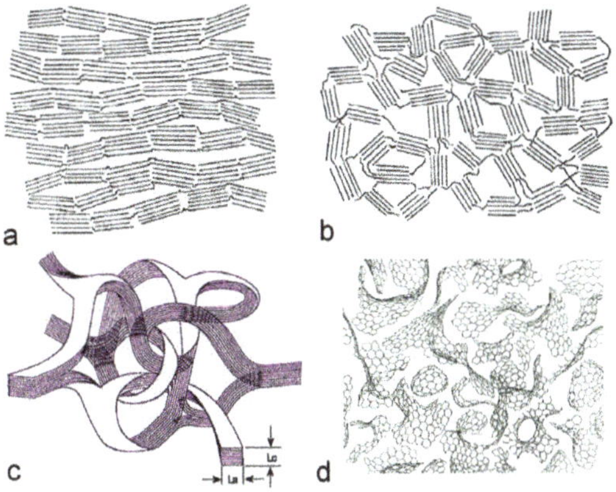

Figure 11.1.3: Evolution of the models describing the graphitic order and amorphous disorder in GC, starting from Franklin's representation of (a) graphitizing and (b) non-graphitizing amorphous carbons followed by Jenkins' model (1972) of non-graphitizing GC correlating crystallinity with L_a and L_c domains (c). (d) Harris' model (early 2000s), where non-graphitizing carbons are represented as consisting of intermixed graphitic and fullerene-like motifs. Reproduced with permission from ref. [15].

All GCs generally compose of disorganized flakes with graphitic layers of sp^2-hybridized carbon atoms arranged in hexagonal pattern (Figure 11.1.3), which are randomly arranged and relatively closely packed. The size and graphitic nature of these crystallites clearly depend on the precursors and pyrolysis conditions. The crystals typically possess low values of L_c (height) and L_a (width) and a turbostratic structure in which the individual carbon layers are rendered in a random fashion along the a axis and rotated in relation to each other with respect to the c axis. This deformation also results in a significantly larger average interlayer spacing as compared with graphite [16]. Due to cavities created by the crystallite packing, GC exhibits a lower density of around 1.5 g cm^{-3} compared with graphite and some porosity, albeit with a limited accessible surface area [14].

11.1.3.2 Synthesis

Today, GC is synthesized via pyrolysis of carbon-rich precursors (e.g. phenol formaldehyde, polyfurfuryl alcohol, polynaphthalene, polyimide, polyviniylidenechloride and perylene tetracarboxylic acid) under an inert atmosphere such as N_2 or Ar [15]. At the beginning of this process, small carbon-based molecules of low molecular weight are released into the gas phase from the solid precursor resin. This is followed by some aromatization of aliphatic species and polycondensation of aromatic molecules.

Above 600 °C, large amounts of heteroatoms (oxygen, nitrogen, etc.) are removed from the precursor in the form of gaseous carbon oxides (CO, CO_2), nitrogen oxides (NO, NO_2), methane (CH_4) and hydrogen (H_2) leading to a residual material mainly composed of carbon at 1300 °C. GC can be classified into "Type 1" which is produced at low temperatures (~1000 °C) and into "Type 2" which is produced at higher temperatures (>2000 °C) [17]. GC produced at higher temperatures tends to have fewer structural defects with a higher degree of organization compared to low-temperature GC.

11.1.3.3 Properties
GC exhibits a relatively high hardness (7 in Mohs scale), low toughness, high thermal resistance, low density and impermeability to gases and liquids as well as extreme durability against corrosion. GC can also be processed into fine particles, thin films, sheets and fibers. The most important feature of GC is its low electrical resistivity of 4.5×10^{-5} Ωm at 30 °C [18].

11.1.3.4 Key applications
GC mostly finds utilization in electrochemical applications, in particular, as inert and reliable electrodes [19]. GC fulfills several electrode requirements as it has a low background response, a fast, reversible and stable electron transport, a wide electrochemical window (especially at reductive potentials) and no toxicity. Combined with its low chemical reactivity, low permeability, ease of cleaning and high hardness, GC has evolved as a standard electrode material. GC surface can be functionalized upon oxidative treatments and can thus be utilized in biochemical sensing applications [15]. Reticulated, cellular and monolithic GC exist as the substructures and find different applications: Reticulated GC (RGC) is a microporous material with high open porosity and large surface area which is formed by struts that are connected by nodes. Cellular GC (CGC) also shows a porous structure, but unlike RCG, its pores are made of small circles called windows. Due to having high surface area, both RGC and CGC are used in water treatment [20, 21], electrochemical sensors [22, 23], energy storage devices [24] and in tissue engineering as scaffolds [25, 26]. Monolithic GC (MGC) does not present a uniform pore structure as in the case of RGC and CGC. However, fissures, tensions and pores can be found in the bulk of MGC due to the diffusion of volatile molecules and shrinkage of the precursor resin. MGC is commonly used as materials in crucibles [27], heart valves [28], dental implants [29] and tools for precision molding [30].

11.1.4 Graphene

11.1.4.1 Structural characteristics
Graphene refers to a two-dimensional (2D) material made of a single sp^2-bonded monolayer carbon sheet, i.e. an individual layer in graphite (Figure 11.1.4) [31–33].

Most research work has been concentrated on such monolayer graphene, although bi-layer and few-layer graphene have been widely investigated [33].

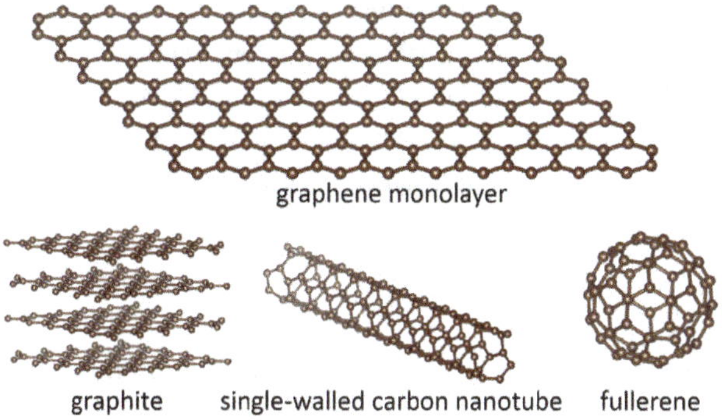

Figure 11.1.4: Structural models of a graphene monolayer, graphite, a single-walled carbon nanotube and a fullerene.

11.1.4.2 Synthesis

Graphene can be produced either top-down by exfoliation from graphite or bottom-up via direct growth from molecular precursors [34]. For top-down exfoliation, individual layers can be mechanically separated by the "scotch-tape" method introduced by Novoselov et al. [31], which made graphene experimentally accessible in 2004. This process, however, yields relatively small (i.e. micrometer range) monolayer graphene flakes for research purposes only and is not scalable. For large amounts of top-down-produced graphene, wet chemical liquid phase exfoliation (LPE) is preferred. Here, graphite is delaminated by shear forces in a liquid suspension medium and then size-selected (by, e.g. centrifugation) with individual exfoliated layers forming stable suspensions in appropriate liquid media [34]. LPE typically yields 2D crystals with a layer number distribution from 1 to ~10 layers and lateral sizes in the small micrometer range.

Alternatively, a graphene-type material can also be obtained by the first delamination of graphite via chemical oxidation (e.g. with potassium permanganate), which results in the formation of graphene oxide (GO) [35], followed by chemical reduction (e.g. with hydrazine) to reduced graphene oxide (rGO) [36]. This process typically yields large amounts of few-layer flakes in small micrometer lateral size range, albeit at the cost of a fairly defective crystalline quality [34]. Their size and defect characteristics somewhat limit the properties of LPE graphene and rGO, but the low cost of their preparation makes LPE graphene and rGO interesting for, e.g. energy applications and as additives/fillers in composite materials (*graphene powders*).

CVD is the method of choice for producing large-sized graphene films with controlled layer numbers and high crystallinity [34]. It is possible to grow monolayer graphene single crystal films with macroscopic lateral dimensions up to millimeters with very low defect levels on appropriate metal catalyst substrates such as copper substrates. These graphene films can then be transferred via removable scaffolds onto desired target substrates. CVD-based graphene films are of predominant interest for (opto-)electronic and barrier applications of graphene. Finally, graphene can also be synthesized in wet chemistry organic synthesis, yielding typically not extended graphene layers but rather defined graphene nanoribbons [34].

11.1.4.3 Properties

By removing a graphene layer from its bulk graphite environment, the properties of the graphene layer change significantly and in an usual and beneficial way. Normalized to its atomic thickness (normally described as 3.354 Å), monolayer graphene exhibits many "record" properties: It is a semi-metal with a linear dispersion near the Dirac point [37], can exhibit ballistic transport over micrometer-scales at room temperature with extremely high electron mobility of up to 200000 cm^2/Vs [38] and has been shown to sustain extreme current densities (10^{12} A/cm^2) [39]. These properties make graphene an excellent conductor at atomic thickness. Notably, its electronic properties, such as charge carrier densities or charge neutrality points, are readily adjustable and highly sensitive towards adsorption by molecules or other matter on the graphene surface [33, 40]. This opens the possibility to electronically tune graphene by so-called hybrid formation or use it as a sensor material. Another possibility is to narrow the lateral size of the graphene film in one dimension and to produce the so-called nanoribbons (width of a new nm), in which the semimetallic graphene transforms into a semiconducting material with a tunable bandgap [33]. Graphene also has intriguing optical properties, as it has a very low absorption cross-section (~2.3%) over the entire visible range [41]. This makes graphene highly interesting for transparent devices. Mechanically, graphene is a very strong atomically thin material (tensile strength ~130 GPa) and has extremely high Young's modulus (~1 TPa) [42]. Additionally, defect-free graphene layers are impermeable to a wide range of atoms, molecules and ions [43, 44], suggesting graphene as an ultimately thin barrier material. In terms of thermal properties, in-plane thermal conductivity in graphene monolayers is also remarkably high (~5000 W/mK) [45]. Being essentially surface of both sides with hardly any volume in between, graphene also has high specific surface area and thus lends itself as a building block for high surface area materials [33]. Finally, graphene is environmentally benign, does not pose health concerns and is potentially cheap to produce, making graphene intrinsically scalable [33, 34]. Notably, many of the record properties of graphene such as electronic and thermal transport were experimentally measured on a micrometer-scale, high-quality monolayer graphene flakes [38, 39, 43–45], but drastically reduce when translated to larger, polycrystalline and restacked graphene assemblies [33]. Despite their lower performance values, nevertheless, such assemblies from graphene building blocks are also of high application potential in, e.g. electrical, energy or composite applications [33, 46].

11.1.4.4 Key applications

This set of unique properties have made graphene a material of considerable research interest in a large number of fields [33], e.g. in cheap, transparent and flexible (2D) electronics (with graphene being an ideal transparent conductive electrode material [47]), in "more-than-Moore" (opto-)electronics, as ultimately thin functional coatings and barrier materials in electronics and metallurgy [48], as electrode materials in supercapacitors and batteries [19], as catalyst supports in (photo-)catalysis [49] and as lightweight high strength fillers in composites, to just name a few. Graphene was also the first experimentally realized, freestanding 2D material and thus has laid the foundation for many research activities into other 2D materials in which graphene often still plays a key component in the so-called van der Waals heterostructure hybrids [33, 50–52].

11.1.5 Carbon nanotubes

11.1.5.1 Structural characteristics

The structure of carbon nanotubes (CNTs) can be derived from graphene by cutting a sheet out of a graphene plane and rolling it into a seamless cylinder (Figure 11.1.4) [53, 54]. CNTs were first reported by Iijima in 1991 [53]. The transition from the 2D graphene sheet to the rolled-up 1D carbon nanotube can be accomplished with an infinite number of cutting lines through the graphene structure that join on equivalent positions to close the nanotube cylinder [54]. The cutting direction, and thus resulting tube structure and diameter, is completely described by a circumferential two-dimensional vector that connects two crystallographically equivalent sites on the cylinder. This vector is defined by the integer chiral indices (n,m). In the case of closed tubes, the ends of this cylinder are then closed by hemispherical fullerene caps. The second distinction can be made between single graphene cylinders, which are termed as single-walled nanotubes (SWNTs), and multiple, concentrically stacked and regularly spaced cylinders, composed of multiple SWNTs with ever increasing diameters, which are termed as multiwalled nanotubes (MWNTs).

11.1.5.2 Structural characteristics

For the synthesis of CNTs, several techniques have been successfully developed [55, 56]. Arc-discharge, solar furnace and laser vaporization of graphite targets all use high temperatures (>1000 °C) to sublime carbon which then condenses into CNTs "powders", often with large amounts of other carbonaceous species. These synthesis techniques can produce nanotubes at reasonable quantities and of decent structural quality, but do not allow localized growth or sufficient control of their aspect ratio and number of walls.

A considerably better control of the CNT characteristics can be achieved by CVD, which has evolved as the dominant synthesis technique for SWCNTs and MWCNTs both on a lab scale and for industrial production. CVD uses many different types of carbon precursors (such as hydrocarbon gases) and relies on catalyst nanoparticles (often metallic Fe, Ni, Co and often supported on metal oxide materials) to catalyze breakdown of the precursor gases, redistribute carbon species and facilitate nucleation and sustained growth of CNTs. Via controlled deposition of catalyst nanoparticles on target substrates, this process allows localized CNT growth at predefined positions on samples with comparably lower temperatures, including control over SWNT/MWNT type [57–59]. A detailed description of the advantages and disadvantages of the various synthesis techniques, catalyst materials and process conditions can be found in ref. [56].

11.1.5.3 Properties

SWNTs can be either semiconducting or metallic, depending on their chirality (n,m) [54–56]. For example, all armchair tubes and chiral tubes, where $n - m$ is divisible by 3, are metallic. The band gap of semiconducting tubes decreases with increasing diameter. Similar to graphene, CNTs can exhibit ballistic electron transport over micrometer scales [60], sustain very high current densities ($\sim 10^{10}$ A/m^2) [61], show high electron mobility (>100000 cm^2/Vs) [62], have very high thermal conductivity ($\sim$3000 W/mK) [63] and yield strength and Young's modulus ($\sim$1 TPa) [64].

11.1.5.4 Key applications

The above detailed remarkable electronic, electron transport and thermal properties have spurred intense research interest across various disciplines, e.g. semiconducting SWNTs are considered as transistor elements for next-generation field-effect transistors [60]. As their electronic structure can easily be modified by adsorbates, such devices are also used as chemical sensors [65]. Metallic nanotubes are also considered as building blocks for current transport in next-generation interconnects [58, 59]. The exceptional high thermal conductivity of carbon nanotubes suggests them as thermal transport materials both on their own as well as embedded in a composite matrix. The high-surface area of CNTs has led to much research for using them as potential capacitor materials in double-layer supercapacitors [66] as well as using them as scaffolds [67] or delivery vehicles [68], in particular when CNTs can be controllably opened at their ends. The high-aspect nature of isolated, stiff CNTs (and the resulting electric field enhancement factor) combined with their high current-carrying capability has led to them being implemented for field emission devices [69]. The exceptional tensile strength and Young's modulus of CNTs influenced significant research efforts to use CNTs as structural materials [70] or active filler materials in composites [71]. Many of these applications benefit from the implementation of CNTs with inorganic functional materials into hybrid materials [72, 73]. The excellent capability of CNTs to accept electrons from the hybrid's components allows for a more efficient charge separation and extraction as well as enhanced charge lifetimes, which commends these hybrid materials photoelectrodes and photocatalysts in energy conversion technologies [74, 75].

11.1.6 Fullerenes

Fullerenes are hollow sphere-like 0D objects, formed by spherical wrapping of a graphene sheet via introduction of isolated pentagons, similar to the corannulene molecule (Figure 11.1.4) [76]. The archetypical fullerene is C_{60}, discovered by Kroto et al. [77] in 1985. Yet, the term "fullerene" applies to all spherical carbon allotropes that contain exactly 12 pentagons, irrespective of their size (i.e. number of hexagons), shape (i.e. spherical, ellipsoid) and defect level (i.e. presence of seven-rings). Fullerenes are typically produced by arc discharge or laser ablation of graphite and often contain mixtures (most prominently C_{60} and C_{70}). Due to their mixed sp^2 (hexagons) and sp^3 (pentagons) character and the resulting angle strain, fullerenes are stable, but more reactive than the other nanocarbons, as represented by a low sublimation temperature, increased solubility in organic media (e.g. carbon disulfide; CS_2) and a high susceptibility towards functionalization (e.g. via electrophilic addition reactions) that tends to focus on the pentagons. Moreover, in contrast to purely graphitic materials, fullerenes do not exhibit a (super)aromatic behavior as the electrons are not fully "delocalized" over the entire molecule. As a result, fullerenes are excellent electron acceptors, which makes them interesting for functionalization and as catalyst for organic reactions. Furthermore, this makes fullerenes electrical semiconductors (e.g. C_{60} has a band-gap of ~1.6 eV, yet the band-gap decreases with the addition of hexagons) [78]. However, when crystallized with alkali and alkaline earth metals, fullerene crystals (aka fullerites) can be turned into either superconductors with high critical temperatures above 30 K (use of up to three alkali metal ions) or insulators (uptake of 4–6 metal ions) [78]. Due to their electrophilicity, fullerenes are often engineered into hybrid systems by anchoring metal ions on their surface (exohedral modification) or into nanocontainers by enclosing metal ions and gas molecules in their interior (endohedral modification). These properties commend fullerenes for use as carrier systems in medicine and biology (e.g. antibiotics and cancer therapy), in cosmetics (e.g. anti-aging) and as functional layers in hybrid organic electronics [76, 79]. Other commercialized applications utilize their small spherical morphology in lubricants and filling material for mechanical composites (e.g. in sports equipment).

11.1.7 Carbyne

Carbyne is a sp^1-hybridized carbon chain with alternating single and triple bonding between the carbon atoms [80–82]. Its exact nature and properties are still an area of active research [80, 81]. Carbyne appears to be difficult to obtain in isolated form but has been reported to be obtainable via, e.g. CNT encapsulation [83].

11.1.8 Diamond

11.1.8.1 Structural characteristics

Two carbon allotropes can be built from sp^3-hybridized carbon atoms in a tetrahedral configuration (Figure 11.1.5). Diamond, the most common, has a cubic structure derived from the face-centered cubic lattice (space group $Fd\bar{3}m$). Each carbon atom is linked to its neighbors by covalent bonds (360 kJ/mol) with a carbon-carbon distance of 0.154 nm. The diamond structure exhibits one of the highest atomic density (1.76×10^{23} atoms cm^{-3}) corresponding to a mass density of 3.52 g cm^{-3}. Lonsdaleite is the second allotrope exhibiting a hexagonal crystallographic structure (albeit lonsdaleite's structural independence has recently come under debate [84]).

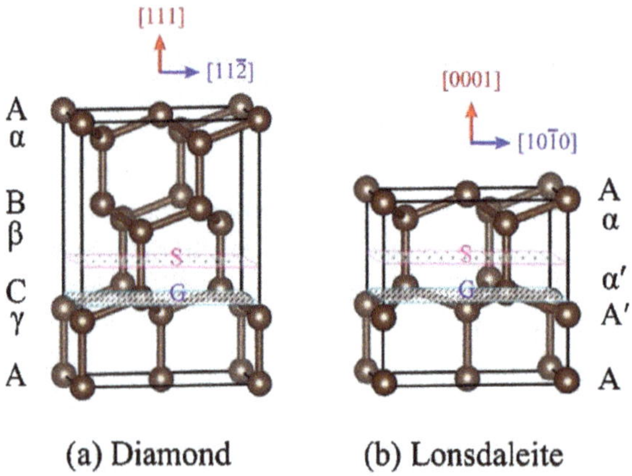

(a) Diamond   (b) Lonsdaleite

Figure 11.1.5: Crystalline structures of diamond and lonsdaleite. Reproduced with permission from ref. [85].

Diamond materials can possess different microstructures from single crystal to polycrystalline films with grain size ranging from microns to nanometers. These microstructures strongly impact the chemicophysical properties. Among the hosted impurities in natural or synthetic diamond, the most abundant one is nitrogen [86]. Depending on its concentration (higher or lower than 1–2 ppm) and its distribution within the structure, two main types (I and II) are defined. Boron and phosphorous atoms are intentionally incorporated in substitution in the diamond structure to confer p and n-type doping, with activation energies of 0.37 and 0.59 eV, respectively [87]. More than 500 optically active defects are reported in the diamond crystalline structure [88]. Recently, impurities from the IV column like silicon, germanium, tin or lead are inserted in diamond to generate SiV, GeV, SnV or PbV photoluminescent color centers by combination with neighboring vacancies [89].

11.1.8.2 Synthesis

According to thermodynamics (Figure 11.1.6), the stability domain of diamond corresponds to high pressure (>13 GPa) and high temperature (>1700 °C). The high-pressure high temperature (HPHT) synthesis started in 1955 [90]. Later, the use of metallic catalysts allowed to lower HPHT synthesis conditions. This technique is compatible with mass production and allows to produce boron-doped diamond of high crystalline quality [91].

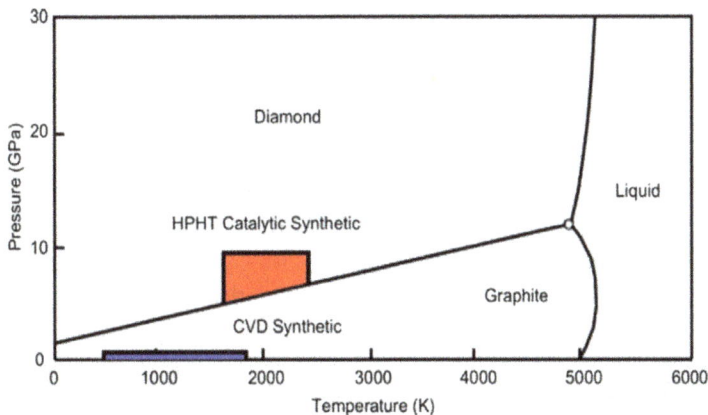

Figure 11.1.6: Phase diagram of carbon. Reproduced with permission from ref. [86].

In 1962, the pioneer work of Eversole [92] revealed the formation of diamond by CVD despite metastable conditions (Figure 11.1.6). CVD synthesis activated by microwave or hot filament quickly spread across research laboratories. Nowadays, this common synthesis route allows to grow diamond films of high purity with controlled microstructure on various substrates. Polycrystalline diamond can be grown on large substrates (up to 4 inches) [93], while one-inch samples are reached for single crystal diamond [94].

For the past 10 years, diamond nanoparticles or nanodiamonds (ND) are actively investigated. ND can be obtained by a top-down approach by milling of bulk diamond, natural, HPHT or CVD grown. This pathway allows to get intrinsic or boron-doped ND but with a size dispersion. With a bottom-up strategy, ND of nanometric size can be obtained by decomposition of explosive mixtures, and this detonation method was initiated in 1963 [95]. The second way is the laser ablation [96]. In the last years, novel CVD [97] and HPHT [98] bottom-up approaches emerged to better control the ND characteristics (size, shape and doping).

11.1.8.3 Properties

Diamond shows outstanding mechanical properties with a Young's modulus of 1050 GPa, a hardness of 10 on the Mohs scale and a very low friction coefficient (0.05–0.15) [99]. The sound velocity through the diamond structure, 18 k ms^{-1}, is one of the highest known. Its remarkable thermal conductivity can reach 2200 Wm^{-1} K^{-1} for diamond single crystals [100], six times higher than copper. It drops to smaller values for nanocrystalline diamond. Diamond is resistant to harsh environment (radiation, corrosive atmosphere, high pressure and temperature). The diamond surface can be efficiently functionalized via the versatile carbon chemistry. Optically, it is transparent over a wide wavelength range from 220 nm to the far IR. It possesses a large refractive index of 2.42. The diamond structure can also host color centers (part a) which possess a very stable photoluminescence. In addition, NV centers can bring spin properties depending on its charge state. Diamond is a biocompatible material, and nanodiamonds can be considered as negative reference for in vitro nanogenotoxicity studies [101].

Single crystal diamond is a promising wide-bandgap semiconductor (5.45 eV) for future electronics. Its electrical and thermal properties outperform those of other wide bandgap semiconductors like 4 H-SiC, GaN or Ga$_2$O$_3$ [102]. Its breakdown field can reach 13 MV/cm. Holes and electrons possess exceptional intrinsic mobilities in diamond up to 3800 and 4500 cm V^{-1} s^{-1} at room temperature, respectively. The intrinsic resistivity of diamond is 10^{16} Ωcm. p-Doping of single crystal diamond with boron is now well controlled over a wide concentration range, whereas n-type doping with phosphorous is still under progress. Boron-doped polycrystalline diamond exhibits noticeable electrochemical properties with a wide working potential window and low background currents [103].

Nanodiamonds (ND) combines properties of bulk diamond with new ones conferred by nanoscale. Especially, their surface-related properties can be exalted as ND can exhibit specific surface area up to 300–400 m^2 g^{-1}. A recently published book focused on specific characterizations and properties of ND [104].

11.1.8.4 Key applications

Diamond is nowadays used for industrial applications. Its huge mechanical properties are exploited in cutting tools, drilling heads, laser cutting and scalpels. Optical windows are made with polished polycrystalline diamond materials for X-rays, microwave, terahertz, CO$_2$ lasers and ultrahigh vacuum [105]. Based on single crystal diamond, X-ray beam position monitors, X-ray monochromators, splitters and detectors (charged high energy particles, neutrons, UV light), dosimeters in radiotherapy are under development [94]. Diamond is also an excellent medium for Raman lasers due to high Raman gain and shift, simple single-line Raman spectrum, wide transparency range, high thermal conductivity and low thermal expansion coefficient [106]. According to its biocompatibility, its mechanical properties and its surface functionalization, diamond is suitable for photonic crystals usable for biosensors [107]. Boron-doped diamond electrodes are used for in vivo bioelectrochemistry and supercapacitors [108]. In biomedical applications, diamond coatings are under consideration for implants and tissue engineering. Functionalized ND is currently used as a carrier for drug deliver or as a biomarker taking advantage of fluorescent color centers [109]. Polymer/nanodiamonds composites are actively investigated, ND bringing its mechanical and thermal assets [110]. Advanced lubricants using detonation of ND are under development to lower the friction and wear properties [111]. The current trend concerns quantum applications based on spin properties of color centers hosted in diamond: thermometry at nanoscale [112] and optical hyperpolarization for MRI applications [113]. The sensitivity of color centers to external magnetic or electric fields, and mechanical strain will be exploited in the near future for nanosensing. In the field of energy applications, the potential of nanodiamonds is under investigation for photocatalysis and photo(electro)catalysis [114].

11.1.9 Conclusion

Carbon is one of the most versatile elements. Its unique flexibility in binding modes creates allotropes with an unprecedented variety in structure and properties, and their utilization in nanomaterials and engineering components is key to advancing next-generation technologies. This renders carbon the perhaps most interdisciplinary element and a game-changer in addressing today's global challenges.

References

[1]   Pisanty A. J Chem Educ, 1991, 68, 804.
[2]   McCreery RL. Chem Rev, 2008, 108, 2646–87.
[3]   Coulson CA, Taylor R. Proc Phys Soc A, 1952, 65, 815–25.
[4]   van Zuilen MA, Mathew K, Wopenka B, Lepland A, Marti K, Arrhenius G. Geochim Cosmochim Acta, 2005, 69, 1241–52.
[5]   Gold V (Ed). The IUPAC Compendium of Chemical Terminology: The Gold Book, International Union Of Pure And Applied Chemistry (IUPAC), Research Triangle Park, NC, 2019.
[6]   (a) Acheson EG. Process of Making Graphite, US 711031, published 1902-10-14; (b) Acheson EG. Method of manufacturing graphite articles, US 617979, published 1899-01-17; (c) Holleman AF, Wiberg N. Anorganische Chemie, 103. Auflage, De Gruyter, Berlin, 2016. ISBN 978-3-11-051854-2.

[7]   Fitzer E, Muller K, Schaefer W. Chemistry and Physics of Carbon, Marcel Dekker, New York, 1971, 237. ISBN-10: 0824712099.

[8]   Burchell TD. Comprehensive Nuclear Materials, Elsevier, Amsterdam, 2012, 285–305.

[9]   Pierson HO. Handbook of Carbon, Graphite, Diamonds and Fullerenes: Processing, Properties and Applications, Noyes Publication, Park Ridge, NJ, 1994. ISBN: 9780815513391.

[10]   Swain GM. Handbook of Electrochemistry. In: Zoski C (Ed). Elsevier, Amsterdam, 2007.

[11]   Alkire RC, Bartlett PN, Lipkowski J. (Eds). Electrochemistry of Carbon Electrodes, Wiley-VCH, Weinheim, 2015. ISBN: 9783527337323.

[12]   Crevillen AG, Pumera M, Gonzalez MC, Escarpa A. Analyst, 2009, 134, 657–62.

[13]   Jouikov V, Simonet J. Langmuir, 2012, 28, 931–8.

[14]   Serp P. Comprehensive Inorganic Chemistry II, Elsevier, Amsterdam, 2013, 323–69.

[15]   Uskoković V. Carbon Trends, 2021, 5, 100116.

[16]   McEnaney B. Carbon Materials for Advanced Technologies. In: Burchell TD. (Ed). Pergamon Press, New York, 1999, 1–34. ISBN 978-0-08-042683-9.

[17]   Ferrer-Argemi L, Aliabadi ES, Cisquella-Serra A, Salazar A, Madou M, Lee J. Carbon, 2018, 130, 87–93.

[18]   de S, Vieira L. Carbon, 2022, 186, 282–302.

[19]   Fuchs D, Bayer BC, Gupta T, Szabo GL, Wilhelm RA, Eder D, Meyer JC, Steiner S, Gollas B. ACS Appl Mater Interf, 2020, 2, 40937–48.

[20]   Jin Y, Shi Y, Chen R, Chen X, Zheng X, Liu Y. Chemosphere, 2019, 215, 380–7.

[21]   Jin Y, Shi Y, Chen Z, Chen R, Chen X, Zheng X, Liu Y. J Cleaner Prod, 2020, 252, 119794.

[22]   Ahmadi-Kashani M, Dehghani H. J Pharmac Biomed Anal, 2021, 194, 113653.

[23]   Gao Y, Li H, Wang L, Gao Y, Ye B. Int J Environ Anal Chem, 2019, 99, 1471–83.

[24]   Vazquez-Samperio J, Acevedo-Peña P, Guzmán-Vargas A, Reguera E, Cordoba-Tuta E. Int J Energy Res, 2021, 45, 6383–94.

[25]   Tadyszak K, Litowczenko J, Majchrzycki Ł, Jeżowski P, Załęski K, Scheibe B. Mater Chem Phys, 2020, 239, 122033.

[26]   Acuña NT, Güiza-Argüello V, Córdoba-Tuta E. Macromol Res, 2020, 28, 888–95.

[27]   de Oliveira CEM, de Carvalho MMG, Miskys CR. J Cryst Growth, 1997, 173, 214–7.

[28]   Jenkins GM, Ila D, Maleki H. MRS Online Pro Libr, 1995, 394, 181–5.

[29]   Jenkins GM, Grigson CJ. J Biomed Mater Res, 1979, 13, 371–94.

[30]   Haq MR, Kim YK, Kim J, Ju J, Kim S. J Micromech Microeng, 2019, 29, 75010.

[31]   (a) Boehm HP, Clauss A, Fischer GO, Hofmann U. Z Anorg Allg Chem, 1962, 316, 119–27; (b) Boehm HP. Angew Chem, 2010, 122, 9520–3; (c) Novoselov KS, Geim AK, Morozov SV, Jiang D, Zhang Y, Dubonos SV, Grigorieva IV, Firsov AA. Science, 2004, 306, 666–9.

[32]   Geim AK, Novoselov KS. Nature Mater, 2007, 6, 183–91.

[33]   Ferrari AC, Bonaccorso F, Fal'ko V, Novoselov KS, Roche S, Bøggild P, Borini S, Koppens FHL, Palermo V, Pugno N, Garrido JA, Sordan R, Bianco A, Ballerini L, Prato M, Lidorikis E, Kivioja J, Marinelli C, Ryhänen T, Morpurgo A, Coleman JN, Nicolosi V, Colombo L, Fert A, Garcia-Hernandez M, Bachtold A, Schneider GF, Guinea F, Dekker C, Barbone M, Sun Z, Galiotis C, Grigorenko AN, Konstantatos G, Kis A, Katsnelson M, Vandersypen L, Loiseau A, Morandi V, Neumaier D, Treossi E, Pellegrini V, Polini M, Tredicucci A, Williams GM, Hee Hong B, Ahn J-H, Min Kim J, Zirath H, van Wees BJ, van der Zant H, Occhipinti L, Di Matteo A, Kinloch IA, Seyller T, Quesnel E, Feng X, Teo K, Rupesinghe N, Hakonen P, Neil SRT, Tannock Q, Löfwander T, Kinaret J. Nanoscale, 2015, 7, 4598–810.

[34]   Backes C, Abdelkader AM, Alonso C, Andrieux-Ledier A, Arenal R, Azpeitia J, Balakrishnan N, Banszerus L, Barjon J, Bartali R. 2D Mater, 2020, 7, 022001.

[35]   Hummers WS, Offeman RE. J Am Chem Soc, 1958, 80, 1339.

[36]   Boehm H, Clauss A, Fischer G, Hofmann U. Z Naturforsch, 1962, 17b, 150–3.

[37] Neto AC, Guinea F, Peres NM, Novoselov KS, Geim AK. Rev Modern Phys, 2009, 81, 109.

[38] Bolotin KI, Sikes KJ, Jiang Z, Klima M, Fudenberg G, Hone J, Kim P, Stormer HL. Solid State Commun, 2008, 146, 351–5.

[39] Gruber E, Wilhelm RA, Pétuya R, Smejkal V, Kozubek R, Hierzenberger A, Bayer BC, Aldazabal I, Kazansky AK, Libisch F, Krasheninnikov AV, Schleberger M, Facsko S, Borisov AG, Arnau A, Aumayr F. Nature Commun, 2016, 7, 13948.

[40] Jariwala D, Marks TJ, Hersam MC. Nature Mater, 2016, 16, 170.

[41] Bonaccorso F, Sun Z, Hasan T, Ferrari AC. Nat Photon, 2010, 4, 611–22.

[42] Lee C, Wei X, Kysar JW, Hone J. Science, 2008, 321, 385–8.

[43] Bunch JS, Verbridge SS, Alden JS, van der Zande AM, Parpia JM, Craighead HG, McEuen PL. Nano Lett, 2008, 8, 2458–62.

[44] Sun PZ, Yang Q, Kuang WJ, Stebunov YV, Xiong WQ, Yu J, Nair RR, Katsnelson MI, Yuan SJ, Grigorieva IV, Lozada-Hidalgo M, Wang FC, Geim AK. Nature, 2020, 579, 229–32.

[45] Balandin AA, Ghosh S, Bao W, Calizo I, Teweldebrhan D, Miao F, Lau CN. Nano Lett, 2008, 8, 902–7.

[46] Kaindl R, Gupta T, Blümel A, Pei S, Hou P-X, Du J, Liu C, Patter P, Popovic K, Dergez D, Elibol K, Schafler E, Liu J, Eder D, Kieslinger D, Ren W, Hartmann P, Waldhauser W, Bayer BC. ACS Omega, 2021, 6, 50, 34301–34313.

[47] Meyer J, Kidambi PR, Bayer BC, Weijtens C, Kuhn A, Centeno A, Pesquera A, Zurutuza A, Robertson J, Hofmann S. Sci Rep, 2014, 4, 5380.

[48] Camilli L, Yu F, Cassidy A, Hornekær L, Bøggild P. 2D Mater, 2019, 6, 022002.

[49] Ren Z, Kim E, Pattinson S, Subrahmanyam K, Rao C, Cheetham A, Eder D. Chem Sci, 2012, 3, 209–16.

[50] Bayer BC, Kaindl R, Reza Ahmadpour Monazam M, Susi T, Kotakoski J, Gupta T, Eder D, Waldhauser W, Meyer JC. ACS Nano, 2018, 12, 8758–69.

[51] Elibol K, Mangler C, Gupta T, Zagler G, Eder D, Meyer JC, Kotakoski J, Bayer BC. Adv Funct Mater, 2020, 30, 2003300.

[52] Gupta T, Elibol K, Hummel S, Stöger-Pollach M, Mangler C, Habler G, Meyer JC, Eder D, Bayer BC. Npj 2D Mater Appl, 2021, 5, 53.

[53] Iijima S. Nature, 1991, 354, 56–8.

[54] Charlier J-C, Blase X, Roche S. Rev Modern Phys, 2007, 79, 677.

[55] Terrones M. Ann Rev Mater Res, 2003, 33, 419–501.

[56] Gebhardt P, Eder D. Nanocarbon-Inorganic Hybrids. In: Eder D, Schlögl R (Eds). De Gruyter, Berlin, 2014, 3–24.

[57] Dupuis A-C. Progr Mater Sci, 2005, 50, 929–61.

[58] Robertson J, Zhong G, Hofmann S, Bayer B, Esconjauregui C, Telg H, Thomsen C. Diamond Rel Mater, 2009, 18, 957–62.

[59] Bayer BC, Hofmann S, Castellarin-Cudia C, Blume R, Baehtz C, Esconjauregui S, Wirth CT, Oliver RA, Ducati C, Knop-Gericke A, Schlögl R, Goldoni A, Cepek C, Robertson J. J Phys Chem C, 2011, 115, 4359–69.

[60] Javey A, Guo J, Wang Q, Lundstrom M, Dai H. Nature, 2003, 424, 654–7.

[61] Wei B, Vajtai R, Ajayan P. Appl Phys Lett, 2001, 79, 1172–4.

[62] Dürkop T, Getty SA, Cobas E, Fuhrer M. Nano Lett, 2004, 4, 35–9.

[63] Kim P, Shi L, Majumdar A, McEuen PL. Phys Rev Lett, 2001, 87, 215502.

[64] Treacy MJ, Ebbesen TW, Gibson JM. Nature, 1996, 381, 678–80.

[65] Dai H. Acc Chem Res, 2002, 35, 1035–44.

[66] Su DS, Schlögl R. ChemSusChem Chem, 2010, 3, 136–68.

[67] Harrison BS, Atala A. Biomater, 2007, 28, 344–53.

[68] Bianco A, Kostarelos K, Prato M. Curr Op Chem Biol, 2005, 9, 674–9.

[69]   Cole MT, Mann M, Teo KBK, Milne WI. Emerging Nanotechnologies for Manufacturing, 2nd edition. In: Ahmed W, Jackson MJ (Eds). William Andrew Publishing, Boston, 2015, 125–86.
[70]   Koziol K, Vilatela J, Moisala A, Motta M, Cunniff P, Sennett M, Windle A. Science, 2007, 318, 1892–5.
[71]   Qian H, Greenhalgh ES, Shaffer MS, Bismarck A. J Mater Chem, 2010, 20, 4751–62.
[72]   Eder D. Chem Rev, 2010, 110, 1348–85.
[73]   Vilatela JJ, Eder D. ChemSusChem, 2012, 5, 2–25.
[74]   Shearer CJ, Cherevan A, Eder D. Adv Mater, 2014, 26, 2295–318.
[75]   Cherevan AS, Gebhardt P, Shearer CJ, Matsukawa M, Domen K, Eder D. Energy Environ Sci, 2014, 7, 791–6.
[76]   Kadish KM, Ruoff RS. Fullerenes: Chemistry, Physics, and Technology, John Wiley & Sons, Hoboken, 2000.
[77]   Kroto HW, Heath JR, O'Brien SC, Curl RF, Smalley RE. Nature, 1985, 318, 162–3.
[78]   Soga T. Nanostructured Materials for Solar Energy Conversion, Elsevier, Amsterdam, 2006.
[79]   Li C-Z, Yip H-L, Jen AK-Y. J Mater Chem, 2012, 22, 4161–77.
[80]   Smith P, Buseck PR. Science, 1982, 216, 984–6.
[81]   Chalifoux WA, Tykwinski RR. Nature Chem, 2010, 2, 967.
[82]   Banhart F. ChemTexts, 2020, 6, 1–10.
[83]   Shi L, Rohringer P, Suenaga K, Niimi Y, Kotakoski J, Meyer JC, Peterlik H, Wanko M, Cahangirov S, Rubio A, Lapin ZJ, Novotny L, Ayala P, Pichler T. Nature Mater, 2016, 15, 634–9.
[84]   Németh P, Garvie LAJ, Aoki T, Dubrovinskaia N, Dubrovinsky L, Buseck PR. Nature Commun, 2014, 5, 5447.
[85]   Yang B, Peng X, Zhao Y, Yin D, Fu T, Huang C. J Mater Sci Technol, 2020, 48, 114–22.
[86]   Ashfold MNR, Goss JP, Green BL, May PW, Newton ME, Peaker CV. Chem Rev, 2020, 120, 5745–94.
[87]   Nemanich RJ, Carlisle JA, Hirata A, Haenen K. MRS Bull, 2014, 39, 490–4.
[88]   Zaitsev AM. Optical Properties of Diamond: A Data Handbook, Springer, Heidelberg, 2001.
[89]   Smith JM, Meynell SA, Jayich ACB, Meijer J. Nanophotonics, 2019, 8, 1889–906.
[90]   Bundy FP, Hall HT, Strong HM, Wentorfjun RH. Nature, 1955, 176, 51–5.
[91]   Bormashov VS, Tarelkin SA, Buga SG, Kuznetsov MS, Terentiev SA, Semenov AN, Blank VD. Diamond Rel Mater, 2013, 35, 19–23.
[92]   Eversole WG. Synthesis of Diamond, 1962, US3030188A.
[93]   Knittel P, Buchner F, Hadzifejzovic E, Giese C, Quellmalz P, Seidel R, Petit T, Iliev B, Schubert TJS, Nebel CE, Foord JS. ChemCatChem, 2020, 12, 5548–57.
[94]   Arnault J-C, Saada S, Ralchenko V. Phys Status Solidi (RRL), 2022, 16, 2100354. DOI: 10.1002/pssr.202100354.
[95]   Volkov KV, Danilenko VV, Elin VI. Combust Explos Shock Waves, 1990, 26, 366–8.
[96]   Hao J, Pan L, Pan L, Gao S, Fan H, Gao B, Gao B. Opt Mater Expr OME, 2019, 9, 4734–41.
[97]   Tallaire A, Brinza O, De Feudis M, Ferrier A, Touati N, Binet L, Nicolas L, Delord T, Hétet G, Herzig T, Pezzagna S, Goldner P, Achard J. ACS Appl Nano Mater, 2019, 2, 5952–62.
[98]   Ekimov EA, Kudryavtsev OS, Mordvinova NE, Lebedev OI, Vlasov II. ChemNanoMat, 2018, 4, 269–73.
[99]   Field JE, Pickles CSJ. Diamond Rel Mater, 1996, 5, 625–34.
[100] Che J, Çağin T, Deng W, Goddard WA. J Chem Phys, 2000, 113, 6888–900.
[101] Moche H, Paget V, Chevalier D, Lorge E, Claude N, Girard HA, Arnault JC, Chevillard S, Nesslany F. J Appl Toxicol, 2017, 37, 954–61.
[102] Donato N, Rouger N, Pernot J, Longobardi G, Udrea F. J Phys D: Appl Phys, 2019, 53, 093001.
[103] Einaga Y, Foord JS, Swain GM. MRS Bull, 2014, 39, 525–32.

[104] Arnault JC (Ed). Nanodiamonds: Advanced Material Analysis, Properties and Applications, Elsevier, Amsterdam, 2017. ISBN: 9780323430296.

[105] Diamond Materials, "Diamond Vacuum Windows," can be found under https://www.dia mond-materials.com/en/products/optical-windows/uhv-vacuum-windows/, accessed december 26, 2021.

[106] Antipov S, Sabella A, Williams RJ, Kitzler O, Spence DJ, Mildren RP. Opt Lett, 2019, 44, 2506–9.

[107] Blin C, Checoury X, Girard HA, Gesset C, Saada S, Boucaud P, Bergonzo P. Adv Opt Mater, 2013, 1, 963–70.

[108] Cobb SJ, Ayres ZJ, Macpherson JV. Ann Rev Anal Chem, 2018, 11, 463–84.

[109] Shenderova OA, Shames AI, Nunn NA, Torelli MD, Vlasov I, Zaitsev A. J Vacuum Sci Techn B, 2019, 37, 030802.

[110] Mochalin VN, Gogotsi Y. Diamond Rel Mater, 2015, 58, 161–71.

[111] Ivanov M, Shenderova O. Curr Op Solid State Mater Sci, 2017, 21, 17–24.

[112] Nguyen CT, Evans RE, Sipahigil A, Bhaskar MK, Sukachev DD, Agafonov VN, Davydov VA, Kulikova LF, Jelezko F, Lukin MD. Appl Phys Lett, 2018, 112, 203102.

[113] Ajoy A, Nazaryan R, Druga E, Liu K, Aguilar A, Han B, Gierth M, Oon JT, Safvati B, Tsang R, Walton JH, Suter D, Meriles CA, Reimer JA, Pines A. Rev Sci Instr, 2020, 91, 023106.

[114] Li S, Bandy JA, Hamers RJ. ACS Appl Mater Interf, 2018, 10, 5395–403.

11.2 Sulfur

Cristian A. Strassert, Rainer Pöttgen

11.2.1 Sulfur as a basic chemical

Elemental sulfur is one of the key precursors that is used on a high tonnage scale for the synthesis of a vast number of basic inorganic chemicals. In the past, it was gained from natural deposits, i.e. sulfur of volcanic origin or from underground reservoirs stemming from anaerobic sulfate-reducing bacterial activity (Frasch process, now outdated) [1]. As a constituent of proteins, sulfur is also found in fossil fuels including anthracite, lignite, tar sands, shale oil and peat. Sulfidic gases in natural gas sources or from bioreactors, on the other hand, stem from anaerobic bacterial activity and can be separated from methane to be further used in the Claus process for the production of elemental sulfur (*vide infra*).

Another large source of sulfur for technical use is based on sulfidic minerals such as ZnS (blende and wurtzite), PbS (galenite), FeS_2 (pyrite), Cu_2S (chalcocite) and CuS (covellite), CdS (greenockit), As_2S_3 (orpiment), Sb_2S_3 (stibnite), Bi_2S_3 (bismuthinite), HgS (cinnabar) or $CuFeS_2$ (chalcopyrite). The roasting processes involving such sulfides (e.g. $PbS + 3/2\ O_2 \rightarrow PbO + SO_2$) lead to the production of vast amounts of sulfur dioxide.

In the past, the direct SO_2 emission stemming from the diverse roasting reactions and from the combustion of fossil fuels led to acid rain, due to the fact that sulfurous acid and sulfuric acid are produced upon reaction with humidity and oxygen in the atmosphere. Nowadays, sulfur dioxide is used as a valuable chemical precursor for the production of plaster as well as sulfuric acid.

Yearly, the production of elemental sulfur surpasses the 50 megaton scale, and roughly 90% is consumed for the production of sulfuric acid. The rest is used for other inorganic products (e.g. SO_2, CS_2 and P_2S_5), pigments such as As_2S_3 and ultramarine blue (see Chapter 1.4), as well as in fire matches, in gun powder, in the construction of streets (together with asphalt), in the rubber-vulcanization process (*vide infra*), in the synthesis of insecticides and in the production of pharmaceuticals. The vast chemistry of sulfur is presented in great detail in the excellent textbooks by Steudel [2] and Holleman, Wiberg and Wiberg [3]; in this chapter, we only summarize the main technical aspects that are relevant for the chemical industry.

11.2.2 SO$_2$ and H$_2$S from desulfurization of combustion exhaust (flue) gases, natural gas and bioreactor methane

Sulfur dioxide (SO_2) and hydrogen sulfide (H_2S) constitute large-scale side products from roasting processes, fossil fuel combustion and methane purification (both

https://doi.org/10.1515/9783110733471-019

from natural gas and from bioreactors). Since the amount of both gases is very large, they have cut out the sulfur extraction from natural deposits completely (i.e. by means of the Frasch technique) and are presently used as the main educts for sulfur chemistry. More than 50% of the industrially used sulfur originates from desulfurization processes.

SO_2 is a valuable precursor in the synthesis of diverse chemicals (mainly SO_3 and sulfuric acid, but also sulfites, thiosulfates, dithionites, sulfinates, etc., *vide infra*) as well as for the sulfochlorination or sulfoxidation of hydrocarbons. It can be transported as a liquid in steel containers and is even used as a cooling agent and non-aqueous solvent due to its high dielectric constant and aprotic character (see Chapter 4.10).

Combustion gases can be treated with aqueous solutions (or suspensions) of calcium hydroxide (or calcium carbonate) to neutralize the resulting sulfurous acid. Calcium sulfite forms in a first step, which is finally oxidized to calcium sulfate by oxygen from the air:

$$Ca(OH)_2 + SO_2 \rightarrow CaSO_3 \cdot \frac{1}{2} H_2O + \frac{1}{2} H_2O$$

$$CaSO_3 \cdot \frac{1}{2} H_2O + \frac{1}{2} O_2 + 3/2 H_2O \rightarrow CaSO_4 \cdot 2 H_2O$$

The calcium sulfate dihydrate is used as the so-called industrial plaster for various large-scale applications, e.g. gypsum plaster, plasterboards or plaster screed. This looks like a win-win situation at first sight; however, it has to be kept in mind that the production of calcium hydroxide starts with calcium carbonate as a natural source, and the endothermic process of lime burning ($CaCO_3 \rightarrow CaO + CO_2$) consumes heat and further liberates carbon dioxide (one equivalent of CO_2 per equivalent of bound SO_2, besides the financial and environmental costs of heat production).

When burning anthracite or lignite, $CaCO_3$ (limestone) is directly added to the fossil fuel (already in the combustion chamber); while its thermal decomposition to CaO constitutes an endothermic step, it is compensated by the exothermic formation of the thermodynamically stable $CaSO_4$, as follows: $CaO + SO_2 + \frac{1}{2} O_2 \rightarrow CaSO_4$. Hence, the overall power output remains unaffected, while entrapping harmful yet valuable SO_2 as solid $CaSO_4$, which is conveniently separated from the exhaust gases (e.g. by means of electrostatic filters).

As mentioned above, an aqueous suspension of limestone can be finely dispersed and mixed with the flue gases, leading to the oxidation of SO_2 to yield $CaSO_4$ (*vide supra*). When burning mineral oil or natural gas, the produced SO_2 can be also trapped by washing the exhaust gases with aqueous ammonia or amine solutions as well as with alkaline (earth) hydroxide solutions or weaker bases ($CaCO_3$, Na_2CO_3, Na_2SO_3, Na-citrate). Alternatively, SO_2 can be condensed with pressure at the temperature of liquid ammonia.

The desulfurization processes of flue gas allows the recovery of vast amounts of sulfur dioxide produced during the combustion of fossil fuels (e.g. in thermoelectric power plants) and particularly high concentrations along the so-called roasting processes (work-up of sulfidic minerals). During the thermal treatment of sulfidic ores with atmospheric oxygen (e.g. CuS, ZnS, PbS or FeS_2), the exothermic character of the reactions sustains these reactions. If needed with higher purity and in larger amounts, SO_2 is also produced by direct combustion of elemental sulfur or H_2S. Usually, the oxidation stops at SO_2 since the exothermic oxidation to SO_3 is kinetically inhibited and thermodynamically unfavorable at high temperatures.

H_2S is a highly toxic gas with a very low boiling point due to its lack of hydrogen bonding abilities; it is a weak acid, but can act as a corrosive agent since most transition metal sulfides are rather insoluble (with S^{2-} being a soft ligand). Hydrogen sulfide is the major by-product from natural gas sources. The so-called sour gases can contain up to 35% hydrogen sulfide, which is entrapped by means of the amine gas treatment, using aqueous solutions of alkylamines (diethanolamine (DEA), monoethanolamine (MEA) and methyldiethanolamine (MDEA) are most commonly used). In the first step, the amine binds the hydrogen sulfide as an alkylammonium salt in water through a simple acid-base reaction (for a general amine: $R\text{-}NH_2 + H_2S \rightarrow R\text{-}NH_3^+ + HS^-$). The hydrogen sulfide is finally recovered by heating the aqueous solution (releasing H_2S into the gas phase); the aqueous alkylamine solution can be reused.

The second large source of hydrogen sulfide stems from the hydrogenating desulfurization of mineral oil, a strictly necessary step before cracking towards the production of low-sulfur diesel or gasoline fuels as well as in the production of organic synthetic precursors. While crude oil from the Arabic peninsula typically has a lower sulfur content, other reservoirs (e.g. from Venezuela) present a significantly higher demand for desulfurization and can only be processed at specialized plants.

In any case, the hydrogen sulfide is then used in the Claus process. In the first step, part of the hydrogen sulfide is partially oxidized (burnt under controlled hypoxic conditions) to sulfur dioxide ($H_2S + 3/2\ O_2 \rightarrow SO_2 + H_2O$). The second step involves the comproportionation towards elemental sulfur ($2\ H_2S + SO_2 \rightarrow 3/8\ S_8 + 2\ H_2O$).

Besides H_2S adsorption on activated carbon, the amine gas treatment can be used for desulfurization of gas produced in biogas plants. Alternatively, it is possible to bind the sulfide with $Fe(OH)_3$ or $FeCl_2$ as highly insoluble FeS directly within the fermenter. The FeS remains in the digestate, which is mainly used as a soil conditioner.

11.2.3 SO_3 production and sulfuric acid

Most of the SO_2 produced worldwide is oxidized to SO_3 in a catalyzed reaction (heterogeneous catalysis, see Chapter 7.2, with V_2O_5) due to the fact that the strongly exothermic (and exergonic) reaction is kinetically inhibited:

$$SO_2 + \tfrac{1}{2}O_2 \rightarrow SO_3 + 98.98\,kJ$$

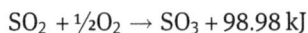

Due to its exothermic character, the yields would drop in the absence of an efficient heat removal, which is why a series of cascaded catalysis with cooling steps is employed in the so-called contact process (before and after the final catalytic stage, most of the SO_3 is entrapped in concentrated sulfuric acid towards oleum to further boost the yields, which therefore constitutes the so-called double-contact-double-absorption process). In the absorption steps, the SO_3 subsequently reacts with water by bubbling in concentrated sulfuric acid towards higher condensed species:

$$SO_3 + H_2O \rightarrow H_2SO_4 + 176.6\,kJ \rightarrow\rightarrow H_2S_2O_7 \rightarrow\rightarrow H_2S_3O_{10} \rightarrow\rightarrow \text{etc.}$$

The obtained oleum is afterwards diluted with water to yield concentrated sulfuric acid, which also constitutes an exothermic reaction. In all cases (including the primary combustion of S_8 and H_2S, the oxidation of SO_2 to SO_3, the reaction of SO_3 with water and the dilution of H_2SO_4), the resulting heat is employed to gain steam for further chemical processes in coupled plants. It is important to notice that a direct reaction of SO_3 with pure water would be very exothermic, which reduces the solubility of the gas with a concomitant drop in yield and difficult cooling management.

If the SO_2 results from direct burning of S_8 or H_2S, it can be directly fed into the SO_3 production; this is not the case for flue gases originated in roasting processes, which have to be treated in order to remove particulate impurities (dust) and potential catalyst poisons (such as As or Sb, present in the exhaust dusts, mostly as oxides).

After the production of chlorine, the synthesis of sulfuric acid constitutes the most reliable indicator for the strength of the chemical industry in a country; more than 90% of the sulfur produced yearly is used to obtain this fundamental chemical. H_2SO_4 is a strong yet weakly oxidizing acid (as compared to $HClO_4$ or HNO_3), and it is highly hygroscopic and possesses a high boiling point. Most of the H_2SO_4 (ca. 60%) is used to process apatite towards phosphate fertilizers (see Chapter 6.6) and to produce $(NH_4)_2SO_4$ (*vide infra*).

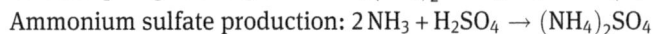

Calcium phosphate digestion: $Ca_3(PO_4)_2 + 2\,H_2SO_4 \rightarrow Ca(H_2PO_4)_2 + 2\,CaSO_4$
Ammonium sulfate production: $2\,NH_3 + H_2SO_4 \rightarrow (NH_4)_2SO_4$

Sulfuric acid is also used to obtain other mineral acids (such as HCl, H_3PO_4 and H_2CrO_4, from their corresponding salts), sulfates (e.g. $Al_2(SO_4)_3$) as well as in the production of TiO_2, and in the processing of uranium or copper ores. Lead accumulators (see Chapter 5.1) also require the use of sulfuric acid, and it constitutes a must-have chemical in every decent chemical lab. Being a strong acid and dehydrating agent, it is able to catalyze and to thermodynamically drive organic reactions as well. In organic chemistry, it is used for the sulfonation or nitration of organic compounds to yield sulfonic acids (aromatic and aliphatic compounds), aromatic nitroderivatives (such as nitrobenzenes, nitrotoluenes, nitrophenols and organic dyes) or nitric esters of polyols (such as nitroglycerin or nitrocellulose). However, for the sulfonation (or sulfatation) of organic compounds, also milder SO_3 derivatives are frequently used, such as chlorosulfonic acid, thionyl chloride and dimethylsulfate (particularly in the production of tensioactive compounds, where also aminosulfonates play a role, *vide infra*).

11.2.4 Sulfur halides and oxyhalides

Sulfur halides (in general, S_2X_2 and SX_2), such as S_2Cl_2 (dichlorodisulfane, a yellow-orange liquid), are toxic and irritating compounds that hydrolyze easily if exposed to moisture while possessing a characteristic stench. S_2Cl_2 is obtained by direct reaction between chlorine gas and molten sulfur at ca. 250 °C and is converted slowly to SCl_2 (dichlorosulfane, always containing some Cl_2 and S_2Cl_2 in equilibrium) over time with catalytic amounts of $FeCl_3$ or I_2 (yearly production up to several kilotons).

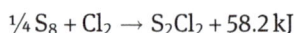

$$\tfrac{1}{4}\,S_8 + Cl_2 \rightarrow S_2Cl_2 + 58.2\,kJ$$

$$S_2Cl_2 + Cl_2 \rightarrow 2\,SCl_2 + 40.6\,kJ$$

S_2Cl_2 is an excellent solvent for elemental sulfur, while yielding polysulfanedichlorides, S_nCl_2, to a minor extent. This solution is used for the vulcanization of rubber, as the S_nCl_2 reacts with and crosslinks the polyisoprene units from natural rubber (*vide infra*). Both S_2Cl_2 and SCl_2 are used to synthesize inorganic derivatives (e.g. $SOCl_2$ and SF_4) and as chlorination or sulfidation agents. SCl_2 easily reacts with alkenes by addition; in the past, it has been used in the production of warfare gas named mustard ($S(CH_2CH_2Cl)_2$) from $H_2CH=CH_2$.

Sulfur hexafluoride (SF_6) is obtained by reaction of elemental sulfur with fluorine:

$$1/8\,S_8 + 3\,F_2 \rightarrow SF_6 + 1220\,kJ$$

The gaseous reaction mixture is heated to 400 °C and afterwards washed with aqueous base to separate lower fluorides (e.g. SF_4 and S_2F_{10}):

$$SF_4 + 6\,OH^- \rightarrow SO_3{}^{2-} + 4\,F^- + 3\,H_2O$$

This is followed by distillation at reduced pressure.

SF_6 has an excellent thermal and electrical insulating ability due to its relative thermodynamic stability paired with very high kinetic inertness towards hydrolysis and other reactions (the latter is related to a closed coordination sphere around the central atom without possibilities of nucleophilic attack, i.e. by an additive mechanism involving an expanded coordination on the intermediary). Hence, SF_6 is used as an insulator and dielectric material with fire and spark-dampening character, which is particularly useful in high-voltage devices and arrays. It is found in transformers and other high-voltage electrical installations and is also used as a protective inert gas when processing molten metals. Due to its low conductivities, it also finds application as a thermal and acoustic insulator (e.g. in multilayered energy-saving windows); this is related to its high molecular mass with concomitantly slow diffusion and transport rates. It is a very efficient greenhouse gas though (20000 times more efficient than CO_2).

Sulfur tetrafluoride (SF_4) is a colorless, suffocating gas that is highly toxic. It is synthesized at the technical scale by the fluorination of elemental sulfur or lower sulfur compounds (e.g. S_nCl_2). It can be transported in steel bottles, but it is very sensitive to hydrolysis ($SF_4 + 2\,H_2O \rightarrow SO_2 + 4\,HF$). Nowadays, it is used as a fluorination agent (e.g. $R_1R_2C{=}O + SF_4 \rightarrow R_1R_2CF_2 + O{=}SF_2$ or $I_2O_5 + 5\,SF_4 \rightarrow 2\,IF_5 + 5\,SOF_2$).

Thionyl chloride ($SOCl_2$) is produced at the scale of 100 kilotons per year by gas phase reaction of SO_2 or SO_3 with Cl_2 in the presence of SCl_2 or S_2Cl_2 with active charcoal as a catalyst:

$$SO_2 + SCl_2 + Cl_2 \rightarrow 2\,SOCl_2$$

$$SO_3 + 2\,SCl_2 + Cl_2 \rightarrow 3\,SOCl_2$$

It is purified by distillation at low temperature since it otherwise decomposes to S_2Cl_2 and Cl_2 above 76 °C. In general, thionyl halides are vigorous dehydrating agents since they are hydrolyzed to SO_2:

$$SOX_2 + H_2O \rightarrow SO_2 + 2\,HX$$

Hence, $SOCl_2$ is used in inorganic chemistry for the synthesis of anhydrous metal halides from the corresponding hydrated precursors, which would otherwise decompose if a thermal dehydration is attempted (as opposed to the gaseous by-products from the reaction with $SOCl_2$, which can be easily separated). In organic chemistry, it is employed for the dehydration in synthetic procedures (e.g. to produce nitriles from primary amides) and for the insertion of the thionyl moiety to yield sulfoxides ($R_1R_2S{=}O$); it plays a role in the production of crop protection agents, pharmaceuticals, dyes and pigments. It is also used for the chlorination of metal oxides and in galvanic elements.

Chlorosulfonic acid (HSO_3Cl) is produced at the technical scale by reaction of anhydrous HCl and liquid SO_3. It is a very strong sulfonation agent (e.g. for the synthesis of organic sulfonic acids) and is yearly produced at the 100 kiloton scale:

$$RH + HSO_3Cl \rightarrow RSO_3H + HCl$$

It is also useful in cases where the sulfonation with fuming sulfuric acid fails and can also be employed as a dehydrating condensation agent.

Fluorosulfonic acid (HSO_3F) is synthesized analogously by the reaction of SO_3 with dry HF in fluid HSO_3F (the product functions as a solvent). It is colorless, a very strong acid and a fluorination or sulfonation agent; it is also used as a catalyst for polymerization or alkylation reactions.

Sulfuryl chloride (SO_2Cl_2) is industrially produced by the direct reaction of chlorine and SO_2 employing a cooled active charcoal catalyst or by the reaction of S_2Cl_2 and chlorine in the presence of oxygen with an active charcoal catalyst.

$$SO_2 + Cl_2 \rightarrow SO_2Cl_2$$

Alternatively, it can be produced by the dismutation of chlorosulfonic acid:

$$2\,HSO_3Cl \rightarrow H_2SO_4 + SO_2Cl_2$$

Sulfuryl chloride (SO_2Cl_2) is used in organic chemistry for dehydration reactions, as a sulfonation or chlorosulfonation agent as well as for chlorination.

Sulfuryl fluoride (SO_2F_2) is obtained by reacting SO_2 and F_2 or by thermal decomposition of fluorosulfonic acid (and its salts as well, e.g. $Ba(SO_3F)_2 \rightarrow BaSO_4 + SO_2F_2$). It is a color- and odorless gas that is surprisingly inert (similar to SF_6), especially if compared with SO_2Cl_2 (it can even be heated with H_2O without hydrolysis and is only slowly hydrolyzed by bases; it does not react with the surface of molten Na). Unlike SF_6, it is toxic for organisms and can be used to combat woodworms (e.g. in furniture and sculptures).

11.2.5 Oxoanionic species beyond sulfates and sulfites

$NaHSO_3$ and $Ca(HSO_3)_2$ (synthesized from SO_2 and NaOH, Na_2CO_3 or $CaCO_3$ in water) are used to process wood towards cellulose paste since they enable the separation of incrusted lignins, leaving the cellulose unaffected. Na_2SO_3 and $Na_2S_2O_5$ can be used as reducing agents (e.g. to protect thiosulfate solutions from oxidation or to neutralize chlorine during the processing of textiles, leather and cellulose); they are precursors in the synthesis of sodium thiosulfate and are employed for the treatment of aqueous effluents.

Sodium dithionite ($Na_2S_2O_4 \cdot 2\,H_2O$) is yearly produced at the 100 kiloton scale. It is synthesized by suspending zinc dust in water and bubbling SO_2 (at 40 °C), followed by the exchange of cations

$$2\,SO_2 + Zn \rightarrow ZnS_2O_4 \rightarrow \rightarrow Na_2S_2O_4$$

as well as by reaction of SO_2 with sodium formiate or borohydride:

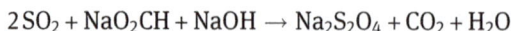

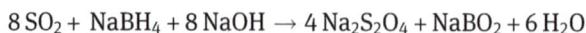

$$2\,SO_2 + NaO_2CH + NaOH \rightarrow Na_2S_2O_4 + CO_2 + H_2O$$

$$8\,SO_2 + NaBH_4 + 8\,NaOH \rightarrow 4\,Na_2S_2O_4 + NaBO_2 + 6\,H_2O$$

Due to its reducing ability, it is employed in staining and printing processes, as a decolorizing agent in the textile and paper industry, to produce Rongalit C®, which is used in the textile industry:

$$Na_2S_2O_4 + 2H_2C{=}O + H_2O \rightarrow NaO_2S{-}CH_2OH + NaO_3S{-}CH_2OH$$

Sodium thiosulfate pentahydrate ($Na_2S_2O_3 \cdot 5\,H_2O$) is also produced at the 100 kilo-ton scale per year and finds many applications beyond the classic photography (where it dissolves the unreacted Ag(I) cations by formation of the water-soluble complex $[Ag(S_2O_3)_2]^{3-}$). In the textile and paper industries, it is used to neutralize (reduce) oxidizing chlorine species (previously used for bleaching) while yielding water-soluble chlorides and sulfates. With iodine, it reacts to tetrathionate, $S_4O_6^{2-}$ and colorless iodide anions, which constitutes the basis for iodometric analysis.

The technical production of peroxodisulfuric acid, a.k.a. Marshall's acid, $H_2S_2O_8$, and its salts, peroxodisulfates, stems from the electrolysis with platinum anodes (at high current densities) of aqueous sulfuric acid or soluble sulfate salts (potassium and ammonium are particularly useful since they crystallize out of the reaction mixture as insoluble peroxodisulfates). Highly concentrated aqueous solutions are used to facili-tate the precipitation of the product and to ensure the oxidation of sulfate (as opposed to the oxidation of water). The commercially available triple salt, $(KHSO_5)_2 \cdot KHSO_4 \cdot K_2SO_4$ (Oxone®), is used as a powerful oxidizing agent. Peroxymonosulfuric acid (Caro's acid, H_2SO_5) is technically synthesized from chlorosulfonic acid and hydrogen peroxide; its salts also constitute strong oxidizing agents.

11.2.6 Vulcanization process, MoS_2 and CS_2

Natural rubber, also called latex, is the polymer cis-1,4-polyisoprene. It needs to be chemically modified in order to become a proper material with desirable mechanical properties and suitable stability as well as adequate processability and durability. Vul-canization involves the cross-linking of the constituting polyisoprene chains with polysulfide chains (at the so-called cure sites, mostly allylic carbon atoms), in analogy to the disulfide bridges that stabilize the structural integrity of proteins while yielding an adequate functionality [4]. In any case, the ability of sulfur to form stable single bonds towards polymeric or cyclic units of variable length constitutes the basis of this helpful strategy. Originally, elemental sulfur was mixed as a powder with latex and then thermally treated to induce the cross-linking sequence; however, the reaction is slow, requiring extensive time periods at high temperatures. Due to the high solubility of sulfur in latex, a crystallization of this yellow element cannot be avoided upon cool-ing. Nowadays, elemental sulfur is dissolved in S_2Cl_2, leading to S_nCl_2 (polysulfane di-chlorides), a key intermediate class that can cross-link the polyisoprene chains under controlled yet milder conditions ("cold vulcanization"). Diverse other accelerants (mostly sulfur-containing organic molecules, peroxides and even metal oxides such as ZnO) as well as catalysts can be part of the so-called cure-package; the details (compo-sition, temperature, time, pressure and even ionizing radiation) constitute well-kept in-house secrets and are protected by patents of the manufacturers.

Molybdenum(IV)disulfide (MoS_2) is used as an additive for lubricants, due to its thermal stability and desirable mechanical properties [5]. Having a layered structure, the lateral displacement with low frictional losses is possible upon application of a shear force. It is a chemically inert transition metal dichalcogenide that withstands high temperatures. The weak van der Waals interactions between the sheets of sulfur atoms are responsible for the low friction coefficients (and even for superlubricity). It can be used as a dry powder, providing good lubricity even at high temperatures in oxidizing atmospheres. It can be blended to yield a composite with graphite or with oils and greases that retain lubricity even if the oil is lost (which is important for critical applications). If added to organic polymers (such as nylon or teflon), improved mechanical and frictional characteristics can be attained.

Carbon disulfide (CS_2) plays a fundamental role in technical processes as a useful solvent despite its high toxicity and flammability. It plays a role in the vulcanization process and is also used as a precursor in the synthesis of the solvent CCl_4:

$$CS_2 + 2\,Cl_2 \rightarrow CCl_4 + \tfrac{1}{4}\,S_8.$$

CS_2 is produced at the megaton scale yearly by the direct reaction of CH_4 with elemental sulfur at ca. 600 °C in the presence of heterogeneous catalysts such as Al_2O_3 or *Kieselgel*:

$$CS_2 + 2\,Cl_2 \rightarrow CCl_4 + \tfrac{1}{4}\,S_8.$$

11.2.7 Phytosanitary and food protection agents

Besides sulfur-containing organic phytosanitary agents [6–8] and sulfate-based fertilizers (Chapter 6.6), elemental sulfur is also used in crop protection, the oldest of all pesticides [9, 10]. The so-called wettable sulfur is simply fine-grained sulfur powder, which is used in combination with a wetting agent acting as fungicide against mildew, scab and botrytis. Wettable sulfur is commercialized under a vast number of trade names. A further formulation is represented by the so-called *Schwefelleber* (i.e. "liver of sulfur" or *hepar sulfuris*, due to its brownish color), a melted (and later pulverized) mixture of potash (K_2CO_3) and elemental sulfur; this product contains potassium polysulfides and potassium thiosulfate, among other species. Also, granulates of sulfur/sodium hydroxide mixtures are available. Lime sulfur is a mixture of different compounds obtained from a boiling solution of $Ca(OH)_2$ and sulfur. It predominantly contains calcium polysulfide (CaS_x mainly with $x = 2$–7), besides minor amounts of calcium thiosulfate (CaS_2O_3), calcium sulfite ($CaSO_3$), residual sulfur and calcium sulfate ($CaSO_4$).

Two different antifungal activities are relevant: (i) sulfur vapor can enter a fungal cell and is enzymatically converted to hydrogen sulfide (H_2S) to exhibit the antifungal activity and (ii) the fine sulfur particles can react with light, water and oxygen to form sulfur dioxide (SO_2), which then constitutes the actual fungicide. Both the reactions and the effectiveness strongly depend on temperature (i.e. inactivity at too low temperature and burn damage on the leaves at excessively high temperatures). The highest effectivity seems to be reached at around 27.5 °C with 75% relative humidity [11].

In the production of wine, sulfites and SO_2 still play a significant role and cannot be replaced. The reducing power of S(IV)-species precludes the oxidation of polyphenols in the grape juice, in the grape must and in the wine. Most importantly, its activity suppresses the replication of unwanted bacteria or fungi and the fermentation processes they cause; at lower concentrations, the yeast employed for wine production is more resistant to sulfites than the detrimental species. After harvesting and to avoid the oxidation of exudates, grapes are typically preserved with potassium metabisulfite, $K_2(O_2S\text{-}SO_3)$ or $K_2S_2O_5$, which is also called potassium pyrosulfite or potassium disulfite; it is prepared by bubbling SO_2 in an aqueous KOH solution. Young wine is treated with gaseous SO_2, a procedure that is being progressively replaced by aqueous solutions of potassium or ammonium sulfites for safety reasons (typically as bisulfites, also called hydrogensulfites, i.e. HSO_3^- salts). For the conservation of wooden barrels, usually an aqueous solution of tartaric acid is combined with SO_2. For the prompt disinfection of barrels, the so-called sulfur strips can be burned to deliver SO_2. In general, S(IV)-species can also be employed for the conservation of other fruits, nuts and juices to suppress unwanted fermentation processes or putrefaction. The reducing power of gaseous SO_2 is also useful for the elimination of insects and for the bleaching of hay, wool and silk (especially if chlorine derivatives are not tolerated) [12].

References

[1] Bertau M, Müller A, Fröhlich P, Katzberg M. Industrielle Anorganische Chemie, 4. Auflage, Wiley-VCH, Weinheim, Germany, 2013.
[2] Steudel R. Chemie der Nichtmetalle, 4. Auflage, Walter de Gruyter GmbH, Berlin / Boston, Germany, 2014.
[3] Wiberg N, Wiberg E, Holleman AF. Anorganische Chemie Band 1: Grundlagen und Hauptgruppenelemente, 103. Auflage, Walter de Gruyter GmbH, Berlin / Boston, Germany, 2017.
[4] Mark JE, Erman B, Roland CM. The Science and Technology of Rubber, 4th edition, Elsevier Inc, Amsterdam / Boston / Heidelberg / London / New York / Oxford / Paris / San Diego / San Francisco / Singapore / Sydney / Tokyo, 2013.
[5] Vazirisereshk MR, Martini A, Strubbe DA, Baykara MZ. Solid lubrication with MoS_2: A review. Lubricants, 2019, 7, 57–92.
[6] Scherer HW. Eur J Agron, 2001, 14, 81–111.
[7] Lamberth C. J Sulfur Chem, 2004, 25, 39–62.
[8] Aula L, Dhillon JS, Omara P, Wehmeyer GB, Freeman KW, Raun WR. Agron J, 2019, 111, 2485–92.

[9]   Tweedy BG. Inorganic sulfur as a fungicide. In: Gunther FA, Gunther JD. (Eds). Res Rev, Vol 78, Springer, New York, NY. https://doi.org/10.1007/978-1-4612-5910-7_3.

[10]   Cooper RM, Williams JS. J Exp Bot, 2004, 55, 1947–53. DOI: 10.1093/jxb/erh179.

[11]   Auger P, Guichou S, Kreiter S. Pest Manage Sci, 2003, 59, 559–65.

[12]   Schandelmaier B. Das Ende der Schwefelbombe. 69. Weinbautage 2016, 1–4. www.weinbau. rlp.de/Internet/global/inetcntr.nsf/suche.xsp?src=W16597ZJCZ&p1=H466PP81UJ&p3= 6XU1060977&p4=0XE18F8V0Z. Accessed February 16 2022.

12 Water, mineral acids and bases

Jens Haberkamp, Thomas Jüstel, Rainer Pöttgen

The natural water cycle is the Earth's most important dynamic physicochemical system. Liquid water is the ubiquitous solvent of biological systems due to its abundance, thermodynamic stability, large surface tension and high polarity. The latter is the origin of its capability to dissolve ions and polar organic molecules. Water vapour and liquid water are also important for the Earth's climate system due to the strong absorption of IR radiation, high standard enthalpy of vaporization (43.990 kJ mol^{-1}) and large heat capacity (4.18 J g^{-1} K^{-1}). While liquid water has a rather large optical window ranging from 200 to 900 nm, ice and clouds (nano to microscale water droplets) cause cooling by the diffuse reflection of visible light, which is known as the global albedo (whiteness) varying around 0.30 [1]. Finally, water is also an important transport carrier for people and goods (from rafts to container ships) as well as for naturally driven mass transport processes, e.g. of dissolved or suspended matter from the source of a small brook via river systems to the oceans.

Freshwater is the basis of all life on Earth. Water is also essential for a multitude of technical or industrial processes, e.g. as solvent for most technologically used acids and bases. The present chapter focuses on common quality features regarding natural water resources and the consequently required treatment to meet the quality standards for drinking water and process water. At the end, we summarize the technologically important aqueous solutions of mineral acids and bases.

12.1 Resources for drinking water and process water

Water treatment for the production of drinking water or process water for industrial purposes includes various technical processes. Depending on the raw water resource as well as the quality required for the intended application, water treatment may focus on a variety of water quality parameters, e.g.:

- pH value, alkalinity and acidity (i.e. acid and base buffer capacity, respectively)
- hardness (predominantly Ca^{2+} and Mg^{2+} concentration)
- salinity including the concentrations of cations (e.g. Na^+, K^+, Ca^{2+} and Mg^{2+}) and anions (e.g. HCO_3^-, Cl^-, NO_3^- and SO_4^{2-})
- heavy metals (e.g. iron, manganese, chromium, lead, uranium and zinc)
- organic substances including natural organic matter (e.g. humic substances) and anthropogenic organic compounds (e.g. pesticides, plasticizers and pharmaceutical residues)
- suspended particulate and colloidal matter including inorganic (e.g. clay minerals and precipitates) and organic particles (e.g. microorganisms, algae, organic detritus and microplastics)

https://doi.org/10.1515/9783110733471-020

In many cases, the raw water used for the production of drinking or process water is abstracted from groundwater. Especially companies in the food and beverage industry often cover their water demand by the abstraction of groundwater, which subsequently has to be treated on-site to meet the water quality requirements for the specific application. In general, groundwater from pore aquifers has a high quality, provided that the aquifer is covered by sufficiently deep ground layers protecting the groundwater from superficial impacts and that the groundwater is not affected by harmful substances of geogenic (e.g. arsenic and uranium) or anthropogenic (e.g. nitrate and aromatic hydrocarbons) origin.

Infiltration of rainwater is a crucial process for groundwater replenishment. Due to the intensive contact of clouds and raindrops with atmospheric air, rainwater contains dissolved carbon dioxide, resulting in slightly acidic pH values of clean rainwater (pH ≈ 5.6 at 420 ppm atmospheric CO_2). During the percolation of precipitation water through soil layers, microbial degradation of organic matter does not only result in the consumption of dissolved oxygen but also results in further increasing carbon dioxide concentrations. In case of calcareous aquifers, the slightly acidic groundwater dissolves calcium carbonate (calcite, $CaCO_3$), resulting in increased concentrations of calcium (Ca^{2+}) as well as bicarbonate (HCO_3^-), which is the predominant buffer component in natural waters and corresponds roughly to the alkalinity:

$$CaCO_3(s) + H_2O + CO_2(aq.) \rightarrow Ca^{2+} + 2\,HCO_3^- \qquad (12.1)$$

The concentration of dissolved calcium (and also magnesium, which is analogously dissolved from geogenic minerals such as dolomite, $CaMg(CO_3)_2$) is commonly referred to as water hardness. In case of 'hard' water, calcite may readily precipitate with increasing calcium and/or carbonate concentrations due to its low solubility. Calcite's low solubility further decreases with increasing water temperature, which may also result in calcite precipitation when hard water is heated – this phenomenon is well known from encrustations in electric kettles and coffee machines. While encrustations in household devices may be annoying for the user, calcite precipitation in industrial processes including hot water generation or cooling processes would cause tremendous costs due to the scaling of pipes and machinery as well as heat exchangers. Therefore, softening of water containing dissolved calcium is a commonly applied industrial water treatment process, e.g. in plants requiring cooling water.

Since groundwater often contains very low oxygen concentrations due to previous microbial oxygen consumption, it is usually characterized by low redox potentials. As a consequence, groundwater often contains increased concentrations of dissolved ferrous iron (Fe^{2+}) and manganese (Mn^{2+}) deriving from the dissolution of natural minerals in the aquifer. When such an anoxic (reduced) water gets in contact with dissolved oxygen, ferrous iron is oxidized to ferric iron (Fe^{3+}) and subsequently precipitates as amorphous ferric hydroxide ($Fe(OH)_3$), whereas dissolved manganese is oxidized to Mn^{4+} and precipitates as manganese dioxide (MnO_2). If dissolved ferrous iron and manganese were not removed from groundwater prior to drinking water distribution or industrial use, these precipitation processes would result in significant encrustations in pipes, household installations and devices (e.g. washing machines) as well as aesthetic problems (e.g. brownish-black staining in laundry and metallic taste). Therefore, the removal of dissolved ferrous iron and manganese in combination with aeration and depth filtration is a common groundwater treatment process for the production of drinking water as well as process water.

In case of insufficient groundwater availability, the raw water for the generation of drinking or process water may be withdrawn from surface water resources, i.e. rivers or lakes. If the soil close to the river or lake provides good properties in terms of composition and permeability, the surface water may be abstracted by bank filtration or subsequent to artificial aquifer replenishment by infiltration of pretreated surface water (the so-called soil aquifer treatment). In both cases, biological and physical purification processes during the soil passage provide for a natural pretreatment, including the degradation of dissolved organic substances and the elimination of pathogens (depending on the hydraulic retention time in the aquifer) prior to technical water treatment. On the other hand, the soil passage often results again in the dissolution of minerals, e.g. ferrous iron, manganese and calcium (cf. above). In case of direct surface water abstraction without any soil passage, the requirements in terms of water treatment are generally high due to a variety of impacts on surface waters without any protection by soil covers.

12.2 Selected water treatment processes

This chapter focuses exemplarily on an overview of processes for the removal of dissolved iron and manganese as well as softening due to their widespread application in the production of drinking and process water. However, water resources may contain additional substances being harmful to health and/or technical systems (cf. Chapter 12.1), thus requiring further and more specific treatment processes, which cannot be considered within the scope of this textbook (for further reading on water treatment, cf. [2, 3]; for textbooks on water chemistry, cf. [4, 5]).

12.2.1 Iron and manganese removal

Chemical elements may occur as oxidized or reduced species, e.g. ferric iron (Fe^{3+}) or ferrous iron (Fe^{2+}). The stability of oxidized or reduced species of the same element in aqueous solution depends on the solution's redox potential E_H (in mV) at a given pH value. This correlation is illustrated by E_H–pH predominance area diagrams (at constant temperature and pressure; Figure 12.1): At redox potentials above the boundary lines of a redox couple (e.g. Mn^{2+}/MnO_2), the oxidized species (e.g. MnO_2 containing Mn(IV)), is predominant, whereas the corresponding reduced species (e.g. Mn^{2+}) is stable at redox potentials below the boundary line. Therefore, the occurrence of oxidized or reduced species can be derived from the knowledge of the redox potential at a given pH value.

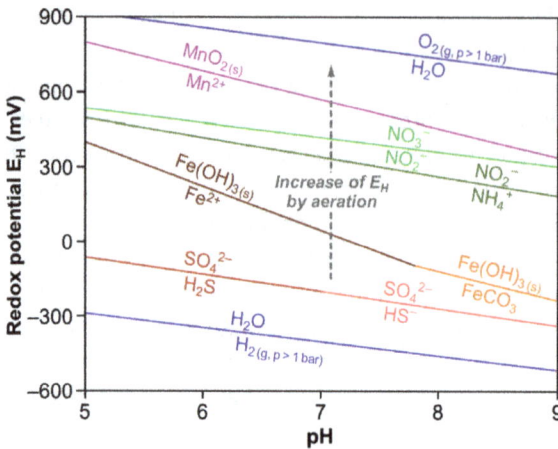

Figure 12.1: E_H–pH predominance area diagram for selected water constituents (boundary lines for iron and manganese concentrations of 10^{-6} mol/L at 25 °C).

Natural groundwaters are often characterized by pH values in the neutral to slightly basic range, i.e. 7 < pH < 8.5. In the absence of dissolved oxygen, nitrate and nitrite, the redox potential is comparatively low, i.e. around $E_H \approx 0$ mV or below. Under these redox conditions, the reduced species ferrous iron (Fe^{2+}) and manganese(II) (Mn^{2+}) are the stable iron and manganese species, respectively. Since both iron and manganese are common soil mineral matter, reduced groundwaters usually contain dissolved ferrous iron and manganese. This may also be observed in case of reduced surface waters, e.g. water from deeper layers of an eutrophic lake.

Iron concentrations in reduced groundwaters may be up to 10 mg/L (usually ≤ 5 mg/L), while manganese concentrations are often in the range of 0.1 to 1 mg/L [2]. Maximum concentrations required by the German Drinking Water Ordinance (2020) are 0.2 mg/L iron and 0.05 mg/L manganese. However, lower target concentrations for water treatment are recommended by the German Technical Standard for iron and manganese removal in order to minimize encrustations in water distribution networks (≤ 0.02 mg/L Fe, ≤ 0.01 mg/L Mn; [6]).

The principle of iron and manganese removal in water treatment is based on the following two steps:

1. increase of the redox potential and subsequent oxidation of the dissolved ferrous iron and manganese ions into insoluble precipitates, i.e. particulate ferric hydroxide ($Fe(OH)_3$) and manganese dioxide (MnO_2) and
2. removal of solid oxidation products by a particle separation process, i.e. filtration.

A sufficiently high redox potential is necessary for the oxidation of the dissolved ferrous iron and manganese ions. The boundary line of the redox couple $Fe^{2+}/Fe(OH)_3$ in the E_H–pH predominance area diagram (cf. Figure 12.1) is reached at moderate redox potential (around $E_H \approx 0$ mV at medium pH), whereas the oxidation of Mn^{2+} requires a higher redox potential. As long as the reduced species requiring a lower redox potential (i.e. Fe^{2+}) has not been oxidized, the redox potential will not increase any further so that the reduced species being stable at higher redox potential (i.e. Mn^{2+}) cannot be oxidized. Therefore, the sequence of iron and manganese removal from reduced waters is thermodynamically determined. The oxidation of dissolved manganese can only proceed subsequent to the oxidation of dissolved ferrous iron. If the raw water also contains ammonia (NH_4^+), this reduced nitrogen compound has to be oxidized via nitrite (NO_2^-) to nitrate (NO_3^-) in a two-stage process called nitrification before the oxidation of Mn^{2+} becomes possible.

The required increase of the redox potential is usually technically achieved by aeration of the raw water using prefiltered ambient air. This is often accomplished by open aeration applying the so-called waterfall aerators, e.g. spray aerators or cascade aerators. In case of spray aeration (Figure 12.2), the water is sprinkled into the air inside aeration tanks, thus forming droplets providing a high specific surface area for the efficient uptake of atmospheric oxygen. In cascade aerators, oxygen uptake from ambient air is achieved by the water falling over several overflow weirs. Open aeration additionally results in the stripping of dissolved carbon dioxide (as well as other potentially dissolved gases), thus contributing to the deacidification of groundwater containing increased carbon dioxide concentrations. Alternatively, the water can be aerated in closed systems where compressed air or pure oxygen

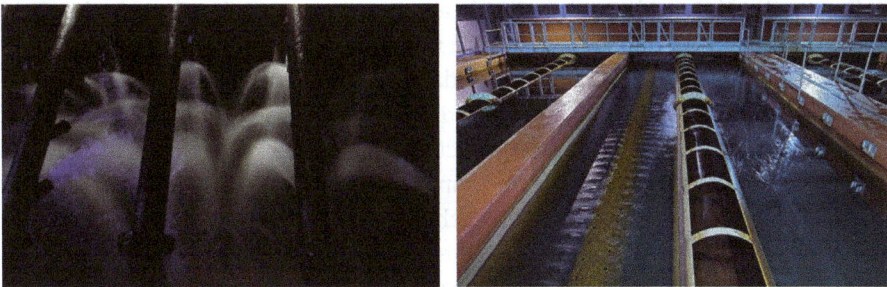

Figure 12.2: Spray aeration of reduced groundwater (left) and open rapid sand filters (right) for drinking water production (© photographs: Berliner Wasserbetriebe).

gas is dissolved in the water. Other oxidants, such as ozone, potassium permanganate or hydrogen peroxide, are rather rarely applied.

The solid oxidation products are commonly removed from the water by the so-called rapid sand filtration, i.e. depth filtration through ca. 1–3 m deep, open or closed sand filter beds at 5–15 m/h filter velocity (cf. Figure 12.2). In most cases of typical drinking water production from groundwater, mono-media filters are applied, where the filter medium has grain sizes of a defined narrow size distribution (e.g. 0.71–1.25 mm). Due to these grain sizes, the flow channels (pores) in the filter bed are relatively large compared to the fine particles to be removed from the water. Therefore, the particles are only to a minor proportion retained by size exclusion, but mostly by other mechanisms resulting in the adhesion of particles to the filter medium. Since the pores of the filter bed become successively constricted by particles deposited during the filtration process, the filter resistance increases gradually. In order to remove deposited material from the filter bed and, therefore, restore the filter capacity, depth filters have to be backwashed discontinuously with a small proportion of the previously filtered water (in drinking water production usually with a frequency of once in up to 10 days).

Even when the thermodynamically required redox potential is sufficiently high subsequent to aeration, the oxidation reactions of ferrous iron and manganese in homogeneous solution are rather slow at medium pH values. While the oxidation rate of ferrous iron increases by 100-fold per pH unit at pH > 5.5, thus providing moderate oxidation rates at medium pH values, the oxidation rate of manganese in homogeneous solution is extremely low at pH < 9 [2]. However, both oxidation reactions are very efficiently accelerated by autocatalytic processes if products of the oxidation reactions are already present. These autocatalytic effects are significant in depth filters when the filter medium is covered by oxidation products, so that iron and manganese removal takes place within only 10–30 min hydraulic retention time in consecutive zones of the filter bed (with ferrous iron oxidation often already beginning in front of the filter; Figure 12.3).

The oxidation reactions are further enhanced by bacteria, which develop in a biofilm on the filter medium and provide additional biocatalytic effects for iron and manganese oxidation due to their specific metabolism. Due to these important autocatalytic and microbial effects on the oxidation rates, efficient iron and manganese removal is only achievable with filter media being covered by iron and manganese oxides as well as iron and manganese bacteria. When clean filter sand is applied, it takes several weeks (regarding iron) to months (in case of manganese) until iron and manganese removal are efficiently achieved due to the slow formation of oxide coatings and biofilms. Therefore, used filter sand with oxide coatings is often added to new filters for iron and manganese removal in order to accelerate this process.

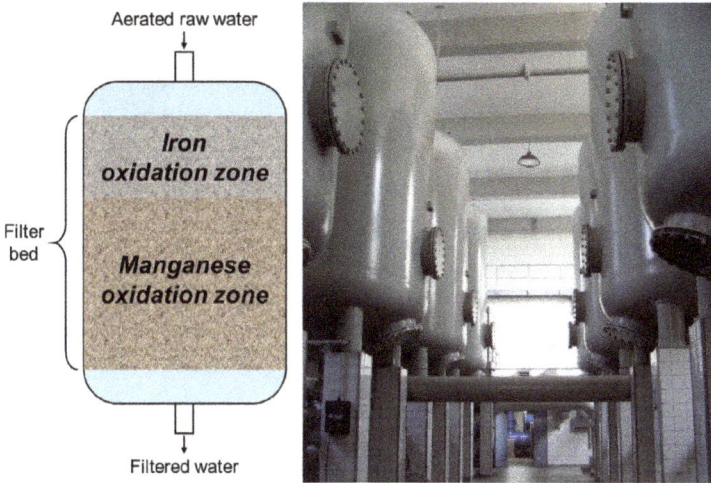

Figure 12.3: Schematic oxidation zones within a mono-media depth filter for iron and manganese removal (left) and closed depth filters for drinking water production at Münster-Hornheide water treatment plant (© photograph: J. Haberkamp).

12.2.2 Softening

As previously described in Chapter 12.1, water hardness may cause severe problems due to scaling, especially in heating and cooling systems where the water temperature rises at least temporarily. Usually, the most significant impact is due to the precipitation of calcite (CaCO$_3$), so that chemical softening processes mainly focus on calcium removal. Magnesium concentrations are commonly lower in natural freshwater, but magnesium may additionally be precipitated in chemical softening as Mg(OH)$_2$ at sufficiently high pH value (pH $\approx$ 10; [3]).

The minor solubility of calcite is reflected by its low solubility product K$_{SO}$:

$$K_{SO} = c\left(Ca^{2+}\right) \cdot c\left(CO_3^{2-}\right) \tag{12.2}$$

$$= 10^{-8.48} \, mol^2/L^2 \left(at\ 25\,°C; [4]\right)$$

Therefore, the solubility and precipitation of calcite depend on the concentrations of both calcium (Ca^{2+}) as well as carbonate (CO$_3^{2-}$). However, it makes no difference in terms of solubility if the concentrations of only calcium or carbonate or both of them increase: As long as the product of both concentrations exceeds the solubility product at the given temperature, calcite will precipitate until the solubility product is met, resulting in a saturated solution.

While the calcium concentration is independent of the pH value, the carbonate concentration strongly depends on the solution's pH value. This is exemplarily illustrated in Figure 12.4 for a closed system with a constant (and, in this example, rather high) total carbon concentration: The higher the pH value, the more bicarbonate (HCO_3^-) ions dissociate so that the carbonate concentration grows with increasing pH value. As a consequence, the maximum allowable calcium concentration decreases with increasing pH value due to the increasing carbonate concentration (cf. calcium solubility curve in Figure 12.4), resulting in calcite precipitation as soon as the solubility product is exceeded.

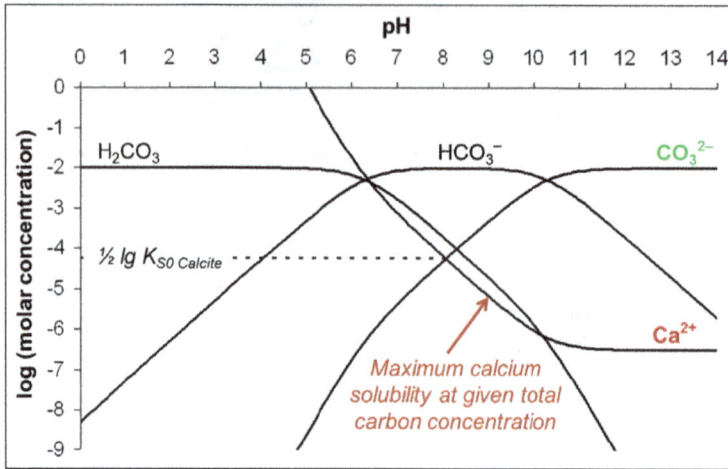

Figure 12.4: Maximum allowable calcium concentration in a solution with constant total carbon concentration ($c_T = 10^{-2}$ mol/L) as a function of the solubility product of calcite at 25 °C in a closed system.

Chemical softening processes are based on the intentional exceedance of the solubility product and thus precipitation of calcite by pH increase and/or addition of calcium or carbonate. In most applications, chemical softening is achieved by the addition of calcium hydroxide (hydrated lime, $Ca(OH)_2$), resulting in a pH increase as well as the following precipitation reactions:

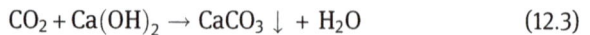

$$CO_2 + Ca(OH)_2 \rightarrow CaCO_3 \downarrow + H_2O \tag{12.3}$$

$$Ca^{2+} + 2HCO_3^- + Ca(OH)_2 \rightarrow 2CaCO_3 \downarrow + 2H_2O \tag{12.4}$$

Equation (12.3) describes the conversion of dissolved carbon dioxide to carbonate due to the increasing pH value, thus inducing calcite precipitation without any reduction of the original solution's hardness. However, dissolved calcium is removed in addition to the added calcium hydroxide as shown in eq. (12.4), resulting in a net calcium removal.

An increase of the pH value, and therefore, formation of carbonate and precipitation of calcite, can also be achieved by the addition of caustic soda:

$$Ca^{2+} + HCO_3^- + NaOH \rightarrow CaCO_3 \downarrow + Na^+ + H_2O \qquad (12.5)$$

Compared to the application of hydrated lime (eq. (12.4)), the addition of caustic soda removes less bicarbonate (HCO_3^-), i.e. alkalinity, at the same calcium removal rate, which is an important aspect with regard to the corrosivity of the water. On the other hand, the sodium concentration in the treated water increases in proportion to the addition of caustic soda therefore limiting the application of caustic soda in drinking water production due to negative impacts on human health and the corresponding drinking water quality.

For efficient chemical softening, the pH value has to be increased to pH > 9. Therefore, the pH value must be readjusted to the neutral pH range subsequent to chemical softening. This is often accomplished by the addition of carbon dioxide (the so-called recarbonation), resulting in the following reactions:

$$2\,OH^- + CO_2 \rightarrow CO_3^{2-} + H_2O \qquad (12.6)$$

$$CO_3^{2-} + CO_2 + H_2O \rightarrow 2\,HCO_3^- \qquad (12.7)$$

In many water treatment processes, only part of the water is treated by softening and subsequently blended with the non-softened portion of the water, which contributes to pH adjustment.

Chemical softening can technically be implemented by precipitation reactors and subsequent sedimentation tanks. However, due to the relatively slow precipitation of calcite in homogenous solution, large reactor volumes are required. The process is significantly more efficient in fluidized-bed reactors, where calcite precipitation is accelerated by the contact with pellet-like calcite precipitates acting as crystal nucleus and being suspended in a sludge blanket in high upflow reactors. Since fine precipitation products are not efficiently eliminated in both configurations, the softened water is usually further treated by depth filtration for particle removal (cf. Chapter 12.2.1).

Apart from chemical softening, the reduction of water hardness can also be accomplished by nanofiltration as well as ion exchange. *Nanofiltration* is a type of pressure-driven membrane filtration where polymeric membranes with ca. 1–10 nm pore diameter are operated at up to 5 bar pressure. Nanofiltration membranes carry negative charges, which causes the electrostatic repulsion especially of polyvalent anions (e.g. SO_4^{2-}). However, in order to maintain electroneutrality on both sides of the membrane, polyvalent cations, such as calcium and magnesium, are retained to a large extent on the concentrate side of the membrane as well, thus resulting in an efficient reduction of hardness (in addition to the elimination of dissolved organic compounds, e.g. pharmaceutical residues or pesticides, due to size exclusion; cf. [7]). Nanofiltration for the purpose of softening is applied on a large scale, e.g. in drinking water treatment plants as well as in small-scale applications for decentralized removal of hardness from drinking water in buildings.

Ion exchangers are usually based on synthetic resins carrying positively or negatively charged, covalently bonded functional groups, which are coupled with negatively or positively charged counter ions, respectively, for reasons of electroneutrality. For softening applications, the cations, calcium and magnesium, can be removed by cation exchangers carrying negative surface charges. At the beginning of the cation exchange process, the ion exchange resin carries monovalent counter ions, e.g. Na^+ or H^+ cations. When water containing divalent calcium and magnesium cations passes through the cation exchanger, these Ca^{2+} and Mg^{2+} ions replace the resin's counter ions due to their higher charge density, e.g. one Ca^{2+} ion replaces two Na^+ ions. Therefore, hardness is efficiently eliminated until the capacity of the cation exchanger is exhausted, i.e. until most of the original monovalent counter ions have been replaced by multivalent cations. At that point, the cation exchanger has to be regenerated. This is accomplished by a concentrated regeneration solution of the original monovalent counter ions, which replace the previously retained multivalent cations due to the high regenerant concentration. Thus, a highly concentrated brine to be discharged results from the regeneration process.

Since the softened water contains all the previously replaced counter ions, the application of cation exchangers with sodium counter ions is restricted in central drinking water production for health reasons. However, household softening devices are often based on cation exchangers with sodium counter ions, resulting in the accumulation of sodium in the final drinking water (which is rather doubtful with regard to the consumer for health reasons). Dishwashers also include ion exchangers for softening purposes in order to avoid lime deposits on dishes, thus requiring the frequent application of significant amounts of 'regenerating salt', which is actually sodium chloride and results in the further accumulation of salts in domestic wastewater.

12.3 Mineral acids for industrial applications

Water is the most important solvent for inorganic acids. Its extraordinary importance is readily underlined by the use of many acidic and basic chemicals on a large tonne scale. In Table 12.1, we summarize some important inorganic acids and their use for diverse processes. These acids are available in technical quality or in p.a. (*pro analysi*) food grade quality. The diluting water for technical quality is frequently just freshwater. Also completely desalinated water is used (ion exchange or reverse osmosis).

Table 12.1: Important mineral and other inorganic acids and selected applications in industry.

amido sulfonic acid – H_2NSO_3H	precursor for the sweetener sodium cyclamate; acidic cleaning agent for ceramics and metals; limescale removal; descaling agent
boric acid – H_3BO_3	precursor for borosilicate glasses; porcelain and enamel; neutron absorber; mild disinfection agent; precursor for fungicides and flame retardants
hydrofluoric acid – HF	etching of glass, quartz and wavers, oil refining, (organo)fluorine chemistry, cleaner
hydrochloric acid – HCl	basic acid in mining and petrochemistry, chemical and pharmaceutical industry; cleaning agent (cement, rust or lime stain removal)
hexafluorosilic acid – H_2SiF_6	synthesis of synthetic cryolite; wood preservation agent; cleaning of copper and brass vessels in breweries
hypophosphorous acid – H_3PO_2	reducing agent (reduction of aryldiazonium salts); electroless nickel plating
nitric acid – HNO_3	synthesis of fertilizers and explosives; nitriding acid for organic synthesis; aqua regia for dissolution of noble metals
phosphoric acid – H_3PO_4	fertilizer production; conservation, acidification and acid regulation in food industry; corrosion protection via zinc phosphates; detergents; etching of wavers
phosphorous acid – H_3PO_3	reducing agent; synthesis of lead phosphonate for PVC stabilization; retardant in concrete
sulfuric acid – H_2SO_4	synthesis of fertilizers; battery acid; TiO_2 and zinc production; CaF_2 digestion

12.4 Hydrogen cyanide and cyanide leaching

Hydrogen cyanide, the so-called prussic acid (German: *Blausäure*), is the basic source for all industrially used cyanides. The educts for the platinum catalyzed synthesis are methane, ammonia and oxygen.

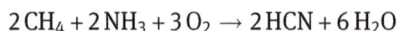

$$2\,CH_4 + 2\,NH_3 + 3\,O_2 \rightarrow 2\,HCN + 6\,H_2O$$

While HCN is an important synthesis building block for many organic compounds, even on a large scale, there are less applications in the field of inorganic chemistry. The important inorganic cyanides are NaCN and KCN. They are mainly used for cyanide leaching, replacing the amalgam process. The cyanide leaching is exemplarily shown for gold with sodium cyanide, but silver ores and potassium cyanide also run the same way. The dicyanidoaurate(I) or dicyanidoargentate(I) solutions are finally reduced with zinc dust, leading to fine gold (silver) particles and the more stable tetracyanidozincate(II) complex.

$$4\,Au + 8\,NaCN + O_2 + 2\,H_2O \rightarrow 4\,Na\left[Au(CN)_2\right] + 4\,NaOH$$

$$2\,Na\left[Au(CN)_2\right] + Zn \rightarrow 2\,Au + Na_2\left[Zn(CN)_4\right]$$

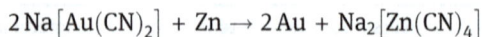

The cyanide-contaminated wastewater can finally be purified through oxidative destruction ($CN^- + H_2O_2 \rightarrow OCN^- + H_2O$) with hydrogen peroxide (Chapter 13.1.1).

Besides the cyanide leaching process, thin silver and gold layers can be deposited on high-quality electronic components by the so-called electroplating procedure (silver-on-copper or gold-on-copper plating), using NaCN or KCN, in order to generate corrosion-resistant coatings. Moreover, potassium cyanide is the precursor for the pigment Prussian Blue (German: *Berliner Blau*), $Fe_4^{III}[Fe^{II}(CN)_6]_3$ (Chapter 1.4).

12.5 Inorganic bases for industrial applications

Similar to the acids discussed in Chapter 12.3, also the inorganic bases have water as solvent. In Table 12.2, we summarize some important inorganic bases and their use for diverse processes.

Table 12.2: Important inorganic bases and selected applications in industry [8].

ammonia – NH_3	nitriding agent in metallurgy; cooling agent (high heat of vaporization); precursor for nitric acid; one of the basic chemicals in inorganic synthesis (amides, amines, nitriles)
lime milk – $Ca(OH)_2$	SO_2 capture; basic construction chemical; synthesis of chlorinated lime; fungicide; acid regulator in food production
sodium hydroxide – NaOH	bauxite digestion; detergent synthesis; etching in microelectronics and metallurgy; precursor for sodium-based chemicals; basic drain cleaner
potassium hydroxide – KOH	selective etching in microelectronics; oxidative alkaline digestion of ores; detergent industry; glass industry; electrolyte in accumulators
sodium hypochlorite – NaOCl	bleaching and disinfection (pools); mold remover; drain cleaner; disinfection in tooth medicine

References

[1] Hanslmeier A. Water in the Universe, Springer, Dordrecht, Heidelberg, London, New York, 2011.

[2] Crittenden JC, Trussell RR, Hand DW, Howe KJ, Tchobanoglous G. (Eds). MWH's Water Treatment: Principles and Design, 3rd edition, John Wiley & Sons, Inc, Hoboken, NJ, USA, 2012.

[3] Worch E. Drinking Water Treatment – An Introduction, Walter de Gruyter, Berlin, Germany/
     Boston,MA, USA, 2019.
[4] Benjamin MM. Water Chemistry, 2nd edition, Waveland Press, Inc., Long Grove, IL, USA, 2015.
[5] Worch E. Hydrochemistry – Basic Concepts and Exercises, Walter de Gruyter, Berlin,
     Germany/ Boston,MA, USA, 2015.
[6] DVGW: Work Sheet 223-1: Deironing and Demanganization – Principles and Processes (in
     German). Technical Standard of the German Technical and Scientific Association for Gas and
     Water (DVGW). Bonn, Germany, 2005.
[7] Kucera J (Ed). Desalination – Water from Water, 2nd edition, John Wiley & Sons, Inc, Hoboken,
     NJ, USA, and Scrivener Publishing LLC, Beverly, MA, USA, 2019.
[8] Büchel KH, Moretto HH, Woditsch P. Industrial Inorganic Chemistry, 2nd edition, Wiley-VCH,
     Weinheim, Germany, 2000.

13 Biomedical and bioactive materials

13.1 Disinfection and inorganic biocides

13.1.1 Hydrogen peroxide and ozone

Rainer Pöttgen, Thomas Fickenscher

Efficient disinfectants are those which release clean and pollutant-free decomposition products. Hydrogen peroxide and ozone are among those compounds since they solely release oxygen and water. The use of hydrogen peroxide and ozone as environmental-friendly oxydation agents for application in disinfection, as oxidation reagents or as educts in inorganic synthesis is summarized in the present chapter. Climate-relevant questions [1–3] of ozone are not considered.

13.1.1.1 Hydrogen peroxide

Hydrogen peroxide belongs to the worldwide-produced basic chemicals with an annual amount of >4 million tons [4] with still increasing requirement.

The very first synthesis of hydrogen peroxide proceeded through decomposition of barium peroxide with hydrochloric or nitric acid, yielding only diluted aqueous solutions. Later, the use of sulfuric acid led to better yields (formation of insoluble $BaSO_4$). This peroxide process was superseded by the anodic oxidation of sulfuric acid with Caro's acid (H_2SO_5) as intermediate product during hydrolysis:

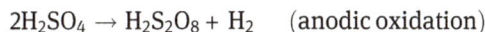

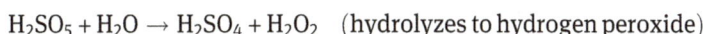

$$2H_2SO_4 \rightarrow H_2S_2O_8 + H_2 \quad \text{(anodic oxidation)}$$

$$H_2S_2O_8 + H_2O \rightarrow H_2SO_4 + H_2SO_5 \quad \text{(hydrolyzes to Caro's acid)}$$

$$H_2SO_5 + H_2O \rightarrow H_2SO_4 + H_2O_2 \quad \text{(hydrolyzes to hydrogen peroxide)}$$

This electrolytically produced hydrogen peroxide is pure and could be obtained with high yields; however, this process is energy-intensive. Nowadays, hydrogen peroxide is exclusively produced via the anthrachinone process (alternating hydrogenation and oxidation) using a 2-alkylanthrachinone. The oxidation step proceeds without catalyst and is also called *autooxidation process* [4]. The 2-alkylanthrachinone solely serves as reactionary and is not consumed during the process. The real reaction is then:

$$H_2(g) + O_2(g) \rightarrow H_2O_2(l)$$

The resulting hydrogen peroxide solution can be concentrated by vacuum distillation. The typical commercially available concentrations are 25, 50 and 70% which are then diluted for the specific applications. A problem of hydrogen peroxide is its decomposition (exothermic under oxygen release!) that proceeds with traces of

https://doi.org/10.1515/9783110733471-021

metal cations, iodide or hydroxide anions and light. This decomposition is prevented by small additions of stabilizing agents like phosphates, phosphoric acid, phosphonic acids or different carboxylic acid esters. As an example, we show the hydrogen peroxide decomposition on a platinum and brownstone (MnO_2) surface in Figure 13.1.1.1.

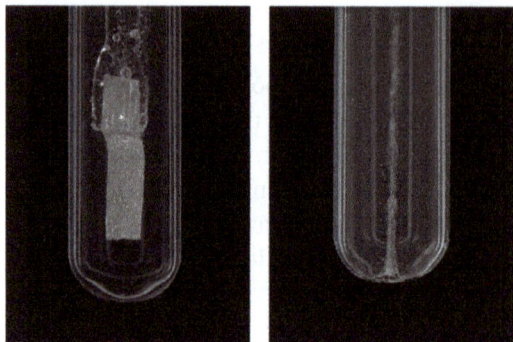

Figure 13.1.1.1: Catalytic decomposition of a 5% hydrogen peroxide solution on the surface of a platinum sheet (left) and brownstone (MnO_2) pieces (right).

The production, transport and stock of hydrogen peroxide requests suitable container materials: stainless steel, aluminum, polyethylene or polypropylene (for low concentrations). The 3% solutions are available on the market in brown glass bottles.

Depending on the application (Table 13.1.1.1), hydrogen peroxide is industrially available in technical quality as well as in high-purity food-grade quality.

Hydrogen peroxide is broadly used since it is one of the environmental-friendly oxidation reagents. Its decomposition ($H_2O_2 \rightarrow H_2O+O$) just releases water and active oxygen. Since meanwhile hydrogen peroxide is broadly available, precursor compounds like sodium peroxide (for in-situ generation of H_2O_2) as oxidation agent today find only moderate use.

The oxidation potential of hydrogen peroxide can be increased by the application of UV light. In the UVOX process (**UV** light and **ox**idation), the O–O bond is cleaved by UV light ($H_2O_2 \rightarrow 2HO$) leaving highly reactive hydroxyle radicals. This procedure finds application in wastewater treatment, e.g. the destruction of herbicides and their metabolites.

A typical technical application of hydrogen peroxide is the oxidative destruction of pollutants in exhaust gas (e.g. $H_2S + 4H_2O_2 \rightarrow H_2SO_4 + 4H_2O$) or contaminated wastewater (e. g. $CN^- + H_2O_2 \rightarrow OCN^- + H_2O$).

Besides the aqueous hydrogen peroxide solutions, several salts of hydrogen peroxide find broad technical application. Of the bleaching agents in detergents (generation of active oxygen), sodium percarbonate has replaced sodium perborate in order to avoid boron-contaminated wastewater.

Magnesium and calcium peroxide show a slow delivery of oxygen in solutions with decreasing pH. Typical applications for these peroxides are groundwater remediation, the baking industry (dough conditioning), tooth cleaning, oral hygiene and hair bleaching. Strontium and barium peroxide are additional oxygen sources in fireworks. Potassium peroxomonosulfate and H_2SO_4/H_2O_2 mixtures (both with high purity) are used for etching printed circuits.

Table 13.1.1.1: Selected applications of hydrogen peroxide for technical use.

Branch	Application
basic chemical synthesis	propene oxide, cyclohexanone oxime/caprolactame/nylon process, sodium perborate $(Na_2[B_2(O_2)_2(OH)_4]\cdot6H_2O)$, sodium percarbonate $(Na_2CO_3\cdot1.5H_2O_2)$
paper industry	bleaching of cellulose and paper
mining	forming soluble hexavalent uranium (from U^{IV}), peroxide-assisted leaching of gold ores, cyanide removal in wastewater (effluent treatment)
household	mold removal
textile industry	bleaching of textile fibers
hairdressing	bleaching of hair (3% solution)
nutrition industry	disinfection of food packaging (35% solution)
aquaculture	treatment of sea lices
medicine	wound disinfection (3% solution)
medical technique	disinfection of contact lenses (3% solution)
space and aviation	fuel (>85%)

13.1.1.2 Ozone

Ozone forms by reaction of oxygen atoms with oxygen molecules. The cleavage of the O_2 bond requires 249 kJ mol^{-1}, and the reaction $O + O_2 \rightarrow O_3$ delivers only 106 kJ mol^{-1} [5]. Thus, ozone is an endothermic compound, and its synthesis $3/2O_2 \rightarrow O_3$ needs 143 kJ mol^{-1}.

The technical production of ozone proceeds through silent electrical discharge (corona discharge) under water-cooling using the so-called Siemens ozonizer. Typical conditions for the low-frequency voltage are 50–500 Hz and 10–20 kV. Usually, pure and dry oxygen is used as educt since the use leads to small amounts of nitrogen oxides as by-products. The high reactivity of ozone limits the materials for construction. Suitable materials are stainless steel (quality 316 L), glass, polytetrafluorethylene, polyvinylidenfluoride or perfluorinated rubber. The ozonization always yields O_2/O_3 mixtures with a maximum ozone content of 15%. An alternative procedure uses UV light, similar to stratospheric chemistry. This technique can work with air; however, the yield is much lower than that with the Siemens ozonizer, and such small UV-based ozonizers can be used for portable applications, also as flow reactors (the optical components of such UV lamps are discussed in Chapter 8.7). This is also possible with portable reactors that work with ceramic plates as electrodes (nickel-tungsten paths embedded in a ceramic matrix). Such electrodes (Figure 13.1.1.2) with 12×5 cm size allow ozone production in the order of 3–5 g h^{-1}. The use of dry air/dry oxygen is always recommended since ozone formation can be quenched by humidity. Since, depending on the purity and the temperature of the surrounding, ozone slowly decomposes (this is the striking difference to hydrogen peroxide), for any application, ozone is freshly prepared. The half-life is ca. 40 min at 293 K and ca. 140 min at 273 K [6].

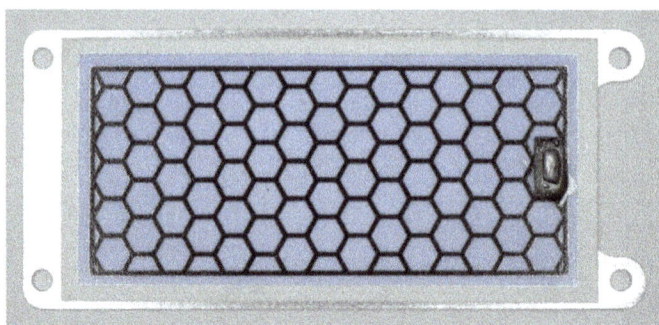

Figure 13.1.1.2: A ceramic electrode for ozone production.

The broad use of ozone relies on its high oxidation capacity. Upon decomposition ($O_3 \rightarrow O_2 + O$), ozone generates molecular oxygen besides a highly reactive O radical. As an example, we discuss a typical inorganic lecture experiment. An O_2/O_3 mixture readily reacts with a clean silver sheet under ambient conditions (Figure 13.1.1.3). The high oxidation potential is used in a manifold of technical applications which are summarized in Table 13.1.1.2.

Besides the many technical applications, ozone finds application in medicine [7, 8], mainly with respect to inactivation of bacteria (Gram-positive and Gram-negative ones), viruses, fungi, yeast and protozoa.

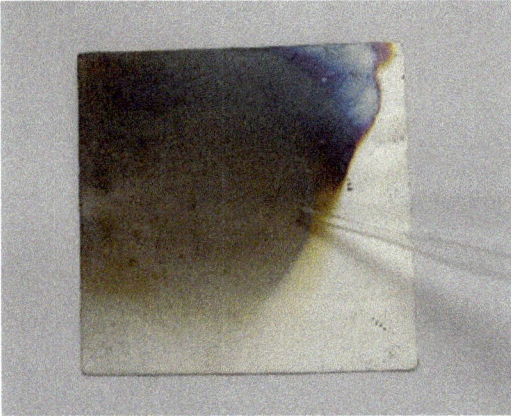

Figure 13.1.1.3: Oxidation of a silver sheet with an O_2/O_3 mixture under ambient conditions.

Table 13.1.1.2: Selected applications of ozone for technical use.

Branch	Application
agriculture and food industry	kill bacteria, water disinfection, generation of germ-free water, sterilization packaging, preservative for the cold storage of eggs and fruits
semiconducting devices	wafer cleaning (oxidation of organic and metallic impurities)
medical sector	disinfection of medical clothes, tools, footware, instruments
sauna, spa, swimming pool	water and surface disinfection
air cleaning/odor removal	air disinfection in conference rooms, removal of bad smell (nicotine smell, corpse odor, fire damage restauration, animal smell, slaughterhouse smell), reduction of droplet infection risk
wine industry	sterilization of barrels and bottle cork
drinking water	sterilization of drinking water, elimination of microbacterial germs, elimination of hospital germs
cooling water treatment	elimination of germs
wastewater treatment	elimination of oxidizable contaminants

📖 References

[1]   Thompson DWJ, Solomon S, Kushner PJ, England MH, Grise KM, Karoly DJ. Nature Geosci,
      2011, 4, 741–9. DOI: 10.1038/NGEO1296.
[2]   Simpson WR, von Glasow R, Riedel K, Anderson P, Ariya P, Bottenheim J, Burrows J, Carpenter
      L, Frieß U, Goodsite ME, Heard D, Hutterli M, Jacobi H-W, Kaleschke L, Neff B, Plane J, Platt U,
      Richter A, Roscoe H, Sander R, Shepson P, Sodeau J, Steffen A, Wagner T, Wolff E. Atmos
      Chem Phys Discuss, 2007, 7, 4285–403. www.atmos-chem-phys-discuss.net/7/4285/2007/.
[3]   Karentz D, Marchi M, Bastianoni S. Encycl Ecol, 2019, 4, 383–90. DOI: 10.1016/B978-
      008045405-4.00865-X.
[4]   Bertau M, Müller A, Fröhlich P, Katzberg M. Industrielle Anorganische Chemie, 4. Auflage,
      Wiley-VCH, Weinheim, Germany, 2013.
[5]   Holleman AF, Wiberg N. Anorganische Chemie, 103. Auflage, De Gruyter, Berlin, 2016. ISBN
      978-3-11-051854-2.
[6]   Mandhare MN, Jagdale DM, Gaikwad PL, Gandhi PS, Kadam VJ. Int J Pharm Biol Sci, 2012, 2,
      63–71.
[7]   Seidler V, Linetskiy I, Hubálková H, Staňková H, Šmucler R, Mazánek J. Prague Med Rep,
      2008, 109, 5–13.
[8]   Tiwari S, Avinash A, Katiyar S, Iyer AA, Jain S. Saudi J Dental Res, 2017, 8, 105–11.

13.1.2 Chlorine-based disinfectants

Rainer Pöttgen, Cristian A. Strassert

Chlorine is one of the most important basic chemicals and also finds a broad range of applications in inorganic and organic syntheses. Most of the technologically used chlorine (around 65 Mio. tons a year worldwide [1]) is generated by means of the electrolytic chloralkali process [2, 3]. An important application of chlorine itself and of its derivatives concerns the production of disinfectants. Selected applications for chlorine, chlorinated lime, sodium hypochlorite, chlorine dioxide and chloramines are summarized hereafter. Organic chlorine-based disinfectants (chlorhexidine, among others) are not part of the scope of the present chapter.

Chlorine-based treatment of drinking water played a decisive role at the end of the nineteenth century and provided readily available water without pathogenic bacteria, parasites and viruses. This is very clearly reviewed in Michael J. McGuire's book *The Chlorine Revolution: Water Disinfection and the Fight to Save Lives* [4]. It has been repeatedly stated that the use of chlorine for drinking water disinfection is probably the most significant public health advancement of the millennium.

Chlorine quickly kills pathogenic bacteria or parasites and inactivates threatening viruses [5]; hence, it dramatically reduced the prevalence of hepatitis, cholera, typhoid fever, legionnaires' disease, and other waterborne affections, such as those caused by *Campylobacter spp.*, *Noroviridae* and amoebae. In general, the disinfecting activity is related to the oxidizing power of the rather hydrophobic Cl_2 molecule, which is able to diffuse across biomembranes while remaining hydrated in aqueous environments. Therefore, it is capable of damaging functional (bio)organic components of pathogens (such as biologically active structures, including proteins, lipids and nucleic acids) or dyes (recall its bleaching capacity).

Besides the difficulties related to the handling of gaseous elemental chlorine possessing a high reactivity and toxicity, it is clear that chlorination of drinking water leads to a characteristic odor and taste. Its relatively low aqueous solubility and instability in water limits its application to the primary disinfection process. For this reason, several derivatives have been implemented as secondary disinfection agents (*i.e.* longer-lived) preventing the proliferation of pathogens in transport pipes and hindering the formation of biofilms. Furthermore, solid salts containing active chlorine derivatives have simplified the transport and handling of disinfecting agents. Hence, chlorine is formed *in situ* by reduction, disproportionation or comproportionation of its chemical precursors, which are used as disinfecting agents and include hypochlorites, chlorine dioxide or chloramines in water.

https://doi.org/10.1515/9783110733471-022

13.1.2.1 Elemental chlorine gas

Around 5% of the worldwide-produced chlorine is used for water sanitation as a primary disinfection agent able to quickly inactivate a broad range of pathogens. However, chlorine is used as sparingly as possible (depending on the environmental or food regulatory agencies and the accepted limits in water analysis). Thus, the trend goes towards the use of drinking water sources with the lowest possible load of pathogens and organic mass. The upper limit for the content of active chlorine according to the German drinking water ordinance amounts to 0.3 mg L^{-1}.

Elemental chlorine is usually transported in its liquid form contained in steel pressure cylinders or pressure tanks (corrosion of the containers is prevented due to surface passivation by $FeCl_3$). For safety reasons, only indirect chlorination finds practical applications [6]. The whole installation operates with a reduced pressure of approximately 850 mbar of chlorine (after decompression from 6 bar from the storage cylinder). Thus, even in the case of a small leakage, no dangerous chlorine gas breakout occurs. The gaseous chlorine is then disseminated in water as finely dispersed bubbles by means of an injector tube (Figure 13.1.2.1) and the resulting aqueous chlorine solution is the final disinfecting agent that is dosed through membrane pumps.

Figure 13.1.2.1: Injector tube for gaseous elemental chlorine dispersion in water. With courtesy and copyright agreement by Lutz-Jesco GmbH.

The process of indirect chlorination with elemental chlorine is fast and very efficient for both drinking water and pool water disinfection. Nevertheless, due to the unwanted formation of potentially health-threatening side products (e.g. chloroalkanes, i.e. $CH_{4-x}Cl_x$) and for safety concerns related to the handling of elemental chlorine, the trend is shifting towards the progressive replacement by sodium hypochlorite (Chapters 13.2.1.2 and 13.2.1.3) and chlorine dioxide (Chapter 13.2.1.4) or monochloramine (Chapter 13.2.1.5) as active agents (*vide infra*).

13.1.2.2 Chlorinated lime

Chlorinated lime is obtained by the reaction of calcium hydroxide (slaked lime) with chlorine under ambient conditions:

$$2Ca(OH)_2 + 2Cl_2 \rightarrow Ca(OCl)_2 \cdot 2H_2O + CaCl_2$$

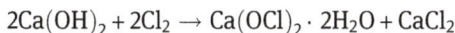

In the product, calcium chloride remains largely in solution and calcium hypochlorite (as an insoluble dihydrate) can be filtered off. However, in the technical process, the reaction is incomplete and the product contains residual chloride and hydroxide. This is often expressed by the following chemical equation:

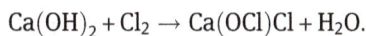

$$Ca(OH)_2 + Cl_2 \rightarrow Ca(OCl)Cl + H_2O.$$

Further technical processes are directed via so-called triple salts. This is discussed and summarized in the bibliographic literature [2].

Ca(OCl)Cl is commercially available in granular form and as convenient tablets. It is much more stable than sodium hypochlorite (*vide infra*) and allows for longer storage intervals. An inconvenience related to chlorinated lime is its limited solubility. It is difficult to prepare clear solutions and this hampers application for practical applications such as disinfection of swimming pool water. Chlorinated lime solutions contain around 3% of effectively active chlorine.

Chlorinated lime (up to 36% of effectively active chlorine content, *vide infra*) was the first chemical that allowed chlorine storage and transport and subsequent recovery by treatment with hydrochloric acid:

$$Ca(OCl)Cl + 2HCl \rightarrow CaCl_2 + Cl_2 + H_2O$$

The decomposition of chlorinated lime proceeds also with sulfuric acid (precipitation of plaster) and occurs slowly in the presence of carbon dioxide from the air (the reaction is driven by the precipitation of $CaCO_3$):

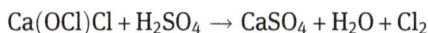

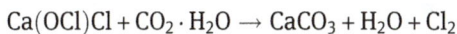

$$Ca(OCl)Cl + H_2SO_4 \rightarrow CaSO_4 + H_2O + Cl_2$$

$$Ca(OCl)Cl + CO_2 \cdot H_2O \rightarrow CaCO_3 + H_2O + Cl_2$$

As in the case of sodium hypochlorite, also chlorinated lime is associated with higher chemical costs if compared with elemental chlorine. Solid chlorinated lime or its suspensions find applications in fields where the complex composition of the mixture involving hypochlorite, chloride and hydroxide, does not constitute a drawback. Typical applications are the use as algicide, for decontamination of warfare agents or large area disinfection of contaminated surfaces containing larger amounts of biological remnants.

13.1.2.3 Sodium hypochlorite

Sodium hypochlorite solution constitutes an alternative to elemental chlorine, since it is easier to handle and due to the fact that these aqueous formulations can be readily transported in large containers. The solution is formed by direct reaction of chlorine gas with caustic soda below 40 °C (to avoid chlorate formation through disproportionation):

$$2NaOH + Cl_2 \rightarrow NaOCl + NaCl + H_2O$$

The hypochlorite solution has a relatively low stability due to photochemical, thermal or chemical decomposition (especially in combination with contaminants if technical-grade materials are employed) and is usually freshly prepared. A slight excess of caustic soda (to keep the pH in the alkaline regime) increases the stability of the solution [2]. The commercially available sodium hypochlorite solutions typically contain 12–15% of effective chlorine: this is defined as the amount of chlorine released (comproportionation $Cl^+ + Cl^- \rightarrow Cl_2$) by the reaction of the hypochlorite solution with hydrochloric acid, based on the mass of the product:

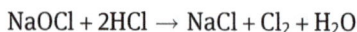

$$NaOCl + 2HCl \rightarrow NaCl + Cl_2 + H_2O$$

Aqueous solutions of alkali-metal hypochlorites are among the oldest commercialized disinfecting agents. The aqueous potassium hypochlorite solution is well known as *Eau de Javelle,* and was already produced in France by the end of the eighteenth century and is still commercialized today. The sodium-based product is known as *Eau de Labarraque.*

Caution! Hypochlorite solutions cannot be used in the presence of acids, since they release toxic chlorine gas.

Sodium hypochlorite solutions find a broad field of application spanning from (i) drinking and (ii) swimming pool water treatment to (iii) mold control, (iv) cyanide destruction and (v) sanitary disinfection (including decontamination of the entire root canal system in endodontics) [7]. Due to the limited shelf-life of the aqueous solutions, on-site generation by electrolysis (minimal chemical storage and transport, no handling of elemental chlorine) is widely used, especially for swimming pool water. However, one limitation for hypochlorite application is related to its high oxidizing ability: even bromide and iodide containing waters can release bromine and iodine by oxidation.

13.1.2.4 Chlorine dioxide
Chlorine dioxide has become an increasingly relevant alternative to elemental chlorine as a broad-spectrum disinfectant. The technological processes for the production of chlorine dioxide start from sodium chlorite, $NaClO_2$, or sodium chlorate, $NaClO_3$, depending on the amounts needed for successive consumption:

(i) Acidification of sodium chlorate on a technical scale:
$NaClO_3 + 2HCl \rightarrow ClO_2 + \frac{1}{2}Cl_2 + NaCl + H_2O$ (alternatively with sulfuric acid)
The separation of chlorine and chlorine dioxide is possible by washing with water (ClO_2 has the higher solubility).

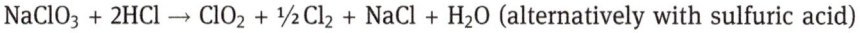

(ii) Acidification of sodium chlorite (hydrogen chloride-chlorite process):

$$5NaClO_2 + 4HCl \rightarrow 4ClO_2 + 5NaCl + 2H_2O.$$

(iii) Oxidation of sodium chlorite (for small quantities), i.e. the chlorine-chlorite process:

$$2NaClO_2 + Cl_2 \rightarrow 2ClO_2 + 2NaCl.$$

(iv) Chlorite-peroxodisulfate process:

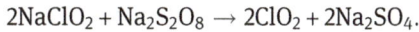

$$2NaClO_2 + Na_2S_2O_8 \rightarrow 2ClO_2 + 2Na_2SO_4.$$

Nowadays, chlorine dioxide is broadly used as a chlorine substitute for diverse disinfection applications (elemental-chlorine-free processes), since it can be obtained *in-situ* from stable salts *via* the reactions listed above. The commercialized products usually include (i) ready-made solutions with typical ClO_2 contents of around 2000 ppm, (ii) two-component systems, where the chlorine dioxide is generated *in situ* by the persulfate-chlorite process (roughly 0.5% solutions) and (iii) one-component systems based on the generation of ClO_2 from sodium chlorite by oxidation. Pure ClO_2 is explosive, which is why the *in situ* formation of diluted products for on-site use is offered. Pure ClO_2 is handled in low concentrations with either nitrogen or carbon dioxide as a diluting, inert gas. The yellow aqueous solutions of chlorine dioxide (Figure 13.1.2.2) are relatively stable.

Figure 13.1.2.2: Photography of an aqueous solution of ClO_2, obtained by the reaction of sodium chlorite solution with hydrochloric acid. Photo by Thomas Fickenscher.

Chlorine dioxide is not used for direct applications on humans and animals. It finds use in diverse surface disinfection processes, for drinking water treatment, as algae-control agent (in pools and aquaria), as a protective agent for liquids in cooling and process systems, for slime control preparations (to avoid biofilm formation), as well as for disinfection in the veterinary, food and feed industry. Further applications concern the disinfection of chemical toilets, sewage fluids, hospital waste and contaminated soil [8–10].

13.1.2.5 Chloramines

Ammonia, alkyl- or arylamines, urea derivatives and other nucleophilic *NH*-containing organic compounds react with chlorine or with hypochlorite in aqueous solution to successively yield *N*-chlorinated mono-, di- or trichloramines. In swimming pools and swimming halls, these products are responsible for the irritation of eyes and mucosa tissues, particularly when urine and skin debris react with the aqueous hypochlorite employed for disinfection. To reduce this unwanted effect, cyanuric acid (Figure 13.1.2.3) is reacted with hypochlorites (or chlorine in alkaline solutions), yielding less reactive *N*-chlorinated intermediates ensuring a steady concentration of active chlorine that guarantees an adequately hygienic disinfection level.

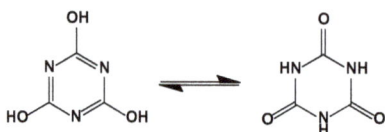

Figure 13.1.2.3: Structural formulae of two relevant tautomers of cyanuric acid.

However, if produced and implemented under controlled conditions, chloramines can be also useful secondary disinfection agents with a longer half-life than hypochlorite [5, 11, 12]. For instance, in the USA, small amounts of gaseous ammonia and chlorine are used in combination to assure a minimum steady concentration of active chlorine in drinking water, as requested by the US EPA. In the aqueous phase, monochloramine is formed :

$$NH_3 + HClO \rightarrow NH_2Cl + H_2O$$

$$NH_3 + Cl_2 \rightarrow NH_2Cl + HCl$$

If high concentrations of chlorinating agents were available, the successive reaction of monochloramine and dichloramine would ultimately lead to trichloramine; however, this is usually avoided, since its volatile nature drops its effectiveness, besides being an irritating species for biological tissues.

Chloramines are particularly useful additives to keep a steady disinfecting level during water circulation in transport pipes, where the formation of bacterial biofilms could lead to severe contamination events. However, if the degree of bacterial contamination or the organic load is high, formation of organic mono- or dichloramines leads to bad organoleptic properties, which are usually not accepted by the consumers and regulatory agencies. Since organic chloramines could be also formed during the primary chlorination process, it is crucial to ensure that drinking water and the corresponding pipes have low initial bacterial and organic loads. Interestingly, the formation of organic chloramines is reduced if the less reactive monochloramine is used as a primary disinfectant instead of elemental chlorine gas, which also minimizes the formation of potentially harmful chloroalkanes during the primary treatment (*vide supra*). However, if compared with elemental chlorine, monochloramine is a slower and less efficient primary disinfecting agent, thus requiring impractically higher dosages. Therefore, its use is restricted to the prevention of bacterial proliferation (*i.e.* as a secondary disinfectant) [5, 11, 12].

All in all, the slight chlorine taste in drinking water is a minor side effect that can be tolerated to guarantee a good level of residual active chlorine ensuring a low level of pathogenic contamination.

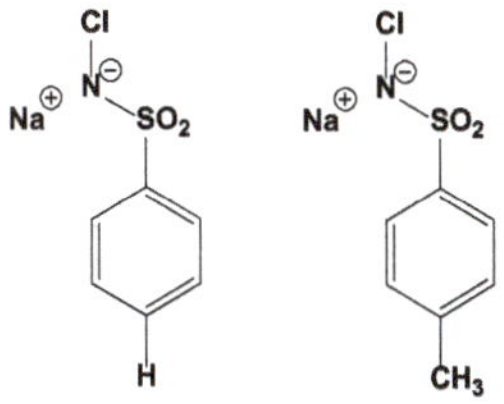

Figure 13.1.2.4: Structural formulae of Chloramine B (left) and Chloramine T (right) as their sodium salts.

Even though they are not chloramines in the strictest sense, the *N*-chlorinated sulfonamides named Chloramine T and Chloramine B (Figure 13.1.2.4) will be briefly discussed herein; these useful derivatives are usually stored, sold and dissolved in water as their water-soluble sodium salts. Besides being good antiseptic agents for disinfection in veterinary or human medical applications (if used at low concentrations), they also constitute practical sources of active chlorine for water and surface decontamination. Hence, they provide steady concentrations of active chlorine with longer half-lives than the rather reactive and relatively unstable hypochlorite anion or elemental chlorine. They are synthesized by reaction of the corresponding sulfonamides with chlorinating agents and typically stabilized and stored as their solid (yet photosensitive) sodium salts.

References

[1]   https://www.chlorineinstitute.org/. accessed on November 9th 2021.
[2]   Bertau M, Müller A, Fröhlich P, Katzberg M. Industrielle Anorganische Chemie, 4. Auflage, Wiley-VCH, Weinheim, Germany, 2013.
[3]   Holleman AF, Wiberg N. Anorganische Chemie, 103. Auflage, De Gruyter, Berlin, 2016. ISBN 978-3-11-051854-2.

[4]   McGuire MJ. The Chlorine Revolution: Water Disinfection and the Fight to Save Lives, American Water Works Association, Denver, CO, 2013, 350. ISBN: 9781583219201.

[5]   Drinking water chlorination – A review on disinfection practices and issues, Chlorine Chemistry Council®, Arlington, VA, United States, 2003. http://c3.org, accessed on November 9th 2021.

[6]   Roeske W, Müller C. Brauwelt, 2003, 287–92.

[7]   Mohammadi Z. Int Dent J, 2008, 58, 329–41.

[8]   Wie M, Lai J, Zhan P. Acta Mikrobiol Sin, 2012, 52, 429–34.

[9]   Liester MB. Int J Med Med Sci, 2021, 13, 13–21.

[10] Chen T-L, Chen Y-H, Zhao Y-L, Chiang P-C. Aerosol Air Qual Res, 2020, 20, 2289–98.

[11] Drinking water requirements for states and public water systems – chloramines in drinking water, United States Environmental Protection Agency®, Washington, DC, United States, 2021. https://www.epa.gov/dwreginfo/chloramines-drinking-water, accessed on November 17th 2021.

[12] Water disinfection with chlorine and chloramine, centers for disease control and prevention®, Washington, DC, United States, 2020. https://www.cdc.gov/healthywater/drinking/public/water_disinfection.html, accessed on November 17th 2021.

13.1.3 Iodine-based disinfection materials

Jonas R. Schmid, Christian Matschke, Sebastian Riedel

Iodine, which belongs to the group of halogens, is one of the least abundant non-metallic elements in earth's composition. Although it does not occur in large quantities, it is found almost everywhere, for example in rocks, soils, waters, plants, animal tissues and in our daily food. The largest reserve of iodine is seawater but only in a concentration of less than 0.05 ppm [1]. Seaweed is one of the most important organisms naturally accumulating iodine and was the source for the discovery of the element by the French chemist Curtois in 1811 while producing niter from seaweed ashes [2]. In the following year, Gay-Lussac and Davey named it after the Greek word "iodes" for the color violet and recognized it as a new element [3, 4]. The industrial production started in the same year and its properties as medical sterilizing agent were applied since 1816 [5]. Seminal work describing the disinfecting efficiency of iodine was published by Davaine between 1874 and 1881. In fact, it was already shown in 1874 that iodine is one of the most effective antiseptics [6]. Today, iodine is still broadly used as a potent disinfectant in a complexed form or as a polyiodide species.

13.1.3.1 Polyiodides and their properties

The chemistry of polyiodides is linked to their use as disinfectants and started with the discovery of the element. Pelletier and Caventou showed that the addition of elemental iodine to strychnine (a highly toxic alkaloid) resulted in the formation of a crystalline compound, strychnine triiodide, which resembles the first trihalide [4, 7]. The increased solubility of iodine in various solvents, when potassium iodide is added, attracted large attention in the first half of the nineteenth century, and several researchers assumed the formation of a triiodide ion. In 1839, Jørgensen investigated the composition of a mixture of iodine and potassium iodide and proposed that the mixture contains iodine as well as iodide supporting the hypothesis for the existence of the triiodide ion $[I_3]^-$ [8]. However, the structure of $K[I_3]$ was only proven unambiguously by X-ray crystallography as shown in Figure 13.1.3.1 [9].

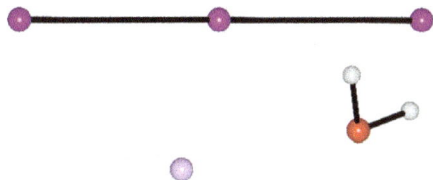

Figure 13.1.3.1: Molecular structure in the solid state of $K[I_3]xH_2O$ [9].

https://doi.org/10.1515/9783110733471-023

Such polyiodides are the basis for various aqueous disinfection solutions where I_2, I^- and $[I_3]^-$ can play an active role, depending on the stoichiometry and the pH value. Higher concentrated solutions, such as Lugol's solution, also contains the species $[I_5]^-$ and $[I_6]^{2-}$ shown in Figure 13.1.3.2 and are well known to detect starch [10]. Interestingly, the tendency of halogens to form polyhalides is found for all stable halogens but is most pronounced for iodine [11].

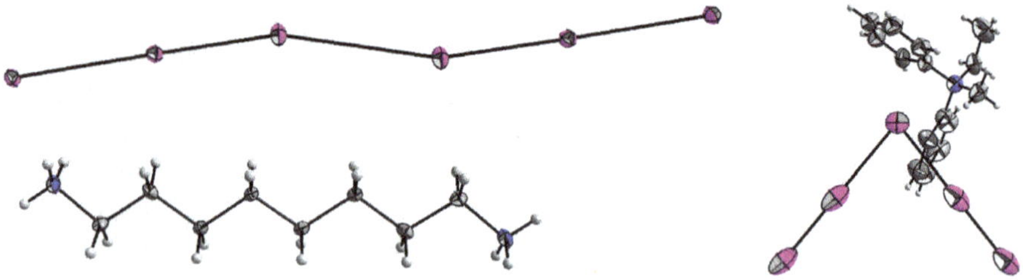

Figure 13.1.3.2: Molecular structure in the solid state of $[H_3N(CH_2)_8NH_3][I_6]$ (left) [12] and $[NPh_2MeEt][I_5]$ (right) [13]. Displacement ellipsoids are shown at the 50% probability level.

If Lugol's solution is dropped onto a test medium (in this case a potato), a discoloration is observed ranging from blue to blue-violet and, depending on the concentration, even black as shown in Figure 13.1.3.3.

Figure 13.1.3.3: Potato with and without potassium triiodide dropped onto the cutting surface. Picture taken by the authors.

Recent scientific investigations indicate that this color change corresponds to the formation of infinite polyiodide chains inside the amylose [14]. Such polyiodides are raging from $[I_2]^-$ over $[I_{29}]^{3-}$ to the above-mentioned infinite polyiodide chains [4]. The bonding situation in these polyhalides can be described by the halogen bonding concept [15]. According to IUPAC it is defined as a "net attractive interaction between an electrophilic region associated with a halogen atom in a molecular entity and a nucleophilic region in another, or the same, molecular entity" in analogy to hydrogen bonding. The halogen bonding is strongly directional and has an optimal angle of 180° in contrast to hydrogen bonding [15]. The halogen bonding donor is accepting electron density (Lewis acid, I_2) from the halogen bonding acceptor (Lewis base, I^-) according to the IUPAC definition. This concept can be applied to describe the structure of polyhalides. In this concept, they are separated into fragments of halogen bonding donors X_2 (e.g., I_2) and halogen bonding acceptors X^- (e.g., I^-) forming the species $[X_3]^-$ (e.g., $[I_3]^-$).

When the electrostatic potential of a dihalogen is plotted on iso-surfaces of the electron density, the potential is anisotropic and not equally distributed over the entire molecule (Figure 13.1.3.4) [11]. A more positive electrostatic potential (blue) is found for the homoatomic dihalogens along the bonding axis, the so called σ-hole. This acts as a Lewis acidic region, whereas the region with negative electronic potential (red) has a Lewis basic character. For the homoatomic dihalogens, the magnitude of the σ-hole increases with the polarizability of the halogens, i.e., with atomic number, which can be seen by comparing F_2 and I_2 (Figure 13.1.3.4) [16].

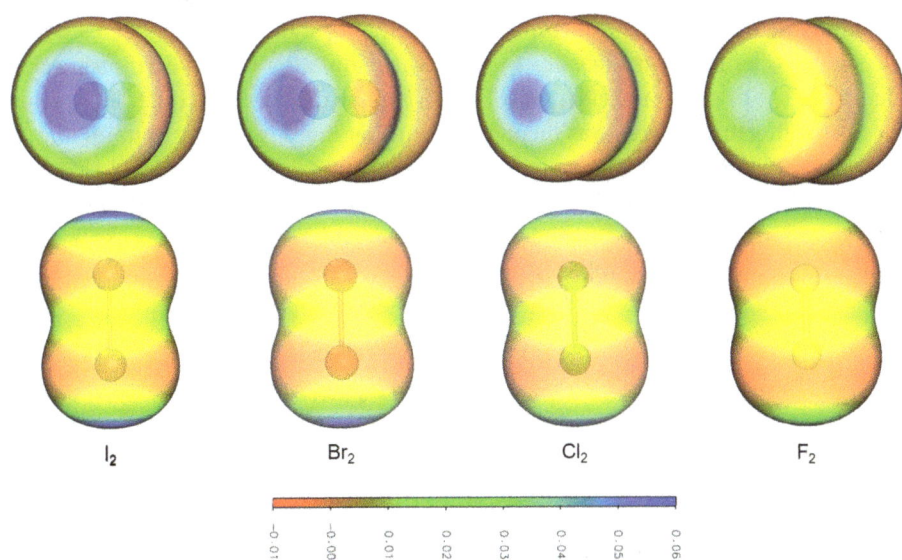

Figure 13.1.3.4: The electrostatic potentials of the halogens F_2, Cl_2, Br_2 and I_2 in the range of −0.01 a.u. (red) to 0.06 a.u. (blue) have been mapped onto their electron densities (iso-surface value 0.0035 a.u.); calculated at the B3LYP-D3/def2-TZVPP level of theory.

The electrostatic potential plot on the iso-surfaces of the electron density for a trihalide is also anisotropic as shown for the dihalogens. In this way, trihalides can interact with nucleophiles (X^-) via their σ-hole along the bond axis and with electrophiles via the annular accumulation of a negative electronic potential perpendicular to the bond axis (X_2). Based on this concept, various polyiodides are formed ranging from $[I_3]^-$, $[I_5]^-$, $[I_6]^{2-}$ and even infinite chains like in the pyrroloperylene–iodine complex and build therefore the fundament for a variety of disinfectants.

13.1.3.2 Iodine as disinfectant

Iodine has been used to cure or prevent infections of wounds for almost 150 years. In the course of time, different treatments were developed based on solutions with and without stabilizers and solid polyiodides. They can be differentiated based on their formulation.

13.1.3.2.1 Aqueous solutions

There are two kinds of formulations of aqueous iodine solutions: (i) without organic compounds and (ii) with organic compounds which coordinate and stabilize the iodine species (iodophors; see Chapter 13.1.3.2.3). Investigations of the water/iodine system revealed that at least nine different equilibria exist where 10 iodine species, like I^-, I_2, $[I_3]^-$, $[I_5]^-$, $[I_6]^{2-}$, HOI, $[OI]^-$ and $[H_2OI]^+$ play a role depending on concentration and pH value [17]. The corresponding equilibria have already been studied focusing on the follow-up reactions and the accumulation of radio-iodine species emerging due to nuclear accidents [10, 17]. The relevance of these different species for the disinfecting process was theoretically further investigated by Gottardi due to inconsistent statements in literature, mainly concerning hypoiodous acid HOI and the cationic iodine species $[H_2OI]^+$. In addition, additives such as an alcohol must be considered, if not a pure, fresh iodine solution, which is not decomposed by disproportionation reactions [10].

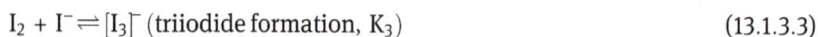

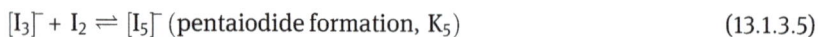

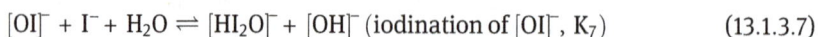

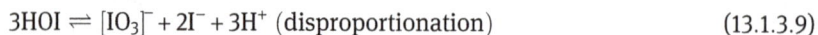

$$I_2 + H_2O \rightleftharpoons HOI + I^- + H^+ \text{ (hydrolysis, } K_1) \tag{13.1.3.1}$$

$$HOI \rightleftharpoons [OI]^- + H^+ \text{ (dissociation of HOI, } K_2) \tag{13.1.3.2}$$

$$I_2 + I^- \rightleftharpoons [I_3]^- \text{ (triiodide formation, } K_3) \tag{13.1.3.3}$$

$$HOI + H^+ \rightleftharpoons [H_2OI]^+ \text{ (protonization of HOI, } K_4) \tag{13.1.3.4}$$

$$[I_3]^- + I_2 \rightleftharpoons [I_5]^- \text{ (pentaiodide formation, } K_5) \tag{13.1.3.5}$$

$$2[I_3]^- \rightleftharpoons [I_6]^{2-} \text{ (dimerization of } [I_3]^-, K_6) \tag{13.1.3.6}$$

$$[OI]^- + I^- + H_2O \rightleftharpoons [HI_2O]^- + [OH]^- \text{ (iodination of } [OI]^-, K_7) \tag{13.1.3.7}$$

$$[HI_2O]^- \rightleftharpoons [I_2O]^{2-} + H^+ \text{ (dissociation of } [HI_2O]^-, K_8) \tag{13.1.3.8}$$

$$3HOI \rightleftharpoons [IO_3]^- + 2I^- + 3H^+ \text{ (disproportionation)} \tag{13.1.3.9}$$

The disproportionation to iodate (eq. (13.1.3.9)) proceeds rather slowly in contrast to the very fast reactions described by eqs. (13.1.3.1)–(13.1.3.8). Gottardi showed that the iodate formation considerably changes with the conditions (manly pH and $c(I^-)$) as described by eqs. (13.1.3.1)–(13.1.3.3) [18]. They studied the main properties of aqueous iodine solutions by calculating the equilibrium constants considering the change in dependency of $c(I_2)$, $c(I^-)$, pH value and dilution.

(Case 1) iodine solutions (10^{-3}–10^{-6} M) without any additional iodide, (Case 2) 0.001 M iodine in the presence of additional iodide, and (Case 3) Lugol's solution (an aqueous solution of 5% iodine and 10% KI according to the United States Pharmacopeia XXI).

(Case 1) In the case of a pure iodine solution a decrease of the number of active species contributing to 99% of the oxidation capacity is observed with dilution and change of the pH value. At a concentration of $c(I_2) = 10^{-3}$ M the five species I_2, HOI, $[I_3]^-$, $[HI_2O]^-$ and $[OI]^-$ are present, while at lower concentrations (10^{-5} M) only I_2, HOI and $[OI]^-$ are primarily existing. The concentration of active iodine in solution is constant up to a pH value of 6 independent of the dilution.

(Case 2) In the presence of iodide, a significantly different composition of the formed species is observed. At a concentration of $c(I^-) = 10^{-1}$ M the range where the concentrations of I_2, $[I_3]^-$, $[I_5]^-$ and $[I_6]^{2-}$ are constant is extended to a pH value of 10 and the formation of other species is suppressed until a pH value of 5. Therefore, only eqs. (13.1.3.3), (13.1.3.5) and (13.1.3.6) are of importance for this case.

(Case 3) In undiluted Lugol's solution the concentration of the main species I^-, $[I_3]^-$, $[I_6]^{2-}$, $[I_5]^-$ and I_2 are constant until a pH value of 10 and the suppression of other species is extended until a pH value of 6. This will become important for the stability of these disinfections in Chapter 13.1.3.4.

In summary, the pH value as well as the concentration of iodide has a very strong influence on the individual equilibrium concentrations. Only at a pH value above 8 and at high dilution, HOI accounts for over 90% of the oxidation capacity (Case 1). If additional iodide is present and a certain pH value (pH 6) is not exceeded only I^-, I_2 and $[I_3]^-$ play a role (Case 2). In Lugol's solution the higher polyiodides $[I_5]^-$ and $[I_6]^{2-}$ make up for 8.2% of the oxidation capacity (Case 3). The addition of iodide not only increases the solubility of iodine but also forms the more stable polyiodides.

13.1.3.2.2 Alcoholic solutions

Alcoholic solutions of iodine undergo different equilibria resulting in the formation of triiodide, which is equilibrated after approximately 24 h.

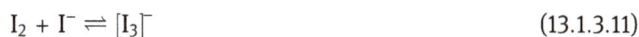

$$ROH + I_2 \rightleftharpoons ROH \cdot I_2 \leftrightarrow [ROHI]^+ \; I^- \qquad (13.1.3.10)$$

$$I_2 + I^- \rightleftharpoons [I_3]^- \qquad (13.1.3.11)$$

$$[I_3]^- + ROH \rightleftharpoons ROH \cdot [I_3]^- \qquad\qquad (13.1.3.12)$$

This system contains several oxidizing iodine species (eqs. (13.1.3.10)–(13.1.3.12)), ranging from I_2 to triiodide-alcohol adducts. Their distribution was not further investigated, as the solvents (alcohols) are themselves strong disinfectants [5, 19].

13.1.3.2.3 Ion exchange resin-based polyiodides and iodophor solutions

Ion exchange resins (IER) represent a unique solid disinfectant without the need of a solution or the possibility of decreasing iodine concentration by sublimation of iodine. They consist of, e.g., a quaternary ammonium ion inside a polymer resin (Amberlite®) which can be charged with elemental iodine to obtain the disinfectant. Iodine carrier materials such as IER release iodine to microorganisms after coming in contact to them and are therefore generally considered demand-release-disinfectants. Iodophors are polymeric organic molecules (such as alcohols, amides or sugars) which coordinate the iodine species. This results in a reduction of the equilibrium concentration compared to aqueous solutions with the same concentration of iodine and iodide as shown in Chapter 13.1.3.1.1. In this case, the relevant species are restricted to I^-, I_2 and $[I_3]^-$ and are simplified to the reaction eqs. (13.1.3.13)–(13.1.3.15).

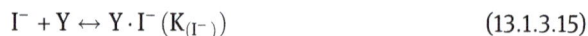

$$I_2 + Y \leftrightarrow Y \cdot I_2 (K_{I_2}) \qquad\qquad (13.1.3.13)$$

$$[I_3]^- + Y \leftrightarrow Y \cdot [I_3]^- \; (K_{I_3^-}) \qquad\qquad (13.1.3.14)$$

$$I^- + Y \leftrightarrow Y \cdot I^- \; (K_{(I^-)}) \qquad\qquad (13.1.3.15)$$

Y = represents a coordination sphere complexing the iodine species in the molecular structure of the iodophor.

Povidone-iodine is a commonly used antimicrobial agent consisting of polyvinylpyrrolidone iodine which is water-soluble containing elemental iodine bound to a synthetic polymer [20]. The amide groups in this polymer could act as electron donors to coordinate elemental iodine, as, for comparison, it is the case for the dioxane iodine complex shown in Figure 13.1.3.5 in the molecular solid-state structure.

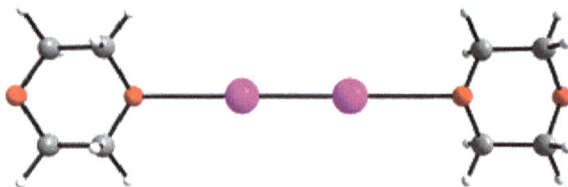

Figure 13.1.3.5: Molecular structure in the solid state of dioxane iodine adduct [21].

Schenck and coworkers investigated analogs of the povidone iodine complex by X-ray diffraction and suggested an amide-bridged proton inside the polymer with intercalated triiodide as shown in Figure 13.1.3.6 [22].

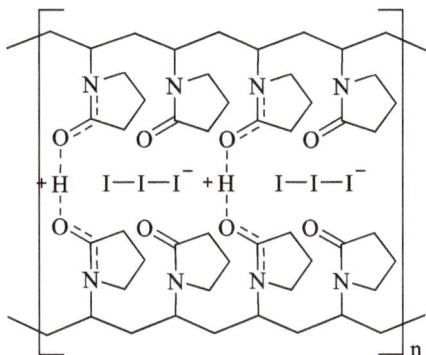

Figure 13.1.3.6: Possible structure of Povidone-triiodide.

13.1.3.2.4 Organic iodine (III) as disinfectants

The unique properties of iodine allow it to form stable polycoordinated multivalent compounds, such as Ph-ICl$_2$, which represents the first organic polyvalent iodine compound [23]. Among the various structural types of polyvalent organic iodine(III) species, the iodonium salts have shown considerable biological activities and many related species have been prepared since their discovery over 120 years ago [24]. There properties can be tailor-made by synthetic methods depending on their biological activities as shown in Table 13.1.3.1 [25–33].

Table 13.1.3.1: Biological activity of bis(diaryl)iodonium salts. Ar = aryl. X = H, F, Cl, Br, alkyl or alkoxy group [25].

Structure	Biological activity	References
	Antimicrobial agents against bacteria, fungi and organisms that attack seeds, roots and above-ground portions of plants, viricidal against small RNA viruses.	[26]
	Antimicrobials for inhibition of the growth of many bacterial and fungal organisms that attack seeds, roots and above-ground portions of plants.	[33]
	Control several bacterial organisms as well as molds, mildews, fungi and slimes.	[31]

Table 13.1.3.1 (continued)

Structure	Biological activity	References
	Anthelmintic agents that are active upon oral administration, active in vitro against gram negative and gram positive organisms.	[28]
	Toxicity against growth of bacteria, fungi, organisms that attack seeds, roots and above-ground portions of plants, micro-organisms responsible for mold, mildew, rot, decay.	[27]
	Active against *Bacillus subtilis, Pseudomonas* species, *Escherichia coli, Staphylococcus* species, fungi, algae, yeasts.	[32]

In addition, a low toxicity towards mammals (LD_{50} in mice) can be observed in the case of (*p*-chlorophenyl)(thienyl)iodonium chloride with a value of >4000 mg/kg [31]. A relatively recent study demonstrated that iodonium salts could be a viable alternative and even surpass the properties of chlorhexidine in some cases, which is commonly used as a disinfectant and antiseptic in dental medicine [34].

13.1.3.3 Mechanisms of the disinfecting activity

In the case of povidone-iodine, the antimicrobial efficiency of iodine is increased through the delivery of the iodine directly to the cell membrane [35]. The iodine rapidly penetrates the micro-organisms [36] and targets key groups of proteins and fatty acids in the cytoplasm, cytoplasmic membrane and nucleotides [37]. Selected reactions that might occur are the following:

1. The S-H group in amino acids such as cysteine could be oxidatively coupled to their disulfides with iodine [38] and resulting in their loss of the ability to connect protein chains by disulfide bridges (Scheme 13.1.3.1) [5].

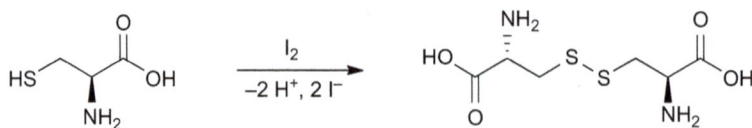

Scheme 13.1.3.1: Iodine-mediated dimerization of amino acids bearing a S-H group.

Additionally, oxidation reactions of the S-H groups to sulfonic acids could take place preventing the formation of disulfide chains (eq. (13.1.3.16)).

$$R\text{–SH} \xrightarrow[-\,2H^+,\,2I^-]{+\,I_2,\,+\,H_2O} R\text{–SOH} \xrightarrow[-\,2H^+,\,2I^-]{+\,I_2,\,+\,H_2O} R\text{–SO}_2H \xrightarrow[-\,2H^+,\,2I^-]{+\,I_2,\,+\,H_2O} R\text{–SO}_3H$$

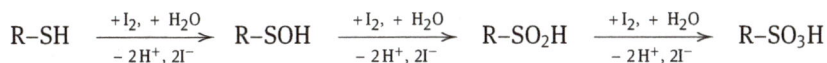

2. Iodination of activated aromatic compounds (amino acids such as tyrosine or histidine) and nucleobases (e.g., cytosine or uracil) could increase the bulk of these molecules and lead to steric hindrance in the formation of hydrogen bonds (Scheme 13.1.3.2).

Scheme 13.1.3.2: Iodination of C-H groups of activated aromatic compounds (tyrosine, histidine, cytosine or uracil). R = activating groups.

3. Iodine could react with the double bonds present in unsaturated fatty acids and lead to a change in the overall physical properties of the lipids and to membrane immobilization (Scheme 13.1.3.3) [39, 40].

Scheme 13.1.3.3: Addition of iodine to double bonds.

13.1.3.4 Stability of iodine-based disinfectants

The stability of the above-mentioned disinfectants solution and solid iodine disinfections depends on the use of additives and pH values. An aqueous solution of iodine will not be stable over time as decomposition into iodide, iodate and many more species is observed. In contrast, no decrease in the disinfecting effectivity is observed in a concentrated iodine, iodide solution with a pH value below 6 or if additives are present. To stabilize these solutions and suppress the formation of iodate, a pH buffer (5–6) is used for the various formulations, usually phosphate or citrate salts [5].

13.1.3.5 Typical disinfectants containing or releasing iodine and their applications

As mentioned above it has to be differentiated between the various formulations (aqueous solutions with or without iodide, additives or IER). A great variety of formulations belong to the group of iodine and iodide (NaI or KI) solutions in water, ethanol, glycerol or in mixtures of these solvents. Formulations according to the

United States Pharmacopeia XXIII are: (I) Iodine Topical Solution (2.0% iodine and 2.4% sodium iodide in an aqueous solution); (II) Strong Iodine Solution (or Lugol's solution, 5% iodine and 10% potassium iodide in an aqueous solution); (III) Iodine Tincture containing (2.0% iodine and 2.4% sodium iodide in aqueous 50% ethanol solution); and (IV) Strong Iodine Tincture (7% iodine and 5% potassium iodide in 95% ethanol). Due to the large amount of iodide present in these formulations, the triiodide equilibrium is of particular importance as shown in Chapter 13.1.3.1.1. The most important application of iodine in human medicine is the disinfection of skin as a preoperative step for surgical disinfection of hands, equipment and patients [39]. They can also be used for treatment of infected or burned skin. Highly concentrated aqueous and alcoholic iodine formulations used until the 1960s were replaced by iodophors as they show minor side reactions and staining of skin [5].

Iodine tinctures are also widely adopted in veterinary medicine for the disinfection of cow's udders before and after milking since 1985 due the reduction of the numbers of staphylococci [41]. However, nowadays, they are replaced by the better-tolerated iodophoric formulations [42].

As discussed above, iodophor-based disinfectants are often related to povidone-iodine and depend on the producer and the formulation sold under different names. The available iodine concentration in povidone-iodine (an iodine 1-vinyl-2-pyrrolidinone polymer complex) according to USP XXIII contains a minimum of 9.0% and maximum of 12.0% available iodine [22]. One well-known example for a commercially available formulation is Betaisodona®. The different formulations are used for disinfection of skin and wounds as shown in Figure 13.1.3.7.

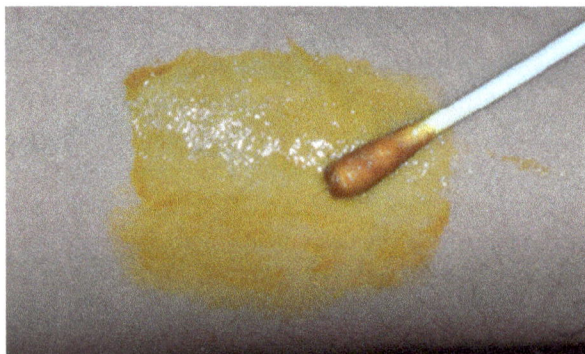

Figure 13.1.3.7: Application of Povidone-Iodine on skin for disinfection.
Picture taken by the authors.

Since 1952, the US Army has used such formulations in form of tablets containing iodine, a carrier and a stabilizing agent to increase the solubility providing an easy way for water treatment. Furthermore, recent studies showed that povidone-iodine is effective against SARS-CoV-2 [43].

Table 13.1.3.2: Overview of medical applications of iodine derivatives.

Product category	Indication	Iodine-Form	Dosage form	Examples (German market)
Disinfection	Skin disinfection, wound disinfection	Povidone-Iodine complex	Ointment, drops, gel, cream, wound gauze, vaginal ovula	Betaisodona® solution, ointment, wound gauze, oral antiseptic Betadona Emra ointment Betaseptic Mundipharma® Solution for application on the skin Jod-Polyvidon Wound and burn ointment Inter Pharm Braunoderm® Traumasept® Vaginal-Ovula
		I_2 – KI (Lugol's solution) in ethanol	Tincture	Jodtinktur Hetterich
Pharmaceuticals	Thyroid drug for goiter treatment and iodine substitution (prevention and therapy of thyroid diseases).	KI	Tablets for oral use	Jodid-ratiopharm® 100 μg Jodid 100 μg HEXAL® Jodid dura® 200 μg Jodinat® 100 μg
	Radioactive therapeutic agent for thyroid tumors or overactive thyroid tissue.	NaI (^{123}I) radioactive	Oral hard capsule or oral solution	Theracap131 37–5550 MBq hard capsule MAH: GE Healthcare Buchler GmbH & Co. KG MONIYOT-131 for therapy 14,8–3700 MBq/ml solution for ingestion MAH: MONROL EUROPE SRL

(continued)

Table 13.1.3.2 (continued)

Product category	Indication	Iodine-Form	Dosage form	Examples (German market)
Radiation protection agent	Reactor accidents	KI	Oral tablets/ solutions	potassium iodide "Lannacher" 65 mg-tablets
Food supplements	Substitution during pregnancy and in case of iodine deficiency	I_2 or KI	Tablets/ capsules/ oral solutions	Folio® SteriPharm GmbH & Co. KG
Diagnostic	Thyroid gland: determination of activity, iodine-storing metastases (labeling), functional status, etc.	NaI (^{123}I) radioactive	Injection, capsules	n.a.
	Thyroid gland: Pretherapeutic determination of activity	NaI (^{131}I) radioactive	Solution	n.a.

IER are in general iodine demand-release-disinfectant and can be used as solid carrier of active iodine. Recent data on iodine resins indicate that they are effective against bacteria, viruses and some protozoa (group of single-celled eukaryotes) [44–46]. The release of elemental iodine under ambient conditions into water is considerably low what makes them a promising agent not only for treating water but also for air [47]. One of the most prominent examples for the application of such an IER is used by the National Aeronautics and Space Administration (NASA), based on an iodine-polyvinyl pyrrolidone complex for their water systems for example in the International Space Station (ISS) [45]. Iodine is a safe alternative to chlorine (often used in water treating facilities) because of its physical state (solid instead of a gas) and ease of handling, in addition to the potential hazards of transporting chlorine gas into space. These charged resins can also be used to eliminate microbes such as bacteria, viruses, parasites and mold. Furthermore, they are effective against bacteria such as anthrax or viruses like SARS, influenza, Newcastle disease, viruses and *Aspergillus niger*, which rarely becomes resistant to iodine, and against weaponized chemical gases (e.g., mustard gas or VX gas). Therefore, gas masks were equipped with IER charged with iodine during the times of war and chemical terrorism [47–50].

The versatile properties of iodine and its derivatives as disinfectant and medicine are summarized in Table 13.1.3.2.

References

[1]   Hora K. Iodine production and industrial applications, IDD Newsletter, August issue, 2016.
[2]   Weeks ME. In Discovery, Curtouis B. Ann Chim, 1813, 91, 304.
[3]   Gay Lussac JL. Ann Chim, 1814, 91, 5.
[4]   Svensson PH, Kloo L. Chem Rev, 2003, 103, 1649–84.
[5]   Gottardi W. In: Tatsuo K. (Ed). Iodine Chemistry and Applications, Wiley, Hoboken, NJ, USA, 2014, 375–410.
[6]   Vallin E-A. Traité Des Désinfectants Et de la Désinfection / Par Vallin E, Paris, 1882.
[7]   Pelletier P, Caventou JB. Ann Chim, 1819, 10, 164.
[8]   Jörgensen SM. J Prakt Chem, 1870, 2, 347–60.
[9]   Thomas R, Moore FH. Acta Crystallogr B, 1980, 36, 2869–73.
[10] Gottardi W. Archiv der Pharmazie, 1999, 332, 151–7.
[11] Sonnenberg K, Mann L, Redeker FA, Schmidt B, Riedel S. Angew Chem Int Ed, 2020, 59, 5464–93.
[12] Van Megen M, Reiss GJ. Inorganics, 2013, 1, 3–13.
[13] Tebbe K-F, Loukili R. Z Anorg Allg Chem, 1999, 625, 820–6.
[14] Madhu S, Evans HA, Doan-Nguyen VVT, Labram JG, Wu G, Chabinyc ML, Seshadri R, Wudl F. Angew Chem Int Ed, 2016, 55, 8032–5.
[15] Desiraju Gautam R, Ho PS, Kloo L, Legon Anthony C, Marquardt R, Metrangolo P, Politzer P, Resnati G, Rissanen K. Pure Appl Chem, 2013, 85, 1711.
[16] Redeker FA, Kropman A, Müller C, Zewge SE, Beckers H, Paulus B, Riedel S. J Fluor Chem, 2018, 216, 81–8.

[17]   Clough P, Starkie H. Eur Appl Res Rep, 1985, 6, 631–776.
[18]   Gottardi W. Zentralbl Bakteriol Mikrobiol Hyg B, 1981, 172, 498–507.
[19]   Bhattacharjee B, Varshney A, Bhat S. J Indian Chem Soc, 1983, 60, 842–4.
[20]   Kramer SA. J Vasc Nurs, 1999, 17, 17–23.
[21]   Marshall WG, Jones RH, Knight KS. CrystEngComm, 2019, 21, 5269–77.
[22]   Schenck H-U, Simak P, Haedicke E. J Pharm Sci, 1979, 68, 1505–9.
[23]   Willgerodt C. J Prakt Chem, 1886, 33, 154–60.
[24]   Hartmann C. Chem Ber, 1894, 27, 426–32.
[25]   Stang PJ, Zhdankin VV. Chem Rev, 1996, 96, 1123–78.
[26]   Doub L. U.S. patent, 3422152, USA, 1969.
[27]   Jezic Z. U.S. Patent, 3622586, USA, 1972.
[28]   Jezic Z. U.S. patent, 3734928, USA, 1970.
[29]   Jezic Z. U.S. patent, 3759989, USA, 1973.
[30]   Jezic Z, Plepys RA. U.S. patent, 3896140A, USA, 1976.
[31]   Moyle CL. U.S. patent, 3944498A, USA, 1976.
[32]   Relenyi AG, Koser GF, Walter RWJ, Kruper WJJ, Shankar RB, Zelinko AP. U.S. patent, 5106407, USA, 1992.
[33]   Jezic Z. U.S. patent, 3712920A, USA, 1973.
[34]   Goldstein EJC, Citron DM, Warren Y, Merriam CV, Tyrrell K, Fernandez H, Radhakrishnan U, Stang PJ, Conrads G. Antimicrob Agents Chemother, 2004, 48, 2766–70.
[35]   Zamora JL. Am J Surg, 1986, 151, 400–6.
[36]   McDonnell G, Russell AD. Clin Microbiol Rev, 1999, 12, 147–79.
[37]   Durani P, Leaper D. Int Wound J, 2008, 5, 376–87.
[38]   Zeynizadeh B. J Chem Res, 2002, 2002, 564–6.
[39]   Horn H, Privora M, Weuffen W. In: Weuffen W. (Ed). Handbuch der Desinfektion und Sterilisation. Volume I, Chapter 7: Halogene und Halogenverbindungen, VEB Verlag Volk und Gesundheit, Berlin, 1972.
[40]   Apostolov K. J Hyg, 1980, 84, 381–8.
[41]   Newbould FHS, Barnum DA. J Milk Food Technol, 1958, 21, 348–9.
[42]   Boddie RL, Nickerson SC. J Dairy Sci, 1997, 80, 1846–50.
[43]   Anderson DE, Sivalingam V, Kang AEZ, Ananthanarayanan A, Arumugam H, Jenkins TM, Hadjiat Y, Eggers M. Infectious Diseases and Therapy, 2020, 9, 669–75.
[44]   Vasudevan P, Tandon M. J Sci Ind Res, 2010, 69, 376–83.
[45]   Punyani S, Narayana P, Singh H, Vasudevan P. J Sci Ind Res, 2006, 65, 116–20.
[46]   Bevan R, Nocker A, Asami M, Bhat V, Boisson S, Brown J, Calderon E, Callan P, Cotruvo J, Cunliffe D, D'Anglada L, de Roda Husman AM, Eckhardt A, Fawell J, Fewtrell L, Gerba C, Hunter P, MacAulay J, Majuru B, Marsden P, Medema G, Meek B, Montgomery M, Nam Ong C, Ramasamy S, Regli S, Robertson W, Rogers L, Snyder S, Sobsey M, Wahabi R, Weisman R. Iodine as a Drinking-water Disinfectant, World Health Organization, 2018. ISBN 978-92-4-151369-2.
[47]   Kaiho T. ARKIVOC, 2021, 2021, 66–78.
[48]   Messier P. U.S. patent, 8091551B2, USA, 2012.
[49]   Bourget S, Ohayon D, Tanelli J, Gendron AM, Messier PJ. Microbiocidal filtering media for individual protection: Facemasks and canisters. 8th International Symposium on Protection against Chemical and Biological Warfare Agents, Gothenburg, Sweden, 2004.
[50]   Ohayon D, Bourget S, Gendron AM, Tanelli J, Low K, Messier PJ. Triosyn technology on the breakdown of chemical warfare agents and industrial chemicals in the workplace. The 9th Symposium on Iodine Utilization, Chiba University, Chiba, Japan, 2006.

13.1.4 Copper-based biocides

Rainer Pöttgen, Thomas Fickenscher

Copper is essential in many biochemical processes and thus a trace element for human beings: Our body contains approximately 3 mg kg^{-1}, along with a daily uptake and delivery of 3–5 mg [1, 2]. This also holds for higher animals, whereas copper solutions are poisonous for lower organisms, including algae, bacteria, or fungi. We use this antimicrobial property daily by employing copper coins, brass doorknobs, copper water pipes or copper flower vases. The present chapter deals with various copper salts that are used in agriculture as fungicides. Many of such fungicides are still on the market, even though with strictly limited admission ranges.

13.1.4.1 Copper corrosion, patina and *Grünspan*

We start herein with the simple copper corrosion on copper sheets (copper roofing or gutters). A fresh copper surface reacts with oxygen from the air within the first few hours and forms an almost transparent film of Cu_2O of 2–4 μm thickness. With ongoing oxidation, copper loses its luster and becomes dark brownish. The oxidized copper surface can then react with reactive species of the air, i.e. H_2O, CO_2 (formation of carbonate), SO_2 (formation of sulfites/sulfates) and with chlorides near the sea, forming the so-called natural copper patina. This patina is always a complex mixture of copper carbonate/sulfate/chloride/hydroxide, and its color can be dull green, turquoise or green blue. As an example, we present the copper roofing and an entrance door of the Münster cathedral in Figure 13.1.4.1.

These patina surfaces are closely related to the secondary minerals azurite (Cu_3 $(OH)_2(CO_3)_2 \equiv 2CuCO_3 \cdot Cu(OH)_2$), malachite ($Cu_2(OH)_2CO_3 \equiv CuCO_3 \cdot Cu(OH)_2$), chalcanthite ($CuSO_4 \cdot 5H_2O$) and atacamite ($Cu_2(OH)_3Cl \equiv Cu(OH)_2 \cdot Cu(OH)Cl$) (atacamite is a biomineral resulting from jaw hardening [$Cu_2(OH)_3Cl$] fibers as found in marine blood worms, e.g. *Glycera dibranchiate* [3]).

Color changes (from brownish to golden) can be achieved by alloying. Typical materials are brass, bronze (Chapter 2.4) or alloys with aluminum and/or zinc [4]. Corrosion leads to different patina colors. The insoluble patina is an excellent surface passivation layer, which additionally serves as a fungicide and keeping the surface homogeneously clean.

One should not confound patina with the so-called *Grünspan* (*verdigris*). The latter is the trivial name for copper acetate monohydrate, the mineral hoganite, Cu $(CH_3COO)_2 \cdot H_2O$. *Grünspan* forms when vinegar-containing food is stored in copper vessels. The approximate composition of this corrosion product is $Cu(CH_3COO)_2 \cdot 3Cu(OH)_2 \cdot 2H_2O$. This observation led to the use of copper acetate as a fungicide. The industrial synthesis of copper acetate proceeds with copper(II) oxide (CuO + 2 CH_3COOH

https://doi.org/10.1515/9783110733471-024

Figure 13.1.4.1: An entrance door and the copper roofing of St.-Paulus Dom in Münster (Westfalen), Germany.

→ $Cu(CH_3COO)_2 + H_2O$) or basic copper carbonate ($CuCO_3 \cdot Cu(OH)_2$): $CuCO_3 \cdot Cu(OH)_2 + 4\ CH_3COOH \rightarrow 2\ Cu(CH_3COO)_2 + 2\ H_2O + CO_2$.

13.1.4.2 *Bouillie bordelaise* (Bordeaux mixture)

One of the oldest broadly used fungicides is *bouillie bordelaise* (Bordeaux mixture). It is a mixture of $CuSO_4 \cdot 5H_2O$ and slaked lime (calcium hydroxide, $Ca(OH)_2$). Different premixed mixtures with varying copper sulfate/calcium hydroxide ratios are commercially available. Also, diverse recipes for custom mixings can be found on the internet.

The Bordeaux mixture was invented in the Haut-Médoc vineyards in the Bordeaux region, France, by the end of the nineteenth century and used most significantly for the protection (preventive treatment) of the grapes primarily from powdery mildew but also from other fungi. The large success of the Bordeaux mixture in the vineyards led then to its use in gardening and in fruit farms. The mixture is sprayed on the whole plant, leaving a stable blueish coating.

In former times, the Bordeaux mixture was directly prepared in the vineyards. The winegrowers used small huts with concrete basins for dispersion of the salt mixture (Figure 13.1.4.2) and repeatedly treated the vines during the vegetation cycle.

13.1.4.3 Copper salts in crop protection

The high effectiveness of the Bordeaux mixture led to the development of other copper-based crop protection products [5–8]. They are broadly used in the cultivation of apples, potatoes, grapes, hops, coffee berries and other vegetables. In the 1960s, a huge number of different copper-based formulations were available [9]. Today's relevant salts are copper(II) hydroxide ($Cu(OH)_2$), basic copper carbonate ($Cu(OH)_2$:$CuCO_3$ 1:1), copper(II) oxychloride ($CuCl_2 \cdot 3Cu(OH)_2$), tribasic copper sulfate ($CuSO_4 \cdot 3Cu(OH)_2 \cdot xH_2O$)

Figure 13.1.4.2: A concrete basin for the preparation of *bouillie bordelaise* (Bordeaux mixture) in a vineyard in the Cerdon Valley (Bugey wine region) in Leymiat, France.

and copper(II) octanoate ($C_{16}H_{30}CuO_4$), which act as contact fungicides. Depending on the company, the copper-based products are sold under different trade names. Representative samples are presented in Figure 13.1.4.3.

In the beginnings of copper-based crop protection, the amount of copper used per hectare was large. The current formulations are more effective, and the amount of copper used per hectare has been drastically reduced meanwhile. Important parameters for an effective fungicidal impact are: (i) a homogeneous distribution of the copper salt along with a permanent contact of the product crystals on the treated surface and (ii) a sufficient rain-fastness. The copper-based fungicides are

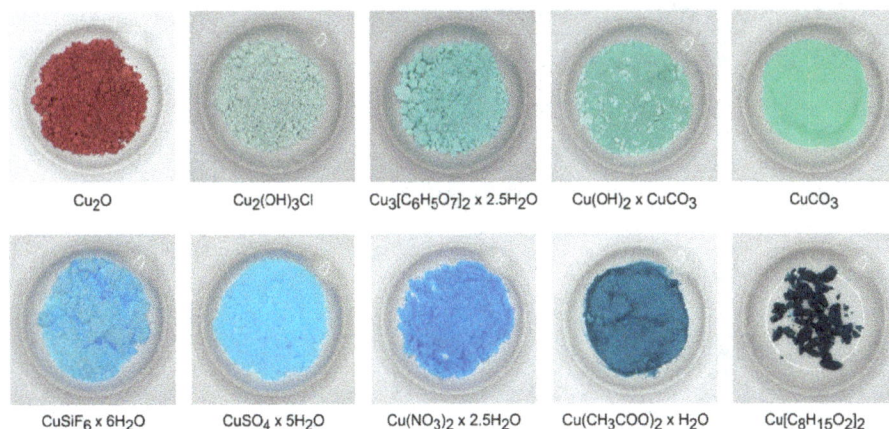

Cu_2O	$Cu_2(OH)_3Cl$	$Cu_3[C_6H_5O_7]_2 \times 2.5H_2O$	$Cu(OH)_2 \times CuCO_3$	$CuCO_3$

$CuSiF_6 \times 6H_2O$	$CuSO_4 \times 5H_2O$	$Cu(NO_3)_2 \times 2.5H_2O$	$Cu(CH_3COO)_2 \times H_2O$	$Cu[C_8H_{15}O_2]_2$

Figure 13.1.4.3: The main copper salts used in applications as fungicides. The samples of copper nitrate and copper octanoate (from an evaporated aqueous solution) were kindly provided by the Eimermacher group and Neudorff, respectively.

available as WP (wettable powder), WG (wettable granules) and SC (suspension concentrate) formulation. It is worthwhile to note that these copper-based crop protections are even used in ecological (bioorganic) agriculture!

13.1.4.4 Wood protection

The impregnation of wood with fungicidal inorganic salts is meanwhile well established for 200 years. The first technical procedure was the so-called kyanization, the surface conservation of wood with diluted $HgCl_2$ solution. This technique was utilized for telegraph poles, railway sleepers, fence posts or vine stakes (any application requesting a long lifetime of the wood); however, due to the toxicity of mercury salts, this conservation technique was abandoned already many decades ago. Later, the same wood products were conserved with chromium salts and/or arsenates. Again, due to toxicity concerns, also this technique is no longer used.

Nowadays, the vacuum/pressure impregnation technique is broadly used for preserving wood (fence posts and bridle gates, cladding timbers, decking, etc.). The active copper salts in this process are copper(II) hydroxide carbonate 1:1 (so-called basic copper carbonate $CuCO_3 \cdot Cu(OH)_2$ 1:1), copper hydroxide ($Cu(OH)_2$), copper(II) hexafluorosilicate ($CuSiF_6$; several hydrates are commercially available) and bis-(N-cyclohexyldiazeniumdioxy)-copper (so-called copper HDO or bis-copper). Depending on the process used, further additives might be 2-aminoethanol, boric acid or didecylpolyoxethylammoniumborate (technical quality). This treatment provides fungicidal, insecticidal and bactericidal protection of wooden construction products with a main prevention of fungal infestation. A copper salt impregnated fence post is shown in Figure 13.1.4.4.

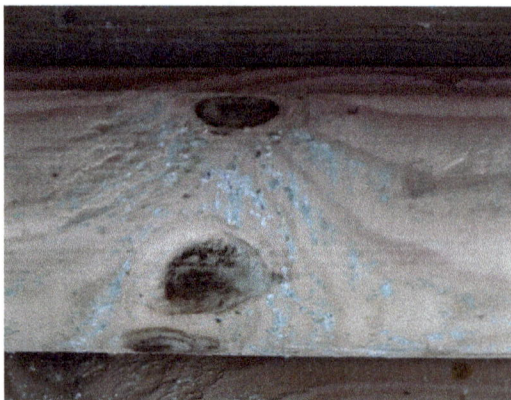

Figure 13.1.4.4: A pressure impregnated fence post. The typical efflorescence of residual blue green copper salts is visible.

13.1.4.5 Further applications of copper salts

Copper sulfate pentahydrate ($CuSO_4 \cdot 5H_2O$) and copper citrate ($C_{12}H_{10}Cu_3O_{14}$) find application for the reduction of sulfide aromas (the so-called *Böckser*) in wine (elimination of sulfide ions through hardly soluble CuS) [10].

Copper nitrate ($Cu(NO_3)_2$) in sulfuric acid solution is used for hoof trimming/claw hardening for cows. Foot rot (hoof infection, so-called *Moderhinke*) of sheeps is treated with copper sulfate solution.

Copper salts play an important role as pigments for antifouling coatings / colors (protection against biological growth) for (i) ship hulls for sea and fresh water use, (ii) for buoys, (iii) for immersed structures or (iv) aquaculture nets. Copper(I) oxide (Cu_2O, the reddish painting of hulls of container ships), copper pyrithione ($C_{10}H_8CuN_2O_2S_2$, the copper salt of pyridine-2-thiol-1-oxide) and copper thiocyanate (CuSCN) are the active components. The technical synthesis of copper thiocyanate proceeds either via synproportionation ($CuSO_4 + Cu + 2\ KSCN \rightarrow 2\ CuSCN + K_2SO_4$) or salt metathesis ($CuCl + KSCN \rightarrow CuSCN + KCl$; driven by the higher lattice energy of potassium chloride). For a broader discussion of this topic, we refer to Chapter 13.1.5, which concentrates on technical antifouling and antimicrobial coatings.

Finally, we shortly comment on fertilizers (for more details see Chapter 6.6), additives in feeding stuff and nutritional supplements. As stated at the beginning, copper is an essential trace element and there is a delicate interplay between deficiency and toxicity [11]; for example the reaction of copper aerosols from copper smelters with sulfur oxides [12]. In this sense, copper(II) hydroxide ($Cu(OH)_2$), copper(I) oxide (Cu_2O), basic copper(II) carbonate ($CuCO_3 \cdot Cu(OH)_2$), copper(II) diacetate monohydrate ($Cu(CH_3COO)_2 \cdot H_2O$), copper(II) chloride dihydrate ($CuCl_2 \cdot 2H_2O$), copper(II) oxide (CuO), copper(II) sulfate pentahydrate ($CuSO_4 \cdot 5H_2O$), copper(II) chelates with amino acids hydrate and copper(II) chelates with protein hydrolysates are used.

References

[1] Marquardt H, Schäfer SG (Eds). Lehrbuch der Toxikologie, BI Wissenschaftsverlag, Mannheim, 1994.

[2] Holleman AF, Wiberg N. Anorganische Chemie, 103. Auflage, De Gruyter, Berlin, 2016. ISBN 978-3-11-051854-2.

[3] Chen LC, Peoples SM, McCarthy JF, Amdur MO. Atmos Environ, 1989, 23, 149–54.

[4] Homepage of Deutsches Kupferinstitut Berufsverband e.V., Heinrichstraße 24, 40239 Düsseldorf, Germany, https://www.kupferinstitut.de/

[5] Homepage of Julius Kühn-Institut, Bundesforschungsinstitut für Kulturpflanzen (JKI), Institut für Strategien und Folgenabschätzung im Pflanzenschutz, Stahnsdorfer Damm 81, 14532 Kleinmachnow, Germany, https://kupfer.julius-kuehn.de/

[6] Kühne S, Roßberg D, Röhrig P, von Mering F, Weihrauch F, Kanthak S, Kienzle J, Patzwahl W, Reiners E. J Kulturpflanzen, 2016, 68, 189–96. DOI: 10.5073/JFK.2016.07.01.

[7] Wilbois K-P, Kauer R, Fader B, Kienzle J, Haug P, Fritzsche-Martin A, Drescher N, Reiners E, Röhrig P. J Kulturpflanzen, 2009, 61, 140–52.

[8] Kühne S, Roßberg D, Röhrig P, von Mering F, Weihrauch F, Kanthak S, Kienzle J, Patzwahl W, Reiners E, Gitzel J. Org Farm, 2017, 3, 66–75. DOI: 10.12924/of2017.03010066.

[9] Rebschutzmittel Verzeichnis, Biologische Bundesanstalt für Land- und Forstwirtschaft in Braunschweig, Merkblatt Nr. 4 (17. Auflage), Februar 1962. www.openagrar.de/merkblatt_004_auflage_17_1962

[10] Renner H, Pour Nikfardjam M. Der Winzer, 2017, 34–8.

[11] López-Alonso M, Miranda M. Animals, 2020, 10, 1890. DOI: 10.3390/ani10101890.

[12] Lichtenegger HC, Schöberl T, Bartl MH, Waite H, Stucky G. Science, 2002, 298, 389–92.

13.1.5 Antifouling coatings

Thomas Schupp

13.1.5.1 Introduction

Fouling is a process where surfaces are colonized by organisms. In a narrower sense, fouling defines the colonization of submerged surfaces by aquatic organisms, where the colonization of hulls of boots and ships has the most commercial impact and raises greatest technological interest.

The fouling progresses in several steps, a succession of organisms in the buildup of the ecosystem [1]. In the first phase, dispersed organic matter adheres to the surface of the boat. This organic matter is made up by extra-cellular biomolecules (polysaccharides, lipids, proteins) and leftovers of biodegradation processes. The buildup of this first, very thin layer occurs immediately after immersion. It forms the basis for the first colonizers, bacteria and diatoms, which start colonization within hours. Within a week, secondary colonizers are observable like macroalgae and protozoa. The tertiary colonizers start to appear within three weeks, and they are represented by crustacea, mollusks and others. These organisms excrete an extracellular matrix which crosslinks on and with the boat hull which makes removal very difficult (Figure 13.1.5.1). Fouling increases the weight of the boat and friction to water, so an increase in fuel consumption of up to 40% may result [1].

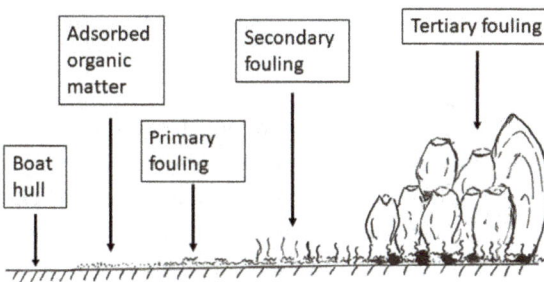

Figure 13.1.5.1: Stages of fouling. Dissolved organic matter in the water adsorbs at the boat hull immediately. Bacteria and diatoms follow within minutes to hours and can form biofilms by cross-linking extracellular matrix. Macroalgae are an important part of secondary fouling, and barnacles and shellfish/bilvalves are examples for organisms of the tertiary fouling (sketched are several *Balanus amphitriti* and one *Mytilus edilis*).

https://doi.org/10.1515/9783110733471-025

13.1.5.2 Some history

Fouling on ships is reported since ancient times, and excellent summaries can be found e.g. in [1, 2]. Here is a brief summary.

Aristotle reported about the Echeneis, who was claimed to be able to slow down or even stop ships (*Remora brachyptera*, "ship stopper" in English, "Schiffshalter" in German). The back fin is mutated to a plate which allows the fish to adhere on larger fish like sharks (and ships) and use them as transporters. However, already Plutarch argued that the weeds on the hull are a more likely a reason for the retardation. The Phoenicians and the Greek recommended to paint the ship hull with tar, asphalt, etc. for protection. It is difficult to disentangle today the prime purpose of this recommendation: tightening the hull, prevent fouling or fight the ship worm, which can pierce the wooden hull so the ship finally sinks. Most likely, the recipes were meant to help in all aspects. Metal sheeting with lead plates was introduced in the late middle ages, but it was noted that these resulted in more severe corrosion of iron parts.

During the eighteenth century, diverse organic coatings like shellac were in use, doped with toxic metal compounds like $CuSO_4$, As_2O_3, or HgO. There was no really satisfying solution; copper sheets were quite helpful as long as hulls could be made of wood, but the steam engines called for iron hulls, which prohibited the use of copper sheets as this would have created a large galvanic element. Therefore, during the nineteenth century, the interest in coatings doped with biocides increased again, while there was a desire to get rid off the toxic and difficult to handle HgO.

13.1.5.3 Coatings with antifouling biocides

Coatings on steel hulls are aimed to prevent corrosion, and biocides in the coating can prevent fouling. For coatings doped with biocides, an as constant as possible release rate is desired [1]. For that reason, the coating should be slowly degraded by seawater as otherwise due to exhaustion of the outer layers and increased distances for diffusion, the release rate can become too small to exert an effect on the target organisms. Compensation of this decreasing release rate by higher initial concentrations is in general no solution as costs for biocides, technical feasibility and toxicity to non-target organisms set limits to this approach. A few examples of biocides that have been used are presented in Table 13.1.5.1. The LC50, EC50 or IC50 values show the average concentration which causes 50% mortality, immobility or growth inhibition in fish, crustacea and algae, respectively, for a given exposure period. Large concentration ranges for some biocides are attributable to different sensitivities within the species and variations in experimental design.

The technical requirements for the biocides are that they are released on time, that is, neither washed out too early nor imbedded too tightly. For optimum processing, they should be soluble in the coating precursors, or at least form emulsions

Table 13.1.5.1: Environmental data for biocides and some precursors; concentrations as µg/L.

Substance	LC50/96 h (fish)	EC50/48 h (crustacea)	IC 50/72 h (algae)	Persistent	Reference[a]
Zinc sulfate	110–2940	155–2910	126–150	not applicable[b]	https://echa.europa.eu/brief-profile/-/brief profile/100.028.904
Zinc pyrithione (zinc-bis(2-sufanyl-pyridinium-N-oxide)	2.6–400	8.2	1.2	no	https://echa.europa.eu/brief-profile/-/brief profile/100.033.324
Ziram (zinc bis(N,N-diethyl-dithiocarbamate)	9.7	48	53–93	no	https://echa.europa.eu/brief-profile/-/brief profile/100.004.808
Pyrithione sodium	2.6–11	11–600	220	no	https://echa.europa.eu/brief-profile/-/brief profile/100.021.179
Sodium diethyl dithiocarbamate	1600–3900	910	1400	no	https://echa.europa.eu/brief-profile/-/brief profile/100.005.192
Tributyltin chloride	2.9	9.8	12.4	yes	https://echa.europa.eu/brief-profile/-/brief profile/100.014.508
Dibutyltin chloride	4000	843	2800	yes	https://echa.europa.eu/brief-profile/-/brief profile/100.010.610
Copper(I) chloride	3–9150	0.5–303	16.5–987	not applicable[b]	https://echa.europa.eu/brief-profile/-/brief profile/100.028.948

[a]Accessed on 10 August 2021.
[b]Zinc and copper are also essential elements.

easily. Finally, they shall be active on target organisms, but not be persistent once released in the environment. For those biocides in Table 13.1.5.1 where there is a "no" entry for persistence, the substance shows rapid biodegradability, rapid photolytic decay and/or rapid hydrolysis in water.

One biocidal additive used since long is Cu_2O; in the presence of oxygenated water, it becomes oxidized and solubilized to the more active form Cu^{2+}.

As can be deduced from data for zinc sulfate, pyrithione sodium and pyrithione zinc, there is obviously some synergism between the metal zinc and its ligand pyrithione.

In the 1960s, tri-alkylated tin compounds were introduced, for example bis(tributyltin)oxide. These substances turned out to be very efficient antifouling agents with a remarkable biocidal activity. Di-alkylated tin compounds are already much less active (Table 13.1.5.1). However, these organotin compounds are persistent in the environment and can accumulate in organisms. For tributyltin (TBT) it was demonstrated that levels as low as about 2 ng/L ($\equiv$0.002 µg/L!) caused imposex in marine snails like *Nupellus lapillus* [3]; TBT inhibits the enzyme aromatase so the snails can no longer convert testosterone in estrogen. Effects on local ecosystems around the harbors became evident, and tri-organotin compounds in antifouling paints were first prohibited for leisure boats in the EU, but finally for ship hulls and other uses in general (ANNEX XVII of Regulation (EC) No. 1907(2006)) [1].

Zinc pyrithione and Ziram (Figure 13.1.5.2) are biocides which are used in polymers, Ziram also in shampoo.

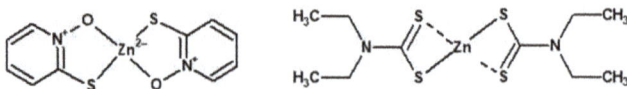

Figure 13.1.5.2: Zinc pyrithione (left) and Ziram (right).

In the European Union, biocides are subject to authorization. Producers and users need to apply for permission for production, import and use, and this permission has to be renewed every 10 years. The aim of this regulation is to have good control over these substances which are designed to kill (or at least control) organisms and which have dispersive use. This shall build up an incentive to develop more targeted biocides with as little impact on non-target organisms as possible [4]; some developments are targeting at biocide reduction by combination with physical effects against fouling (see next chapter).

13.1.5.4 Other, non-biocidal approaches

To synthesize a biocide which is very active against target organisms but harmless for a non-target organism remains being a huge challenge. Other approaches were tried and are still under development.

Hydrophobic surfaces shall retard the colonialization of the hull [1, 5, 6]. Hydrophobicity can be introduced by fluorinated polymers, for example hexafluoro-acetone phenol condensate, reaction product with epichlorohydrine and crosslinked with hexamethylene-diisocyanate oligomers. This coating is capable of wetting polytetrafluoro ethylene (PTFE) which can be dispersed as additive, generating a very-low-energy surface which shows increased resistance against colonization. Methoxy silicon (silicones) is another attempt, but because of abrasion sensitivity they are grafted as loops on polyurethane coatings. These hydrophobic surfaces delay the colonization by tertiary organisms, and cleaning of the hull is comparatively easy before these organisms have settled.

Electric pulses on the steel hull were tried against fouling. Some effect observed was likely attributable to the formation of chlorine, but this caused also problems in terms of corrosion and production of poorly defined, chlorinated organic compounds. Therefore, this technique was not followed up [1].

Hydrophilic surfaces can also prevent colonization by minimizing interface energy [6]. Polyethylene glycol was shown to be effective, but suffers from limited environmental stability.

Other attempts try to mimic the shark or the whale skin [6], as these organisms are not prone to fouling. It turned out that the special structure of the skin is responsible for the antifouling effect. Bacteria and protozoa – the first colonizers – have difficulties to attach to structures which are smaller in size than these organisms. Nanosized particles embedded in the polymer coating may perhaps generate a similar effect, and superhydrophobic nanoscale TiO_2 showed initial promising results. Together with probably synergistic copper, photoactive TiO_2 nanoparticles deposited on copper showed strong antifouling activity under sunlight irradiation [7].

13.1.5.5 Conclusions

In times of worldwide trade, fouling is still an issue of high cost and presenting technical challenges for research and development. To keep away the target organisms is not only an economical question; in terms of invasive species, good antifouling precautions could as well help to control this issue. However, repellence of target organisms must not be paid off by inacceptable damage to non-target organisms and ecosystems. As a result, the very efficient tri-organotin compounds had to be replaced. Emerging technologies (nanotechnology, photo-chemistry, etc.) provide a basis to face the challenges.

References

[1]   Yebra D, Kiil S, Dam-Johansen K. Antifouling technology – past, present and future steps towards efficient and environmentally friendly antifouling coatings. Prog Org Coat, 2004, 50, 75–104.

[2]   U.S. Naval Institute, Annapolis, Maryland. (George Banta publishiung company, Menasha, WI 1952). Marine fouling and its prevention. Chapter 11: The history of the prevention of fouling. Contribution N. 580 from the woods hole oceanographic institute: https://darchive.mblwhoilibrary.org/bitstream/handle/1912/191/chapter%2011.pdf?sequence=20, download 14. 09.2021

[3]   Bettin C, Oehlmann J, Stroben E. TBT-induced imposex in marine neogastropods is mediated by an increasing androgen level. Helgoländer Meeresunters, 1996, 50, 299–317.

[4]   Kyei S, Darko G, Akaranta O. Chemistry and application of emerging ecofriendly antifouling paints: A review. J Coat Technol Res, 2020, 17, 315–32.

[5]   Braay RF. Clean hulls without poisons: Devising and testing nontoxic marine coatings. J Coat Technol Res, 2000, 72, 45–56.

[6]   Tian L, Yin Y, Bing W, Jin E. Antifouling technology trends in marine environmental protection. J Bionic Eng, 2021, 18, 239–63.

[7]   Liu H, Raza A, Aili A, Lu J, Al-Ghaferi A, Zhang T. Sunlight-sensitive anti-fouling nanostructured TiO_2 coated Cu meshes for ultrafast oily water treatment. Nature Sci Rep, 2016, 6, 25414.

13.2 Radiotracers for diagnostic imaging

Andreas Faust

Since the first publication of a medical X-ray image by the physicist Wilhelm Conrad Röntgen displaying his wife's left hand in 1895, and the following discovery of radioactive elements by Henri Becquerel, Marie and Pierre Curie a few years later, medical imaging has become one of the most important diagnostic techniques [1, 3]. In 1923, Georg Karl von Hevesy investigated the absorption and translocation of ^{212}Pb in plants (horse beans) and later established the radiotracer principle, which was honored with the Nobel Prize in Chemistry in 1943: "By making use of radioactive indicators we can label atoms (ions), molecules and even larger units [. . .] subsequently, their path and fate in the living organism can be followed" [2]. During the last 50 to 60 years, several imaging techniques have been developed. Today, the most common are PET (positron emission tomography) [4] and SPECT (single photon emission computed tomography) [5] for diagnostic imaging, based on the radiotracer principle by Hevesy; they are thematically domiciled in nuclear medicine [6]. The decision whether a radionuclide is dangerous or beneficial for the patient depends on the dose of radiation measured in Sievert (Sv). Due to the necessity of only very low concentrations (picomolar) when used for diagnostic applications and the short half-lives of the administered radionuclides, the overall radiation dose is comparable to CT-measurements (Computed Tomography).

13.2.1 Radiotracer principle

A significant part of the highly interdisciplinary field involving nuclear medicine is represented by radiochemistry, or, more specifically, radiopharmaceutical chemistry, which provides appropriate radionuclides or radioisotopes and develops new radiolabeled drugs for diagnostic and therapeutic (see 13.3.) applications. To have an international consensus on nomenclature in radiopharmaceutical chemistry a working group was established in 2015 defining standardized rules pertinent to this field [7]. The following chapters contemplate this consensus. For diagnostic imaging, β^+ emitters are used for PET (where annihilation of β^+ particles with electrons yields two entangled γ-photons of 511 keV, Figure 13.2.1), whereas pure γ-emitters for SPECT are needed. Hence, the effective half-lives (combination of physical and biological half-life) have to be long enough to address the medical question but not too long in order to minimize the overall radiation dose administered to the patient. Only a few radionuclides fulfill these requirements (Table 13.2.1).

A radiotracer consists of a drug having a certain affinity for the biological target of interest, as well as a radioactive label, which is covalently bound or installed *via* labeled chelators. The biological target could be a receptor, an enzyme or a transport

https://doi.org/10.1515/9783110733471-026

PET-detector

$p \rightarrow n + e^+ (\beta^+ \text{-emission})$

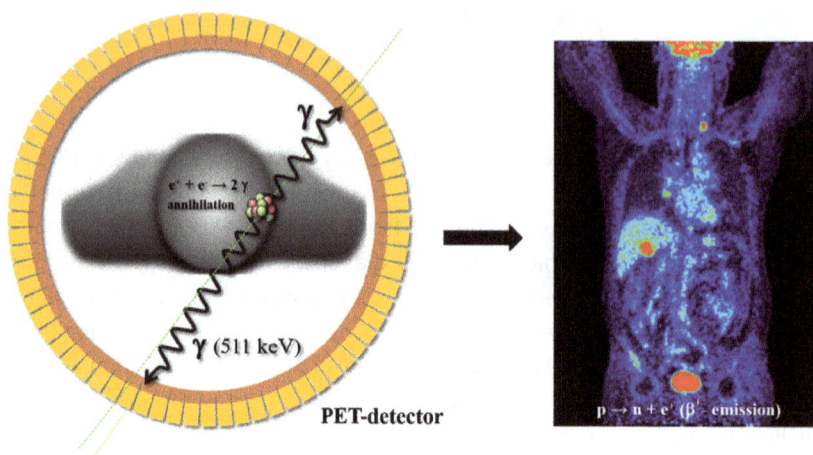

Figure 13.2.1: After injection of a positron-emitter into a patient, the emitted β^+-particles annihilate with electrons from the surrounding and resulting two entangled γ-photons of 511 keV, which can be detected by a PET-detector ring in a human PET-scanner. After mathematical reconstruction of the acquired data, three-dimensional distribution images of the radiotracer are provided for the medical assessment. © European Institute for Molecular Imaging (EIMI), University of Münster, Germany.

Table 13.2.1: Prominent radionuclides used in nuclear medicine (not only radiometals), their decay parameters, methods of production and application [1].

radio-nuclide	half-life	decay		method of production	application
^{11}C	20.4 min	β^+	cyclotron	^{10}B(d,n)^{11}C $\rightarrow$ ^{11}B ^{11}B(p,n)^{11}C $\rightarrow$ ^{11}B ^{14}N(p,α)^{11}C $\rightarrow$ ^{11}B	PET
^{13}N	9.96 min	β^+	cyclotron	^{12}C(d,n)^{13}N $\rightarrow$ ^{13}C ^{13}C(p,n)^{13}N $\rightarrow$ ^{13}C ^{16}O(p,α)^{13}N $\rightarrow$ ^{13}C	PET
^{15}O	2.03 min	β^+	cyclotron	^{14}N(d,n)^{15}O $\rightarrow$ ^{15}N ^{15}N(p,n)^{15}O $\rightarrow$ ^{15}N ^{16}O(p,pn)^{15}O $\rightarrow$ ^{15}N	PET
^{18}F	109.7 min	β^+	cyclotron	^{18}O(p,n)^{18}F $\rightarrow$ ^{18}O ^{20}Ne(d,α)^{18}F $\rightarrow$ ^{18}O	PET
^{64}Cu	12.4 h	β^+, β^-	cyclotron	^{64}Ni(p,n)^{64}Cu $\rightarrow$ ^{64}Ni/^{64}Zn ^{66}Zn(d,α)^{64}Cu $\rightarrow$ ^{64}Ni/^{64}Zn	PET
^{67}Ga	78.3 h	EC, γ	cyclotron	^{68}Zn(p,2n)^{67}Ga $\rightarrow$ ^{64}Ni	SPECT
^{68}Ga	68 min	β^+, γ	generator	^{68}Ge / ^{68}Ga	PET
^{81m}Kr	13.1 s	γ	generator	^{81}Rb / ^{81m}Kr	SPECT

Table 13.2.1 (continued)

radio-nuclide	half-life	decay	method of production		application
^{82}Rb	75 s	β^+, γ	generator	^{82}Sr / ^{82}Rb	PET
^{89}Zr	78.5 h	β^+, γ	cyclotron	^{89}Y(p,n)^{89}Zr $\rightarrow^{89}$Y	PET
^{99m}Tc	6.02 h	γ	generator	^{99}Mo / ^{99m}Tc	SPECT
^{111}In	2.81 d	EC, γ	cyclotron	^{109}Ag(α,2n)^{111}In $\rightarrow^{111}$Cd ^{111}Cd(p,n)^{111}In $\rightarrow^{111}$Cd	SPECT
^{123}I	13.2 h	EC, γ	cyclotron	^{122}Te(d,n)^{123}I $\rightarrow^{123}$Te ^{124}Xe(p,2n)^{123}Cs $\rightarrow^{123}$Xe $\rightarrow$ ^{123}I $\rightarrow^{123}$Te ^{127}I(p,5n)^{123}Xe $\rightarrow$ ^{123}I $\rightarrow^{123}$Te	SPECT
^{124}I	4.15 d	β^+, γ	cyclotron	^{124}Te(p,n)^{124}I $\rightarrow^{124}$Te ^{125}Te(p,2n)^{124}I $\rightarrow^{124}$Te	PET
^{201}Tl	73.1 h	EC, γ	cyclotron	^{203}Tl(p,3n)^{201}Tl $\rightarrow^{201}$Hg	SPECT

system (e.g. a metabolic pathway). To get a good signal to background ratio, the radiotracer should have a good affinity and specificity for the target and a pharmacokinetic profile ensuring a suitable enrichment. In this regard, polarity, molecular size, unspecific protein binding and (metabolic) stability are important aspects to be considered while developing new radiotracers.

Mechanisms for *in vivo* enrichment could be, among others, passive transport by diffusion (lung ventilation), perfusion (discrimination between healthy and dead tissue), adsorption effects (e.g. diphosphonate complexes chelating Ca^{2+} in bones), ion transport ([^{99m}Tc][TcO$_4$]$^-$ in the thyroid gland), active transport (e.g. [^{99m}Tc]Tc-MAG$_3$ for kidney function), and metabolic trapping (non-degradable radio-metabolites that accumulate over time upon uptake, e.g [^{18}F]fluorodeoxyglucose ([^{18}F]FDG)), as well as targeting of disease-related overexpressed receptors or enzymes, which are addressed in a highly specific manner. Typically injected activities of radiotracers only involve amounts of substance ranging from nmol down to pmol, which means that the biological system is accompanied, but not affected by the radiotracer; hence, the trace and specific accumulation of radioactivity in the body is followed.

13.2.2 Generator-based radioisotopes

There are two main origins of radioisotopes used in nuclear medicine, namely the generator-based production and the (mainly on-site) use of cyclotrons. In comparison with generator-based radioisotopes, the production with a cyclotron is very expensive and personnel-intensive. Therefore, the generator-based radioisotope technetium-99m became the most important representative in nuclear medicine. Generally, the structure of a radioisotope-generator is very simple. In the middle of a shielded box, a small column is placed, where the parent nuclide is fixed at the top. Over time, it decays and produces the daughter nuclide, which can be washed out by an adequate eluent and further used in coordination-chemical derivatization procedures or organic-chemical syntheses. Regarding technetium-99m, the mother nuclide is molybdenum-99 fixed as the dianionic, hence less mobile, molybdate $[^{99}\text{Mo}][\text{MoO}_4]^{2-}$ adsorbed on an Al_2O_3-column; upon radioactive decay, the more mobile monoanionic pertechnetate species $[^{99\text{m}}\text{Tc}][\text{TcO}_4]^-$ is accumulated, which can be eluted with saline solution (0.9% NaCl in water, see Figure 13.2.2).

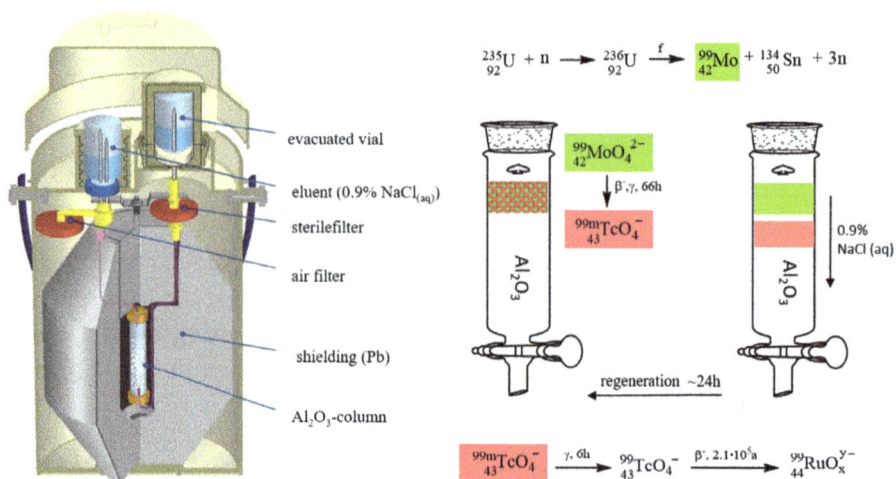

Figure 13.2.2: Left: Schematic view on a $^{99\text{m}}$Tc-generator. Right: reaction scheme of the ^{99}Mo-production in a nuclear reactor. The alumina-fixed $[^{99}\text{Mo}][\text{MoO}_4]^{2-}$ decays to $[^{99\text{m}}\text{Tc}][\text{TcO}_4]^-$ and can be eluted with saline and further processed. After 24 h, an equilibrium between rising and decaying $[^{99\text{m}}\text{Tc}][\text{TcO}_4]^-$ is reached yielding the highest possible activity concentration for medical use, which is optimal for the daily clinical routine. Schematic picture: adapted with permission of the National Centre for Nuclear Research, Radioisotope Centre POLATOM, Poland.

How did technetium, a biochemically irrelevant metal, become so important over time? The parent nuclide molybdenum-99 is produced by nuclear fission of uranium-235 and is therefore readily available from commercial reactors used in electricity power or research plants. By releasing electrons with a half-life of 66 h, it generates technetium-99m, which itself decays with a half-life of 6 h upon emission of γ-photons (E_γ = 141 keV). For the SPECT-cameras based on the frequently used NaI(Tl) or $\text{Cd}_{1-x}\text{Zn}_x\text{Te}$-detectors, this is an optimal energy range for high resolution imaging (spatial resolution 1–2 mm), while the radiation dose for the patient is comparatively low [6]. Technetium-99m represents a metastable excited state and forms technetium-99 by isomeric transition, a radioactive decay releasing γ-photons. With a half-life of 211000 years, technetium-99 (also named technetium-99 g with "g" for "ground state") is nearly stable for practical purposes; at the usually low concentrations employed, it cannot be detected above the natural background activity permanently available everywhere (see also 13.2.4).

Another important generator-based nuclide is gallium-68 (half-life: 68 min.), which is used as a positron emitter for PET. Here, the parent nuclide is germanium-68 (half-life: 271 d) fixed on a modified silica column, which is eluted with 0.05 M HCl while collecting $^{68}Ga^{3+}$ as the complex anion tetrachloridogallate $[^{68}Ga][GaCl_4]^-$. For subsequent chelation, $^{68}Ga^{3+}$ is mostly used under mild acidic conditions (pH 3–4) that are ensured by buffering with citrate or HEPES (4-(2-hydroxyethyl)-1-piperazine-ethanesulfonic acid). These weakly coordinating buffers protect Ga^{3+} from immediate precipitation as colloidal $[^{68}Ga]Ga(OH)_3$. This is important for the subsequent ligand exchange and chelation process to yield the desired ^{68}Ga-complexes; this procedure is mainly carried out between mildly acidic and up to neutral pH-range, often at higher temperatures [8].

13.2.3 Cyclotron-based radioisotopes

At many research and therapy centers (e.g. university hospitals), cyclotrons for positron emitting radioisotopes are available. They can accelerate different charged particles, such as protons, deuterons, and α-particles at different energies to enable the production of a wide variety of radioisotopes. Compared to nuclear reactors, the daily on-demand availability of these radioisotopes renders them an important cornerstone in nuclear medicine. Small cyclotrons with a proton energy below 20 MeV are widely distributed as in-house production facilities providing the routinely used short-lived positron emitters. These are mainly fluorine-18 and carbon-11, but also radiometals like zirconium-89 are produced. Higher proton energy cyclotrons are usually run in academic research institutes or at commercial facilities to produce intermediate half-life SPECT and PET radioisotopes as well as radioisotopes for therapeutic applications and parent nuclides that are loaded onto generator systems [9]. In most cases, an ion source is placed at the center of the cyclotron, as in the case of the production of high-energy protons. Hence, hydrogen gas forms a plasma upon heating, and the statistically formed hydrogen anions (H^-) are forced onto circular paths by a homogeneous magnetic field. For acceleration between the hollow "D"-shaped sheet metal electrodes ("dees"), an oscillating electric field (72 MHz for the Eclipse RD Cyclotron by Siemens) is applied. As a result, the protons in the gap between the two electrodes are accelerated inside the vacuum chamber (Figure 13.2.3). At the exit point in the periphery, a carbon foil strips the electrons from the hydrogen anions, transforming them into positively charged high-energy protons [10]. The target can be a gas, a liquid or a solid. For instance, for the production of zirconium-89, a solid target consisting of a natural yttrium-89 foil is often used to generate zirconium-89 in a

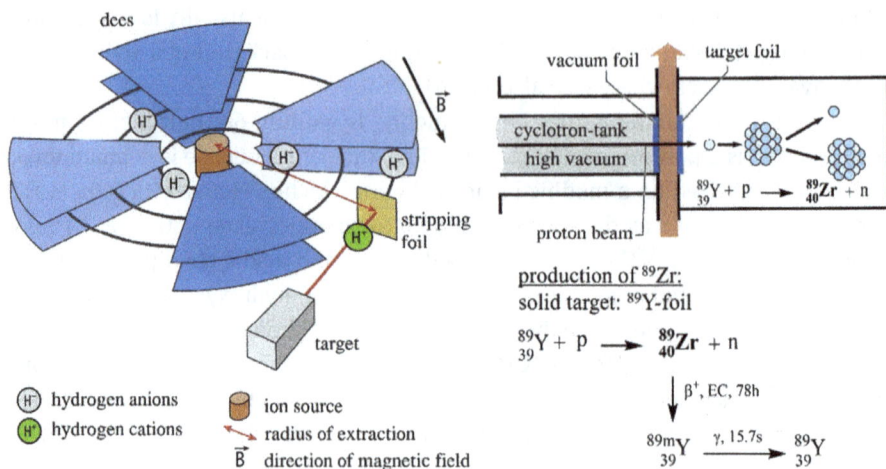

Figure 13.2.3: Functional insight into a cyclotron typically used on-site in university hospitals. Normally, fluorine-18 is produced with $[^{18}O]H_2O$ as a liquid target. For the production of zirconium-89, a solid yttrium-89-target is needed, which provides zirconium-89 after bombardment with high energy protons (modified from [1] with permission of Thieme-Verlag, Stuttgart, Germany).

p,n-nuclear reaction; after purification, finally, $[^{89}Zr]Zr$-oxalate is obtained, which can be transformed to other Zr-complexes (see 13.2.7).

13.2.4 Technetium-99m for SPECT

The highly advantageous decay properties of technetium-99m coupled with the availability of a practical generator system along with the facile radiolabeling are the main reasons for the high variety of ^{99m}Tc-labeled radiopharmaceuticals used for many diagnostic targets. Its position at the center of the transition metal block in the periodic table of the elements provides a wide range of oxidation states (−1 to 7) and diverse coordination numbers (mostly 4–7) [11].

The generator-produced $[^{99m}Tc][TcO_4]^-$ is commonly used for thyroid gland scintigraphy. The specific molecular pathway targeted therein is represented by the sodium/iodide symporter (NIS) located on the basolateral membrane of the thyroid follicular cell and transports beside iodide, also the monoanionic $[^{99m}Tc]TcO_4]^-$ due to its similar anionic radius and charge density [12]. For all other diagnostic radiotracers, pertechnetate must be reduced to form other complexes or chelates. In nuclear medicine, commercially available and legally approved preparation kits are used. In an evacuated flask equipped with a reducing agent (normally Sn(II) salts) and incorporating the corresponding ligand to be inserted, the aqueous pertechnetate solution from the generator is placed and heated following a defined protocol. After quality control, it is used for injection as obtained (Figure 13.2.4)

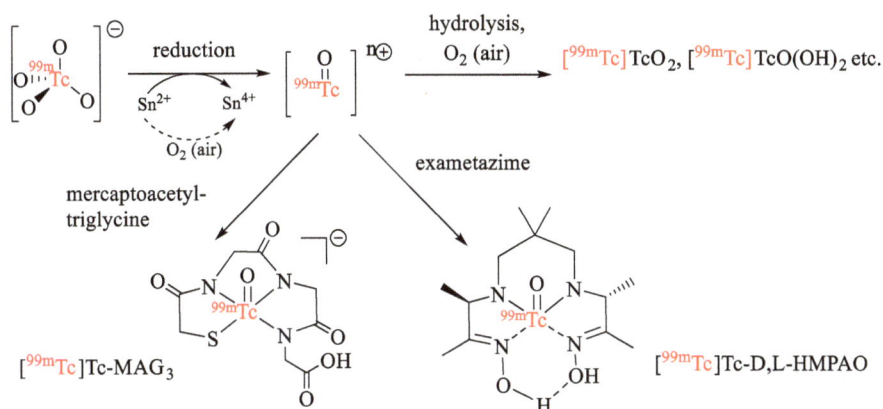

Figure 13.2.4: General synthesis protocol for many ^{99m}Tc-based radiotracers used in nuclear medicine (preparation kit). The first step is the reduction of $[^{99m}Tc][TcO_4]^-$, e.g. with Sn(II) salts, followed by the complexation with different chelators (in the figure: MAG$_3$ and HMPAO). It is important to have evacuated kit-vials to elute the $[^{99m}Tc][TcO_4]^-$ solution from the generator directly into the vial while avoiding oxygen-mediated inactivation of the reducing agent and preventing the production of ^{99m}Tc-oxides/hydroxides (modified from [1] with permission of Thieme-Verlag, Stuttgart, Germany).

In clinical routine, many different ^{99m}Tc-radiotracers are used. The examples described in this chapter are illustrated in Figure 13.2.5. One family are the oxygen-containing complexes with a TcO^{3+}-core, which can be coordinated with N$_4$, N$_x$O$_{4-x}$, N$_x$S$_{4-x}$-donor ligands. One example is $[^{99m}Tc]$Tc-HMPAO, used as an imaging agent for cerebral blood flow (CBF). The neutral complex overcomes the blood brain barrier (BBB) and after enzymatic conversion to a more hydrophilic species, it is trapped while resulting in an amplification of the signal. Frequently, the N$_x$O$_{4-x}$-ligand DTPA (diethylenetriaminepentaacetic acid) is used in ^{99m}Tc-chemistry. The resulting complex is useful to monitor the kidney function by evaluation of glomerular filtration rates. One of the most prominent N$_x$S$_{4-x}$-donor ligand is MAG$_3$ (mercaptoacetyltriglycine). The resulting ^{99m}Tc-complex is also used to evaluate the kidney function, particularly the tubular secretion. Also dithiolato ligands like DMSA (2,3-dimercaptosuccinic acid) form defined and highly stable complexes. $[^{99m}Tc]$Tc-DMSA (imaging of kidney morphology and function) is excreted mostly in an intact form into the bladder [13]. Cationic Tc(V)-dioxido complexes, which are stabilized with N$_4$-ligands, are not used for medical applications. However, when using diphosphanes as ligands, the resulting *trans*-dioxido complex with technetium-99m is readily obtained and clinically used for myocardial imaging ($[^{99m}Tc]$Tc-tetrofosmin, also known as Myoview^{TM}).

Tc-diazenide complexes had their breakthrough towards medical applications with the introduction of carboxylated hydrazinopyridines in the early 1990s; they contain a stable metal-nitrogen involving a multiple bond between the cation and the so-called HYNIC (6-hydrazino-4-nicotinic acid) [14]. This HYNIC-ligand is conjugated

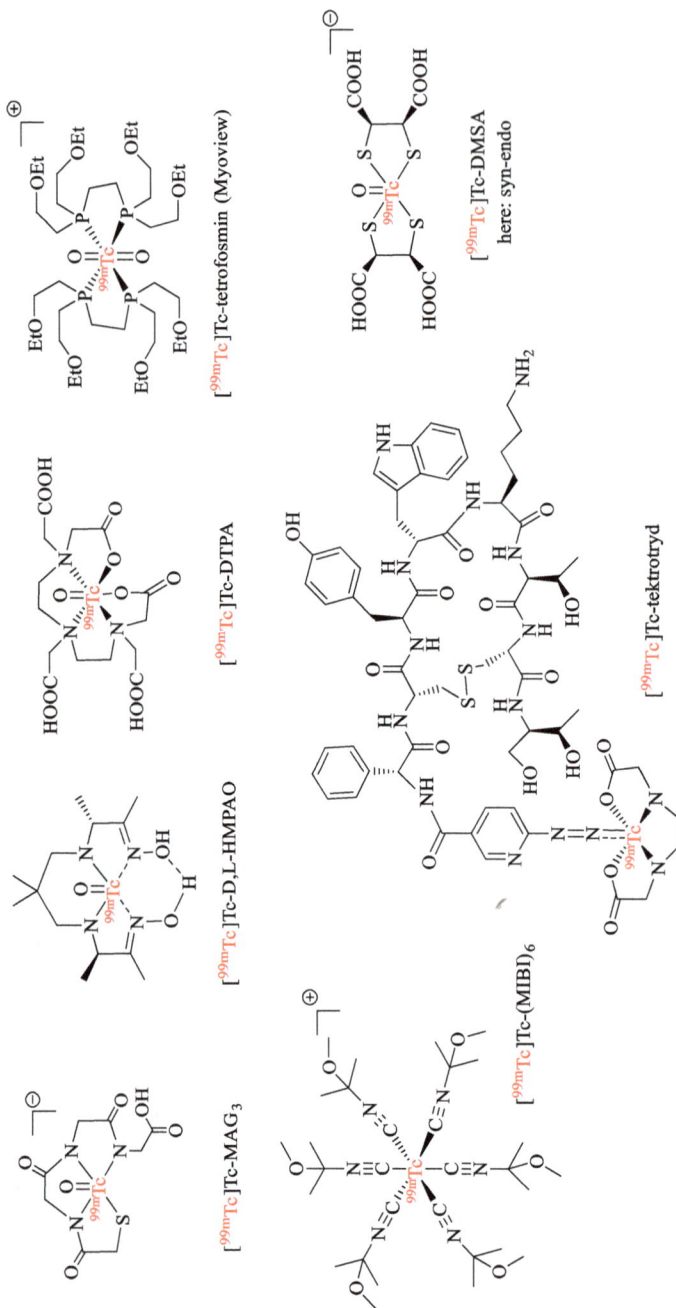

Figure 13.2.5: A small selection of ^{99m}Tc-radiotracers that are used in medical applications and as described in this chapter.

to the targeting molecule (small molecules, peptides or proteins) as an amide *via* the carboxylic acid. This results in a versatile variety of ^{99m}Tc-tracers for imaging, including different tumor classes, infections or apoptotic processes [15]. However, the coordination mode of HYNIC with technetium-99m can be either as a monodentate or as a bidentate ligand. Since hydrazines are reducing agents, it is proposed that a reduction takes place upon coordination with technetium-99m, and therefore HYNIC binds as a diazene (neutral) or as a diazenido (anionic) ligand. Additional co-ligands, such as tricine or EDDA (1,2-ethylenediamine-diacetic acid) complete the coordination sphere. The exact nature of the coordination environment has not been fully elucidated so far [16]. One prominent example of this type of complexes in nuclear medicine is $[^{99m}$Tc]Tc-tektrotyd, an imaging agent addressing the somatostatin receptor (neuroendocrine tumors); it involves a $[^{99m}$Tc]Tc(HYNIC)(EDDA)-complex conjugated to Tyr^3-Octreotide as a targeting cyclic peptide (Figure 13.2.5).

Organometallic technetium(I) complexes for medical application came into account in the early 1980s with the synthesis of the cationic $[^{99m}$Tc]Tc-hexakis(2-methoxy-isobutyl-isocyanide), also known as $[^{99m}$Tc][Tc-(MIBI)$_6$]$^+$. It is extensively used in nuclear cardiology and supplanted $[^{201}$Tl]TlCl for imaging of myocardial perfusion. During the last 20 years, the development of new diagnostic ^{99m}Tc-radiotracers has been mostly driven by the use of Tc-tricarbonyl complexes [17]. They contain three carbon monoxide molecules arranged as monodentate ligands in a facial coordination mode, whereas other donors can occupy the other three coordination sites. The preparation starts with the formation of the precursor *fac*-$[^{99m}$Tc(H$_2$O)$_3$(CO)$_3$]$^+$, which then reacts with ligands that replace the labile aqua ligand. Although there is no Tc-tricarbonyl complex in clinical use yet, its high *in vivo* stability renders it an attractive platform for the development of new ^{99m}Tc-radiotracers [16].

13.2.5 Gallium-68, copper-64 and zirconium-89 for PET

Besides fluorine-18, carbon-11, nitrogen-13 and oxygen-15, which are very common radionuclides in radiopharmaceutical chemistry and PET-diagnostics, radiometals are also in use. The most prominent one is the generator-based radionuclide gallium-68 (see also 13.2.2). Gallium-68 has a well-established coordination chemistry with a short half-life of 68 min, which makes it ideal for rapidly localizing and clearing radiotracers while minimizing patient-absorbed radiation doses. Due to the fact that Ga(III) and the biologically relevant Fe(III) species have similar properties, the use of thermodynamically stable and kinetically inert chelators is required for the proper application of gallium. It is an acidic metal ion with a high affinity for hydroxide anions (pH 4–7), forming the insoluble Ga(OH)$_3$ at pH = 7 and yielding the soluble tetrahydroxidogallate anion in strongly alkaline media. Ga(III) prefers to form complexes with coordination number of 4 to 6 while reaching the highest complex stability with hexadentate chelating agents yielding a distorted octahedral

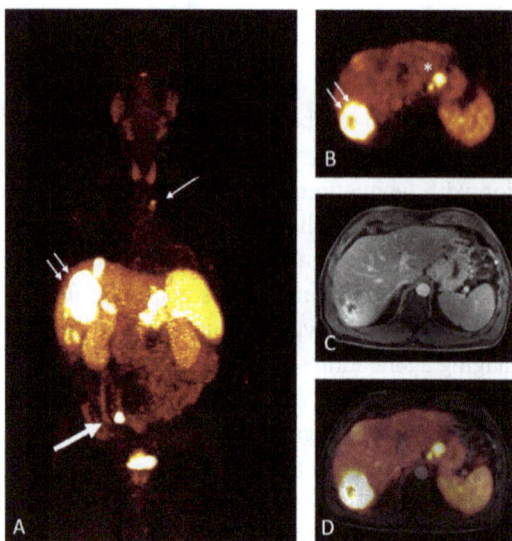

Figure 13.2.6: A: [^{68}Ga]Ga-DOTA-TATE-PET of a 65y old patient with neuroendocrine tumor in the terminal ileum (big arrow), liver metastases (doubled arrow) and bone metastasis (small arrow). B: transversal PET, C: MRI (magnetic resonance imaging) and D: PET-MRI-fusion a [^{68}Ga]Ga-DOTA-TATE positive central necrotic liver metastasis and a small retroperitoneal lymph node metastasis. © Department of Nuclear Medicine, University Hospital, Münster, Germany.

geometry. Hard donor atoms like carboxylates, phenols and hydroxamates as well as amines are preferred by the hard Ga(III) cation [18].

The most prominent ligand is represented by the DOTA-chelator (1,4,7,10-tetraazacyclododecane-1,4,7,10-tetraacetic acid), which is used in PET-radiotracers like [^{68}Ga]Ga-DOTA-TATE (DOTA-Tyr^3-octreotate) or [^{68}Ga]Ga-DOTA-TOC (DOTA-Phe^1-Tyr^3-octreotate) and fulfills the required characteristics to be used in PET-diagnostics and radiotherapy (Figure 13.2.7). However, other chelators like NOTA, NODAGA or TRAP are objects of ongoing developmental steps towards higher yields at lower labeling temperatures and more favorable *in vivo* characteristics, especially if they are conjugated with heat-sensitive targeting units like peptides, nano, affi or antibodies. Nowadays, the most extensively used ^{68}Ga-radiotracers are [^{68}Ga]Ga-DOTA-TATE (used to image the overexpression of the somatostatin receptor, for example in neuroendocrine tumors) and [^{68}Ga]Ga-PSMA-617 or [^{68}Ga]Ga-PSMA-11 (for the imaging of prostate cancer and its metastasis while targeting the prostate-specific membrane antigen PSMA) (Figure 13.2.7). In Figure 13.2.6 is an example of [^{68}Ga]Ga-DOTA-TATE positive metastases with the corresponding PET-MRI-fusion showing the possibilities to detect even very small lesions. Due to their high specificity in the diagnostic setup, the use of yttrium-90 or lutetium-177 for treatment is also possible. Their comparable coordination chemistry results in the same *in vivo*-characteristics and specificity with low side effects (see Chapter 13.3.1).

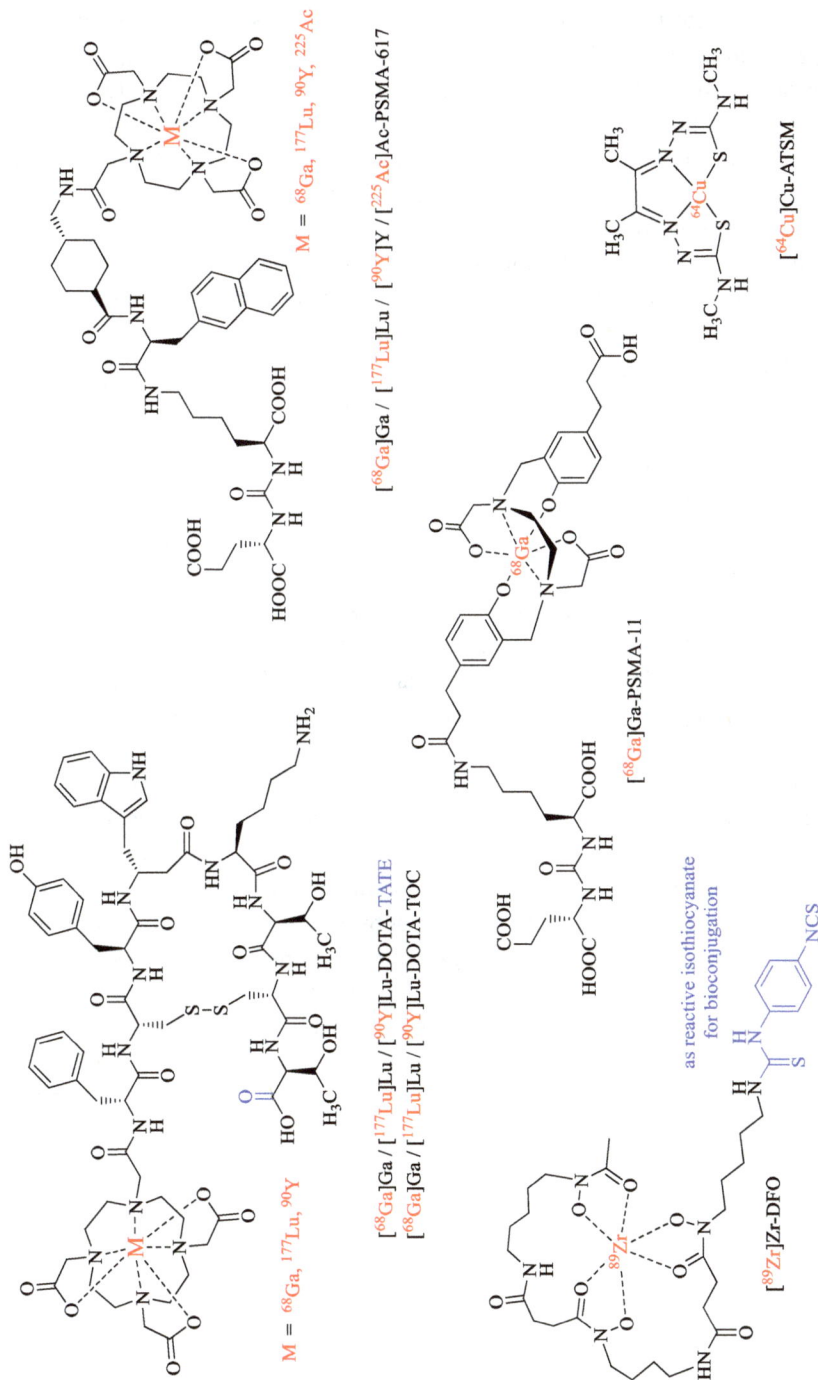

Figure 13.2.7: A selection of radiotracers labeled with radiometals, which are used in medical applications and are described in this chapter. [^{89}Zr]Zr-DFO as reactive isothiocyanate is one example of different reactive groups for labeling peptides, proteins or antibodies.

Copper-64 is not in routine clinical use so far but has become the object of ongoing clinical studies. It is typically used in the oxidation state +II, while the use of hexadentate chelators saturates its coordination sphere to yield a stable and inert complex minimizing interactions with exogenous ligands and biological binders. In comparison to Ga(III), the Cu(II)-cation prefers softer donor ligands, such as amines, imines or thiols [18]. One example is [^{64}Cu]Cu-ATSM (Cu(II)-diacetyl-bis(N^4-methylthiosemicarbazone) as marker of hypoxia) (Figure 13.2.7). It has a high affinity for hypoxic tumor tissues and can therefore be used to measure the degree of oxygen depletion in tumors [19]. The emission of electrons (β$^-$) besides positrons (β$^+$) renders copper-64 also to an isotope for therapeutic use. It enables the diagnostic imaging and therapeutic intervention from the same isotope avoiding the need for a second nuclide as theranostic pair (see ^{68}Ga/^{177}Lu).

Zirconium-89 has become increasingly visible in the literature during the last years. Particularly for the use with antibodies, which have a rather long biological half-life and slow distribution kinetics, it perfectly matches with its intrinsic half-life of 78.5 h. The circulation of targeting antibodies can be monitored for several days up to a few weeks. Although the radiation dose is comparatively high in comparison with other clinically used radioisotopes, there are increased efforts to translate zirconium-89-based radiotracers to the clinic. Here, the additional emission of high-energy γ-photons in combination with its long half-life has to be considered. One major advantage is that it is retained in cells after being internalized (e.g. as ^{89}Zr-antibody-conjugate), leading to irreversible cellular uptake [20]. Zr(IV) is a highly charged and very hard cation preferring hard donors like carboxylates or hydroxamate-oxygen anions. Nearly all ^{89}Zr-radiotracers rely on derivatives of the bacterial siderophore desferrioxamine (DFO), which binds the zirconium-89 with the three hydroxamate moieties in a hexadentate coordination mode (Figure 13.2.7). This ligand is then conjugated *via* amino-reactive groups, e.g. activated esters, iso(thio)cyanates or thiol-reactive maleinimides, to peptides, proteins, or antibodies. The first example was [^{89}Zr]Zr-DFO-zevalin, which was used as a PET-surrogate for [^{90}Y]Y-zevalin (see Chapter 13.3.1). Other antibody conjugates followed, such as [^{89}Zr]Zr-DFO-bevacizumab (VEGF-receptor imaging, overexpressed in many cancers), [^{89}Zr]Zr-DFO-trastuzumab (imaging of breast cancer) or [^{89}Zr]Zr-DFO-cetuximab for the imaging of head, neck and colorectal cancer [18].

References

[1] Dietlein M, Kopka K, Schmidt M. Nuklearmedizin, Basiswissen und klinische Anwendung, Schattauer Verlag, Stuttgart, 2017.

[2] Hevesy G. The absorption and translocation of lead by plants: A contribution to the application of the method of radioactive indicators in the investigation of the change of substance in plants. Biochem J, 1923, 17, 439–45. Hahn LA, von Hevesy, Lundsgard EC. The circulation of phosphorus in the body revealed by application of radioactive phosphorus as indicator. Biochem J, 1937, 31, 1705.

[3] Doi K. Diagnostic imaging over the last 50 years: Research and development in medical imaging science and technology. Phys Med Biol, 2006, 51, R5–28.

[4] Muehllehner G, Karp JS. Positron emission tomography. Phys Med Biol, 2006, 51, R117–38.

[5] Jaszczak RJ. The early years of single photon emission computed tomography (SPECT): An anthology of selected reminiscences. Phys Med Biol, 2006, 51, R99–116.

[6] Schlesinger TE, Toney JE, Yoon H, Lee EY, Brunett BA, Franks L, James RB. Cadmium zinc telluride and its use as a nuclear radiation detector material. Mat Sci Eng, 2001, 32, 103–89.

[7] Coenen HH, Ad G, Adam M, Antoni G, Cutler CS, Fujibayashi Y, Jeong JM, Mach RH, Mindt TL, Pike VW, Windhorst AD. Consensus nomenclature rules for radiopharmaceutical chemistry – setting the record straight. Nucl Med & Biol, 2017, 55, V–XI.

[8] Roesch F, Riss PJ. The renaissance of the ^{68}Ge/^{68}Ga radionuclide generator initiates new developments in ^{68}Ga radiopharmaceutical chemistry. Curr Top Med Chem, 2010, 10, 1633–68.

[9] Boschi A, Martini P, Costa V, Pagnoni A, Uccelli L. Interdisciplinary tasks in the cyclotron production of radiometals for medical applications. The case of ^{47}Sc as example. Molecules, 2019, 24, 444–58.

[10] Cyclotron, from Wikipedia, the free encyclopedia (accessed September 21, 2021, at https://en.wikipedia.org/wiki/Cyclotron.htm.)

[11] Abram U, Alberto R. Technetium and rhenium: Coordination chemistry and nuclear medical applications. J Braz Chem Soc, 2006, 17, 1486–1500.

[12] Singh N, Lewington V. Molecular radiotheragnostics in thyroid disease. Clin Med (London), 2017, 17, 453–7.

[13] Delange M, Piers D, Kosterink J, van Luijk WH, Meijer S, de Zeeuw D, van der Hem GK. Renal handling of technetium-99m DMSA: Evidence for glomerular filtration and peritubular uptake. J Nucl Med, 1989, 30, 1219–23.

[14] Schwartz D, Abrams M, Hauser M, Gaul FE, Larsen SK, Rauh D, Zubieta JA. Preparation of hydrazino-modified proteins and their use for the synthesis of 99mTc-protein conjugates. Bioconj Chem, 1991, 2, 333–6.

[15] Meszaros LK, Dose A, Biagini SCG, Blower PJ. Hydrazinonicotinic acid (HYNIC) – Coordination chemistry and applications in radiopharmaceutical chemistry. Inorg Chim Acta, 2010, 363, 1050–69.

[16] Papagiannopoulou D. Technetium-99m radiochemistry for pharmaceutical applications. J Label Compd Radiopharm, 2017, 60, 502–20.

[17] Alberto R, Schibli R, Egli A, Schubiger PA, Herrmann WA, Artus G, Abram U, Kaden TA. Metal carbonyl syntheses XXII. Low pressure carbonylation of [MOCl$_4$]$^-$ and [MO$_4$]$^-$: The technetium (I) and rhenium(I) complexes [NEt$_4$]$_2$[MCl$_3$(CO)$_3$]. J Organomet Chem, 1995, 493, 119–27.

[18] Long N, Wong WT. The Chemistry of Molecular Imaging, John Wiley & Sons, NJ, USA, 2015.

[19] Pasquali M, Martini P, Shahi A, Jalilian AR, Oss JA Jr, Boschi A. Copper-64 based radiopharmaceuticals for brain tumors and hypoxia imaging. Q J Nucl Med Mol Imag, 2020, 64, 371–81.

[20] van Dongen G, Visser G, Lub-de Hooge M, de Vries EG, Perk LR. Immuno-PET: A navigator in monoclonal antibody development and applications. Oncologist, 2007, 12, 1379–89.

13.3 Radiotherapeutics

Andreas Faust

Besides examining organ function and structure on a molecular basis by means of radioactive nuclides for PET or SPECT, radioactive species can be also used to treat diseases such as arthritis or cancer. The low number of useful therapeutic radionuclides have half-lives ranging from hours to several days and are mainly β⁻-emitters; in a very few cases, also α-emitters with a high linear energy transfer (LET) can be considered [1]. In therapeutic interventions, the radiation dose for the patient is higher as in PET or SPECT diagnostics, but ideally it is focused on the tissue of interest, e.g. in the case of cancer. Hence, the radionuclide should target the neoplastic cells with high specificity while killing them selectively and avoiding unwanted side effects in healthy tissues (low unspecific binding, fast and complete clearance). Relevant therapeutic radionuclides with their particle energies and maximal penetration depth in tissues are listed in Table 13.3.1.

Table 13.3.1: Prominent radionuclides used for therapeutic applications in nuclear medicine, their decay characteristics, particle energies and maximal penetration depths in tissues [2].

radio-nuclide	half-life	decay	β⁻ / α-energies [MeV]		penetration of tissue [mm]	
			maximal	average	maximal	average
^{32}P	14.3 d	β⁻	1.71 (β⁻)	0.7 (β⁻)	8.7 (β⁻)	2.9 (β⁻)
^{90}Y	2.67 d	β⁻	2.28 (β⁻)	0.93 (β⁻)	12.0 (β⁻)	3.9 (β⁻)
^{131}I	8.0 d	β⁻	0.81 (β⁻)	0.18 (β⁻)	2.4 (β⁻)	0.9 (β⁻)
^{153}Sm	1.94 d	β⁻	0.81 (β⁻)	0.22 (β⁻)	3.0 (β⁻)	1.2 (β⁻)
^{166}Ho	26.8 h	β⁻	1.85 (β⁻)	0.67 (β⁻)	10.2 (β⁻)	3.4 (β⁻)
^{169}Er	9.4 d	β⁻	0.35 (β⁻)	0.1 (β⁻)	0.9 (β⁻)	0.5 (β⁻)
^{177}Lu	6.7 d	β⁻	0.5 (β⁻)	0.13 (β⁻)	2.5 (β⁻)	0.7 (β⁻)
^{186}Re	3.72 d	β⁻ / EC	1.07 (β⁻)	0.35 (β⁻)	3.6 (β⁻)	1.8 (β⁻)
^{188}Re	17 h	β⁻	2.12 (β⁻)	0.76 (β⁻)	11.0 (β⁻)	3.5 (β⁻)
^{198}Au	2.69 d	β⁻	0.96 (β⁻)	0.31 (β⁻)	3.8 (β⁻)	1.6 (β⁻)
^{211}At	7.21 h	α / EC	5.87 (α)		0.08 (α)	
^{223}Ra	11.43 d	α	5.82 (α)		<0.1 (α)	
^{225}Ac	10.0 d	α	5.94 (α)		0.05–0.08 (α)	

https://doi.org/10.1515/9783110733471-027

13.3.1 Cancer therapy

The oldest example of radiotherapy involves iodine-131. It was the first radionuclide ever used for clinical treatment in the 1940s and has paved the way for the development of further radiotracers for medical applications. It is the most commonly used radionuclide for the treatment of different kinds of thyroid cancer [3]. Due to the high affinity and selectivity for the thyroid tissue, it is the radionuclide of choice for this treatment until today. For thyroid scintigraphy with SPECT, $[^{99m}Tc][TcO_4]^-$ is used nowadays instead of iodine-isotopes (see Chapter 13.2.4).

During the past decade, the theranostic pair gallium-68 (PET) and lutetium-177 (therapy) became increasingly important in nuclear medicine. One example is PSMA-617, which targets the prostate-specific membrane antigen (PSMA) and which can be labeled with different radionuclides (see Figure 13.2.6). Due to the high and specific tumor uptake, these PSMA-derivatives enable the implementation of radionuclide therapy employing ^{177}Lu along with diagnostic PET-imaging. In Figure 13.3.1, an example of metastatic castration-resistant prostate cancer (mCRPC) is shown, where all other initial treatments including surgery and hormone-therapy failed. After four cycles of approx. 6 GBq $[^{177}Lu]Lu$-PSMA-617, a dramatic decrease of tumor-activity was observed by using $[^{68}Ga]Ga$-PSMA-617-PET. In addition, the PSA-value in blood-samples dropped down to nearly normal values [4]. Recently, also the α-emitter actinium-225 ($[^{225}Ac]Ac$-PSMA-617) was successfully used for the treatment of castration-resistant prostate cancer in a human study [5].

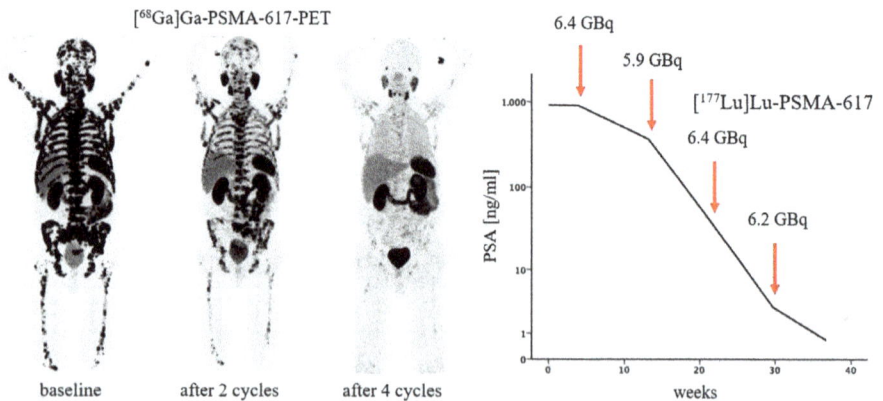

Figure 13.3.1: A clinical example involving a 77-year-old heavily pretreated patient with metastasized castration-resistant prostate cancer. The patient was treated with four cycles of approx. 6 GBq $[^{177}Lu]Lu$-PSMA-617. Tumor-reduction was visualized by $[^{68}Ga]Ga$-PSMA-617-PET and control of the PSA-level in blood. Reprinted with permission from [4]; Wolters Kluwer Health, Inc.

A prominent ^{68}Ga/^{177}Lu theranostic match pair is represented by the peptide-based radiopharmaceuticals (somatostatin receptor (sstr) agonists and antagonists) for imaging and treatment of neuroendocrine tumors. As sstr agonists, ^{68}Ga/^{177}Lu-DOTA-TOC and ^{68}Ga/^{177}Lu-DOTA-TATE (see Figure 13.2.6) are used for the therapy of neuroendocrine tumors.

Another commonly used therapeutic radionuclide is yttrium-90. With its nearly three-day half-life and pure high-energy β^--emission resulting in a deep tissue penetration of 12 mm, it finds wide use in antibody and peptide labeling. Besides DOTA-TATE, DOTA-TOC and PSMA-617, also the labeling of antibodies is frequently carried out. One example is ^{90}Y-ibritumomab tiuxetan (^{90}Y-Zevalin®) as the first radioimmuno-therapeutic drug to treat low-grade or follicular B-cell non-Hodgkin's lymphoma (NHL) [6]. Yttrium-90 is also trapped in so-called microspheres. These small resins are applied in SIRT (selective internal radiation therapy), where the radioactive dose is locally placed *via* catheter and therefore ideal for the treatment of liver tumors, which have shown no response to systemic chemotherapy. The liver, with its dual blood supply system, receives it mostly *via* the hepatic artery. Thus, it is possible to deposit radioactive-labeled microspheres in the tumor (radioembolization) while avoiding harmful side effects for the healthy tissue (Figure 13.3.2). There are different types of microspheres available: TheraSphere® and SIR-Spheres® loaded with yttrium-90 as the radioactive source and QuiremSpheres® loaded with the holmium-166 [7].

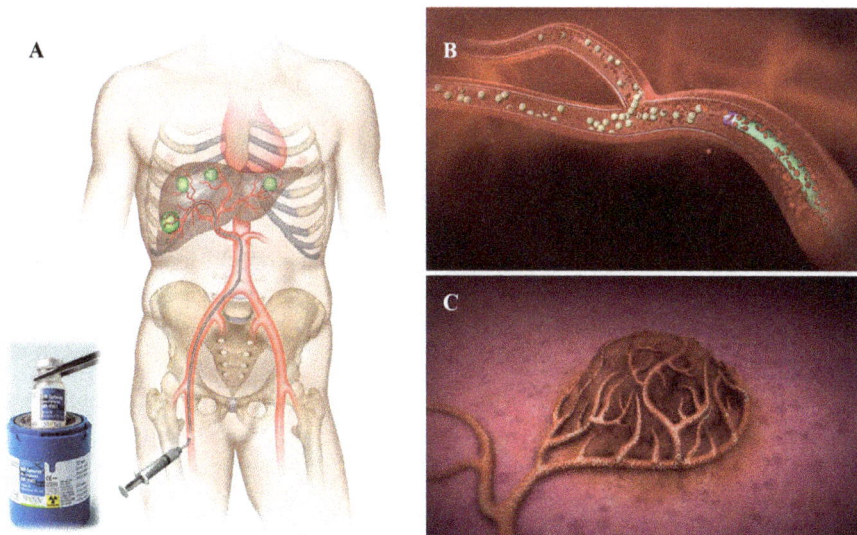

Figure 13.3.2: A: Injection of ^{90}Y-labeled resin microspheres as minimally invasive treatment of liver tumors *via* micro catheter in the hepatic artery; B: microspheres are carried directly to the tumor and trapped. C: Treatment upon β^--emission leads to tumor necrosis. Rearranged with permission from Sirtex© Medical Europe GmbH.

For targeted radionuclide therapy in patients with metastatic pheochromocytoma and paraganglioma, $[^{131}I]$metaiodobenzylguanidine ($[^{131}I]$mIBG) is used as a β^--emitter or, alternatively, the related $[^{211}At]$mABG as an α-emitter. Nonetheless, the evaluation of doses and schedules intending to enhance the effectiveness and to decrease the toxicity are under ongoing development [8]. For patients with strong bone metastasis (e.g. in prostate cancer) and particularly if all other curative approaches have failed, the α-emitter Radium-223 (Xofigo®) can be used. It has shown that treatment with $[^{223}Ra]RaCl_3$ prolongs the survival of patients with bone metastases from prostate carcinoma and additionally, it reduces pain [9].

13.3.2 Palliative radiotherapy

Palliative pain treatment is also possible with β^--emitting radionuclides. Here, metastatic bone pain is frequently the problem faced by cancer patients in terminal stages; in such cases, palliation rather than the cure is the goal of treatment. Bone-specific radiopharmaceuticals have shown favorable results in the relief of bone pain. The most common radiopharmaceutical is $[^{153}Sm]$Sm-EDTMP (Quadramet®). Due to its structure, there is a high affinity for the bone mineral hydroxyapatite, but the exact mechanism of pain reduction is still unknown. Different mechanisms of action are currently being discussed, including the reduction of cytokines, of tumor growth factors or of inflammatory cells as well as several other hypotheses [10]. Structurally related are $[^{186}Re]/[^{188}Re]$Re-HEDP and $[^{117}Sn]$Sn-DTPA (Figure 13.3.3), even though they are not so frequently used in clinical praxis. Also other bone-affine ions, such as $[^{89}Sr]SrCl_2$ or $[^{32}P]PO_4^{3-}$, are known for their use in palliative treatments [2].

13.3.3 Radiosynovectomy

The curative treatment of inflamed tissue (radiosynovectomy) has a long tradition in nuclear medicine and is known as radiosynoviorthesis. Hence, β^--emitters are injected into joint-spaces to treat synovitis. Presently, depending on the joint's size, roughly three different radiocolloids are in clinical use: colloidal $[^{90}Y]$Y-citrate or silicate with particle sizes ranging from 10 nm to 1 μm (treatment of big joints, e.g. knee), colloidal $[^{186}Re]$Re-sulfide (particle sizes between 30 and 50 μm) for medium-sized joints (shoulder, elbow, hips), and colloidal $[^{169}Er]$Er-citrate for small joints (like fingers, using a particle size ranging from 0.1 to 1 μm) [11]. The particle size should correspond to the thickness of the inflamed synovium. In a sterile injection procedure, local skin anesthesia is required before puncture. During the injection of the radiopharmaceutical formulation, the position of the needle is controlled by X-ray radiography. After administration, anesthetics such as prilocaine are injected as well as glucocorticoids to prevent synovitis [12]. Upon emission of β^-radiation, the injected particles cause irreversible damage to the cells of the synovial membrane. Starting with excitation and ionization, a large number of secondary particles such as free radicals are created, resulting in subsequent apoptosis and ablation of the inflamed synovial tissue [13]. Presently, new radiopharmaceuticals are developed for radiosynovectomy, such as the phytate complex of lutetium-177 or holmium-166 (see Figure 13.3.3) [14].

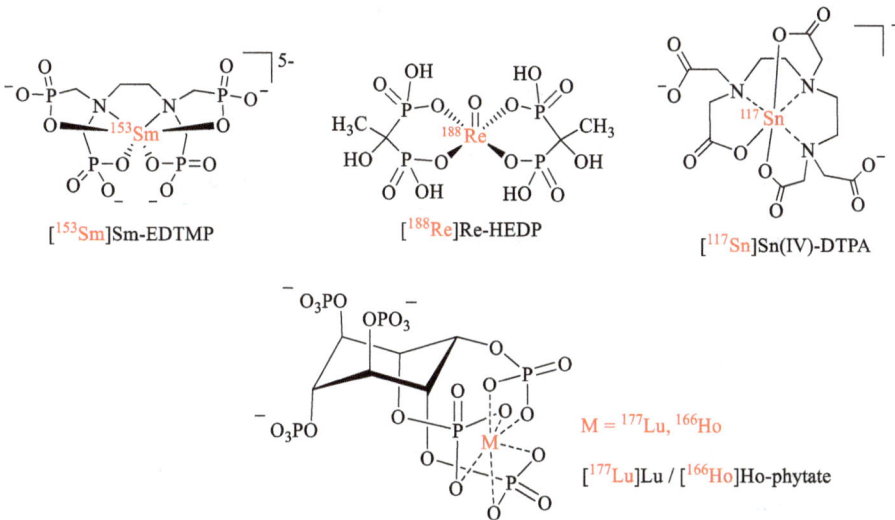

$[^{153}Sm]Sm$-EDTMP

$[^{188}Re]Re$-HEDP

$[^{117}Sn]Sn(IV)$-DTPA

$M = {}^{177}Lu, {}^{166}Ho$

$[^{177}Lu]Lu / [^{166}Ho]Ho$-phytate

Figure 13.3.3: Structures of radiopharmaceuticals used in palliative treatment and radiosynovectomy.

References

[1]   Kostelnik TI, Orvig C. Radioactive main group and rare earth metals for imaging and therapy. Chem Rev, 2019, 119, 902–56.
[2]   Dietlein M, Kopka K, Schmidt M. Nuklearmedizin, Basiswissen und klinische Anwendung, Schattauer Verlag, Stuttgart, 2017.
[3]   IAEA. Nuclear medicine in thyroid cancer management: A practical approach. IAEA-TECDOC-1608, Vienna, 2009.
[4]   Roll W, Bräuer A, Weckesser M, Bögemann M, Rahbar K. Long-term survival and excellent response to repeated ${}^{177}Lu$-prostate-specific membrane antigen 617 radioligand therapy in a patient with advanced metastatic castration-resistant prostate cancer. Clin Nucl Med, 2018, 43, 755–6.
[5]   Sanli Y, Kuyumcu S, Duygu DH, Büyükkaya F, Civan C, Isik EG, Ozkan ZG, Basaran M, Sanli O. ${}^{225}Ac$-prostate-specific membrane antigen therapy for castration-resistant prostate cancer: A single-center experience. Clin Nucl Med, 2021, 46, 943–51.
[6]   Grillo-López AJ. Zevalin: The first radioimmunotherapy approved for the treatment of lymphoma. Expert Rev Anticancer Ther, 2002, 2, 485–93.

[7]   (a) Smits MLJ, Elschot M, van den Bosch MAAJ, van de Maat GH, van het Schip AD, Zonnenberg BA, Seevinck PR, Verkooijen HM, Bakker CJ, de Jong HWAM, Lam MGEH, Nijsen JFW. In vivo dosimetry based on SPECT and MR imaging of [166]Ho-microspheres for treatment of liver malignancies. J Nucl Med, 2013, 54, 2093–100. (b) Pöttgen R, Jüstel T, Strassert CA. Rare Earth Chemistry, de Gruyter, Berlin/Boston, 2020, 444–52.

[8]   Jimenez C, Erwin W, Chasen B. Targeted radionuclide therapy for patients with metastatic pheochromocytoma and paraganglioma: From low-specific-activity to high-specific-activity iodine-131 metaiodobenzylguanidine. Cancers (Basel), 2019, 11, 1018–38.

[9]   Nilsson S. Radionuclide therapies in prostate cancer: Integrating Radium-223 in the treatment of patients with metastatic castration-resistant prostate cancer. Curr Oncol Rep, 2016, 18, 14–26.

[10] Taheri M, Azizmohammadi Z, Ansari M, Dadkhah P, Dehghan K, Valizadeh R, Assadi M. [153]Sm-EDTMP and ^{177}Lu-EDTMP are equally safe and effective in pain palliation from skeletal metastases. Nuklearmedizin, 2018, 57, 174–80.

[11] Kampen WU, Boddenberg-Pätzold B, Fischer M, Gabriel M, Klett R, Konijnenberg M, Kresnik E, Lellouche H, Paycha F, Terslev L, Turkmen C, van der Zant F, Antunovic L, Panagiotidis E, Gnanasegaran G, Kuwert T, van den Wyngaert T. The EANM guideline for radiosynoviorthesis. Eur J Nucl Med Mol Imag, 2021. DOI: 10.1007/s00259-021-05541-7.

[12] Antilgan HI, Sadic M, Koca G, Korkmaz M. Radiosynovectomy: Current status and clinical utility. Int J Health Sci Res, 2016, 6, 324–36.

[13] Ailland J, Kampen WU, Schünke M, Trenntmann J, Kurz B. Beta irradiation decreases collagen type II synthesis and increases nitric oxide production and cell death in articular chondrocytes. Ann Rheum Dis, 2003, 62, 1054–60.

[14] Yousefnia H, Jalilian AR, Bahrami-Samani A, Mazidi M, Maragheh MG, Abbasi-Davani F. Development of ^{177}Lu-phytate complex for radiosynovectomy. Iran J Basic Med Sci, 2013, 16, 705–9.

13.4 Active inorganic drugs and excipients

Fabian Herrmann

Although current medicinal chemistry relies to a large extent on low-molecular-weight organic compounds as therapeutic agents, a variety of inorganic species are still of specific relevance in modern pharmacotherapy. Due to their physicochemical properties, pharmaceutically used inorganics are mainly employed as auxiliaries and excipients in a variety of drug formulations and typically do not exert any pharmacological activity by themselves (e.g. pigments or gel builders). Nonetheless, some specific inorganic compounds (or even isolated metal ions) are of high significance in the treatment of a variety of disorders ranging from psychotic affections to cancer, due to their particular pharmacodynamic and pharmacokinetic properties. With special emphasis on practical relevance, this chapter is focused on auxiliary and active inorganics with current importance in evidence-based therapy, while specifically excluding those that are considered obsolete by modern standards, e.g. the osmotic laxatives mirabilite (also known as Glauber's salt, $Na_2SO_4 \times 10\ H_2O$) and epsomite ($MgSO_4 \times 10\ H_2O$), or the disinfectant $KMnO_4$. Inorganic agents employed for the supplementation of micronutrients or trace elements were also not covered, due to their relatively unspecific pharmacological properties beyond the supply of the ions themselves. Additionally, auxiliaries employed only due to non-pharmacological activity (i.e. purely because of their physicochemical properties) were also not included in the current chapter and are eventually only briefly mentioned (e.g. inorganic pigments, fillers or gelators apart from SiO_2, flow regulators, as well as effervescent decay accelerators).

13.4.1 Lithium salts as antidepressants

Lithium salts (lithium carbonate, lithium acetate, lithium hydrogenaspartate) have an over 70-year-long history in the therapy of psychotic affections with a current focus on the treatment of bipolar disorders, and are classified as so-called mood stabilizers. When the antipsychotic action of lithium was described in 1949 by John Cade for the first time [1], this lightest solid element constituted the only therapeutic option for the treatment of a variety of mental disorders, and has been of major importance in this regard ever since (e.g. as part of the WHO's List of Essential Medicines [2]). Despite the chemical simplicity of the lithium cation (compared to modern low molecular drugs used in psychotherapy), its antipsychotic mechanism of action is still unknown to a significant degree. In inorganic chemistry, a well-known diagonal relation rationalizes the chemical similarities encountered between magnesium and lithium (e.g. insolubility of their phosphates and carbonates as well as comparable solvation enthalpies). Hence, a correlation between the biochemical

https://doi.org/10.1515/9783110733471-028

properties of magnesium and lithium ions has been established. Due to their comparable charge, atomic, ionic and solvated radii, a variety of effects in magnesium- and sodium-dependent physiological processes are found, based on their similar charge density. Even without identifying the specific mechanism of action, biochemical interactions with a variety of sodium- or magnesium-involving target proteins have been described in the literature (e.g. Na^+/K^+-ATPase [3]). Additionally, it is widely accepted that intracellular Li^+ ions affect the phosphatidylinositol turnover, leading to the reduction of intracellular inositol and subsequently to the specific downregulation of neurotransmitter activity [4]. Whatever pharmacodynamic actions may be induced by lithium therapy, they seem to be mainly generated by its cationic form and its similarity with the physiologically relevant species derived from the elements magnesium and sodium.

Since the carbonate salt of lithium has a relatively low aqueous solubility, a fast release upon oral intake leading to a spike in the plasmatic concentration of the cationic species is precluded, which would otherwise lead to an acute neuro-, nephro- or cardiotoxicity (which in turn can be observed for lithium halides and other soluble salts). Once in the plasma, the active form is the solvated lithium cation; it possesses the highest solvation enthalpy of all alkali metal cations as well as several solvent layers that lead to an effective radius and ionic mobility resembling solvated magnesium, sodium and potassium species. Solvated lithium cations permeate biomembranes (comparable to sodium ions), but are far less efficiently excreted by ion transporters, leading to an intracellular enrichment of Li^+ [4].

13.4.2 Gold-based complexes as anti-arthritic drugs

Although gold is mainly used in nanoparticle formulations in current therapy (e.g. for cancer treatment, see 13.4.3), gold salts and complexes themselves have a long history in the treatment of arthritic diseases. Starting with Robert Koch´s identification in 1890 of the *in vitro* activity related to AuCN against the etiologic agent of tuberculosis (*Mycobacterium tuberculosis*) [5], this precious metal gained growing attention concerning its biological activity and potential therapeutic use. In the course of this augmented interest in Au-based compounds, Jacques Forestier described in 1929 the ability of ionic gold compounds to relieve joint pain in his arthritic patients, which in some cases even led to a complete remission [6]. Based on this observation and other promising biological effects, gold(I) complexes have been therapeutically used as standard options to treat arthritic diseases until the late 1990s (so-called chrysotherapy) when more efficient and less toxic therapies progressively replaced the Au-based options. With those alternatives available, gold compounds are currently not used as the initial therapeutic options in the treatment of arthritic diseases [4]. As in the case of lithium therapy (13.4.1), a precise mechanism of action related to the monovalent gold ions acting as anti-arthritic drugs has not been identified so far. It is widely accepted, however, that the beneficial effects of chrysotherapy are mainly caused by the gold ions themselves (Au^+ and its metabolites, namely Au^{3+}, $[Au(CN)_2]^-$ and Au^0). Especially gold(I) ions have a high affinity for thiol groups of proteins (e.g. cysteine residues), forming protein-gold(I)-thiolato complexes (Figure 13.4.1) [7]. In the case of the main pro-inflammatory transcription factors involved in the pathogenesis of arthritic diseases (e.g. NF-kappaB, AP-1), specific interactions between gold(I) ions and cysteine and/or selenocysteine groups reduce the ability of those transcription factors to bind to the corresponding DNA sequences, subsequently lowering the overall inflammatory response [4]. Corresponding to this mechanism, an overall reduction of extravasation processes of immunocompetent blood cells is also mediated by gold(I) ions by the inhibition of the expression of endothelial adhesion molecules (e.g. ICAM-1), leading altogether to a reduction of inflammatory processes in arthritic joints [4].

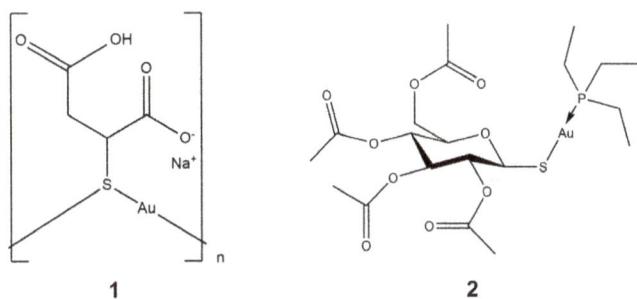

1   **2**

Figure 13.4.1: Chemical formulae of the most prominently used Au(I) drugs in anti-arthritic therapy: sodium aurothiomalate **(1)** and auranofin **(2)**.

13.4.3 Cancer therapy: platinum complexes and nanoparticle formulations (Au, HfO_2 and Fe_3O_4)

Modern cancer chemotherapy relies to a great extent on inorganic cytostatics. Especially the intensively used platinum(II) complexes (cisplatin, carboplatin and oxaliplatin) are of striking relevance in today's treatment of solid tumors. The Pt(II) complexes used in oncology exhibit a coordination-chemical interaction involving mainly the N-donors from bases in DNA. This leads to replication defects during cell division while ultimately causing cell death (typically by apoptosis). Hence, the specific complex-forming ability of Pt-based cytostatic agents is crucial for their DNA damaging capacity. Only *cis*-configured Pt(II) complexes are used as therapeutics and only those possess the ability to form complexes with adjacent bases (from the nucleotides on the same strand) that actually constitute DNA double helix (mainly due to spatial arrangement). In the case of cisplatin possessing two amino and two chlorido ligands, the first step involves hydrolysis with exchange of the anionic donors by aqua ligands, followed by the irreversible formation of complexes with the N-donors of the bases from adjacent nucleosides located at the same strand of the double-stranded DNA. Since the replication of DNA is altered, it mostly affects fast-replicating cells, which are characteristic of neoplastic affections. The introduction of Pt-based cytostatics (derived from cisplatin, such as carboplatin and oxaliplatin, see Figure 13.4.2) as therapeutic options altogether revolutionized the chemotherapy of neoplastic affections, sometimes even leading to total remissions in oncological patients (e.g. testicular cancer) [4].

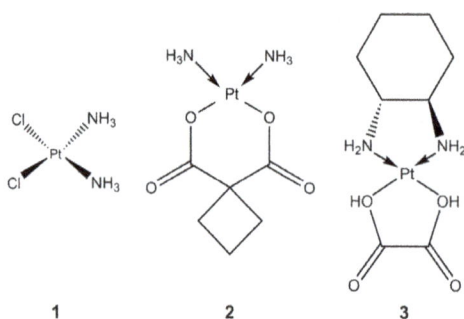

Figure 13.4.2: The most prominently used Pt(II)-based anti-cancer drugs in today's therapy of neoplastic affections: cisplatin (**1**), carboplatin (**2**) and oxaliplatin (**3**).

Gold nanoparticles (AuNPs) are currently extensively investigated as innovative therapeutic or diagnostic options (theranostics) for the treatment of a variety of solid tumors. They comprise promising properties, e.g. as drug delivery vehicles for therapeutic cargos, molecular probes or biosensors, only to name a few. Because of the easy functionalization of AuNP surfaces by the addition of thiolated species, as well as due to their chemically adjustable size and shape, this Au-based theranostics offer a variety of drug delivery options, e.g. in cancer therapy. Additionally, AuNPs show interactions with the time-dependent electromagnetic fields of photons, leading to oscillations of the free electrons (surface plasmon resonance, SPR). The energy absorbed during this process is subsequently converted into heat, which allows to specifically inactivate tumor tissues *via* hyperthermia-related damage. Additionally, AuNPs are also available as radioactive formulations ([198]Au and [199]Au isotopes; see Chapter 13.3) used to specifically deliver beta and gamma radiation for the destruction of tumor tissue [8].

Another innovative approach in the treatment of solid tumors involves the recently emerging therapeutic options by localized inactivation through ionizing radiation. Hafnium oxide (HfO_2) NPs constitute examples of localized tumor therapy, specifically enhancing radiation-mediated damage inside neoplastic tissues. In the case of HfO_2-NPs, the high electron density of the transition metal enhances the local absorption cross-section for the radiation and is therefore injected as a NP-formulation directly into the tumor, where it enhances the local damage when external ionizing sources are applied during radiotherapy [9].

Iron oxides in the form of superparamagnetic magnetite (Fe_3O_4) nanoparticle formulations possess the ability to transduce the energy from oscillating magnetic fields into heat and can therefore also be used as remotely activated heat sources. Because of these peculiar properties of Fe_3O_4-NPs, the use in the treatment of inoperable tumors (e.g. brain tumors such as glioblastomas) is currently emerging as a promising therapeutic option. With the local administration of such NPs directly into the solid tumor, subsequent induction of localized magnetic hyperthermia leads to localized cell damage, allowing to directly inactivate solid tumors that are otherwise inaccessible by standard therapies [10, 11].

13.4.4 Active inorganic excipients: ZnO, TiO$_2$, clays

The *Papyrus Ebers*, one of the oldest medicinal papyri known today (sixteenth century BC), already mentions the advantageous medical properties of zinc oxide (*zinci oxidum*, ZnO, mentioned in the form of calamine, a mixture of ZnO and around 0.5% of Fe_2O_3). This amorphous oxide mixture comprises specific sought-after properties, rendering it one of the most commonly employed pharmaceutical excipients in wound healing treatment. Apart from its physicochemical advantages, such as a considerable hydrophilicity and astringency, it is also known for anti-microbial activities that are particularly beneficial in the treatment of skin disorders. ZnO is therefore widely used in topical applications for the therapy of common skin lesions (e.g. *pasta zinci* for diaper rashes) but also in the care of more complex affections (e.g. leg ulcers). Additionally, ZnO is also extensively used as a pigment in sunscreen lotions, due to its UV-absorbing ability [12].

Titanium dioxide (TiO_2) shares some pharmaceutical applications with ZnO and is also a commonly employed excipient. Besides its use in sunscreen lotions as an UV-blocking agent, it is also a widely used white pigment in the manufacture of tablets and dragées. Just as ZnO, TiO_2 also comprises some beneficial properties for the treatment of exudative skin lesions, such as its high ability to bind excess of water. Because of the missing antimicrobial activity as well as the currently discussed toxic potential of TiO_2, ZnO is typically preferred for this indication. Due to eventual genotoxic effects, the usage of TiO_2 (E 171) was recently re-evaluated by the European Food Safety Authority (EFSA) and considered to be unsafe for use as a food additive [13]. Because of this re-evaluation, TiO_2 is currently disappearing from many typical applications and is gradually replaced by proven non-toxic alternatives (e.g. in the production of tablets or sunscreen lotions) [14].

Clays used for medicinal purposes are typically composed of layered aluminosilicates and mostly administered topically, e.g. in the course of the so-called traditional mud therapy. A variety of naturally occurring clays have been used already in ancient times for different purposes. In modern medicine and pharmacy, only very few of the wide variety of available clays are currently still administered. One relevant clay used for medicinal purposes is kaolinite, a layered silicate with the repeating unit $Al_2(OH)_4Si_2O_5$. Kaolinite, also known as *bolus alba*, is characterized by a 1:1 distribution of silicate tetrahedra-containing layers and corresponding sheets with octahedrally coordinated aluminum; it is used today for external therapies, mainly because of its ability to take up water, as a pharmaceutical excipient in the production of powders, or as dispersant in suspensions. Another medicinally relevant clay is characterized by a 2:1 distribution of layers with silicate tetrahedra and sheets with octahedrally coordinated aluminum centers, the so-called montmorillonite (named after the city Montmorillon in France; Figure 13.4.3). The resulting physicochemical properties are comparable to kaolinite, leading to similar applications in medicine and pharmacy. Altogether, the mentioned medicinal clays are broadly used in topical remedies because of their ability to bind significant amounts of water, besides the desirable rheological properties they can impart to formulations (including viscosity and thixotropy) [15].

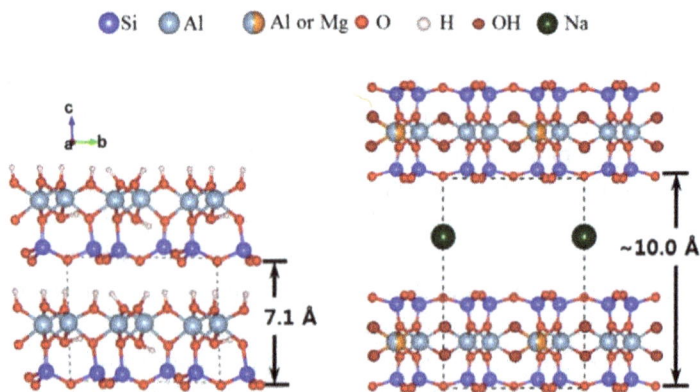

Figure 13.4.3: Structure of the pharmaceutically used clays kaolinite (1) and montmorillonite (2). Adapted with permission from [15] Copyright Elsevier.

13.4.5 Inorganic antipeptic drugs: Hydrotalcite [Al$_2$Mg$_2$(OH)$_{16}$CO$_3$ x 4 H$_2$O] and bismuth salts

Inorganic antacids have a long history in the treatment of gastroesophageal reflux diseases (GERD) as oral preparations of basic sodium, magnesium, calcium or aluminum salts (e.g. carbonates, hydrogen carbonates or hydroxides). Their main benefit is exerted by the direct neutralization of excess stomach protons after oral intake and the subsequent relief of heartburn symptoms. Because of the direct and immediate neutralization of secreted acid, a rapid reduction of the ventricular pH is induced while subsequently leading to increased proton secretion into the lumen. Due to this unspecific mechanism of action, the continuous use of those remedies as self-medication of GERD is considered as problematic nowadays [4].

Hydrotalcite ($[Al_2Mg_2(OH)_{16}CO_3 \times 4\ H_2O]$), a layered aluminum silicate frequently used as a pharmaceutical remedy in the therapy of GERD, is composed of a magnesium hydroxide matrix, showing partial replacement of Mg^{2+} by Al^{3+} ions as well as inter-layer anions, such as OH^- and CO_3^{2-}. Its main benefit in GERD therapy is given by its ability to neutralize excess of gastric acid, pepsin as well as cholic acid, therefore reducing the overall symptoms accompanied by GERD. Because of its direct mechanism of action, the relieving effects are timely exerted, resulting in its typical use in the self-medication of reflux diseases. As opposed to the mentioned non-layered antacids, the additional benefit of hydrotalcite is given by a retarded uptake of gastric protons into the layered silicate and their subsequent delayed neutralization. In contrast to the previously mentioned inorganic antacids, this is not reflexively followed by an overflow of gastric protons secreted due to a dropping ventricular *p*H [16].

Bismuth salts (e.g. bismuth subsalicylate (Pepto-Bismol) or bismuth citrate) constitute another relevant group of inorganic remedies administered for the treatment of GERD, diarrhea or even more complicated gastrointestinal affections such as *Helicobacter pylori*-caused peptic ulcers. Having been administered for almost three centuries, bismuth salts mainly exert their activity in the upper gastrointestinal tract (stomach and duodenum). Besides their ability to remedy typical symptoms of diarrhea as well as of uncomplicated reflux disease, bismuth salts are currently mainly administered due to their specific antibacterial activity against *Helicobacter pylori* (a commonly occurring pathogen leading to stomach ulcers and even neoplastic affections like stomach cancer) within the so-called quadruple therapy for *H. pylori* eradication. The mode of action concerning the activity of bismuth salts in the upper gastrointestinal tract has not been elucidated to a sufficient extent. It is commonly accepted, however, that besides the affinity of bismuth ions to the ventricular mucosa, anti-*Helicobacter* effects include the formation of bismuth complexes in the bacterial cell wall as well as in the periplasmatic space, the inhibition of bacterial enzymes (e.g. urease, catalase and lipase), the inhibition of ATP synthesis as well as inhibitory activity concerning the adhesion of *H. pylori* to the stomach mucosa [17].

13.4.6 Sodium fluoride

Sodium fluoride (NaF) is still widely used today in the management and prophylaxis of caries-induced teeth lesions. Its prophylactic mechanism of action is mainly exerted *via* the replacement of OH^- ions by fluoride anions in the hydroxyapatite complex of tooth enamel ($Ca_5[OH(PO_4)_3]$), resulting in an increased apatite hardness and subsequently in the lower susceptibility to the formation of bacteria-induced caries. It is therefore typically supplemented in tooth paste formulations or topically administered directly on the enamel in higher local doses, during teeth development in minors and as prophylactic treatments for adults [14].

13.4.7 Carbo medicinalis

The term *carbo medicinalis* refers to activated charcoal formulations with a specifically high surface area (up to over 2000 m^2 for 1 g of *carbo medicinalis*) normally containing over 90% of elemental carbon. Because of its enormous specific surface area in combination with its highly lipophilic characteristics, *carbo medicinalis* has been successfully used for decades in the treatment systemic intoxications. Its ability to bind a high variety of organic compounds from the gastrointestinal tract upon oral administration and therefore to inhibit their systemic resorption still renders *carbo medicinalis* one of the most important antidotes in clinical use today [14]).

13.4.8 Sodium nitroprusside

Sodium nitroprusside is an inorganic cyanido complex ($Na_2[Fe(CN)_5NO] \times 2\ H_2O$) consisting of an iron center (Fe^{3+}) surrounded by one nitroso and by five cyanido ligands. It is typically used in the management of cardiovascular diseases, exerting its beneficial effects due to its ability to act as NO donor resulting in vasodilatation involving coronary arteries and peripheral vessels. Besides the release of NO, nitroprusside is also known to liberate cyanide ions to a certain extent, leading to the necessity to co-administrate sodium thiosulfate (13.4.9 as an antidote during long-term therapy [4, 14]).

13.4.9 Sodium thiosulfate

Sodium thiosulfate ($Na_2S_2O_3 \times 5\ H_2O$) is another typical example for an important inorganic antidote in clinical use and is mainly administered by injection because of its ability to detoxify cyanide ions by forming harmless thiocyanate *in vivo*. Hence, the CN$^-$ acts as a nucleophile that reacts with the peripheral S-atom of the $\underline{S}\text{-}SO_3^{2-}$ anion to yield a $\underline{S}\text{-}CN^-$ species and a SO_3^{2-} unit. Besides the therapy of oral cyanide intoxications, it is also commonly used as adjuvant, e.g. during long-term nitroprusside therapy (13.4.8). Apart from the mentioned classical use established for decades, sodium thiosulfate is currently under investigation concerning other promising therapeutic effects (e.g. as adjuvant in the management of cisplatin-induced side effects in cancer chemotherapy [4, 18]).

13.4.10 Silicon dioxide

In pharmaceutical formulations, the use of silicon dioxide (SiO_2) is related to its amorphous character with high structural dispersion (particle size around 20 nm), if compared with naturally occurring silicon dioxide variations (e.g. the different quartz crystals). It is technologically used due to its capability to bind significant amounts of water, as a gel builder or as a filler in tablet formulations [14]. Because of the small particle size of amorphous silicon dioxide and the resulting potential for unwanted inhalation, its toxicological properties are currently under evaluation in the course of the REACH program [19].

13.4.11 Inorganic contrast agents (Gd(III) chelates and BaSO₄)

In the current diagnosis of a plethora of diseases, non-invasive imaging techniques including computer tomography (CT) or magnetic resonance tomography (MRT) allow insights into solid organs or soft tissues and are therefore of specific importance.

The rationale behind the application of contrast agents for diagnostic imaging by CT is altogether the enhancement of image quality, mainly due to the introduced electron density differences between an organ of interest and its surrounding tissues. Considering computer tomography, barium sulfate ($BaSO_4$) is still widely administered as a positive contrasting agent, due to its high potential to absorb X-rays and subsequently enhancing radiodensity and therefore the imaging contrast. Because of the considerably high toxicity of barium ions themselves, only the highly insoluble and therefore non-absorbable barium sulfate ($BaSO_4$) can be safely used. It is mainly administered orally as a suspension for the general visualization of the gastrointestinal tract [4].

In the case of magnetic resonance tomography (MRT) diagnostics, contrast agents altering the magnetic environment of hydrogen atoms are typically employed to improve the contrast. The lanthanide Gd(III) ion, for example, exerts a strong paramagnetic effect due to its seven unpaired electrons on the f-sub-shell, leading to an enhanced magnetic field affecting protons of water in the direct vicinity of the Gd(III) centers. This results in a shortened relaxation time of the affected hydrogen atoms in the externally applied magnetic field during MRT while leading to an overall enhanced contrast (positive contrasting agent) [4].

As in the case of $BaSO_4$, the high toxicity of isolated Gd(III) ions imposes specific hindrances in the application of Gd-based contrast agents (Figure 13.4.4). Therefore, they are typically administered as highly stable chelate complexes involving central Gd-atoms surrounded by tightly bound organic ligands (e.g. DTPA, DOTA, see 13.2.5) in order to prevent the systemic uptake of Gd(III) ions and the resulting toxicity [4].

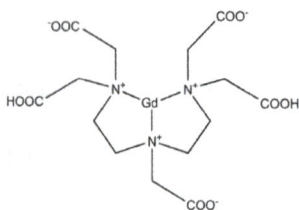

Figure 13.4.4: Gadopentetic acid, a typical Gd-based imaging agent used in magnetic resonance tomography for contrast enhancement.

13.4.12 Aluminum-based adjuvants (ABA) in vaccines

Aluminum salts, e.g. aluminum oxyhydroxide (AlO(OH)) or aluminum hydrogen phosphate (AlHPO$_4$), have been used for almost a century as immune response-enhancing auxiliaries in a variety of vaccines. Despite their widespread use, the mechanism of the exerted immune-stimulating properties is still not precisely known today. It is commonly accepted, however, that the addition of aluminum salts to vaccines leads to the formation of aluminum(III) aggregates, which are phagocytosed by macrophages or dendritic cells after administration of the vaccine, subsequently enhancing immune conversion processes. The improved induction of immunogenicity exerted due to the added aluminum auxiliaries can be attributed at least to some extent to the maintained high intracellular Al(III) concentration in macrophages and dendritic cells, the resulting intracellular pH change, the formation of reactive oxygen species (ROS) as well as by specific effects on the stability of phagosomal membranes [20].

13.4.13 Zinc salts as insulin stabilizers

When Dorothy Crowfoot Hodgkin and co-workers elucidated the structure of pig insulin by X-ray crystallography in 1969, an interaction potential involving zinc ions and insulin monomers (leading to the specific formation of less soluble zinc insulin complexes) had been already known for decades [21]. Physiologically speaking, the presence of zinc and calcium ions are of particular importance in the assembly of stabilized zinc-calcium-insulin complexes within dictyosomes of beta-pancreatic islets before secretion [22]. This complexation is mediated by Zn^{2+} ions binding in a histidine-containing allosteric insulin site (His B10 [22]), altogether resulting in the formation of relatively insoluble insulin hexamers. Because of their low solubility as well as due to the fact that only monomeric insulin shows intrinsic activity at insulin receptors, this physiological insulin assembly mechanism leads to an retardation of insulin activity. This effect has been mimicked for decades by long-lasting zinc-insulin formulations used in the therapy of diabetes. Although zinc-stabilized insulin preparations have lost most of their relevance today (due to more preferable retardation principles), zinc is still used as adjuvant in current insulins employed for the therapy of diabetes as an additional retardation mechanism (e.g. Neutral Protamine Hagedorn (NPH) insulin).

13.4.14 Inorganic antiseptic agents (Ag$^+$ and Thiomersal)

The antimicrobial properties of silver have already been known since ancient times, when the Greeks or Romans used silver bowls and containers to protect beverages such as water or wine from microbiological contamination. These and other empirical findings concerning the activity of the "noble" transition metal silver have led to a broad application as antimicrobial agent throughout human cultural history, with increasing prevalence emerging in the nineteenth century (e.g. as a suture material in surgery or as a germicide in colloidal form). As is the case for other pharmacologically active metal ions (e.g. Li$^+$, see 13.4.1), the precise mechanism of action of isolated Ag$^+$ ions has not been conclusively elucidated to date. Altogether, it is commonly acknowledged that the biologically active agent in silver formulations is the liberated Ag$^+$ ion itself. In general, a series of cytotoxic mechanisms causing specific damages to bacterial or cancer cells have been described and discussed in the past. Besides changes in cellular ion uptake and exchange or occurring Ag$^+$-mediated complex formation with DNA or RNA, the specific protein denaturation activities of isolated silver ions have been discussed as relevant mechanisms for antibacterial and anticancer activities. It is nevertheless noteworthy that the mentioned biological activities are altogether not sufficient to explain the reduced cytotoxicity of Ag$^+$ ions regarding mammalian tissues, if compared to their high activity against bacterial or cancer cells [23]. Because of the relatively easy accessibility of antibacterial silver nitrate formulations, ophthalmic AgNO$_3$ solutions are still of high importance in the prophylaxis of neonatal conjunctivitis in socioeconomically underdeveloped areas of the world, where gonococcal eye infections are one leading cause of neonatal blindness [24].

Thiomersal (a trade name for a sodium ethyl-mercury thiosalicylate, Figure 13.4.5) is another inorganic antimicrobial agent, with some relevance even in modern pharmaceutical chemistry. It was developed in the late 1920s and has since been used as a preservative in cosmetics or pharmaceutical preparations, due to the antibacterial and antifungal properties exerted by the high ethylmercury content (around 49.6%

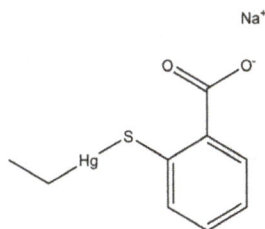

Figure 13.4.5: Sodium ethyl-mercury thiosalicylate (thiomersal).

ethylmercury content by weight). Thiomersal steadily decomposes in solution to form ethylmercury, which is subsequently converted into inorganic mercury and altogether responsible for the antimicrobial activity. Although a variety of clinical studies have shown the harmlessness of the used preservative concentrations (especially in vaccines), thiomersal was progressively removed in the past, especially from pharmaceutical preparations, due to discussions about its safety starting in 1999 [25].

References

[1] Cade J. Lithium salts in the treatment of psychotic excitement. Med J Aust, 1949, 2, 349–52.
[2] Homepage of World Health Organisation, Genf, Switzerland, https://www.who.int/groups/expert-committee-on-selection-and-use-of-essential-medicines/essential-medicines-lists/ (accessed on 05.06.2021).
[3] Yanagita T, Maruta T, Uezono Y, Satoh S, Yoshikawa N, Nemoto T, Kobayashi H, Wada A. Lithium inhibits function of voltage-dependent sodium channels and catecholamine secretion independent of glycogen synthase kinase-3 in adrenal chromaffin cells. Neuropharmacology, 2007, 53, 881–9.
[4] Geisslinger G, Menzel S, Gudermann T, Hinz B, Ruth P. Mutschler Arzneimittelwirkungen – Pharmakologie, Klinische Pharmakologie, Toxikologie, Wissenschaftliche Verlagsgesellschaft Stuttgart, Stuttgart, 2020.
[5] Gibier P. Dr. Koch´s discovery. N Am Rev, 1890, 726–31.
[6] Forestier J. The treatment of rheumatoid arthritis with gold salts injections. Lancet, 1932, 441–4.
[7] Bhabak K, Bhuyan B, Mugesh M. Bioinorganic and medicinal chemistry: Aspects of gold(I)-protein complexes. Dalton Trans, 2011, 40, 2099–111.
[8] Freitas L, Ferreira A, Thipe V, Varca G, Lima C, Batista J, Riello F, Nogueira K, Cruz C, Mendes G, Rodrigues A, Sousa T, Alves V, Lugao A. The state of the art of theranostic nanomaterials for lung, breast, and prostate cancers. Nanomaterials, 2021, 11, 2579.
[9] Scher N, Bonvalot S, Le Tourneau C, Chajon E, Verry C, Thariat J, Calugaru V. Review of the clinical applications of radiation-enhancing nanoparticles. Biotechnol Rep, 2020, 28, e00548.
[10] Kim J, Heo D, Wang J, Kim H, Ota S, Takemura Y, Huh C, Bae S. Pseudo-single domain colloidal superparamagnetic nanoparticles designed at a physiologically tolerable AC magnetic field for clinically safe hyperthermia. Nanoscale, 2021, 13, 19484–92.
[11] Zuk M, Gaweda W, Majkowska-Pilip A, Osial M, Wolski M, Bilewicz A, Krysinski P. Hybrid radiobioconjugated superparamagnetic iron oxide-based nanoparticles for multimodal cancer therapy. Pharmaceutics, 2021, 13, 1843.
[12] Lansdown A, Mirastschijski U, Stubbs N, Scanlon E, Agren M. Zinc in wound healing: Theoretical, experimental and clinical aspects. Wound Repair Regen, 2007, 15, 2–16.
[13] Homepage of European Food Safety Authority, Parma, Italy, https://www.efsa.europa.eu/de/news/titanium-dioxide-e171-no-longer-considered-safe-when-used-food-additive/ (accessed on 12.06.2021).
[14] Unterhalt B. Pharmanorganische Chemie – Anorganisch-chemische Wirkstoffe: Eigenschaften, Herstellung Verwendung, Wissenschaftliche Verlagsgesellschaft Stuttgart, Stuttgart, 2000.

[15] Yang J, Lee J, Ryu H, Elzatahry A, Alothmann Z, Choy J. Drug-clay nanohybrids as sustained delivery systems. Appl Clay Sci, 2016, 130, 20–32.

[16] Holtmeier W, Holtmann G, Caspary W, Weingärtner U. On-demand treatment of acute heartburn with the antacid hydrotalicte compared with famotidine and placebo. J Clin Gastroenterol, 2007, 41, 564–70.

[17] Lambert J, Midolo P. The actions of bismuth in the treatment of *Helicobacter pylori* infection. Aliment Pharmacol Ther, 1997, 11, 27–33.

[18] Kurreck A, Gronau F, Vilchez M, Abels A, Enghard A, Brandl R, Francis B, Föhre C, Lojewski C, Pratschke J, Thuss-Patience P, Modest D, Rau B, Feldbrügge L. Sodium thiosulfate reduces acute kidney injury in patients undergoing cytoreductive surgery lus hyperthermic itraperitoneal hemotherapy with cisplatin: A single-center observational study. Ann Surg Oncol, 2021, 29, 152–62.

[19] Homepage of European Chemicals Agency, Helsinki, Finland, https://echa.europa.eu/de/sub stance-information/-/substanceinfo/100.028.678/ (accessed on 24.06.2021).

[20] Danielsson R, Eriksson H. Aluminum adjuvants in vaccines – A way to modulate the immune response. Semin Cell Dev Biol, 2021, 115, 3–9.

[21] Adams M, Baker E, Blundell T, Harding, M, Dodson E, Hodkin D, Dodson G, Rimmer B, Vijayan M, Sheat S. Structure of rhombohedral 2 zinc insulin crystals. Nature, 1969, 224, 491–5.

[22] Dunn M. Zinc-ligand interactions modulate assembly and stability of the insulin hexamer – A review. BioMetals, 2005, 18, 295–303.

[23] Medici S, Peana M, Nurchi V, Zoroddu M. Medical uses of silver: History, myths, and scientific evidence. J Med Chem, 2019, 62, 5923–43.

[24] Moore D, MacDonald N. Preventing ophthalmia neonatorum. Can J Infect Dis Med Microbiol, 2015, 26, 122–5.

[25] Homepage of National Centre for Immunisation Research and Surveillance, Australia, Sydney, Fact Sheet Thiomersal, https://www.ncirs.org.au/sites/default/files/2018-12/thiomersal-fact -sheet.pdf (accessed on 04.11.2021).

13.5 Inorganic biomaterials

Reinhard Maletz, Matthias Epple

13.5.1 General aspects

Biomaterials are applied to replace or restore lost body functions [1–4]. For this, they have to be biocompatible, i.e. suitable for the intended purpose. The term "biocompatibility" comprises three major aspects, i.e. mechanical, chemical and biological compatibility. Mechanical compatibility means appropriate stability (e.g. for a hip implant) or elasticity (e.g. for an intravascular stent), including the absence of mechanical wear and fatigue during continuous mechanical load. Chemical compatibility involves the absence of adverse chemical reactions, like the release of toxic ions by corrosion or the degradation of a polymer to non-toxic reaction products. Finally, biological compatibility requires the absence of unwanted biological responses in the body, e.g. the avoidance of immunological reactions and an implant surface that is suitable for cell on-growth in the case of bone implants. It must be stressed that the term "biocompatibility" depends on the intended application of a biomaterial, i.e. the different aspects will be variable depending on the medical case and implantation site [5].

Inorganic materials that are used in medicine come from the three main materials classes, i.e. metals, polymers and ceramics. Metallic biomaterials are usually applied for orthopedic implants (e.g. hip endoprostheses, plates, nails and screws), in orthodontics (e.g. brackets, orthodontic wires), in dentistry for tooth filling and tooth replacement (crowns and bridges) and in cardiovascular surgery (stents). Only a few metals and alloys are applied due to biocompatibility considerations; usually corrosion resistance and mechanical strength are the decisive factors. Most implant metals are non-biodegradable, with stainless steel, titanium and its alloys, cobalt-chromium alloys and nickel-titanium shape memory alloys (Nitinol) as most prominent examples. In dentistry, noble metal alloys based on gold, silver, platinum as well as the now less frequently used amalgam (silver-mercury alloy) are applied as well [1]. The biodegradable metals magnesium and iron have also gained some interest in the last decade as they can corrode and dissolve after having served their mechanical purpose, e.g. in intravascular biodegradable stents [6, 7].

Although most polymers are of organic nature, also in biomedicine, there are two inorganic polymers that have found their way into the clinics. Polysiloxanes (silicones) are used in artificial joints and breast implants [8, 9]; polyphosphazenes are applied as surgical suture material [10]. Silicones are usually not-biodegradable whereas polyphosphazenes can degrade to phosphoric acid and ammonia and their derivatives. Furthermore, many organic polymers are applied together with inorganic fillers as demonstrated below for tooth restoration.

https://doi.org/10.1515/9783110733471-029

Ceramics are by definition inorganic non-metallic materials. Most of them can be considered as ionic solids or sparingly inorganic salts, although they can be perceived as covalent giant molecules as well. Of course, the binding has covalent and ionic parts so that they can usually be regarded from both sides (consider, e.g. diamond, silicon nitride, quartz and silicates). In biomedicine, three kinds of ceramics are routinely applied [10]. First, calcium phosphates are used in bone contact due to their inherent similarity to human hard tissue [11]. Second, hard ceramics like zirconia and alumina are used in hip joints and for artificial teeth [12]. Third, silicate-based ceramics are used for tooth restoration, also together with organic polymers (see below).

In general, many developments in modern biomedicine rely on inorganic materials. In the following, typical examples are presented. Metallic biomaterials for prostheses are covered in Chapter 2.8.

13.5.2 Inorganic biomaterials in orthopedic surgery

13.5.2.1 Bone substitution and augmentation

Surgeons and dentists often face a situation where lost bone has to be replaced by a solid filler material to ensure a proper healing [13, 14]. Typical clinical instances are accidents (trauma), tumor extractions and inflamed or necrotic bone. Unfortunately, humans possess only little excess bone for re-implantation (there is some of it in the hip, i.e. the iliac crest, that can be transplanted). Bone from other people can be infectious, and animal bone leads to severe immunogenic effects and subsequent implant rejection. Therefore, artificial materials are of prime importance in the clinics. Calcium phosphate ceramics are most prominent due to their presence in human hard tissue, i.e. bone and teeth. For this purpose, calcium phosphate is employed by nature in most mammals, birds and fish since millions of years [15]. Figure 13.5.1 shows a fossilized shark tooth that consists of calcium phosphate.

Due to its omnipresence in the human body, calcium phosphate is biocompatible and biodegradable. It is therefore ideally suited to fill bone defects and to attract bone-forming cells (osteoblasts) to its surface and make them produce new bone. It is also convenient that calcium phosphates are generally sparingly soluble in water and easily and economically prepared by precipitation, sol-gel-synthesis and also by solid-state reaction (firing). In terms of biocompatibility, it is a clear advantage of synthetically prepared ceramics that they are neither immunogenic nor infectious, unlike materials of biological origin.

Chemically, calcium phosphate constitutes a whole family of different compounds, due to the inclination of the phosphate anion to occur in fully deprotonated (PO_4^{3-}), monohydrogenated (HPO_4^{2-}) and dihydrogenated ($H_2PO_4^-$) form, depending on the pH during precipitation [16]. Together with the cation Ca^{2+}, different combinations are possible. Many calcium phosphates have been given acronyms for quick

Figure 13.5.1: Calcium phosphate as basis of bone and teeth: a fossilized shark tooth, consisting of fluoroapatite (outer part: enamel) and hydroxyapatite (inner part: dentin).

identification, with hydroxyapatite, $Ca_5(PO_4)_3OH$ (HAP), tricalcium phosphate, Ca_3 $(PO_4)_2$ (α-TCP and β-TCP; two polymorphic phases with this stoichiometry are known), octacalcium phosphate $Ca_8H_2(PO_4)_6 \cdot 5\ H_2O$, dicalcium phosphate dihydrate, $CaHPO_4 \cdot 2\ H_2O$ (DCPD), and monocalcium phosphate monohydrate $Ca(HPO_4)_2 \cdot H_2O$ (MCPM) as most prominent ones. The mineral in mammalian hard tissue is always hydroxyapatite which is the least soluble calcium phosphate phase at neutral pH. Interestingly, sharks use a derivative of hydroxyapatite, i.e. fluoroapatite, $Ca_5(PO_4)_3F$ (FAP), as biomineral in the outer part of their teeth, for yet unknown reason [17].

Calcium phosphates are frequently applied to fill bone defects. Figure 13.5.2 gives a few examples. Granulated particles can be prepared by agglomeration of a previously precipitated or ground powder. If a ceramic is compacted, e.g. by sintering, it can then be shaped into objects with defined shape, including pores. This is advantageous to fill bone cavities with a given shape, and also to permit the ingrowth of bone cells, followed by slow resorption of calcium phosphate and its replacement by newly grown bone. Notably, porous calcium phosphate implants can also be prepared from animal bone via high-temperature calcination (>1000 °C). This preserves the porous bone structure and the biomineral calcium phosphate but removes all potentially infectious and immunogenic organic parts of the bone (proteins, bacteria, cells etc.). A related material is a calcium phosphate cement where a calcium phosphate phase is precipitated in situ in a bone defect [18]. Bioglasses are amorphous solids that are based on a silica body which accommodates CaO and P_2O_5 as part of the SiO_2 network structure. At a sufficiently high content of calcium and phosphate, a bioglass is biodegradable in the body [19].

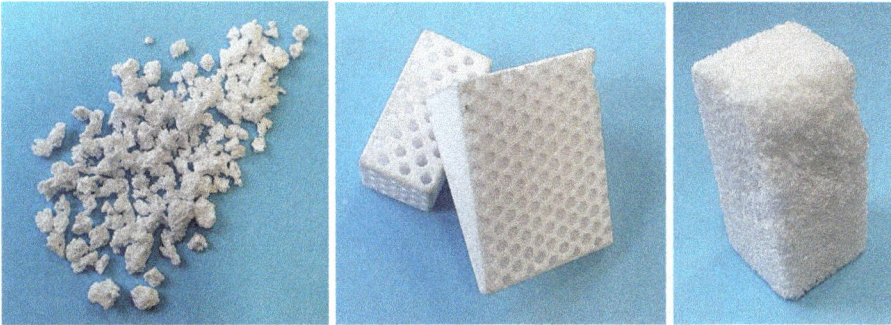

Figure 13.5.2: Calcium phosphate-based bone substitution materials. Left: Granulated β-tricalcium phosphate; center: hot-pressed and machine-cut β-tricalcium phosphate; right: sintered bovine bone, consisting of hydroxyapatite.

13.5.2.2 Endoprostheses

The second major area where ceramics are used in biomedicine are endoprostheses to replace degraded joints, usually in the hip and in the knee [2]. Figure 13.5.3 shows a typical example of an artificial hip joint that is implanted with the long shaft into the thigh (femur) and fixed with the cup in the pelvic bone.

To ensure a good mechanical fixation of the shaft in the femoral bone, it is often coated with calcium phosphate. This enhances the bone-bonding in comparison to the bare metal (usually stainless steel or a titanium alloy) and makes the arrangement more stable for mechanical load during walking. The calcium phosphate layer is usually applied by high-temperature plasma-spraying where a calcium phosphate powder is passed through a plasma torch. The droplets of the molten calcium phosphate are then sprayed onto the metallic implant where they solidify and firmly attach to the metal surface [20].

The joint itself must have a low friction coefficient and a low degree of wear over the two- to three-decade-long lifetime of a hip joint. The most common combination is a ball of stainless steel in combination with a cup of high-molecular weight polyethylene (HMWPE). This combination is easy to manufacture and comparably inexpensive. However, it suffers from a considerable wear of the softer part, i.e. the polyethylene, over time, which leads to a gradual loosening of the ball-cup joint. Furthermore, the debris particles of polyethylene cause a low-level but persistent inflammation around the joint which is held responsible for the eventual failure of such an endoprosthesis after 10 or 20 years. This problem can be overcome by using a hard-ceramic ball-cup pair, e.g. from alumina (Al_2O_3) or zirconia (ZrO_2). This is more expensive to manufacture and to shape but leads to much less wear. It is therefore expected to considerably enhance the lifetime of an artificial hip endoprosthesis [21].

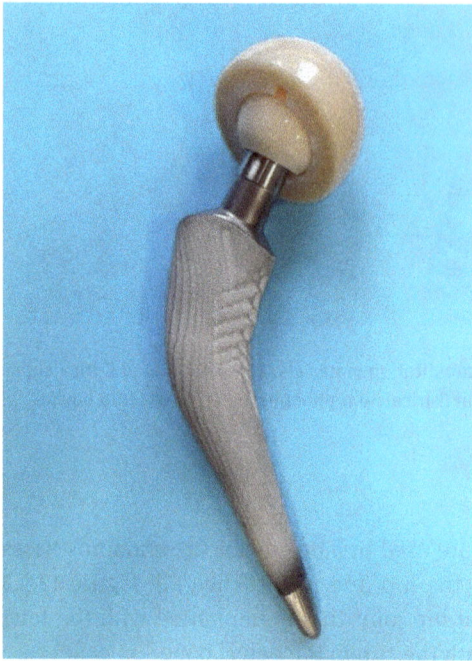

Figure 13.5.3: A modular hip endoprosthesis, consisting of a metal shaft, coated with calcium phosphate (white) for bone on-growth, and an aluminum oxide ball/cup as artificial joint.

13.5.3 Inorganic biomaterials in dentistry

Dentistry with its sub-disciplines of conservative dentistry, prosthodontics, orthodontics, implantology and prophylaxis is a broad field for innovative material solutions that facilitate very good clinical success rates for various indication areas and optimize dental treatment processes. Inorganic materials are frequently the focus of product developments in this area. Restorations of purely inorganic composition are found, for example, in titanium implants or in the use of precious or non-precious metal alloys and zirconia or silicate ceramics for crowns and bridges. In the case of sculpturable restorative composites, the inorganic fillers play a decisive role in defining important properties such as strength, abrasion resistance and gloss. Inorganic chemistry is also the basis for the class of materials known as glass ionomer cements, which were developed exclusively for dentistry. An acid-base setting reaction of fluoroaluminosilicate glass particles with aqueous polycarboxylic acids produces biocompatible cements with a wide range of indications. Furthermore, inorganic fluorides are the basis for effective caries prophylaxis.

13.5.3.1 Dental ceramics for indirect restorations

The widespread use of ceramics in the fabrication of crowns and bridges was enabled by the development of ceramics for veneering metal framework. The coefficient of thermal expansion was adapted to the metal, thereby making it suitable for indications involving masticatory loads up to the posterior teeth.

Ceramic restorations without metal support, so-called monolithic or all-ceramic restorations, became available in the 1980s. The pioneer restorations in this field were initially made of feldspar ceramic, which were supplemented by leucite-reinforced glass-ceramics at the time [22]. However, as both systems are used in all-ceramic restorations, they are only suitable for indications of smaller dimensions, such as inlays or low-load single crowns.

Further development steps for improved strength led to glass-ceramics with a high crystallite content (approx. 70% by volume). These silicate ceramics can be used as monolithic restorations or as framework in combination with feldspar ceramics. An important representative of this material class is the lithium disilicate glass-ceramic from the system $SiO_2–Li_2O–Al_2O_3–K_2O–ZrO_2$, which can be used for bridges up to three-units due to its high strength. Upon heat treatment, the material initially forms lithium metasilicates at 550–750 °C, which crystallize to lithium disilicate crystals by reaching 840 °C.

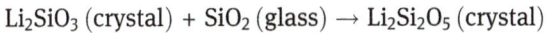

$$Li_2SiO_3 \text{ (crystal)} + SiO_2 \text{ (glass)} \rightarrow Li_2Si_2O_5 \text{ (crystal)}$$

In order to make this technology feasible for subtractive manufacturing using CAD/CAM, easy-to-mill blocks of predominantly lithium metasilicates are produced. Crystallization to $Li_2Si_2O_5$ crystals takes place only after subtractive machining at the user's site in his own furnace. The pre-crystallized form has a bluish cast and acquires the tooth color not before the crystallization firing step is completed [23].

Currently, the most important oxide ceramic in dentistry is polycrystalline zirconia (ZrO_2). The material properties allow the substitution of metal as a framework for long-span bridges as well as for implants. Its color, which is more attractive than that of any metal, even allows for monolithic restorations. These applications were made possible by stabilizing the tetragonal modification of zirconia at room temperature by doping it with oxides such as MgO, CaO, CeO_2 or Y_2O_3. Depending on the amount of dopant, a partially stabilized zirconia (PSZ) or a fully stabilized zirconia (FSZ) can be produced.

The tetragonal partially stabilized phase is of particular importance, since under certain conditions, such as tensile stresses, a partial transformation to the monoclinic phase can occur. This localized transformation is advantageous in terms of the mechanical properties of the zirconia. When a crack in the material is forced, it triggers a spontaneous transformation of the tetragonal to the monoclinic phase, which is associated with a volume increase in the grains, leading to an increase in the resistance to further propagation of the crack. This is the reason why the mechanism is also called "transformation amplification" [24].

If a very fine-grained microstructure is generated by using fine powders, high-pressure and low-temperature sintering options, the metastable zirconia with a tetragonal structure gains exceptionally high strengths. Such zirconia is also called polycrystalline tetragonal zirconia (TZP – tetragonal zirconia polycrystal). A typical example is the commercially available 3 mol% yttria-stabilized zirconia (3Y-TZP) with about 98% tetragonal phase.

If the amount of doping oxides is further increased, the cubic phase configuration can be stabilized at room temperature. While the "classical" 3Y-TZP has very high flexural strengths of over 1000 MPa, but shows low translucency, the 5Y-TZP with cubic-tetragonal stabilized phase (about 53% cubic) has a lower flexural strength of about 600 MPa, but shows high translucency due to significantly reduced light scattering resulting in better aesthetic properties [25] (Figure 13.5.4).

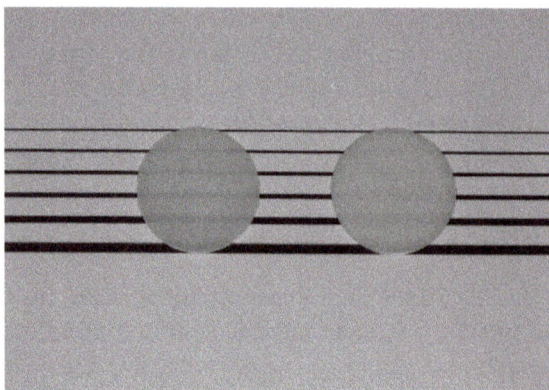

Figure 13.5.4: Comparison of the translucencies of different zirconia: Left: 5Y-TPZ; right: 3Y-TPZ.

The widespread use of densely sintered zirconia in the dental field has been made possible especially by advances in CAD/CAM techniques. State of the art is the use of pre-sintered blocks or blanks, which can be easily processed by a milling machine and only require a second sintering step in a furnace to finish the restoration piece. The CAD/CAM work flow for the manufacturing of zirconia bridge is shown in Figure 13.5.5.

13.5.3.2 Glass-ionomer cements for direct restorations

The history of dental cements began in the nineteenth century with zinc phosphate cement formed by zinc oxide and phosphoric acid. At the beginning of the twentieth century, silicate cement, which hardens by reaction of aluminum silicate glasses with phosphoric acid, followed as a tooth-colored filling. However, the clinical applications of these cements were not very broad due to low strength and lack of adhesion to the tooth structure. The demand for a sculpturable dental filling material with good

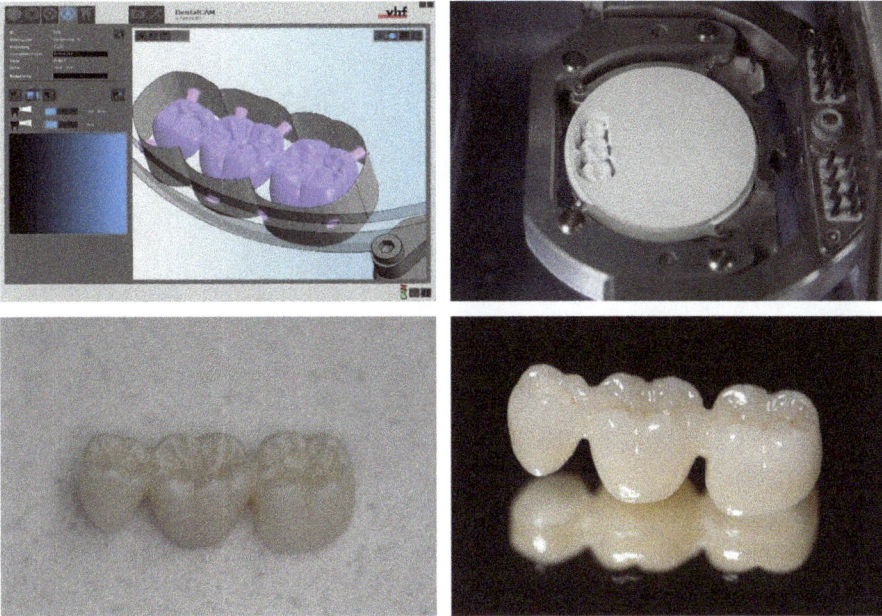

Figure 13.5.5: CAD/CAM manufacturing of a zirconia 3-unit posterior bridge: Top left: Nesting of the bridge in the CAM software; top right: manufacturing of the pre-sintered zirconia bridge in a 5-axis milling machine; bottom left: bridge after sintering; bottom right: completion of the restoration by color individualization and glaze firing.

physical properties, biocompatibility as well as adhesion to the tooth and appealing esthetics initiated the development of glass ionomer cements (GIC) in the late 1960s.

These materials are powder/liquid systems that cure by acid-base reaction. The powder consists of finely ground reactive fluoroaluminosilicate glasses (20–50 µm) made of the three essential constituents SiO_2, Al_2O_3, and CaF_2, to which $AlPO_4$ and cryolite (Na_3AlF_6) may be added, and which are melted at 1100–1500 °C [26]. Calcium is often replaced by strontium to increase radiopacity. The liquid is an aqueous polycarboxylic acid, most commonly polyacrylic acid or its copolymers with maleic or itaconic acid. After mixing the two components, the setting process starts. First the acid attacks the glass network releasing cations, mainly Na^+, K^+, Ca^{2+} and Al^{3+}. With these cations, salt bridges are formed between the polycarboxylic acid chains resulting in a hydrogel matrix. Here, calcium polyacrylate formation shows advantageous reaction kinetics and thus proceeds faster than aluminum polyacrylate formation [27]. Due to the acid attack on the surface of the glass particles, the acid-soluble components are dissolved and the acid-insoluble SiO_2 framework remains on the particle surface. The glass core remains intact and serves as a filler in the cement matrix (Figure 13.5.6).

Figure 13.5.6: SEM image of a cured GIC showing the glass cores with the SiO_2 framework in the hydrogel matrix.

Water plays a crucial role in the setting process. During setting, external moisture must be kept away from the cement to prevent the loss of cations. After setting, water prevents dehydration of the material. The setting process can lead to further post-curing over a period of up to one year [28]. With compressive strengths of 200–300 MPa and flexural strengths up to 50 MPa, glass ionomer cements achieve physical properties making them suitable for posterior applications. GICs also perform well under hot/cold stress in the mouth, as they adapt very well to the surrounding tooth structure with coefficients of thermal expansion of $10–12 \cdot 10^6$ K^{-1} [29].

The good chemical adhesion of GIC to the hydroxyapatite of the tooth (HAP) is of great importance for clinical use. This results from the reaction of the carboxyl groups of the polycarboxylic acid with the phosphate groups of the HAP, while electroneutrality is being restored by Ca^{2+} [28]. Furthermore, the continuous fluoride release of GIC causes a cariostatic effect, preventing secondary caries effectively. Overall, the biocompatibility of this class of materials is considered as very good.

GIC are attractive materials from a clinical point of view, as they are less time-consuming, technique-sensitive and cost-intensive as compared to composites. By varying the powder/liquid ratios, different consistencies can be realized, which are used in restorative therapy in the anterior and posterior regions and for the cementation of prosthetic restorations. In addition, GICs are frequently used for semi-permanent fillings and in pediatric and geriatric dentistry. They are also predestined for atraumatic restorative treatment (ART), a procedure based on excavating carious teeth using hand instruments only, which is preferably used in developing countries.

The application of GICs has been further optimized with the introduction of mixing capsules replacing the error-prone hand-mixing variants. After mechanical activation of the capsule, the contained pre-portioned powder and liquid quantities are mixed for a short time in a high-frequency mixer; thereupon the dentist can apply the homogenized material directly from the capsule into the cavity. The functional principle of a mixing capsule and the clinical application in the context of a posterior filling are shown in Figures 13.5.7 and 13.5.8.

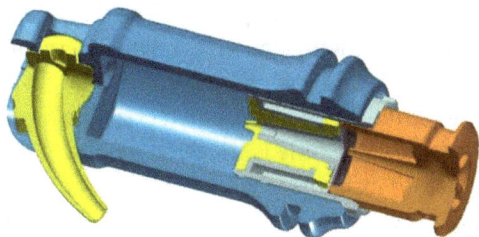

Figure 13.5.7: When activating a GIC mixing capsule by pressing the plunger (orange), the liquid is transferred from the plunger into the chamber containing the powder (blue). Mixing then takes place for 10 s at a frequency of 4000 rpm.

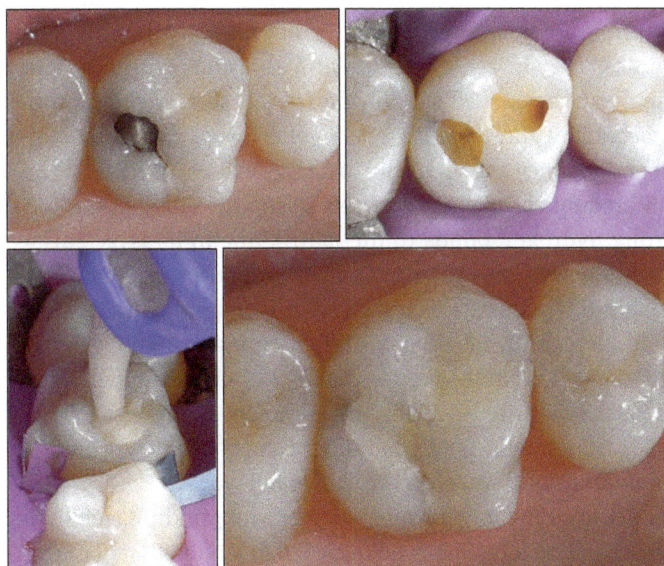

Figure 13.5.8: Replacement of defective restorations with glass ionomer cement.

13.5.3.3 Composite technology

Dental composites, i.e. materials made of polymethacrylates and inorganic fillers, represent the material class with the most comprehensive range of indications in the field of dentistry due to an enormous range of different properties. Indirect composites, as used for CAD/CAM materials or veneering systems, are already cured at the manufacturer or in the dental laboratory. In the case of direct composites, a paste is cured intraorally [22]. In addition to adhesive restorative therapy, fissure sealing, core buildup, cementation of prosthetic restorations and temporary restorations represent further important clinical applications.

The basic monomers of modern composites are bifunctional methacrylates with constitutive parts selected from the groups of functionalized bisphenols, urethanes, ethylene glycols or (cyclo)alkanes. Mixtures of bisphenol A glycidyl methacrylate (BisGMA), urethane dimethacrylate (UDMA) and triethylene glycol dimethacrylate (TEGDMA) are frequently used.

For adhesive fillings or fissure sealants, the free-radical polymerization with the Norrish type II photoinitiator camphorquinone/amine is triggered by blue light [30]. If photopolymerization is not possible, for example when luting indirect composite restorations or making temporary crown and bridges with composites, a chemical initiation system is indicated. In these cases, dibenzoyl peroxide is often used as an initiator in combination with an accelerator amine, which necessitates the use of a two-component system.

The most significant progress in composites has been achieved through continuous improvements of inorganic filler systems. The fillers are not only the main component of a dental composite in terms of quantity; they also significantly determine its relevant properties. They increase strength and abrasion resistance, lower the coefficient of thermal expansion, reduce polymerization shrinkage and water absorption, improve polishability and esthetics, enable effective diagnosis by X-rays and thus define clinical performance and longevity to a large extent.

In addition to small amounts of highly dispersed silicon dioxide (approx. 2–10 wt%), which forms strong hydrogen bonds to the matrix due to its large surface area and thus significantly determines the rheological properties of the material (e.g. thixotropy), the main filler consists of splinter-like inorganic particles in the μm range. The significant increase in the quality of light-curing restorative composites is directly linked to innovations in the field of milling techniques. Initially, macrofiller composites with particle sizes of 10–50 μm were developed. This was followed by midifiller composites (1–10 μm) and finally minifiller composites (0.6–1 μm) [31]. Typical particle size distributions of glass fillers are shown in Figure 13.5.9.

The transition from dry grinding to very sophisticated wet grinding techniques with special design of the grinding aggregates not only led to a reduction of the particle size with an associated narrow particle size distribution but also established the basis for highly translucent composites by avoiding grinding abrasion. Today, glass particles in a size range <200 nm are accessible (Figure 13.5.10). The consistent reduction of filler sizes leads to composites with optimized polishability/gloss and improved compressive strength.

In the early days, quartz was used as a filler. However, special dental glasses were developed soon. Nowadays, many filling materials are based on glasses made of SiO_2, Al_2O_3, B_2O_3 and BaO or SrO. The alkaline earth elements are used to increase radiopacity. One problem is that larger amounts of barium or strontium increase the refractive index of the glass. However, to obtain a composite with high translucency for highly

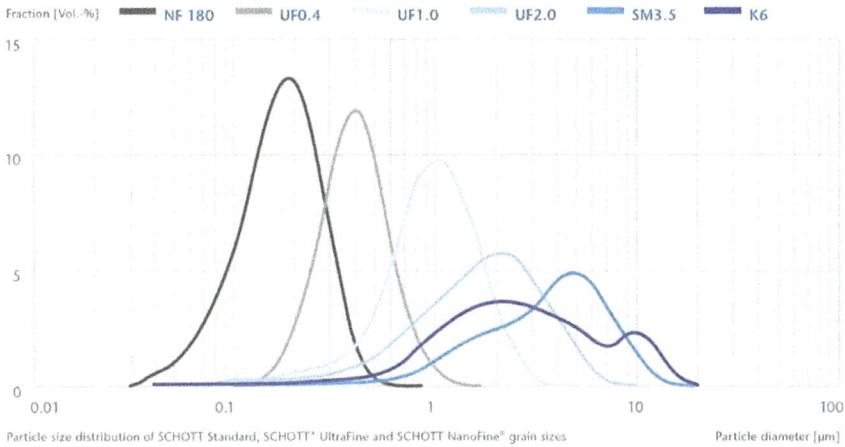

Figure 13.5.9: Particle size distribution of glass filler for dental composites. The standard K and SM grain sizes are produced by means of dry grinding technology, the Ultrafine (UF) and Nanofine (NF) grain sizes are produced by wet grinding technology (© Schott AG).

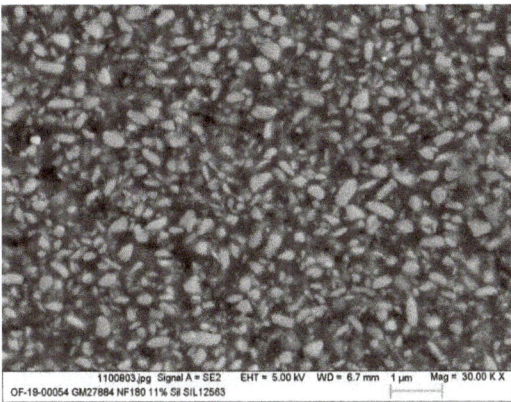

Figure 13.5.10: SEM image of a cured composite with a glass filler of an average particle size of 180 nm. The value 180 nm corresponds to the modal value of the volume distribution of the equivalent diameters of the glass particles (© Schott AG).

esthetic restorations, the refractive indices of the methacrylate polymer, which ranges from 1.50 to 1.58, and the fillers must match as closely as possible to minimize light scattering at the interfaces. From there, one is limited in implementing very high X-ray visibilities solely by the glasses, since increasing the amounts of heavy elements even further would shift the refractive index out of the desired range. A neat trick to improve X-ray diagnostics is the use of YbF_3, a rare earth compound with an amazingly low refractive index of 1.55, which is in the desired range and thus allows high light transmission. A composite filling with high radiopacity is shown in Figure 13.5.11.

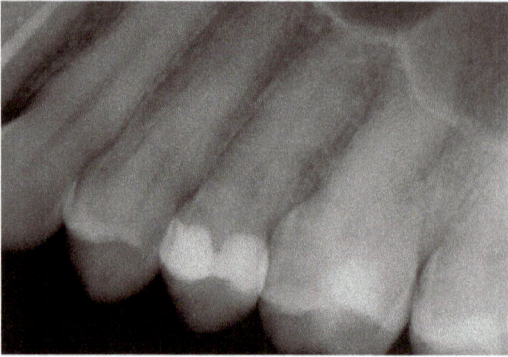

Figure 13.5.11: X-ray diagnostic check of a composite filling (with courtesy of Prof. Dr. J. Manhart, LMU Munich).

In order to covalently bind the silicate-based fillers to the polymer matrix, they are coated with methacryloyl silanes. This results in significantly higher strength values and improved longevity.

There is a constant effort to maximize the inorganic filler content in order to improve surface hardness, abrasion resistance, flexural and compressive strength and to reduce polymerization shrinkage by minimizing the amount of polymerizable matrix at the same time. However, classical composite technology is limited by the fact that decreasing particle size comes along with an increasing particle surface, which would eventually increase the viscosity of the composite too much due to the increased interactions with the monomers. To get around this obstacle, the "nano effect" is exploited. If surface-functionalized, non-agglomerated SiO_2 nanoparticles <50 nm are incorporated in the monomer matrix, its viscosity hardly increases, even at high particle contents of approx. 50 wt%. Now, glass particles can be used to "fill up" the composite to a viscosity suitable for processing thereby reaching a total inorganic filler content of 89 wt%. This approach, also known as nanohybrid technology, microscopically follows the principle of high space filling (Figure 13.5.12) and results in very resistant, low-shrinkage composites. Modern nanohybrid composites achieve very high strengths (flexural strength >180 MPa, Young's modulus >16 GPa, compressive strength >400 MPa) with low polymerization shrinkage of 1.6 vol%.

Unlike amalgam, which can only be anchored in the tooth by creating appropriate macroretentions, composites are adhesively bonded to the tooth structure using strong dentin/enamel bonding agents stabilizing the tooth. Composite filling therapy also enables defect-oriented, minimally invasive cavity preparation that is gentle to the tooth substance. Despite preliminary skepticism, the fillings proved to be durable. In a clinically controlled study, an overall success rate of 97.1% was determined for posterior restorations with composite systems after 12 years [32]. In addition, the fillings are highly esthetic and blend into the natural tooth surrounding because composite systems perfectly imitate both the different shades and the

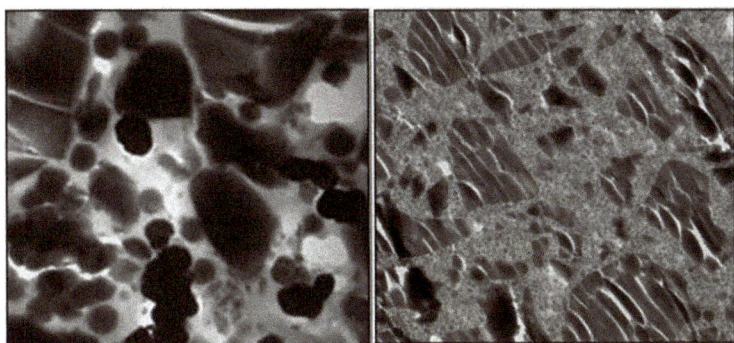

Figure 13.5.12: TEM images of ultra-thin sections of a common minifiller composite (left) and a nanohybrid composite (right) with glass fillers of approx. 1 µm (with courtesy of Prof. Dr. M. Warkentin, University of Rostock).

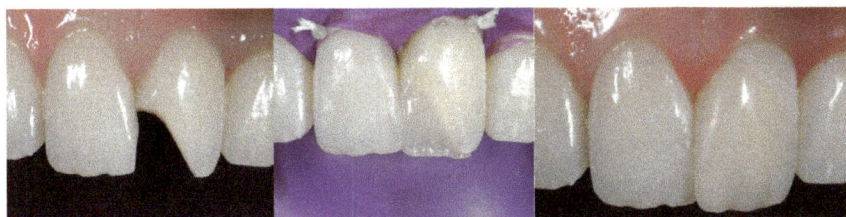

Figure 13.5.13: Restoration of an anterior tooth trauma by means of an adhesive composite filling.

different translucency levels of dentin and enamel. The clinical case in Figure 13.5.13 shows the perfect esthetic result of an adhesive composite filling.

The success in the continuing development of composites into high-strength materials also makes this class of materials attractive for the subtractive fabrication of dental restorations using CAD/CAM techniques. Compared to ceramics, composite restorations are antagonist-friendly, are easy to customize/repair and take less time to fabricate because they do not require sintering. The latest generation of nanohybrid composites actually achieves fracture strengths for crowns or anterior and premolar bridges that are in the range of lithium disilicate ceramics.

13.5.3.4 Inorganic-organic hybrid polymers – ORMOCER

The material class of ORMOCERs has been developed by the Fraunhofer Institute for Silicate Research in recent decades and has been made usable for many applications. These are inorganic-organic hybrid polymers that can be designed according to a modular system. ORMOCERs are based on multifunctional silanes which contain hydrolysable/condensable alkoxysilyl groups, an organic polymerisable functional group (e.g. a methacrylate group), and an organic spacer unit as connecting element. In addition, further modifications can be brought into the network by functionalizing the spacer unit. The variability of this modular system via multi-functional silane precursors as well as the structure of corresponding methacylate-functionalized silanes are shown in Figures 13.5.14 and 13.5.15.

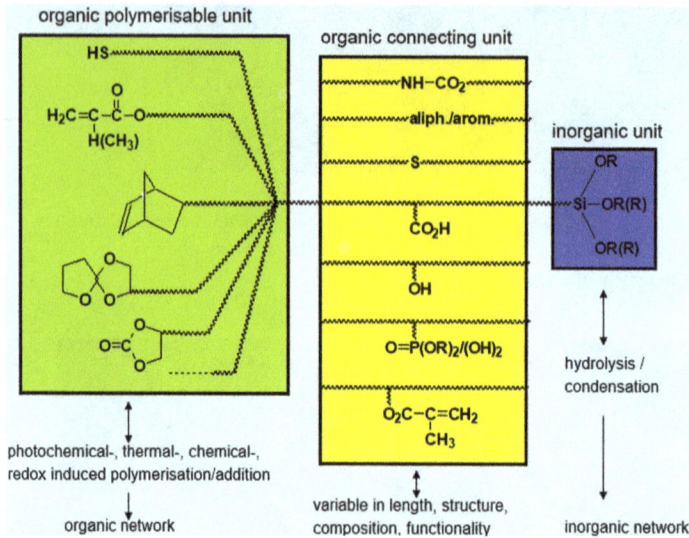

Figure 13.5.14: Various combinations of organic polymerizable units, organic connecting units and inorganic units provide the basis for the modular system of ORMOCERs (with courtesy of Fraunhofer ISC, Würzburg).

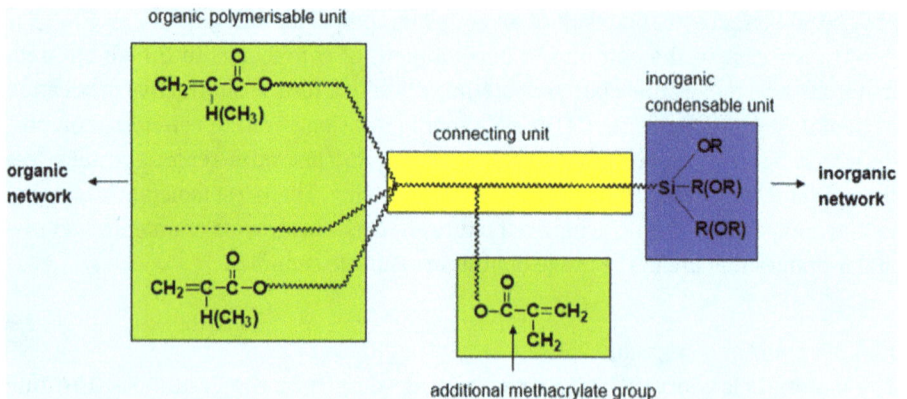

Figure 13.5.15: Methacrylate-functionalized silanes are the starting materials for free-radical polymerizable ORMOCER resins (with courtesy of Fraunhofer ISC, Würzburg).

The inorganic network of ORMOCERs is formed first by classical inorganic sol-gel-processing reactions. Controlled hydrolysis and condensation of the multifunctional silanes lead to an oligomeric Si-O-Si-nanostructure (Figure 13.5.16).

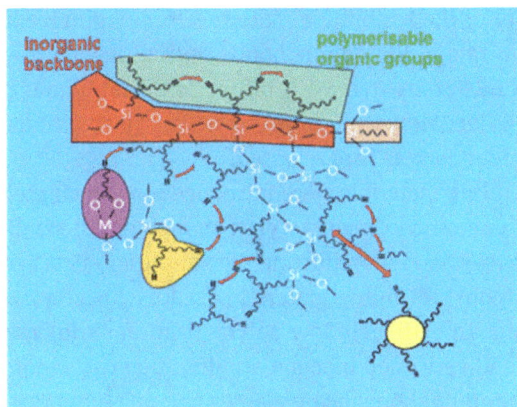

Figure 13.5.16: Schematic illustration of the ORMOCER structure after formation of the Si-O-Si-network. In addition to the polymerizable double bonds, further heteroatoms (M) or organic functional groups can optionally be incorporated (with courtesy of Fraunhofer ISC, Würzburg).

In the second step the organic polymerizable functional groups are crosslinked during the final curing step (e.g. photopolymerization). The concept is to combine properties of organic polymers (functionalization, toughness) with properties of glass-like materials (hardness, transparency) in order to generate new/synergetic properties not accessible otherwise with classical composites. By choosing the desired amount of the different structural elements of ORMOCERs their properties can be adjusted in a wide range between more inorganic or more organic compositions [33]. Materials based on ORMOCER technology are used, for example, in the scratch-resistant coating of eyeglass lenses or as hollow fiber membranes for gas separation processes [34].

The variability of inorganic and organic structural components resulting from the synthesis principles predestines this class of hybrid polymers for use in dental filling composites. The basic idea is to substitute the classical dimethacrylates of composites partially or preferably completely by ORMOCER structures on the level of the already polycondensed, methacrylate-functionalized Si-alkoxides.

Thus, according to the modular system, the silane precursors can be specifically adapted to the desired properties. Due to the variability of organic and inorganic network density and controlled spacer length between both network forming sites, Young's modulus and thermal expansion coefficients can be controlled [34]. Furthermore, the refractive index of the matrix can be perfectly matched to that of the inorganic fillers by selective choice of aliphatic and aromatic spacer groups [35].

The even greater advantage of using an ORMOCER matrix, however, is the significant reduction of polymerization shrinkage, since the photo-induced linkage of the methacrylate groups of the ORMOCER macromolecules leads to lower shrinkage compared to significantly lower molecular weight dimethacrylates. In addition, this approach allows the biocompatibility of the restorative materials to be optimized, since, assuming the same conversion rate of the double bonds during polymerization of the multifunctional ORMOCERs compared with bifunctional monomers, there are, from a statistical point of view alone, no unlinked polymerizable units and thus, in effect, no more "residual monomers" are released.

In a first step, ORMOCER-based filling materials were introduced to the market at the end of the 1990s, which still contained dimethacrylate monomers in small proportions for viscosity adjustment reasons. By optimizing the syntheses, 15 years later it was possible to develop a nanohybrid filling material based exclusively on ORMOCER technology, which exhibits an extremely low polymerization shrinkage of 1.25 vol%, high surface hardness and very good biocompatibility. Due to its inorganic filler structure consisting of aluminum borosilicate glasses and SiO_2 nanoparticles in combination with the Si-O-Si structure of the polymerizable matrix, this system is referred to as "pure silicate technology" (Figure 13.5.17).

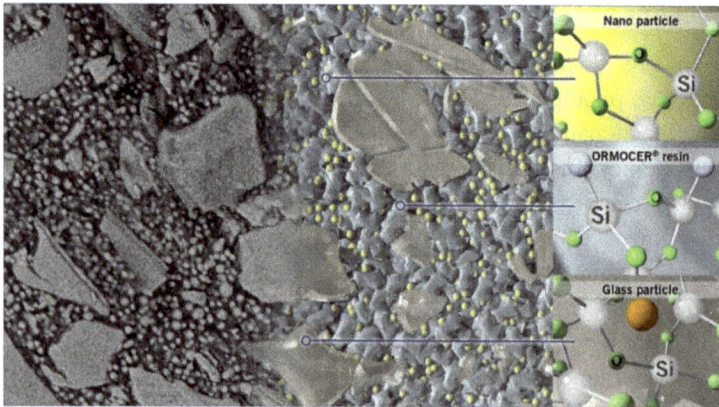

Figure 13.5.17: Illustration of pure silicate technology based on a TEM image of a Nano-Hybrid-ORMOCER filling material (with courtesy of Prof. Dr. D. Behrend, University of Rostock). All components, ORMOCER resin, glass particles and nanoparticles are formed by silicate structures.

References

[1] Epple M. Biomaterialien und Biomineralisation. Eine Einführung für Naturwissenschaftler, Mediziner und Ingenieure, Teubner, Wiesbaden, 2003.

[2] Ratner BD, Hoffman AS, Schoen FJ. Biomaterials Science. An Introduction to Materials in Medicine, Academic Press, 2004.

[3]   Williams DF. On the nature of biomaterials. Biomaterials, 2009, 30, 5897–909.
[4]   Wintermantel E, Ha SW. Medizintechnik – Life Science Engineering, Springer, Berlin
       Heidelberg, 2009.
[5]   Williams DF. On the mechanisms of biocompatibility. Biomaterials, 2008, 29, 2941–53.
[6]   Witte F, Hort N, Vogt C, Cohen S, Kainer KU, Willumeit R, et al. Degradable biomaterials based
       on magnesium corrosion. Curr Opin Solid State Mater Sci, 2008, 12, 63–72.
[7]   Iqbal J, Onuma Y, Ormiston J, Abizaid A, Waksman R, Serruys P. Bioresorbable scaffolds:
       Rationale, current status, challenges, and future. Eur Heart J, 2014, 35, 765–76.
[8]   Berry MG, Davies DM. Breast augmentation: Part I – a review of the silicone prosthesis. J
       Plast Reconstr Aesthet Surg, 2010, 63, 1761–8.
[9]   Sokolova V, Epple M. Brustimplantate. Chemie in unserer Zeit, 2012, 46, 76–9.
[10]   Ulery BD, Nair LS, Laurencin CT. Biomedical applications of biodegradable polymers. J Polym
       Sci Pt B-Polym Phys, 2011, 49, 832–64.
[11]   Habraken W, Habibovic P, Epple M, Bohner M. Calcium phosphates in biomedical
       applications: Materials for the future?. Mater Today, 2016, 19, 69–87.
[12]   Piconi C, Maccauro G. Zirconia as ceramic biomaterial. Biomaterials, 1999, 20, 1–25.
[13]   Dorozhkin SV, Epple M. Die biologische und medizinische Bedeutung von
       Calciumphosphaten. Angew Chem, 2002, 114, 3260–77.
[14]   Heinemann S, Gelinsky M, Worch H, Hanke T. Resorbable bone substitution materials. An
       overview commercially grouting materials and new research approaches in the area of
       composites. Orthopade, 2011, 40, 761–73.
[15]   Baeuerlein E, Behrens P, Epple M (Eds). Handbook of Biomineralization, Wiley-VCH,
       Weinheim, 2007.
[16]   Dorozhkin SV. Bioceramics of calcium orthophosphates. Biomaterials, 2010, 31, 1465–85.
[17]   Enax J, Prymak O, Raabe D, Epple M. Structure, composition, and mechanical properties of
       shark teeth. J Struct Biol, 2012, 178, 290–9.
[18]   Chen ZG, Zhang XL, Kang LZ, Xu F, Wang ZL, Cui FZ, et al. Recent progress in injectable bone
       repair materials research. Front Mater Sci, 2015, 9, 332–45.
[19]   Jones JR. Review of bioactive glass: From Hench to hybrids. Acta Biomater, 2013, 9, 4457–86.
[20]   Heimann RB. Thermal spraying of biomaterials. Surf Coat Technol, 2006, 201, 2012–9.
[21]   Heimke G, Leyen S, Willmann G. Knee arthroplasty: Recently developed ceramics offer new
       solutions. Biomaterials, 2002, 23, 1539–51.
[22]   Ilie N, Stawarczyk B. Direkte und indirekte Komposite. In: Rosentritt M, Ilie N, Lohbauer U.
       (Eds). Werkstoffkunde in der Zahnmedizin, Georg Thieme Verlag, Stuttgart, 2018, 183–237.
[23]   Lohbauer U, Belli R, Wendler M. Keramische materialien (I). In: Rosentritt M, Ilie N, Lohbauer
       U (Eds). Werkstoffkunde in der Zahnmedizin, Georg Thieme Verlag, Stuttgart, 2018.
[24]   Garvie RC, Hannink RH, Pascoe RT. Ceramic steel?. Nature, 1975, 258, 703–4.
[25]   Kieschnik A, Rosentritt M, Kleine BS. Werkstoffkunde für Zahnärzte – Teil 3: Zirkonoxide.
       Zahnärztliche Mitteilungen (Zm), 2019, 109, 448–51.
[26]   Wilson AD, McLean JW. Glasionomerzement, Quintessenz Verlags-GmbH, Berlin, 1988.
[27]   Nicholson JW, Brookman PJ, Lacy OM, Wilson AD. Fourier transform infrared spectroscopic
       study of the role of tartaric acid in glass-ionomer dental cements. J Dent Res, 1988, 67,
       1451–4.
[28]   Lohbauer U. Glasionomerzemente. In: Rosentritt M, Ilie N, Lohbauer U (Eds). Werkstoffkunde
       in der Zahnmedizin, Georg Thieme Verlag, Stuttgart, 2018, 79–90.
[29]   Craig RC. Restaurative Dental Materials, 11th edition, Mosby, London, 2002.
[30]   Ikemura K, Endo T. A review of the development of radical photopolymerization initiators
       used for designing light-curing dental adhesives and resin composites. Dent Mater J, 2010,
       29, 481–501.

[31] Ferracane JL. Resin composite – state of the art. Dent Mater, 2011, 27, 29–38.
[32] Frankenberger R, Reinelt C, Glatthöfer C, Krämer N. Clinical performance and SEM marginal quality of extended posterior resin composite restorations after 12 years. Dent Mater, 2020, 36, e217–e28.
[33] Haas KH. Hybrid inorganic–organic polymers based on organically modified Si-alkoxides. Adv Eng Mater, 2000, 2, 571–82.
[34] Haas KH. Crosslinked heteropolysiloxanes as inorganic-organic polymers: Precursors, synthesis, properties and applications. In: Nalwa HS (Ed). Handbook of Organic-inorganic Hybrid Materials and Nanocomposites. 1, American Scientific Publishers, 2003, 207–39.
[35] Haas KH, Wolter H. Synthesis, properties and applications of inorganic–organic copolymers (ORMOCER®s). Curr Opin Solid State Mater Sci, 1999, 4, 571–80.

14 Criticality of inorganic resources

Michael Binnewies

14.1 Introduction

The standard of living of almost all people is directly linked to the global availability of raw materials. This applies equally to inorganic and organic precursors. Organic compounds originate in living nature. Fossil sources are crude oil, natural gas and the various forms of coal (hard coal, lignite, etc.). They belong to the non-regenerative raw materials and are predominantly used as energy sources. The remaining organic raw materials are mostly agricultural products. These are the nutritional basis of all living beings on earth. In principle, they are regenerative sources. All inorganic raw materials, on the other hand, are non-regenerative in nature. They come predominantly from the lithosphere. For industrialized countries, these raw materials are of particular economic importance, along with energy supply. The secure supply of raw materials of any kind is directly linked to the standard of living of almost all people, even in the less industrialized countries. The reserves of non-regenerative raw materials are not infinite. Moreover, most of them are geographically very unevenly distributed. Thus, it is only logical that industrialized countries in particular most carefully observe the availability of raw materials in the present and the future.

When assessing the geological availability of raw materials, the following terms are used:

- **Reserve**: reserves are deposits that can be exploited economically with the current state of technology.
- **Resource**: resources are deposits that are proven to exist but cannot be exploited economically with the current state of technology.
- **Range**: Duration (in years, (a)) during which the reserves or resources can still be used. Since the future demand for a raw material is unknown and difficult to predict, it is often referred to as **static range** (**static reserve range**, **static resource range**). This refers to the range that arithmetically results if the future demand for a raw material is the same as the current demand. As a rule, however, this will not be the case.

The actual range of a raw material will in most cases be less than the static range, because the demand increases over time in most cases. This is illustrated by one example. The consumption of tellurium has increased more than 400% from 1999 to 2019 [1]. This is due to the increasing use (in the form of cadmium telluride, CdTe) in thin-film solar cells and in Peltier elements (e.g., in the form of bismuth telluride, Bi_2Te_3). In rarer cases, the demand may also decrease. An example of this is the building material asbestos. Global consumption of asbestos decreased by

https://doi.org/10.1515/9783110733471-030

approximately 40% from 1999 to 2019 [2]. The reasons for the declining demand are well known.

14.2 Global availability of raw materials

Table 14.1 compiles data on selected non-regenerative inorganic raw materials (column 1). Columns 2 and 3 contain data on global mine production in 1999 and 2019. In Columns 4 and 5 are data on the size of reserves and resources, respectively, according to the state of knowledge in 2019. Columns 6 and 7 compile the static ranges of reserves (1999/2019) and the static ranges of resources (1999/2019) for the selected raw materials, and column 8 contains the percentage changes in the respective ranges from 1999 to 2019. For the elements gallium, indium and germanium, which are technologically very important today, no meaningful data can be given for the reserves and resources, because no mineable deposits of minerals containing these elements are known. Gallium, indium and germanium are by-elements that may be produced as by-products in the mining of other elements. Gallium is a tramp element of aluminum. Indium and germanium are found in some deposits of zinc ores (ZnS). The available amounts of gallium, indium and germanium are thus linked to the processed amounts of bauxite and sphalerite, respectively.

Looking at the data on the static ranges of reserves for 2019 (column 6), it becomes clear that the reserves of 24 of the 38 raw materials considered will be exhausted in less than 100 years. However, it is commonly assumed that once the raw material reserves are exhausted, the prospective use of raw material resources will set in. In this case, the number of raw materials with a range of less than 100 years is reduced from 24 to 8 (numbers in bold in column 7). It can therefore be assumed for the near future that, from a global perspective, there will not be an acute shortage of the raw materials mentioned within the next few decades. In addition, it is to be expected that, as a result of successes in the exploration, resources will be bigger in the future than is known today. However, this situation can also rapidly change if new types of technology cause the demand for certain raw materials to rise sharply. An example is indium. In the last century, indium was used for certain solders and as a sealing material in vacuum technology, i.e., only in very specific fields of application. Today, indium is essential for display technology (ITO, indium-tin-oxide) and is used much more extensively. From 1970 to today, the production of indium and indium compounds has increased by a factor of about 10. Conversely, new technologies can also reduce the demand for a particular raw material. For example, the use of silver (as AgBr) for the photographic process is virtually non-existent today.

Increasing quantities of certain industrial goods reduce the ranges of the raw materials they contain. The ever shorter service lives of some industrial products are also additionally contributing to the shortening of ranges. On the other hand,

Table 14.1: Mine production, reserves, resources, static ranges of reserves and resources in the years 1999 and 2019.

1	2	3	4	5	6	7	8
Raw material/ element	Mine production 1999 (10^6 t)	Mine production 2019 (10^6 t)	Reserves 2019 (10^6 t)	Resources 2019 (10^6 t)	Static range of reserves 1999/2019 (a)	Static range of resources 1999/2019(a)	Change 1999→2019 (%)
Li	$15 \cdot 10^{-3}$ [3]	$77 \cdot 10^{-3}$ [a] [3]	17 [3]	80 [3]	1133 / 221	5333 / 1039	−81
Be	$336 \cdot 10^{-6}$ [4]	$260 \cdot 10^{-6}$ [4]	n.a. [4]	$>100 \cdot 10^{-3}$ [4]	n.a.	297 / 385	30
Sr	$315 \cdot 10^{-3}$ [5]	$220 \cdot 10^{-3}$ [5]	6.8 [5]	>1000 [5]	22 / 31	3175 / 4545	43
Barite	3.8 [6]	9.5 [6]	320 [6]	740 [6]	84 / 34	195 / 78	−60
Titanium ore	4 [a] [7]	7.6 [7]	820 [7]	>2000 [7]	205 / 108	$>500 / >263$	−47
Zr	$815 \cdot 10^{-3}$ [a] [8]	$1400 \cdot 10^{-3}$ [8]	62 [8]	n.a. [8]	76 / 44		−42
V	$40 \cdot 10^{-3}$ [9]	$73 \cdot 10^{-3}$ [9]	22 [9]	>63 [9]	550 / 301	1575 / 863	−45
Nb	$18.5 \cdot 10^{-3}$ [10]	$74 \cdot 10^{-3}$ [10]	>13 [10]	n.a. [10]	$>703 / >176$		−75
Ta	$473 \cdot 10^{-6}$ [11]	$1,800 \cdot 10^{-6}$ [11]	$>90 \cdot 10^{-3}$ [11]	$>150 \cdot 10^{-3}$ [12]	$>190 / >50$	$>317 / >83$	−75

(continued)

Table 14.1 (continued)

1	2	3	4	5	6	7	8
Raw material/element	Mine production 1999 (10^6 t)	Mine production 2019 (10^6 t)	Reserves 2019 (10^6 t)	Resources 2019 (10^6 t)	Static range of reserves 1999/2019 (a)	Static range of resources 1999/2019(a)	Change 1999→2019 (%)
Cr	12.8 [13]	44 [13]	570 [13]	>12,000 [13]	45 / 13	>938 / >273	−71
Mo	$129 \cdot 10^{-3}$ [14]	$290 \cdot 10^{-3}$ [14]	18 [14]	25.4 [14]	140 / 62	197 / 88	−55
W	$31.3 \cdot 10^{-3}$ [15]	$85 \cdot 10^{-3}$ [15]	3.2 [15]	>6.2 [16]	102 / 38	>198 / >73	−63
Mn	6.7 [17]	19 [17]	810 [17]	>5100 [16]	121 / 43	>761 / >268	−65
Re	$44 \cdot 10^{-6}$ [18]	$49 \cdot 10^{-6}$ [18]	$2.4 \cdot 10^{-3}$ [18]	$11 \cdot 10^{-3}$ [18]	55 / 49	250 / 224	−10
Iron ore	992 [19]	2,500 [19]	170,000 [19]	>800,000 [19]	171 / 68	>806 / >320	−60
Fe (Pig iron)	521 [20]	1300 [20]					
Co	$28.3 \cdot 10^{-3}$ [21]	$140 \cdot 10^{-3}$ [21]	7 [21]	25 [21]	247 / 50	883 / 178	−80
Ni	1.1 [22]	2.7 [22]	89 [22]	>130 [22]	81 / 33	118 / 48	−59
Cu	12.6 [23]	20 [23]	870 [23]	>2100 [23]	69 / 44	>167 / >105	−37

Ag	$15.9 \cdot 10^{-3}$ [24]	$27 \cdot 10^{-3}$ [24]	$560 \cdot 10^{-3}$ [24]	$>570 \cdot 10^{-3}$ [16]	35 / 21	>36 / > **21**	−42
Au	$2.3 \cdot 10^{-3}$ [25]	$3.3 \cdot 10^{-3}$ [25]	$50 \cdot 10^{-3}$ [25]	$90 \cdot 10^{-3}$ [16]	22 / 15	39 / **27**	−31
Zn	7.6 [26]	13 [26]	250 [26]	1,900 [26]	33 /19	250 / **146**	−42
Bauxite	123 [a] [27]	$370^{[a]}$ [27]	30,000 [27]	>55,000 [27]	244 / 81	>447 / >**149**	−67
Al	22.7	64					
Ga	n.a.	$320 \cdot 10^{-6}$ [28]	n.a.	n.a.			
In	$240 \cdot 10^{-6}$ [29]	$760 \cdot 10^{-6}$ [29]	n.a.	n.a.			
Natural graphite	$578 \cdot 10^{-3}$ [30]	1.1 [30]	15 [30]	>800 [30]	519 / 272	>1384 / >**727**	−47
Ge	$58 \cdot 10^{-6}$ [31]	$130 \cdot 10^{-6}$ [31]	n.a.	n.a.			
Sn	0.21 [32]	0.31 [32]	4.7 [32]	n.a.	22 / 15	n.a.	−32
Pb	3 [33]	4.5 [33]	90 [33]	>2000 [33]	30 / 20	>667 / >**444**	−33
Phosphate rock	138 [34]	240 [34]	69,000 [34]	n.a.	500 / 228	n.a.	−42

(continued)

Table 14.1 (continued)

1	2	3	4	5	6	7	8
Raw material/ element	Mine production 1999 (10^6 t)	Mine production 2019 (10^6 t)	Reserves 2019 (10^6 t)	Resources 2019 (10^6 t)	Static range of reserves 1999/2019 (a)	Static range of resources 1999/2019(a)	Change 1999→2019 (%)
As	$41.5 \cdot 10^{-3}$ [35]	$33 \cdot 10^{-3}$ [35]	$750 \cdot 10^{-3}$ [16]	11 [16]	18 / 23	265 / 333	26
Sb	$138 \cdot 10^{-3}$ [36]	$160 \cdot 10^{-3}$ [36]	1.5 [36]	>3.9 [16]	11 / 9	>28 / >24	−14
Se	$1.3 \cdot 10^{-3}$ [37]	$2.8 \cdot 10^{-3}$ [37]	$99 \cdot 10^{-3}$ [37]	n.a.	76 / 35	n.a.	−54
Te	$110 \cdot 10^{-6}$ [38]	$470 \cdot 10^{-6}$ [38]	$31 \cdot 10^{-3}$ [38]	$>47 \cdot 10^{-3}$ [16]	282 / 66	427 / 100	−77
Fluorspar	4.2 [39]	7 [39]	310 [39]	5,000 [39]	74 / 44	1190 / 714	−40
PGM	$275 \cdot 10^{-6}$ (Pd + Pt) [40]	$390 \cdot 10^{-6}$ (Pd + Pt) [40]	$69 \cdot 10^{-3}$ [40]	$100 \cdot 10^{-3}$ [40]	251 / 177	364 / 256	−30
REE	$76.5 \cdot 10^{-3}$ [41]	$210 \cdot 10^{-3}$ [41]	120 [41]	n.a.	1569 / 571	n.a.	−64

PGM: Platinum group metals.
REE: Rare earth elements
n.a.: Not available
aWithout US

the constantly advancing miniaturization of certain components, particularly in electronics, contributes to lower consumption of raw materials. It is therefore very difficult to predict how the ranges of raw materials will develop in the future.

A range of a raw material of, say, 100 or even 500 years, may seem reassuring at first glance. However, in relation to the several hundred thousand years history of homo sapiens, a period of a few hundred years is only a tiny time span. The ranges mentioned also mean that industrial activities will only be possible to a limited extent once natural deposits have been exhausted, unless there is a fundamental change in the way raw materials are handled. Managing raw materials in a cycle must necessarily take on much greater significance in the future than it does today.

14.3 Criticality of raw material supply using the EU as an example

So, from a global point of view, a shortage of any inorganic raw material is not to be expected in the near future. Nevertheless, many countries, especially the industrialized nations, some communities of states like the European Union, banks, insurance companies, industrial companies, investors etc. are following the global situation with utmost attention. The main reason for this is the very uneven geographical distribution of raw materials. The fewer countries that own a certain raw material and also offer it on the world market, the greater the supply risk for the countries that do not own them but need their supply. The same applies to the companies that extract and process raw materials. The fewer the companies provide them, the greater the supply risk. This is also referred to as **country concentration** or **company concentration**, or **country risk** or **company risk**. Risks for the supply also lie in the political circumstances of the producing countries and the general geopolitical situation. Additionally, the pricing of technology raw materials, which is essentially determined by the producing countries and companies, is also a cause of concern for the industrialized nations. All this shows that the concept of criticality can only ever be country-specific. There is no international agreement on which elements or raw materials are "critical" in terms of their supply situation.

Like supply risk, the economic importance of a particular commodity varies from country to country. The literature often plots the supply risk against the economic importance of a commodity in a diagram (Figure 14.1). If a raw material has a high supply risk and a high economic importance for a certain country, it is counted as a critical raw material. Raw materials with high criticality are located in the gray area in Figure 14.1.

Using the example of the European Union, an overview is given below of which raw materials are considered critical. At three-years intervals since 2011, the European Commission has published a list of industrial raw materials whose supply situation is considered critical for the EU (Table 14.2).

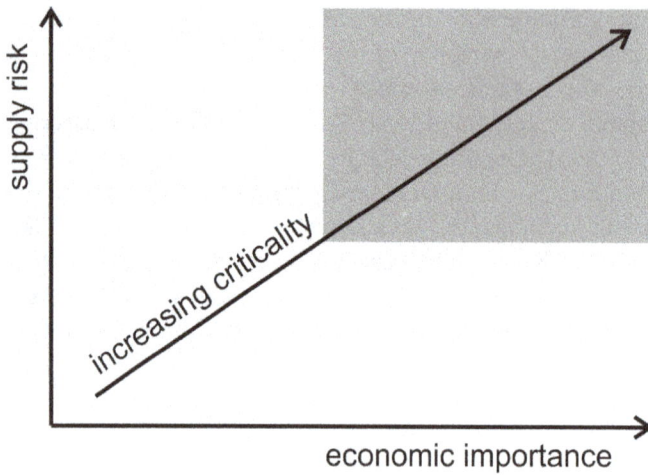

Figure 14.1: Diagram of criticality.

Table 14.2: European lists of critical raw materials 2011–2020 [42–45].

2011	2014	2017	2020	Import reliance* (2020) [45]
			Li	100%
Be	Be	Be	Be	n.a.
	Magnesite			
Mg	Mg	Mg	Mg	100%
			Sr	0%
		Baryte	Baryte	70%
		Sc	Sc	100%
			Ti	100%
		Hf	Hf	0%
		V	V	n.a.
Nb	Nb	Nb	Nb	100%
Ta		Ta	Ta	99%
	Cr			
W	W	W	W	n.a.
Co	Co	Co	Co	86%

Table 14.2 (continued)

2011	2014	2017	2020	Import reliance* (2020) [45]
	Borate	Borate	Borate	100%
			Bauxite	87%
Ga	Ga	Ga	Ga	31%
In	In	In	In	0%
Natural graphite	Natural graphite	Natural graphite	Natural graphite	98%
	Coking coal	Coking coal	Coking coal	62%
	Si (element)	Si (element)	Si (element)	63%
Ge	Ge	Ge	Ge	31%
		P	P	100%
	Phosphate rock	Phosphate rock	Phosphate rock	84%
Sb	Sb	Sb	Sb	100%
		Bi	Bi	100%
Fluorspar	Fluorspar	Fluorspar	Fluorspar	66%
		He		
		Natural rubber	Natural rubber	100%
PGM	PGM	PGM	PGM	100%
REE	REE	REE	REE	100%
Quantity 14	19	26	29	

PGM: Platinum group metals
REE: Rare earth elements
n.a.: Not available
*(Import − Export)/(Domestic production + Import − Export)

It is apparent that the number of critical raw materials has roughly doubled from 2014 to 2020. A reversal of this trend is hardly to be expected. Remarkably, there are quite a few raw materials on this list considered, although there are also a lot of those that have particularly long ranges of well over a hundred years from a global perspective: Li, Be, Sr, V, Co, natural graphite, phosphate rock, fluorspar, PGM, REE. This is mainly due to the high country risk of some of these raw materials. The following selection makes this clear: Li: 78% from Chile [45], Be: 88% from the USA [45], Sr: 100% from Spain [45], Co: 68% from DR Congo [45], natural graphite: 47% from

China [45], phosphate rock: 71% from Kazakhstan [45], PGM: approx. 90% from RSA and Russia [45], REE: 98% from China [45].

14.3.1 Selected examples

14.3.1.1 Lithium

With increasing adoption of electrically powered vehicles, the supply of lithium feedstocks is of growing importance. World annual production from primary feedstocks in 2019 was 77,000 t, 55% of which was in Australia [3]. However, by far the largest reserves are in Chile, which accounted for 23% of the world annual production in 2019. However, the world's largest resources of lithium are in Bolivia, which is currently not even one of the producing countries [3]. Today, it is assumed that the world's resources of lithium are sufficient to power all vehicles worldwide with lithium-ion batteries. In addition, it is expected that in the future significant shares of lithium consumption will be covered by secondary raw materials (battery recycling).

14.3.1.2 Rare earth elements

In particular, the extreme dependence on China for the import of REE raw materials is problematic, as these elements play a central role in electromobility and power generation from regenerative energies (wind turbines). While it is true that there are a number of deposits of raw REE sources worldwide, economically justifiable exploitation is currently practically non-existent. In this context, the attempt by the former President of the USA, Donald Trump, to buy Greenland, which has considerable deposits of these elements, should be mentioned. Understandably, this request was rejected by Denmark, whose territory includes Greenland.

14.3.1.3 Platinum group metals

The country concentration for platinum group metals (PGMs) is also extremely high. An economically significant application of some of these metals is the fabrication of automotive catalytic converters. Worldwide, 72.4 million passenger cars were manufactured in 2019, of which about 25 million were produced by EU automakers [46]. Each of these vehicles contains about 2 g of platinum metals (Pd, Pt, Rh) in its exhaust catalyst. Thus, car manufacturers worldwide consume about 150 t PGM per year (of 390 t annual mine production, see Table 14.1). EU car manufacturers therefore account for about 50 t of PGMs per year. Should the Republic of South Africa and Russia – for whatever reasons – no longer be able or willing to supply these metals, the annual global production of these metals from primary raw materials would amount to only approx. 40–80 tons. The worldwide economic consequences would be incalculable, also for the European automotive industry. Replacing combustion engines with electric motors would solve this problem.

14.3.1.4 Indium

The demand for indium has been growing steadily for years. There are two reasons for this. Since the ban on solders containing lead in the EU from 2007, alloys containing indium have been used in the electronics industry. This has significantly increased the demand for indium. However, it is also required in rapidly growing quantities as one of the transparent conductive oxides (TCO) for the fabrication of displays. For example, the fabrication of a 32'' display requires about 3 g of indium in the form of indium tin oxide (ITO) [47]. Electronic devices have progressively shorter lifetimes due to technological advancements. As a result, the demand for indium can be expected to increase significantly. However, the production of indium from primary raw materials cannot be increased at will. Indium is a by-product of zinc and lead production. It can be obtained as a by-product during the smelting of these metals. It is foreseeable that the demand for indium cannot be met from primary raw materials for very long. One way out of this situation may be the use of substitute materials. In the case of solders for the electronics industry, this could be a return to the situation before 2007. Finding a substitute for ITO is not as easy. It is true that there are several other TCOs known besides ITO. Among these, ITO is so far the most suitable for coating displays. The other alternative is to recycle indium from old equipment containing ITO. However, at present this is still uneconomical.

14.3.1.5 Germanium

The raw material situation for germanium is similar to that of indium. There are no known deposits of germanium worth mining. It is obtained in small quantities as a by-product of zinc smelting. The available quantity of germanium is therefore closely correlated with the production of zinc. Germanium also has a high country risk; about 2/3 of the germanium produced worldwide comes from China. Germanium is technologically of particular importance today (along with erbium and terbium) for high-speed data transmission in fiber optic networks. Since such fiber optic networks are long-lived assets, recovery of germanium from fiber optics is not an option in the foreseeable future.

14.4 Substitution of critical raw materials through development of new technologies

Once an element or raw material is classified as critical, this classification does not necessarily have to be permanent. If new high-yield primary deposits of this substance are discovered, the supply situation for this raw material can improve significantly. The development of new technologies for extracting this substance from primary or secondary raw materials can also lead to a lower criticality. Also, the development of new types of technologies can cause the demand for a particular raw material to decrease, but the demand for another raw material to increase. Table 14.3 gives some examples of how novel technologies that can result in the substitution of one raw material for another.

Table 14.3: New technologies: substitution of raw materials.

Technology (old)	Technology (new)	Raw material(s) (old)	Raw material(s) (new)
Photographic film	Semiconductors	Ag	Si
Light bulb	Light emitting diode (LED)	W	Ga, In, REE
Gas discharge lamp	Light emitting diode (LED)	Hg, Eu	Ga, In, REE
Cathode ray tube	Flat screen	W, Pb, Ba, Sr	In (ITO)
Power station	Wind power plant	Coal, oil, U	Cu, Nd, Co
Combustion engine	Electric motor	Oil, Al	Cu, Nd, Sm
Telephone	Mobile phone	Fe, Cu, Sn	Cu, Ag, Au, Sn, Pd, Ga, In, Si, Ta, Li, Fe, Co, Ni, REE . . .
–	Computer	–	Si
Data transfer old	Data transfer new	Cu	Ge

Table 14.3 lists some technologies that have undergone fundamental changes in recent decades. As a result of these changes, certain raw materials are no longer consumed or are consumed in smaller quantities, while the demand for other raw materials is increasing. This change in technology makes predictions about the future demand for a particular commodity very difficult. An example that each of us has in mind is the development of digital photography, which makes silver bromide coated films obsolete. We are currently experiencing a fundamental change in lighting technology: classic incandescent bulbs are virtually no longer used, and the end of gas discharge lamps is also in sight. Compared to the two old technologies, light-emitting diodes have the advantage of significantly lower energy consumption. For their production, gallium, indium and some REE are needed, elements with a critical supply situation from the point of view of the EU. The classic television set has long discarded. During the production of cathode ray tubes, considerable amounts of heavy metals (Pb, Ba, Sr) were needed to absorb the X-rays formed in the tube. The flat screens used today require indium (ITO) for the display. A modern wind turbine contains several hundred kilograms of neodymium, but saves fossil fuels. Smartphones contain a variety of different metals, including silver, gold and palladium. The content of precious metals makes it profitable to recycle electronic devices such as smartphones. However, many metals are lost in the process.

Modern computer technology has no technological antecedents. Computer chips have become mass products for which large quantities of high-purity silicon are produced. Worldwide, about 7 million tons of elemental silicon were produced in 2019, of which 4.5 million tons were produced in China [48]. Today, optical fibers or, better more precisely, optical waveguides are used for high-speed transmission of large amounts of data. Optical fibers are thin threads (diameter about 0.1 mm) of fused silica, which have a core of germanium dioxide. Optical fibers have a much lower attenuation than electrically conductive coaxial cables made of copper. Most of the germanium produced worldwide is nowadays consumed for the production of optical fibers.

14.5 Use of secondary raw materials: recycling

It is to be hoped and expected that in the future an increasing amount of secondary raw materials will be used to manufacture new industrial goods. For a number of metals, this is already state of the art. For many other metals, however, the recovery of metals from waste is still in its infancy. In this context, one also speaks of urban mining. Table 14.4 shows the proportion of metals and semimetals recovered from waste in a quality that allows them to be reused for their original purpose (end-of-life recycling rate EOL-RR) [49].

If we look at the end-of-life recycling rates for the metals classified as critical raw materials by the European Commission in 2020 (Table 14.4), it becomes clear that only a few of these elements are recovered from secondary raw materials to a greater extent: Li (<1%), Be (<1%), Mg (25–50%), Sr (<1%), Sc (<1%), Ti (>50%), Hf (<1%), V (<1%), Nb (>50%), Ta (<1%), W (10–25%), Co (>50%), Ga (<1%), In (<1%), Ge (<1%), Sb (1–10%), Bi (<1%), PGM (>50%), REE (<1%). In order to ensure a secure supply for the industrial nations in the future, it will be imperative to make increased use of secondary raw materials and to manufacture industrial goods from them. At present, many rare elements are still irretrievably lost in the processing of primary and secondary raw materials because it is not economical to extract them. Two examples may illustrate this.

- Many bauxite deposits contain small amounts of gallium. However, its separation is energy-intensive and requires many processing steps. Gallium is preferably extracted where energy and labor are cheap, in China. If the gallium is not separated, it remains in the red mud. This occurs in large quantities as a waste product in the production of aluminum. Red mud is landfilled or increasingly used in road construction and the ceramics industry.
- Electronic scrap contains appreciable amounts of tantalum (and other rare base elements). When electronic scrap is reprocessed, these elements are transferred to the slag produced during the processes. This is typically used in road construction. Recovery of valuable constituents then hardly seems possible at that point.

Table 14.4: End-of-life recycling rates for metals and semi-metals.

H																	He
Li	Be											B	C	N	O	F	Ne
Na	Mg											Al	Si	P	S	Cl	Ar
K	Ca	Sc	Ti	V	Cr	Mn	Fe	Co	Ni	Cu	Zn	Ga	Ge	As	Se	Br	Kr
Rb	Sr	Y	Zr	Nb	Mo	Tc	Ru	Rh	Pd	Ag	Cd	In	Sn	Sb	Te	I	Xe
Cs	Ba	La	Hf	Ta	W	Re	Os	Ir	Pt	Au	Hg	Tl	Pb	Bi	Po	At	Rn

Ce	Pr	Nd	Pm	Sm	Eu	Gd	Tb	Dy	Ho	Er	Tm	Yb	Lu

Aa	< 1%
Aa	1 – 10 %
Aa	10 – 25 %
Aa	25 – 50 %
Aa	> 50 %

Probably the best example of careful human handling of a non-regenerative raw material is gold. Gold is probably the metal that has been used by humanity for the longest time. It is also one of the rarest elements. It ranks 75th in element abundance, in close proximity to tellurium and platinum. Nevertheless, there is no real shortage of gold. This is essentially due to the fact that this metal, which people have known and used for about 6000 years, has always been held in special esteem. Gold was and is the epitome of value. Consequently, gold has been preserved, it has been used but not consumed. The vast majority of all gold ever mined is still available today. Transferring this way of dealing with non-regenerative raw materials to other rare elements would help to avoid or reduce future supply problems. In the longer term, any industrial activity must involve a circular economy of the raw materials used. For many materials, it may be still uneconomical to use them in cycles. However, these materials should at least be kept available for future generations, even if this involves entails costs.

References

[1]   United States Geological Survey, Tellurium, 1999/2019.
[2]   United States Geological Survey, Asbestos Statistics and Information, 1999/2019.
[3]   United States Geological Survey, Lithium, 1999/2019.
[4]   United States Geological Survey, Beryllium, 1999/2019.
[5]   United States Geological Survey, Strontium, 1999/2019.
[6]   United States Geological Survey, Barite, 1999/2019.
[7]   United States Geological Survey, Titanium mineral concentrates, 1999/2019.
[8]   United States Geological Survey, Zirconium and Hafnium, 1999/2019.
[9]   United States Geological Survey, Vanadium, 1999/2019.
[10]   United States Geological Survey, Niobium (Columbium), 1999/2019.
[11]   United States Geological Survey, Tantalum, 1999/2019.
[12]   Angerer G, Erdmann L, Marscheider-Weidemann F, Scharp M, Lüllmann A, Handke V,
       Marwede M. Rohstoffe für Zukunftstechnologien, Fraunhofer IRB, Stuttgart, 2009.
[13]   United States Geological Survey, Chromium, 1999/2019.
[14]   United States Geological Survey, Molybdenum, 1999/2019.
[15]   United States Geological Survey, Tungsten, 1999/2019.
[16]   Frondel M, Grösche P, Huchtemann D, Oberheitmann A, Peters J, Vance C, Angerer G,
       Sartorius C, Buchholz P, Röhling S, Wagner M. Trends der Angebots- und Nachfragesituation
       bei Mineralischen Rohstoffen, RWI, Essen, 2007.
[17]   United States Geological Survey, Manganese, 1999/2019.
[18]   United States Geological Survey, Rhenium, 1999/2019.
[19]   United States Geological Survey, Iron ore, 1999/2019.
[20]   United States Geological Survey, Iron and steel, 1999/2019.
[21]   United States Geological Survey, Cobalt, 1999/2019.
[22]   United States Geological Survey, Nickel, 1999/2019.
[23]   United States Geological Survey, Copper, 1999/2019.
[24]   United States Geological Survey, Silver, 1999/2019.
[25]   United States Geological Survey, Gold, 1999/2019.
[26]   United States Geological Survey, Zinc, 1999/2019.
[27]   United States Geological Survey, Bauxite and Alumina, 1999/2019.
[28]   United States Geological Survey, Gallium, 1999/2019.
[29]   United States Geological Survey, Indium, 1999/2019.
[30]   United States Geological Survey, Graphite (natural), 1999/2019.
[31]   United States Geological Survey, Germanium, 1999/2019.
[32]   United States Geological Survey, Tin, 1999/2019.
[33]   United States Geological Survey, Lead, 1999/2019.
[34]   United States Geological Survey, Phosphate Rock, 1999/2019.
[35]   United States Geological Survey, Arsenic, 1999/2019.
[36]   United States Geological Survey, Antimony, 1999/2019.
[37]   United States Geological Survey, Selenium, 1999/2019.
[38]   United States Geological Survey, Tellurium, 1999/2019.
[39]   United States Geological Survey, Fluorspar, 1999/2019.
[40]   United States Geological Survey, Platinum-Group Metals, 1999/2019.
[41]   United States Geological Survey, Rare Earths, 1999/2019.
[42]   European Commission, Report on Critical raw materials for the EU, 2011.
[43]   European Commission, Report on Critical raw materials for the EU, 2014.

[44] European Commission, Report on Critical raw materials for the EU, 2017.
[45] European Commission, Report on Critical raw materials for the EU, 2020.
[46] Handelsblatt 9.11.2020.
[47] Niederschlag E, Stelter M. Erzmetall, 2009, 62, 17–22.
[48] United States Geological Survey, Silicon.
[49] United Nations Environment Program, Recycling Rates of Metals, 2011.

15 Toxicology of inorganic compounds

Thomas Schupp

15.1 Introduction

Toxicology is the science of the adverse interaction between an organism and molecules. Since ancient times the knowledge about harmful and non-harmful materials is inherited from generation to generation, whereby hear-saying was (and perhaps still is) a dominant information source. Toxicity is something that can make people scary, and magic was (and partly is) associated with the mode of action of toxins; just think about the fairy-tale of Snow White and The Seven Dwarfs. A milestone in the understanding of toxicology was the statement of the Swiss physician Theophrastus Bombastus von Hohenheim, who used the artist name "Paracelsus", claiming that everything is toxic, there is nothing non-toxic. It is the dosage alone that discriminates whether a substance acts toxic or not[1] [1]. This statement was brought forward in defense against accusations he was going to intoxicate his patients. Note, that in those days mercury, arsenic and other potentially toxic compounds were used as pharmaceuticals (and still are today, for example mercury in 2-(ethylmercury thio) benzoic acid, CAS-No. 54-64-8). In fact, it is well known that pharmaceuticals can have toxic side-effects which become more and more predominant with increasing dose. Therefore, pharmacology and toxicology can be regarded as two sides of a coin for many substances and products. Mathieu Joseph Bonaventur Orfila was a French chemist in the nineteenth century who introduced systematic experiments with animals in toxicology. This created the basis for a better understanding of toxicity and the interaction between dose and response. From the very first notion of critical situations at workplaces, recommendations for precaution were noted. For example, in the sixteenth century Georg Agricola wrote in his book No. 6 that "we have to put more emphasis on the health than on the benefit to keep our ability to perform the daily tasks of work"[1] [2].

Having a cursory look on dose-response relationships, some basic concepts in toxicology can be visualized. With increasing dosage, the effect created increases. However, at low dosages, the effect may not be discernible from the background. For example, a substance may cause a reduction on the number of red blood cells (erythrocytes), which is known as anemia; however, the number of erythrocytes per µL in human blood is not a distinct, sharp value but ranges between about 4 and 6 million. Therefore, if compared to a control group, a slight reduction in erythrocyte count caused by a substance may still be in the normal background range and is not necessarily adverse. However, a more pronounced reduction below the background at a

---

1 Free translation of the author.

https://doi.org/10.1515/9783110733471-031

higher dosage may cause fatigue and, therefore, would be regarded as adverse (see Figure 15.1).

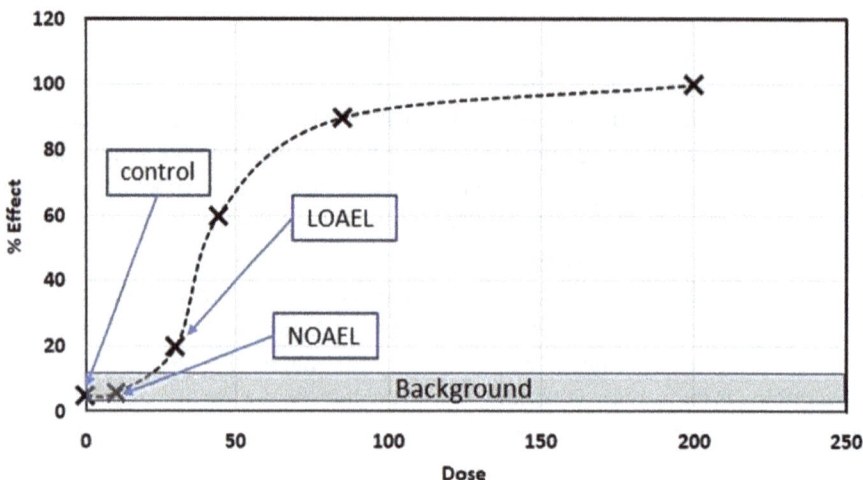

Figure 15.1: Generic dose-response curve for a toxic substance. "Background" describes the range of the natural variability in the population. In a test, where a dose creates an effect in the range of the population background, this is typically indicated as "not adverse".

The highest tested dose, that does not cause an observable adverse effect, is called the "No Observable Adverse Effect Level" (NOAEL); the lowest tested dose that creates an observable adverse effect is called "Lowest Observable Adverse Effect Level" (LOAEL).

Toxicologists may enter in lengthy and tedious discussions onto what is and what is not adverse. These discussions can be left to the experts. In general, an adverse effect is something that affects the fitness, life-span and/or reproduction performance of an organism. The terminus "Observable" pinpoints to the important fact that the design of the experiment chosen determines the accuracy of the observation. With a NOAEL, a threshold can be defined below which the substance is not to be expected to pose a threat to human or wildlife health. This is matching the paradigm of Paracelsus, and this is the basis for the derivation of exposure limits: starting from the NOAEL, certain safety-factors are used to derive tolerable dosages for workers or consumers. As a rule of thumb, a daily dose which is at least 100 times lower than the NOAEL in a chronic animal study is expected as low enough so detrimental effects to the health of consumers are not to be expected.

However, for certain effects, in theory it is not possible to define a threshold or something like a safe level. The typical example is a genotoxic carcinogen. From a purely theoretical point of view, a single molecule of the toxin may change the DNA of a cell, and this might be the final event to transform the cell to a malignant tumor cell. Nevertheless, the effect of very low dosages may be so small that it is blurred by the background, but in a strict sense, it cannot be calculated as "zero", as small as it ever may be (Figure 15.2). These substances with a stochastic activity are an exemption of the paradigm of Paracelsus. However, society could define a dosage where the expected effects are rated as sufficiently small (or unlikely) and tolerable against potential benefits the product provides (p. e. residual carcinogenic benzene in gasoline).

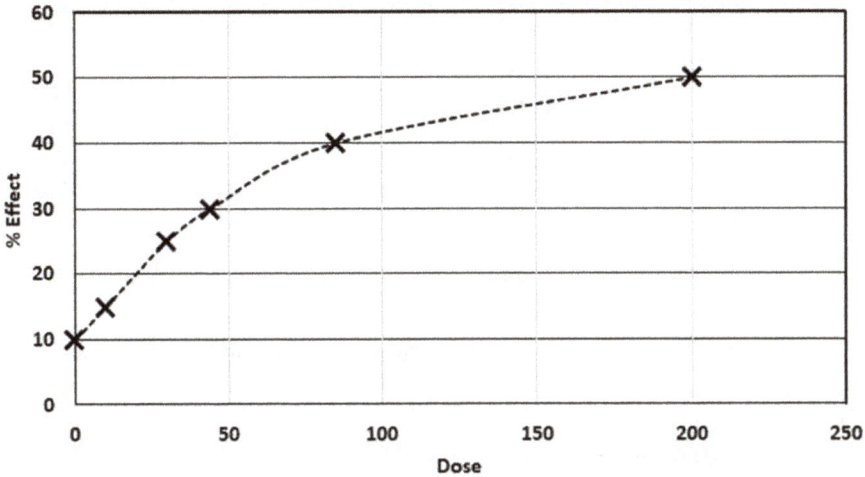

Figure 15.2: Dose-response for a substance without threshold.

A decline in dose is not always beneficial for an organism for every possible element. This is explicitly the case for trace elements. Many cobalt salts, for example, are classified as carcinogens, but cobalt also is the central metal in co-factors like vitamin B_{12}, to name an example. Therefore, a too low daily dose would result in adverse effects as well. This is schematically demonstrated in Figure 15.3.

The dose-response relationship of a substance in a toxicity test is crucial to estimate the risk posed by the substance. The toxicity is an inherent property and determines what kind of effect is triggered in a specific organ or tissue. The concentration of the substance in the respective organ/tissue determines the grade (or severity) of the specific effect and is a function of the exposure. Therefore, risk is defined as

$$RISK = TOXICITY \times EXPOSURE.$$

If the exposure is zero, there is no risk, irrespective what toxic potency is exerted by the substance. However, in case of a high toxic potency, a small exposure can result in critical risk.

After this introduction of some basic concepts in toxicology, the following sections will describe the toxicity of some selected inorganic compounds. The selection is arbitrary, based on the authors perception of historical or current relevance, and/

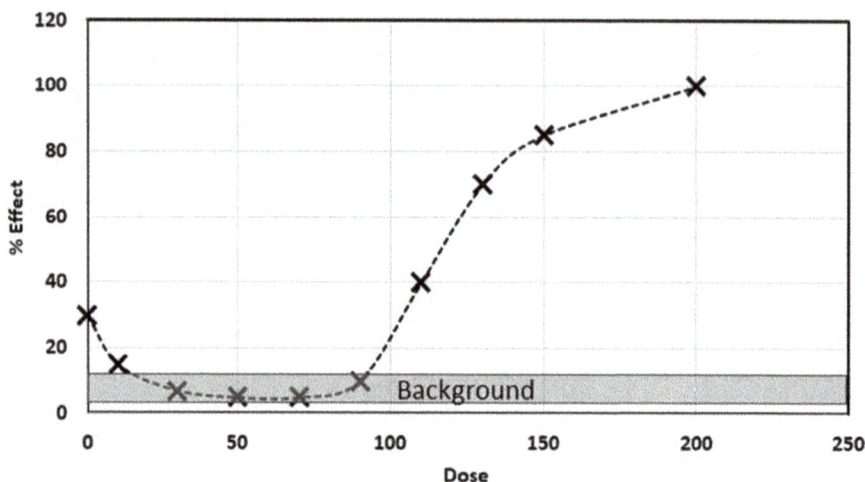

Figure 15.3: Dose-response for a trace element.

or knowledge about the molecular mechanisms of toxicity. Substance groups to be described are:
(i) Heavy metals and
(ii) Toxic and asphyxiant gases

15.2 Heavy metals

Heavy metals presented in this chapter include arsenic, lead, cadmium, and mercury. They have some history as toxins, partly due to workplace intoxication (mining, smelter), partly due to assassination. There is a history of chemical regulation over the last decades, and permitted uses were reduced step by step.

Because of their known toxicity, the use of several heavy metals has been restricted in many regions. Workplace exposure limits and biological limits have been set (p. e. permitted maximum values or guidance values in blood or urine). However, there is a geological background of lead, arsenic, mercury, and cadmium, but also dispersed emissions from incineration plants (power plants run with hard coal or lignite, crematoria, etc.) and partly from industrial processes. These emissions contribute to exposure of the general population via inhalation and via the food chain.

A common feature of these metals is their binding to sulfhydryl groups in proteins and co-factors which results in dysfunction of enzymes. Chelators like ethane-1,2-diamine-N,N,N',N'-tetraacetate (EDTA) or 2,3-dimercapto-propane sulfonate may be used for purging the body from heavy metals under close supervision by physicians.

15.2.1 Lead

Lead is known and used since antique times. To name a few examples, lead was used for piping, for balancing, and as shot by slingers in the Roman Empire. It also has had some use for the manufacturing of vessels for food and beverages. In combination with vinegar, lead metal can be dissolved by air oxidation:

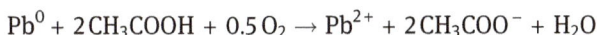

$$Pb^0 + 2\,CH_3COOH + 0.5\,O_2 \rightarrow Pb^{2+} + 2\,CH_3COO^- + H_2O$$

By this reaction, the acidic vinegar is transformed to lead acetate which has a sweet taste. The assumption that romans knew certain pottery being excellent in ameliorating acidic wine opens the door for speculations why the Roman Empire ceased. Misuse of lead acetate to improve wine quality was reported still in the early twentieth century [3]. The German poet Heinrich Heine was probably killed by lead poisoning [4]. That working with molten lead can be detrimental to health is known since long; the German engineer Georg Bauer (published in Latin as Georgius Agricola) recommended to eat butter as remedy for lead intoxication [2].

Lead piping is a known cause for intoxication of residents. Accordingly, this use is prohibited since decades,[2] and solder for drinking water pipes has to be free of Pb. Nevertheless, even in recent times old lead pipes in old buildings can be a cause for elevated lead levels in blood of residents [5].

Pottery can be a source of lead exposure. The European Union Ceramic Articles Directive (Directive 76/893/EEC) defines a migration limit of 0.8 mg Pb per dm^2 surface [6].

Lead pigments (see also Chapter 1.4) are appreciated because of their brilliance and opacity, for example white lead ($2PbCO_3 \cdot Pb(OH)_2$), Naples yellow ($Pb(SbO_3)_2 \cdot Pb_3(SbO_4)_2$), lead chromate ($PbCrO_4$, yellow) and lead molybdate ($PbMoO_4$, red orange). Red lead or minium (Pb_3O_4, deep red) can be used as desiccant in varnishes. These pigments find no longer commercial application.

Tetraethyl lead was used as anti-knock additive in gasoline. However, epidemiological evidence showed a convincing link between lead exposure and retarded mental development of children; living distance to high traffic roads correlated with blood lead levels and cognitive deficits. As a result, the use of tetraethyl lead as anti-knocking agent in gasoline was phased out in industrialized countries in the 1970s–1990s.

Other uses of Pb are lead sheets in roofing, rechargeable lead batteries (see also Chapter 5.1) and lead in ammunition. For the latter case, especially the use of gunshot can present a ticking time bomb as Pb in soil is taken up by plant crops and contributes to the daily Pb intake of consumers [7].

---

2 Germany: since 1973; see https://www.umweltbundesamt.de/umwelttipps-fuer-den-alltag/essen-trinken/blei-im-trinkwasser#unsere-tipps.

To protect consumers and workers in the European Union, the use of Pb and its compounds is subject to authorization according annex XVII of REGULATION (EC) No 1907/2006 [8]. For example, lead pigments for artwork to preserve historic paintings in an original state may be permitted by way of national derogation. According to Directive 2009/48/EC, toys must not contain more than 0.5–23 ppm Pb, depending on the probe [9]. The European Drinking Water Directive [10] defines a limit of 5 µg/L for Pb, to be met by January 12, 2036, the latest; up to that date, 10 µg/L are tolerated, which is identical with the provisional drinking water guideline value of the WHO [11].

Concerning the toxicology of lead and its compounds, the interested reader is referred to the review of the US ATSDR [12].

Lead can be absorbed by inhalation of dusts/aerosols, and from the digestive tract if swallowed; absorption via skin is possible but much less dominant than oral absorption. In the digestive tract, children absorb 20–50% of the water soluble Pb, and adults absorb 3–10%. Cell membrane transporters for Ca^{2+} and Fe^{2+} are involved in the uptake; therefore, the uptake of Pb is saturable and in competition with the uptake of Ca^{2+} and Fe^{2+}. That is, if the body has a high demand for these essential metals (growth, disease, etc.) and/or their content in the diet is low, Pb uptake is increased. In the body, Pb is distributed via blood and deposited primarily in bones, followed by the kidneys. In case of bone injury or disease, stored Pb can be remobilized. Pb is eliminated via urine and feces with a biphasic half-life, the fast process with a half-life of a few weeks to two years and the slow process with a half-life of 10–20 years. Pb can be excreted in human milk.

Several molecular mechanisms for Pb toxicity are discussed. Pb^{2+} disturbs the homeostasis of important ions like Fe^{2+} and Ca^{2+}, but also Zn^{2+} and others. The disturbance of Ca^{2+} transport is associated with decreased nerve cell activity and pulse propagation. Pb^{2+} interferes with the cell energy metabolism so that finally apoptosis (programmed cell-death) can be induced. Pb^{2+} interferes with many enzymes and proteins in general; in some cases, at very low concentrations, Pb^{2+} can activate enzymes, but at elevated concentrations enzymes and proteins are inhibited. To name a few important examples, Pb^{2+} inhibits enzymes responsible for anti-oxidative defense which caused oxidative stress and damage. Pb^{2+} inhibits the δ-amino-levulinic acid dehydrogenase (ALAD), which catalyzes the cyclization of two molecules δ-amino-levulinic acid to porphobilinogen (Figure 15.4).

δ-Amino-levulinic acid (ALA) and porphobilinogen are intermediates in the synthesis of heme. In the final step of heme syntheses (Figure 15.5), chelatase transfers Fe^{2+} to the porphyrin ring; this enzyme is also a target for the toxic action of lead.

The molecular mechanisms identified are helpful for understanding lead toxicity. For acute intoxications, typical symptoms are nausea, vomiting, obstipation, extreme belly pain and colic, oliguria and anuria, hypotonia (low blood pressure) and coma. The gut pain and colic are attributable to an overflow of δ-amino-levulinic acid (ALA) which enters nerve cells, causing paralysis. Starting with numb feeling in the legs, this paralysis can cover the whole body and finally leads to breathing failure. All these

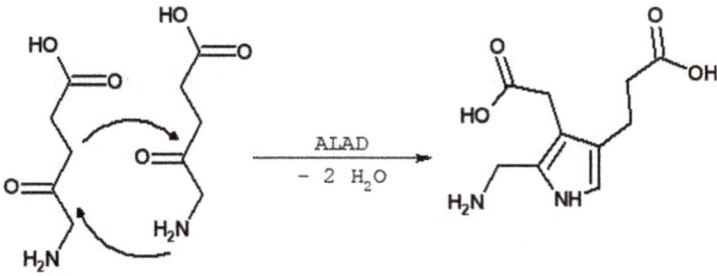

Figure 15.4: Synthesis of porphobilinogen, catalyzed by the Pb-sensitive δ-amino-levulinic acid dehydrogenase (ALAD).

symptoms can be caused by the rare disease Acute Intermittend Porphyria, where an enzyme defect causes the ALA overflow. About 5 to 30 g Pb are expected to be lethal for an adult person.

For chronic exposure, the most critical and sensitive endpoint is developmental neurotoxicity. Epidemiological studies have revealed that low level lead exposure is causative for mental retardation in children and may cause an irreversible reduction in the intelligence quotient. This effect is the reason for the low drinking water guidance levels, and the 10 µg/L issued by the WHO are rated as "preliminary" because they are not regarded as convincingly safe; keeping 10 µg/L pays tribute to two facts: (1) in many regions of the world drinking water with lower levels is simply not available, and (2) reliable and reproducible analytical methods and equipment must be available for drinking water surveys.

Other effects of low-level chronic exposure to lead are kidney failure, hypertension / cardiovascular effects, anemia, reduced male fertility and kidney cancer.

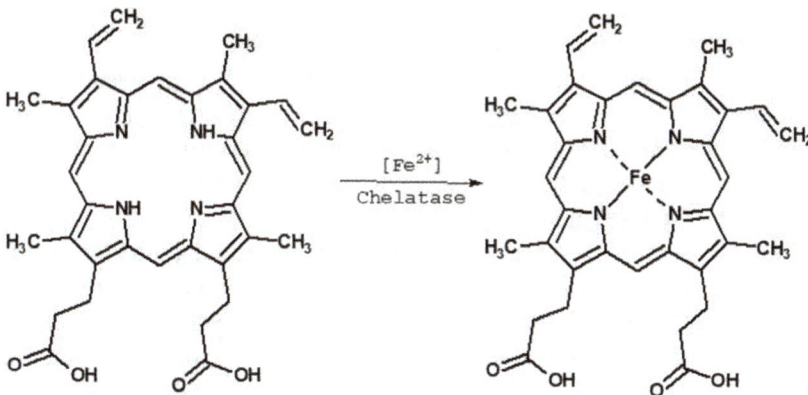

Figure 15.5: Final stage of the physiological heme synthesis: the enzyme chelatase can be inhibited by lead.

15.2.2 Cadmium

Cadmium minerals are a companion to zinc minerals and, therefore, processing of zinc minerals is always linked to cadmium exposure. Specific technical uses do exist, like Cd plating as anti-corrosion measure, CdS as a yellow pigment, Cd soaps as UV stabilizers in PVC and other plastics, or use in Ni/Cd accumulators. However, due to the high toxicity of Cd, most of these uses are meanwhile prohibited or restricted.

Exposure of the general population can occur via stack emissions from smelters, via Cd in soil which is taken up by plants and ends up in crops, via drinking water or via migration from consumer goods. For ceramic articles with intended food contact, the Directive 76/893/EEC sets a migration limit of 0.07 mg Cd per dm^2 surface [6]. Drinking water limits are 5.0 [10] and 3 µg/L [11], and toys must not contain more than 0.3–17 ppm, depending on the probe investigated [9].

At workplaces, exposure can occur primarily via inhalation while performing "hot" processes like smelting, welding or soldering; at elevated temperatures, CdO has a significant volatility and can be inhaled.

A comprehensive overview of the toxicity of cadmium was published by the U.-S. ATSDR [13]. Inhaled dust and particles can be resorbed if they reach the deep airways (alveoli), and tobacco smoking is a significant contributor to the total body burden of Cd. About 1–10% of the orally ingested Cd is resorbed in the intestine; higher resorption rates in case of iron deficiency indicates that Fe-transporters play a role in Cd uptake. Compared to inhalation and oral absorption, skin absorption is insignificant. Cd is accumulated in the kidneys which achieve highest concentrations of all tissues, and to a much lower extent in the liver. In the kidney, Cd is "trapped" by metallo-thioneins, and this process is likely a detoxification. However, if the storage capacity approaches saturation, kidneys will be damaged. Over lifetime, the Cd content in kidneys increases until it achieves a steady state at the age of 50–60 years. Resorbed Cd is excreted equally via urine and feces, with a half-life of more than 20 years. Cd is a competitor of Ca, which is the basis for the toxic effects.

Inhalation of Cd-fumes can cause a fatal pneumonia and lung edema; if survived, such an event may cause chronic bronchitis and lung emphysema. Symptoms usually appear or aggravate with a few hours delay and cover coughing, chest tightness, fever, chilling etc. ("Cd-flue"). A single event of overexposure may cause persistent impaired lung function.

Repeated inhalation overexposure may result in impaired lung function, anosmia (loss of sense of smell), emphysema of the lung and chronic airway obstruction. For chronic inhalation as well as oral exposure, the kidney is the critical target organ for systemic toxicity. Proteinuria is followed by oligouria and finally anuria. Further, by intervention with Ca-transporters, and due to the reduced vitamin D production as a result of the kidney damage, Cd disturbs the homeostasis of bones, causing bone-softening and malformation. In animal studies after oral exposure, Cd can affect male fertility. For developmental toxicity in rats, neurotoxicity in the pups is a sensitive marker of Cd toxicity.

A historical disaster occurred in Japan in the twentieth century in the Toyama region. Polluted water was used to float rice fields, and cereals showed a considerable uptake of Cd. Endemic poisoning resulted in the itai-itai disease, an osteomalacia, osteoporosis and softening of the bone structure which resulted in painful deformations ("itai-itai" is Japanese for 'ouch-ouch').

15.2.3 Arsenic

Arsenic is another metal which compounds are known since ancient times. It was used as pigment (auripigment, As_2S_3 and realgar, As_4S_4). The misuse of arsenic trioxide for murder dates back several centuries, as it can easily be added to food without changing the taste or odor. The symptoms of acute intoxication resemble those of infectious diseases, so the murderer had a real chance to keep his assault undetected. With the introduction of the Marsh test (introduced by James Marsh in 1836) changed the situation, as from now on there was a specific test that could be used to demonstrated elevated levels of arsenic in deceased persons. As an illustrative example, in the German city of Bremen, Ms. Gesche Margarethe Gottfried was executed after being convicted of having killed some husbands and neighbors with "mice-butter" (butter doped with arsenic, marketed as a rodenticide) [4].

In the nineteenth century, a Cu-As pigment known as Schweinfurt green (Germany), Scheele's green, or Paris green was used to color textiles and wallpapers. The formula is about $Cu(CH_3COO)_2 \cdot 3Cu(AsO_2)_2$, and the story of its market penetration and final prohibition is an illustrative example of a historical environmental health debate. The story is told in a German publication [14]. Briefly, Scheele's green was the first, brilliant green pigment that allowed to paint wonderful natural scenes on wallpaper. The economically upcoming middle class (Bourgeoise) was longing for articles and goods to demonstrate their wealth and welcomed these new wallpapers of unprecedented brilliance. However, with market penetration cases of diseases summed up which resembled those of arsenic intoxication. Some scientists required prohibitions, others argued against that due to lack of convincing evidence between wallpaper use and disease cases, which resulted in a kind of ping-pong of introducing and withdrawing market restrictions. As Scheele's green is comparatively insoluble in water, without swallowing the material it was hardly conceivable that the wallpapers could be the reason for diseases; furthermore, if they were accountable, why didn't diseases pop up in a more uniform pattern instead of spotted, isolated cases? Meanwhile, it turned out that cases of diseases obviously were linked to wallpapers in humid chambers and the occurrence of a pungent odor resembling the presence of mice or garlic. Finally, it could be demonstrated that the combination of Scheele's green on wallpapers in humid chambers caused mold fungus growth, and the reduction of arsenic to volatile compounds like di- and tri-methyl arsine which then result in inhalation exposure [15–17].

Chronic intake of low levels of arsenic induces some tolerance, and arsenic in-take was used to create a more healthy, "younger" and stronger appearance of horses, or even of persons by "self-medication", which sometimes was backfiring and fatal [4].

Arsenic can be misused for warfare agents. Diphenyl arsine chloride (CLARK I) is a strong irritant and induces vomiting, and the blistering agent Lewisite was pro-duced by Great Britain and the USA from November 1918 onward, but not used in combat; it smells like Geraniums. The arsenic-based warfare agents (Figure 15.6) can penetrate the skin and block enzymes by reaction with sulfhydryl groups. Dumped chemical warfare agents based on arsenic are an environmental threat and time bomb [18].

Figure 15.6: Arsenic warfare agents: CLARK I (left) and Lewisite (right).

Salvarsan, a (3-amino-4-hydroxyphenyl)arsenic compound, was the first chemo-therapeutic agent that was successfully introduced into the market in 1910 and was the first medication that allowed to treat syphilis [19].

Today, As is still used in special alloys as well as in the production of semi-conductors. At workplaces, exposure may occur via the inhalation route as many processes make use of the relative volatility of As compounds. Arsenic compounds may be used in biocides and wood preservatives, for example chromium-copper-arsenate. Such uses are now restricted under the Biocides Regulation in the EU.

Exposure of the general public occurs primarily via the food chain, and rice can be a significant contributor [20]. Interestingly, a specific As-species is of impor-tance. Arseno-betaine, which is the most important arsenic compound in seafood, is hardly metabolized in human beings and is not regarded as a critical contami-nant. Concerning the oxidation state, As^{3+} compounds exert a more pronounced toxicity than As^{5+} compounds. In some regions, groundwater achieves compara-tively high levels of arsenic which results in drinking water levels much higher than the 10 µg/L recommended by the WHO [11]. Lowering ground water levels shifts certain soils from anoxic to aerobic conditions, and oxidation of insoluble ar-senic sulfides creates the more water-soluble oxides. In Bangladesh, for example, high levels of arsenic in drinking water are a known cause for the comparatively high incidence of lung, bladder and skin cancer [21]. The limit in toys is 0.9–47 ppm, depending on the probe [9]. Marketing and use of arsenic compounds is restricted under Annex XVII, entry 19 of Regulation (EC) 1907/2006 [8].

Several molecular mechanisms for arsenic toxicity have been revealed or are under discussion. For acute toxicity, As^{3+} is the more critical oxidation state. As^{3+} blocks the co-factor lipoic acid by formation of a bis-sulfur-chelate complex, so the cell can no longer generate energy equivalents by degradation of glucose (Figure 15.7). This explains that arsenic compounds which can penetrate the skin act as blistering agents: the cell is cut off from the production of the universal energy carrier adenosine triphosphate (ATP) and falls into necrosis.

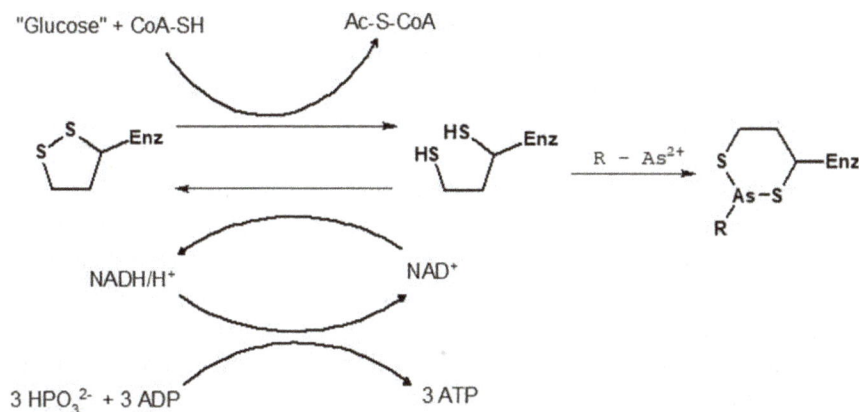

Figure 15.7: Block of lipoic acid due to complexation of its reduced state by arsenic acid cuts the cell off from energy supply: the degradation of glucose produces reduced co-factors; the re-oxidation generates an electrochemical potential which is used for the synthesis of the universal energy carrier adenosine triphosphate (ATP; simplified scheme).

Arsenate, AsO_4^{3-}, mimics/replaces phosphate in cellular processes. This involves signal transduction processes, and due to interference with the relevant enzymes, an external "growth" signal is not switched off, priming the cell for unlimited growth, which finally ends in the formation of tumors. In addition, arsenic can induce direct DNA damage [22].

A comprehensive overview of the toxicological profile of arsenic is published by the U.S. ATSDR [23] and reviewed by Tchounwou et al. [24]. In the body, As is metabolized by oxidation / reduction and methylation reactions. Mono- and dimethyl arsenite are excreted in urine; poor secondary methylation was shown to be linked to increased toxicity. Acute intoxication results in vascular shock, arrhythmia, prolonged vomiting and diarrhea which may be fatal. Chronic overexposure results in skin effects, where hyperkeratosis at hand and feet is a typical hallmark, together with white stripes in finger nails. Vascular capillary defects may cause ischemic necrosis ("blackfoot disease") and hypertension. Vascular damage is deemed responsible for effects at the respiratory system like hemorrhagic bronchitis and lung edema. Arsenic is muta-genic and a known human carcinogen (cancer of the lung, kidney, bladder and skin). Arsenic is fetotoxic and teratogenic (malformation of limbs).

15.2.4 Mercury

Mercury is another heavy metal with a long history of use by mankind. Red mercury oxide (cinnabar) was used as red pigment and mined in antique times, already. Mer-cury was known to be able to extract gold and silver from poor ores as amalgams, and fuming-off the mercury in open fires to isolate the gold or silver, or for gold plat-ing (fire gilding) of articles, was a reason for severe inhalation exposure to elemental mercury. Agricola (Georg Bauer) recommended that workers should stay upwind while running such processes because otherwise they would be at risk to lose their teeth [2].

Mercury salts were used for felt production by hatters, and because of the neuro-toxicity of mercury, "mad as a hatter" became a proverb. Mercury salts are potent biocides and had some use in medication and medicine. To a limited extend, mercury compounds are still in use for pharmaceutical and medical device applications, p. e. Thiomersal (sodium-2-(ethylmercurio-thio)benzoate), which is bacterio- and fungi-static. Mercury alloys are excellent putties for filling gaps in teeth; phenylmercury compounds were/are used as fungicides for seeds.

Because mercury is a liquid metal at ambient temperature and has a high den-sity, there are ample technical applications, for example measuring devices (ther-mometers, barometers, etc.), electricity (switches, electrodes, Hg discharge lamps, etc.) and chemistry (alloys, reactants, electrolysis processes).

In the biosphere, mercury is quasi persistent and runs through cycles without any permanent sinks (Figure 15.8). This persistence in combination with mercury toxicity triggered phase-out programs in the European Union, and the use of mercury com-pounds is restricted [8]. Drinking water guidance levels for inorganic mercury are $1\,\mu g/L$ [10] and $6\,\mu g/L$ [11], the limit in toys is 1.9–90 ppm, depending on the probe [9].

The quasi-persistence of mercury in the biosphere resulted in the Minamata catas-trophe in the middle of the twentieth century [25]. Mercury salts discharged in waste-water ended up in the Minamata bay; in sediments, it was transformed to methyl mercury, enriched in fish and shell-fish, and taken up via seafood by the general popu-lation. About 2300 patients were recognized, and more than 1000 died from poisoning.

A comprehensive summary of the toxicological properties of mercury and its com-pounds was published by the U.S. ATSDR [26]. For mercury, the species has a big influ-ence on the toxicological effects, which may differ between Hg^{2+} salts, Hg^0 and methyl mercury. Metallic mercury has a sufficiently high vapor pressure to cause intoxication by inhalation; therefore, spilled mercury should immediately be taken up by zinc or copper powder (formation of amalgams) or sulfur (slow reaction to insoluble HgS; the

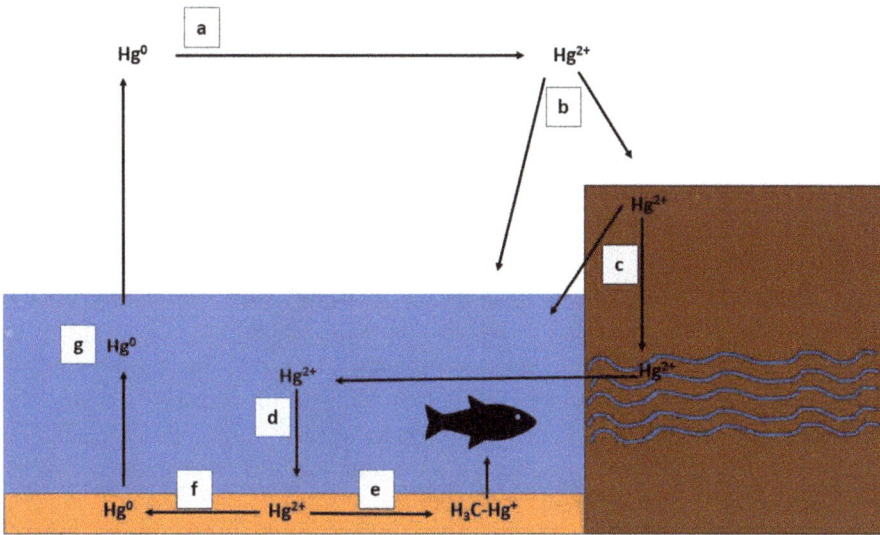

Figure 15.8: Simplistic scheme of the mercury cycle in the biosphere. a) elemental mercury has a considerable vapor pressure, and in the air, it is oxidized to Hg^{2+}; b) Hg^{2+} precipitates and is washed-out by rain on land and water surfaces; c) on land, Hg^{2+} is washed out and finally enters water bodies; d) Hg^{2+} in water equilibrates with sediment; e) in the sediment, microbial methylation may occur, resulting in bio-accumulative methyl mercury wish enters the food-chain; f) reduction in the sediment delivers elemental mercury; g) elemental mercury from the sediment enters the water body and the air; the cycle is closed.

initial main effect of sulfur is preventing the mercury bubbles mechanically from being dispersed and spread). Fatalities due to respiratory failure after acute Hg^0 exposure were reported, but concentrations achieved at these incidents are uncertain. Symptoms are burning pain in the chest, shortness of breath and coughing, but also fever and gastrointestinal disturbance. The most sensitive target organ for elemental mercury and especially for organometallic mercury compounds is the central nervous system. Acute and chronic intoxication can result in gait disturbance, tremor (first in hands, but spreading over the whole body with ongoing intoxication), delayed motor activity, headache, memory loss, mood disturbances and emotional instability ("mad as a hatter"). Psychological, but also neuromuscular disorders may persist even when mercury was phased out. One case of a dramatic poisoning with a tiny amount of dimethyl mercury was recorded [27, 28]: five months after a drop of dimethylmercury hit the disposable latex glove, the researcher realized difficulties with balance and speech; several weeks later she was in coma and died 10 months after the incidence, having elevated levels of mercury in the blood.

Hg^{2+} salts are corrosive to the skin and mucous membranes, and oral uptake results in gastro-intestinal symptoms (gut pain, vomiting).

The kidney is a target organ for repeated exposure to mercury, showing elevated tissue levels and impaired function after repeated and/or prolonged exposure to small dosages; renal failure may also occur after acute overexposure and may be fatal.

Mercury is toxic to reproduction, increasing the number of stillborn in exposed experimental animals and causing delayed neuronal development, mental retardation, and partly visible brain damage.

In the lung, inhaled Hg^0 is oxidized to Hg^{2+}; by action of methyl transferase, but probably also by microbial activity in the colon, Hg^{2+} is transformed to methyl mercury. Methyl mercury as well as Hg^0 can cross the blood-brain barrier easily. In the brain, re-oxidation (de-methylation) creates Hg^{2+} for which the blood-brain barrier is comparatively tight, resulting in an accumulation of mercury. This explains the susceptibility of the central nervous system to mercury intoxication. Due to metallothionein and the affinity to sulfur, the kidney enriches mercury and is the other important target organ for mercury intoxication.

15.3 Toxic and asphyxiant gases

Toxic and asphyxiant gases is the other large group of inorganic chemicals which have triggered the attention of toxicologists. Talking about the toxicity of this group of inorganic compounds can fill up tens of pages, easily. Therefore, a selection is made, and carbon monoxide, carbon dioxide, hydrogen cyanide and chlorine shall be addressed in some more detail. They are important gases in chemistry and technology, but there are also unintended sources of exposure for the general population.

15.3.1 Carbon monoxide

Carbon monoxide is an important gas in the chemical industry, and the production of aldehydes (oxo-synthesis), methanol and gasoline (Fisher-Tropsch Synthesis) are few examples. The latter ones may gain in importance as these processes allow to make use of plant materials and organic wastes to produce basic organic chemicals, reducing the dependence on fossil resources. Further, CO is an important ligand in organometallic chemistry, and it is used for the production of highly purified nickel (Mond process).

With the beginning of industrialization, gas illumination of streets and buildings was a hallmark of technical progress. In those days, town gas (city gas) was produced by the reaction between water steam and coke, producing a mix of mainly CO and H_2. Every larger town has had its gas plant, and the gas was made available in every household via a piping system, offering a much more convenient energy source than hard coal. However, accidental or intended opening of the gas tap was a reason for fatal accidents or a mean of committing murder or suicide. A switch to natural gas (methane) eradicated the threat of intoxication, but left the danger of explosion and fire.

Incomplete combustion of organic fuels because of insufficient oxygen supply results in formation of toxic levels of CO in the air. This is the main cause of fatalities in indoor fires. Similarly, poor ventilation of gas-heaters can be a cause of CO intoxication, be it that crows have built a nest in the chimney, be it that gas- or petrol-driven heaters are used indoors without caring for proper ventilation and clarifying the fitness of the device for indoor heating. Placing and running charcoal barbecue indoors in case of failure of the central heating system is an iterative and unfortunately highly reproducible cause of fatalities in winter time.

> Never run an open fire or charcoal barbecue indoors, and . . .
> . . . never run a gasoline or gas driven power generator indoors,
> . . . unless proper ventilation is checked and guaranteed!

The CO generated in room fires is a special threat for rescue teams. Not only do they need a self-sustained breathing apparatus to avoid CO intoxication; by opening doors while searching for victims, fresh air can ingress the room, and the mix of CO and oxygen in the air is explosive and may ignite at hot surfaces or glowing material. This results in the feared flash-over. Indoor smoke detectors, or specific CO detectors is a comparatively cheap and very effective measure against fire fatalities! Mind regular battery control!

Carbon monoxide binds to the Fe^{2+} in hemoglobin; as it has a much higher affinity than oxygen, the Hb is no longer available for oxygen transport. Due to the cherry-red color of the complex, victims of CO intoxication "look healthy". The toxicological effects of CO are a result of incomplete oxygen supply of tissues and are summarized, p. e., in the German GESTIS database [29]. First signs of intoxication are headache, followed by fatigue, somnolence and impaired vision, and increased uptake of CO finally induces unconsciousness, coma, and cramps. Survival of a severe, acute intoxication may leave mental deficits (memory loss, gait disturbance). These mental deficits may also result from repeated low exposure. Overexposure of pregnant women may result in stillborn, or mental and physical retardation of the child.

15.3.2 Carbon dioxide

Carbon dioxide (see also Chapters 4 and 16) is a product of complete oxidation of organic compounds, released in fermentation processes and by oxidative respiration. In technical applications, it is used as solid "dry ice" for cooling, and to make up beverages like beer and lemonade. Carbon dioxide is an endogenous substance in the body and it has physiological functions; however, overexposure may have hazardous effects.

An overview concerning toxic effects of carbon dioxide is provided by the German GESTIS database [30] and in [31]. As product of respiration, CO_2 is transported to the lung via the blood circulation, where it is released in the alveoli and exhaled. This exhalation is reduced if the concentration in the surrounding air increases. An increasing level of CO_2 in the blood first triggers intravascular receptors, and the breathing rate and circulation are increased. Beginning at 0.1% CO_2 in the air, people may sense headache, and the vigilance starts to be depressed. If the concentration in air exceeds 4%, physiological compensation becomes more and more difficult and finally fails, headache becomes more intense, agitation or dizziness and light-headedness are further symptoms. The body is falling into acidosis, and above 10% cramps and loss of consciousness up to coma will occur.

As CO_2 has a higher density than air (1.98 vs 1.20 g L^{-1} at room temperature), it may form invisible "lakes" in pits and cellars (fermentation and manure pits at farms, wine cellars). People entering these "lakes" become unconscious within seconds and die by asphyxiation. For first responders it is of utmost importance that they call technical help as soon as possible (fire brigade). Victims can be rescued only (!) by rescuers with self-sustained breathing apparatus! The vision "I can hold my breath and bring my friend back to the fresh air" has betrayed these first-responders: it was always mortal to do so! Why this attempt fails is not so easy to understand, but here is my working hypothesis: note that hypercritical CO_2 acts like an organic solvent. As a non-polar, small molecule it can penetrate the skin. Therefore, although the first responder holds breath, diving into 100% CO_2 results in dermal uptake, therefore to increasing blood levels. As the stem-brain reacts to the increasing levels, it triggers an irresistible breathing command to the body. As a result, the "hold back breath" is exhaled and pure CO_2 is inhaled: end of the story! Once in a safe surrounding, providing first aid is no longer critical.

Vulcanic activity results in large emissions of geological CO_2. Such emissions may cause overnight, silent disasters, as happened at lake Nyos in Cameroon in 1986 [32]. A blow-out of CO_2 from the maar killed more than 1700 people.

Finally, it should be mentioned that solid carbon dioxide (dry ice) can cause frost burns.

15.3.3 Hydrogen cyanide

At ambient conditions, hydrogen cyanide is at its edge of physical state between liquid and gas, as the boiling point is about 26 °C [33]. HCN is an important base chemical in organic synthesis, and the cyanide ion is involved in many metallurgical processes and is an important ligand in metal complexes. Prussian blue is a blue pigment in use since centuries. Cyanide can be released unintendedly, for example by digestion of bitter almonds or apricot kernels, or in smoldering fires of organic materials which contain nitrogen, like polyamides, polyurethanes, leather or silk. As a result, HCN can contribute to smoke gas toxicity.

It is the high affinity to metals that makes cyanide a strong toxin. Especially the cytochrome A, an iron-porphyrin enzyme playing a crucial part in the cellular respiration by transferring electrons to molecular oxygen in the cell, is the vulnerable point in cyanide intoxication as in its oxidized state, the Fe^{3+} binds CN^- and is blocked for further oxidation/reduction cycles.

An overview on the HCN toxicity and related symptoms is provided in [34, 35]. Obviously, gaseous HCN can be taken up by inhalation and via the skin. First symptoms are slight irritation of the eyes, increased heart rate, then breathing difficulties, headache, convulsions, increasing unconsciousness and finally coma. In case of intoxication, the first target to be achieved by first responders is the maintenance of the vital functions. Trained emergency staff may apply antidotes (if permitted by regional regulations). Thiosulfate injections provide sulfur for the endogenous transformation of cyanide to the less critical thiocyanate. Redox-active dyes like toluidine blue oxidize a part of the Fe^{2+} in hemoglobin (Hb) to Fe^{3+} (methemoglobin, MetHb), which then competes with cytochrome A for the cyanide ion and, doing so, regenerates part of the cytochrome A pool. However, if Hb is already partly occupied by CO, which is to be expected in fire victims, further reduction of still functioning Hb is counter-productive, and instead of toluidine blue, hydroxo-cobolamin (vitamin B_{12}) is the best choice for reactivation of cytochrome A.

15.3.4 Chlorine

Chlorine is the halogen that has a brought diverse use and application, and many branches of chemistry are based on chlorine. It is even the element, the pale green gas, that has important uses, and chlorine itself as well as its organic compounds has a Janus face if it comes down the threats and benefits. Life is impossible without chloride anions; they play an important role for the regulation of osmotic pressure and signal transduction in organisms, but too high concentrations are incompatible with survival of most aquatic organisms. Chlorine used as a disinfectant certainly has saved millions of lives (and will do so in the future), taken alone the treatment of drinking water and pool water. However, chlorine was also one of the first warfare agents, and its introduction in Flanders 1915 by the German army opened the Pandora Box of chemical warfare. Chlorinated organic compounds shifted from appraisal for pest and disease prevention – p. e. malaria by killing mosquitos with DDT, which was acknowledged with the Nobel Prize for medicine (1948) – to bans because of persistence and accumulation of chlorinated pesticides in nature.

Elemental chlorine is used in sanitation and in chemical industry. Exposure needs to be tightly controlled. Although it has a typical odor, the sensation of chlorine does not guarantee prevention against fatal inhalation. Toxicity and symptoms are summarized by different expert panels [36, 37]. First symptoms of overexposure are irritative sensations in the throat; higher concentrations trigger chest pain and cough. Even if first symptoms were moderate or even mild and then disappeared, within some hours delay a critical lung inflammation may occur, accompanied by a lung edema: inflammatory responses and initial first tissue damage in the alveoli (the end bubbles of the airway, where gas exchange takes place and blood and air are separated by only two layers of cells at both sides of the basal membrane) make them leaky, and blood plasma intrudes the air space. Due to the protein content of the plasma, but primarily by dissolved lung surfactant, agitation of the intruded liquid causes foaming. The foam cuts off an increasing number of alveoli from respiratory air exchange, and shortness of breath may become fatal. If this intoxication is survived, bronchitis may persist for weeks, and it may end up in lung emphysema, which means persisting shortness of breath, sometimes even at limited physical exercise. For that reason, any person with real or presumed overexposure to chlorine needs to be put under medical surveillance in a hospital.

15.4 Dusts and fibers

Inorganic dusts and fibers are an example that toxicity is not necessarily triggered by chemical composition, only; shape, surface and physical behavior as well as resistance to biological attack play an important role.

Naturally, the respiratory tract is exposed to gases and particles inhaled with the air. Particles may deposit on the surface of the airways, and the location of deposition depends on the aerodynamic diameter of the particles. Inhaled particles are differentiated by their "fraction" they are allocated to [38]:
(i)   inhalable fraction: all particles that pass nose/mouth
(ii) extra-thoracic fraction: inhaled particles that fail to pass the larynx
(iii) thoracic fraction: particles that pass the larynx (inhalable)
(iv) respirable fraction: particles from the thoracic fraction that can penetrate into non-ciliated airways

As a rule of thumb, particles with a mean aerodynamic diameter of 20 μm and below 10 μm are respirable. Certainly, there is some variety depending on individual parameters like body size, breathing rate and mouth versus nose-breathing.

By their simple presence, particles can cover the epithelial cells in the respiratory tract and need to be removed so the epithelial cells can be kept viable. This clearance is performed by mucus production and the ciliated epithelial cells, and concerted action of the ciliates transports the mucus with all embedded particles to the larynx where it is swallowed into the digestive tract. In the final bronchioles and alveoli there are no more ciliated cells and macrophages engulf particles and move "upstream" or leave the lung via the lymphatic system.

Besides direct toxic action to the cells of the respiratory tract, particles may act toxic simply by exhausting or hampering the clearance mechanism.

15.4.1 Asbestos and inorganic fibers

Asbestos is a known human carcinogen, and the toxicological profile is summarized by expert panels [39, 40]. Originally, asbestos was introduced in the market because of excellent technical performance, not least to beneficial effects for workplace safety, like resistance to acids, bases and especially due to its non-flammability, as it can be spun and woven to all types of textiles. However, during the 1960s and 1970s it became obvious that workers exposed to asbestos developed a special kind of lung cancer (mesothelioma) after a latency period of about 30 years and with a very poor prognosis. Consistently, asbestos was classified as a proven human carcinogen. Actually, it is the physical properties of asbestos that are responsible for its carcinogenicity. Asbestos fibers undergo longitudinal splitting very easily, whereas orthogonal splitting hardly occurs. The resulting small median aerodynamic diameter ensures that inhaled fibers reach the terminal bronchioles and alveoli. In the alveoli, macrophages start a futile attack on the fibers and finally die without having engulfed or physically changed the fiber, which would be a prerequisite for the fiber removal. These attacks release reactive oxygen species (ROS), which would be very helpful in attacking germs, but leave the fibers unaffected; not so the DNA of nearby cells, which is damaged by the ROS. Slowly the fibers migrate into the mesothelium where they persist further macrophage attack. The longitudinal splitting of the fibers finally results in shapes which are suspected to interfere with the cellular microtubules which are responsible for the proper segregation of the chromosomes during cell mitosis. This is hypothesized to add to the mutagenicity of the ROS, and over time cells will be transformed to cancer cells. Meanwhile, the use of asbestos is restricted in many regions, and deconstruction of contaminated building material or safe disposal of equipment can be performed by authorized companies, only.

After having realized the risk posed by asbestos, other inorganic fibers were taken into focus, being on natural or man-made origin (rock-wool, glass-wool; see also Chapter 1.2) [41]. Usually, particles with a length/diameter ratio more than 3 are regarded as fibers. Besides the shape, water solubility and resistance to macrophage attack are important determinants for the biological effects triggered by fiber inhalation. If the half-life is below about 20 days, the fibers do not seem to result in adverse health effects. Not all fibers are suspected to trigger cancer. However, repeated inflammatory response can cause lung fibrosis, which leads to permanent shortness of breath; this is because functional lung tissue is subsequently replaced by stiff, non-functional connective tissue.

15.4.2 Silica

Granular dust, being morphologically different from fibers discussed above, also shows biological effects in dependence on the physical-chemical speciation. Silica shall be discussed as an example, and amorphous silica is to be distinguished from crystalline silica [42–44]. Overexposure to amorphous silica can result in silicosis, probably because of the exhaustion of the clearance mechanisms of the lung at sufficiently high exposure.

Crystalline silica (e.g. quartz) is more critical. It can result in inflammation and silicosis at lower concentrations than amorphous silica, and it is a proven human carcinogen. One assumed reason for this higher toxicity is the lower water solubility; in addition, crystalline silica has a higher density of radical generating centers than amorphous silica. In that respect, "fresh" quartz surfaces (by fractionation of larger particles) are more active in terms of inflammation than "aged" surfaces; still disordered surface silanol groups in fresh quartz are a discussed reason for the higher activity [45]. The exact / complete mechanism of quartz toxicity still is unclear, but direct interaction with cell membranes seems to be involved in the toxic mode of action. Quartz in the bronchioles and alveoli is attacked by macrophages and causes inflammation and the release of ROS which are hold responsible for the DNA damage, which finally ends up in cell transformation to tumor cells; direct interaction with the DNA is unlikely [46]. Several epidemiological studies have revealed that increased incidence of silicosis is associated with increased incidence of lung cancer. In addition to respiratory effects, quartz exposure in workers was associated with decreased kidney function, probably by inflammation due to deposited silica crystals [42].

15.4.3 Inert dust

Granular dusts are particles with a length/diameter ratio not higher than 3. Amorphous silica (see 15.4.2), but also quartz coated with organic silicones needs much higher loads than quartz to trigger an inflammation in the airways. Like alumina (Al_2O_3), they can be allocated to the "inert dusts", whereby inertness is rated against crystalline silica or asbestos. In general, it is assumed that "inert" particles do not trigger directly a biochemical response in a cell that leads to noticeable cell damage. However, when overwhelming clearance mechanisms, the body may respond with inflammation, and adverse effects may occur in response to this inflammation. Chemists may be aware of the difference between "inert" silica and "active" silica if they think about thin layer chromatography or column chromatography. Inactive silica (diatomaceous earth) does not contribute to substance separation, but it may be used as a concentrating zone. For substance separation (separating zone), crystalline silica is needed.

For inert dusts as well, clearance efficiency in the respiratory tract is determined by the particle mean aerodynamic diameter and by the water solubility. Therefore, these properties will influence to level of acceptable load, and occupational exposure limits need to be referred to these characteristics. If clearance cannot keep up with the particle load, the body will react with inflammation. Repeated and/or prolonged inflammation can end up in fibrosis, emphysema, and/or chronic obstructive lung disease, which will cause shortness of breath and lung decrement [47].

15.4.4 Nanoparticles (NP)

Over the recent decades, much progress has been made in producing and understanding particles of nanoscale size. By definition, nanoparticles (NP) are called those granules which have a mean aerodynamic diameter below 100 nm. At these small sizes, quantum-mechanical effects may become prevalent and materials start to exert properties not observable for the corresponding larger particles. Just to name two examples (out of a plethora), nanosized gold looks red, and nanosized titanium dioxide activates oxygen under UV/VIS-irradiation and is prone to be used as a photo-redox catalyst [48]. Due to their small size which coincides with viruses and cellular macromolecules, many NP can easily penetrate tissue barriers. This, and experiences gained with asbestos, fostered concerns about potential health effects. It needs to be mentioned that human beings always were, and are exposed against nanomaterials; exposure did not start with the market introduction of man-made NP, but abrasive dusts, condensation aerosols and smoke always provided for NP exposure of mankind [49, 50]. Of course, the chemical identity of the NP remains to be an important determinant of the NP toxicity. For example, it is not to be expected that Ag, TiO_2, or carbon NP of the same size show the same biological activity (but they may do so by chance); therefore, the emphasis in the discussion of NP toxicity is to be laid on the question whether or not there are adverse effects that are driven solely by the physical form, if it comes to the comparison between salt, NP or bulk material for a given mass formula and crystal structure.

Due to their small mean aerodynamic diameter, after inhalation NP can achieve higher loads in the terminal airways than larger particles. For TiO_2, up to threefold higher loadings were observable. For more water-dissolvable NP like silver or aluminum, the difference to soluble salts is much less obvious, and distribution in the body can be explained by dissolved ions originating from the NP. After dermal contact or oral intake, in principle NP can be taken up easier into the blood stream than larger particles. The size of the NP makes them prone for phagocytosis; this results in increased intracellular load, which may on the one hand increase toxicity due to higher exposure, and on the other hand this makes NP interesting candidates for targeted drug delivery.

NP that have entered the circulation mainly end up in liver and kidneys, and within the tissues, they are found predominantly in phagocytotic cells. Metal based NP (and also some other) induce oxidative stress. It was shown that 10 nm Ag NP were more active in inducing liver and kidney damage than 40 or 100 nm particles. In terms of oxidative stress, the available surface is of importance, and for a given weight of NP it is evident that the surface increases if the mean diameter decreases (note: in toxicology the effects are measured as [mass material] / [mass of the organism]; 10 mg Ag NP per kg rat means much more Ag surface for 10 nm particles than for 40 nm particles). The role of the surface becomes evident when coatings are applied to the NP, which usually alters the biological effects. Therefore, toxicological results gained with one type of NP cannot easily be transferred to other types of NP [51].

In general, NP sized material may be more critical in terms of toxicity than bulkier material. In any case, the availability and loads in tissues is expected to be increased. Whether additional chemical-biological interactions are to be expected due to the NP size will certainly depend on the exact composition and structure of the NP material. In the European Union, registrants bringing chemicals onto the market need to cover the question of nanoforms and need to incorporate this in their chemical safety assessment and chemical safety report [52].

15.5 Conclusions

In this book chapter, some basic principles of toxicology were introduced and a few classes of inorganic materials were presented. The selection was made as a best try to cover historical interesting and/or currently relevant areas. It is of no surprise that only the surface of the discipline Toxicology could be scratched. However, much is gained if a few messages are taken home and are conveyed to society:

(i)   The dose determines whether a chemical is acting toxic or not.

(ii) As a result, there is nothing that can be claimed to be "non-toxic".

(iii) In advertisements for products, beware of phrases like "non-toxic" or "no chemistry": either the authors are deplorably ignorant or there is some risk that they want to fool you!

(iv) Most chemicals have thresholds below which a detrimental effect is not to be expected.

(v)   For some chemicals, it is difficult, if not impossible, to define a safe level. Practical thresholds might be agreed upon, where risk is levelled off by benefit.

(vi) Typically, chemicals show a dose-response curve, meaning higher dose = more risk.

(vii) Trace elements are an example where too little of the material is also risky for health.

For a deeper insight into toxicology, the interested reader is referred to the following publications:

Schupp, T.: Hazardous Substances – Risks and Regulations, de Gruyter, Berlin, 2020 – This textbook is aimed at postgraduate students and provides an introduction into the basic principles of chemical hazards (physical-chemical, toxicological, ecotoxicological), the environmental behavior of chemicals and the subsequent classification and labeling and relevant regulations for handling and marketing.

Greim, H.; Snyder, R: Toxicology and Risk Assessment: A Comprehensive Introduction, Second edition, John Wiley & Sons, Hoboken / Chichester, 2019 – This book is recommended for readers who like to acquire a deeper and profound understanding in toxicology, toxicological mechanisms and the principles of risk assessment.

References

[1]   Goldhammer K. Paracelsus – Vom Licht der Natur und des Geistes, Reclam, Stuttgart, 1960.
[2]   Agricola G. De Re Metallica Libri XII – Zwölf Bücher vom Berg- und Hüttenwesen., marixverlag, Pößneck, Vol. 1556, 1928.
[3]   Stiefler G. Über Fälle von Bleilähmung nach Genuß bleihaltigen Obstweines (Mostes); nebst Bemerkungen über das Vorkommen chronischer Bleivergiftungen unter der bäuerlichen Bevölkerung Oberösterreichs. Zeitschrift Für Die Gesamte Neurologie Und Psychiatrie, 1922, 25–34.
[4]   Hofmann F. Arsen ohne Spitzenhäubchen. Kleine Geschichte der Gifte, Herder, Freiburg im Breisgau (Germany), 2012.
[5]   Mende A, Blei im Blut: Auch wenig ist giftig 21 08 2012. [Online]. Available: https://www.phar mazeutische-zeitung.de/ausgabe-342012/blei-im-blut-auch-wenig-ist-giftig/.
[6]   EChACeramic Articles Directive 1984. [Online]. Available: https://www.echa.europa.eu/web/ guest/legislation-profile/-/legislationprofile/EU-CERAMIC_FCM.
[7]   Schupp T, Damm G, Foth H, Freyberger A, Gebel T, Gundert-Remy U, Hengstler J, Mangerich A, Partosch F, Röhl C, Wollin K-M. Long-term simulation of lead concentrations in agricultural soils in relation to human adverse health effects. Arch Toxicol, 2020, 94, 2319–29.
[8]   EChA, REGULATION (EC) No 1907/2006 OF THE EUROPEAN PARLIAMENT AND OF THE COUNCIL of 18 December 2006 2006. [Online]. Available: https://eur-lex.europa.eu/legal-content/en/ TXT/HTML/?uri=CELEX:02006R1907-20210215#tocId2.
[9]   EC, Directive 2009/48/EC of the European Parliament and of the Council of 18 June 2009 on the safety of toys., European Commission, 30 06 2009. [Online]. Available: https://eur-lex. europa.eu/eli/dir/2009/48/oj/eng. [Zugriff am 21 07 2021].
[10]   EU, DIRECTIVE (EU) 2020/2184 OF THE EUROPEAN PARLIAMENT AND OF THE COUNCIL of 16 December 2020, European Union, 23 12 2020. [Online]. Available: https://eur-lex.europa. eu/legal-content/EN/TXT/PDF/?uri=CELEX:32020L2184&from=EN. [Zugriff am 21 07 2021].
[11]   WHO, Guidelines for Drinking Water Quality, World Health ORganization, 2017. [Online]. Available: https://www.who.int/publications/i/item/9789241549950. [Zugriff am 21 07 2021].
[12]   ATSDR, Toxicological Profile for Lead, 08 2020. [Online]. Available: https://www.atsdr.cdc. gov/ToxProfiles/tp13.pdf.
[13]   ATSDR, Toxicological Profile for Cadmium, U.S.Department of Health and Human Services, Agency for Toxic Substances and Disease Registry., September 2012. [Online]. Available: https://www.atsdr.cdc.gov/ToxProfiles/tp5.pdf. [Zugriff am 21 July 2021].
[14]   Andreas H. Schweinfurter Grün – das brillante Gift. Chem unserer Zeit, 1996, 3, 23–31.
[15]   Hamberg NF. Chemische Untersuchung der Luft in Wohnzimmern mit arsenikhaltigen Tapeten. Arch Pharm, 1875, 233–54.
[16]   Challenger F, Higginbottom C, Ellis LL. The formation of organo-metalloidal compunds. Part I. J Chem Soc, 1933, 95–101.
[17]   Cernansky S, Urik M, Sevc J, Khun M. Biosorption and Biovolatilization of Arsenic by Heat-Resistant Fungi. Environ Sci Pollut Res, 2007, Special Issue 1, 31–5.
[18]   Haas R, Krippendorf A, Schmidt TC, Steinbach K, Löw EV. Chemisch-analytische Untersuchung von Arsenkampfstoffen und ihren Metaboliten. UWSF – Z Umweltchem Ökotox, 1998, 5, 298–301.

462 15 Toxicology of inorganic compounds

[19] Helmstädter A, Chemisch auf Erreger zielen *Pharmazeutische Zeitung*. https://www.pharma zeutische-zeitung.de/ausgabe-51522010/chemisch-auf-erreger-zielen/, 2010.

[20] Gundert-Remy U, Damm G, Foth H, Freyberger A, Gebel T, Golka K, Röhl C, Schupp T, Wollin K-M, Hengstler JG. High exposure to inorganic arsenic by food: The need for risk reduction. Arch Toxicol, 2015, 2219–27.

[21] Smith AH, Lingas EO, Rahman M. Contamination of drinking-water by arsenic in Bangladesh: A public health emergency. Bull World Health Organ, 2000, 9, 1093–2003.

[22] Wei S, Zhang H, Tao S. Toxicol Res, 2019, 8, 319–27.

[23] ATSDR, Addendum to the Toxicological Profile for Arsenic Agency for Toxic Substances and Disease Registry, February 2016. [Online]. Available: https://www.atsdr.cdc.gov/toxprofiles/ Arsenic_addendum.pdf. [Zugriff am 22 July 2021].

[24] Tchounwou PB, Yedjou CG, Udensi UK, Pacurari M, Stevens JJ, Patlolla AK, Noubissi F, Kumar S. State of the science review of the health effects of inorganic arsenic: Perspectives for future research. Environ Toxicol, 2019, 2, 188–202.

[25] Harada M. Minamata disease: Methylmercury poisoning in Japan caused by environmental pollution. Crit Rev Toxicol, 1995, 1, 1–24.

[26] ATSDR, Toxicological Profile for Mercury, U.S. DEPARTMENT OF HEALTH AND HUMAN SERVICES – Agency for Toxic Substances and Disease Registry., March 1999. [Online]. Available: https://www.atsdr.cdc.gov/ToxProfiles/tp46.pdf. [Zugriff am 23 July 2021].

[27] RCI B, Dimethylmercury poisoning, Berufsgenossenschaft Rohstoffe und Chemische Industrie, 2021. [Online]. Available: https://www.bgrci.de/fachwissen-portal/topic-list/labo ratories/accident-events/dimethylmercury-poisoning. [Zugriff am 23 July 2021].

[28] Ewe T. Titelthema – Mörderische Forschung: Drama im Labor, *Bild der Wissenschaft*. https://www. wissenschaft.de/allgemein/titelthema-moerderische-forschung-drama-im-labor/#, 1999.

[29] DGUV, GESTIS Substance Database, Institut für Arbeitsschutz der Deutschen Gesetzlichen Unfallversicherung., 26 July 2021. [Online]. Available: https://gestis-database.dguv.de/data? name=001110. [Zugriff am 26 July 2021].

[30] DGUV, Carbon dioxide, Institut für Arbeitsschutz der Deutschen Gesetzlichen Unfallversicherung, 2020. [Online]. Available: https://gestis.dguv.de/data?name= 001120&lang=en. [Zugriff am 5 August 2021].

[31] Guais A, Brand G, Jacquot L, Karrer M, Dukan S, Grevillot G, Molina T, Bonte J, Regnier M, Schwartz L. Toxicity of Carbon Dioxide: A Review. Chem Res Toxicol, 2011, 2061–70.

[32] Kling G, Clark M, Compton H, Devine J, Evans W, Humphrey A, Koenigsberg E, Lockwood J, Tuttle M, Wagner G. The 1986 Lake Nyos Gas Disaster in Cameroon, West Africa. Science, 1987, 169–75.

[33] EChA, Hydrogen cyanaide brief profile, European Chemicals Agency, 1 July 2021. [Online]. [Zugriff am 5 August 2021].

[34] DGUV, Hydrogen cyanides, Institut für Arbeitsmedizin der Deutschen Gesetzlichen Unfallversicherung., 5 July 2006. [Online]. Available: https://gestis.dguv.de/data?name= 012450&lang=en. [Zugriff am 5 August 2021].

[35] DFG, MAK value documentation for hydrogen cyanide, MAK-Komission, Deutsche Forschungsgemeinschaft, 2001. [Online]. Available: https://onlinelibrary.wiley.com/doi/ epdf/10.1002/3527600418.mb7490vere0019. [Zugriff am 5 August 2021].

[36] ATSDR, Toxicological Profile for Chlorine, U.S. DEPARTMENT OF HEALTH AND HUMAN SERVICES, Public Health Services, Agency for Toxic Substances and Disease Registry., 2010. [Online]. Available: https://www.atsdr.cdc.gov/ToxProfiles/tp172.pdf. [Zugriff am 5 August 2021].

[37] DFG, MAk Value Documentation for Chlorine., MAK-Kommission, Deutsche Forschungsgemeinschaft., 2004. [Online]. Available: https://onlinelibrary.wiley.com/doi/epdf/10.1002/3527600418.mb778250e3814. [Zugriff am 5 August 2021].

[38] Brown J, Gordon T, Price O, Asgharian B. Thoracic and respirable particle definitions for human health risk assessment. Part Fibre Toxicol, 2013, 12.

[39] ASTDR, Toxicological Profile for Asbestos, U.S. DEPARTMENT OF HEALTH AND HUMAN SERVICES, Public Health Service, Agency for Toxic substances and Disease Registry., September 2001. [Online]. Available: https://www.atsdr.cdc.gov/ToxProfiles/tp61.pdf. [Zugriff am 5 August 2021].

[40] DFG, MAK value documentation for Asbestos, Deutsche Forschungsgemeinschaft, MAK-Kommission, 1991. [Online]. Available: https://onlinelibrary.wiley.com/doi/epdf/10.1002/3527600418.mb133221stae0002. [Zugriff am 5 August 2021].

[41] DFG, Fibrous dust, inorganic. MAK-value documentation, Deutsche Forschungsgemeinschft, MAK-Kommission, 2017. [Online]. Available: https://onlinelibrary.wiley.com/doi/epdf/10.1002/3527600418.mb0243fase6519. [Zugriff am 5 August 2021].

[42] ATSDR, Toxicological profile for silica., U.S. Department of Health and Human Services, Agency for Toxic Substances and Disease Registry., September 2019. [Online]. Available: https://www.atsdr.cdc.gov/ToxProfiles/tp211.pdf. [Zugriff am 5 August 2021].

[43] DFG, Silica amorphous. MAK value Documentation, Deutsche Forschungsgemeinschaft, MAK-Kommission, 1991. [Online]. Available: https://onlinelibrary.wiley.com/doi/epdf/10.1002/3527600418.mb763186e0002. [Zugriff am 5 August 2021].

[44] DFG, Silica, crystalline. MAK value documentation, Deutsche Forschungsgemienschaft, MAK-Kommission., 2000. [Online]. Available: https://onlinelibrary.wiley.com/doi/epdf/10.1002/3527600418.mb0sio2fste0014. [Zugriff am 5 August 2021].

[45] Pavan C, Fubini B. Unveiling the Variability of "Quartz Hazard" in Light of Recent Toxicological Findings. Chem Res Toxicol, 2017, 469–85.

[46] Borm P, Fowler P, Kirkland D. An updated review of the genotoxicity of respirable crystalline silica. Part Fibre Toxicol, 2018, 23.

[47] DFG, General Threshold Limit Value for Dust, Deutsche Forschungsgemeinschaft, MAK-Kommission, 1999. [Online]. Available: https://onlinelibrary.wiley.com/doi/epdf/10.1002/3527600418.mb0230stwe0012. [Zugriff am 6 August 2021].

[48] Noman M, Ashraf M, Ali A. Synthesis and applications of nano-TiO_2: A review. Environ Sci Pollut Res, 2019, 3262–91.

[49] Jeevanandam J, Barhoum A, Chan Y, Dufresne A, Danquah M. Review on nanoparticles and nanostructured materials: History, sources, toxicity and regulations. Beilstein J Nanotechnol, 2018, 1050–74.

[50] Gebel T, Damm G, Foth H, Reyberger A, Gundert-Remy U, Hengstler J, Kramer P-J, Lilienbblum W, Röhl C, Schupp T, Weiss C, Wollin K-M. Manufactured nanomaterials: Categorization and approaches to hazard assessment. Arch Toxicol, 2014, 2191–211.

[51] Najahi-Missaoui W, Arnold R, Cummings B. Safe Nanoparticles: Are We There Yet?. Int J Mol Sci, 2021, 385 ff. DOI: https://doi.org/10.3390/ijms22010385.

[52] EChA, How to prepare registration dossiers covering nanoforms. European Chemicals Agency, April 2021. [Online]. Available: https://echa.europa.eu/documents/10162/1804633/howto_prepare_reg_dossiers_nano_en.pdf/5e994573-6bf9-7040-054e-7ab753bd7fd6. [Zugriff am 6 August 2021].

16 Chemical products: gradients, energy balances, entropy

Martin Bertau, Thomas Jüstel, Rainer Pöttgen, Cristian A. Strassert

The 15 previous chapters impressively demonstrated the width of inorganic chemistry. Our daily power supply and the many functional materials, which we regularly use as a matter of course, are only made possible through creative chemical solutions based on inorganic as well as organic concepts and reactions.

Besides primary mineral sources, these processes require vast amounts of energy. However, energy cannot (!) be *created* or *renewed* (this is the core of the first and second laws of thermodynamics, respectively) and only limited amounts of reliable resources are actually available. In particular, many industrialized countries have nearly no primary sources of minerals and fossil fuels – they mostly depend on the supplies from politically or economically unstable countries, where the production often occurs under questionable standards. In addition, primary sources are not endless and should be strategically employed by maximizing the efficiency of use and processing. In particular, fossil carbon deposits should be spared as key feedstock for the organic chemical industry: besides not being endless, they are way too valuable to be simply combusted.

In view of the problems related to the lack of resource sovereignty, the energetic balances associated to mining and recycling have to be considered as a whole. This final chapter summarizes some thoughts concerning gradients, energy balances and resources. These topics are currently intensively discussed in society, politics and science – unfortunately, in many cases without the required correctness and factuality. We live with the laws of nature and have to respect them soberly.

16.1 Gradients

Most people recall that energy E (or enthalpy H or mass m) cannot be created or destroyed – it can only be interconverted (first law of thermodynamics and special relativity). Ergo, for a closed system, $\Delta E = 0$. Absolute energies are meaningless and need to be defined with respect to a standard state: this integration constant can be arbitrarily defined as 0, which stems from the fact that the change in energy ΔE can be considered to result from the line integral of the product between a particular intensive quantity (ι) and a differential extensive variable ($\delta \varepsilon$) along a certain path: $\Delta E = \int \iota \, \delta \varepsilon$. On the other hand, certain intensive variables (such as force F, pressure P, temperature T, chemical potential μ and electrical potential ϕ) control the "flow direction" (or flux) of extensive quantities (e.g. distance s, volume V, heat Q, number of particles n, and charges q).

https://doi.org/10.1515/9783110733471-032

It is evident that the direction in which the flux of an extensive quantity $\delta\varepsilon$ occurs is always opposite to the direction of a gradient of the intensive variable controlling it ($d\iota/dr$, where r is a unitary vector pointing into a particular direction under consideration), so that the existing gradient is being reduced: gradients of force, pressure, temperature (more precisely, $1/T$), and electrochemical potential cause the differential changes of distance s, volume V, heat Q, number of particles n and charges q. Hence (and nobody would expect such processes to reverse spontaneously!), charged particles flow until electrochemical potentials are equal; the wind blows as long as air pressure differences exist; stones roll downhill until all force vectors are cancelled out; the hot sausage warms up the cold beer standing next to it until thermal equilibrium is reached. In these cases, the total amount of energy remains absolutely invariant: it is just dissipated and becomes useless. Most important, however, is the fact that only if a gradient $d\iota/dr$ is available, the resulting flux $\delta\varepsilon$ can be used to interconvert energy forms and to use part of it (free energy) with a limited efficiency η due to dissipative losses (e.g. friction, overpotentials, activation barriers or ohmic resistances) while the relevant gradients $d\iota/dr$ are always irreversibly reduced on the way. Moreover, the buildup of a new gradient $d\iota'/dr'$ must be coupled to the degradation of an even larger gradient $d\iota''/dr''$ to overcompensate for dissipative losses; therefore, **energy cannot be created or destroyed, and gradients cannot be spontaneously renewed**! Living systems are actually far away from equilibrium and permanently degrade free energy sources to overcompensate for ongoing degenerative processes (which is a kind of dissipative loss); the lack of externally available gradients means equilibrium, i.e. death.

The above-mentioned discussion is summarized in the second law of thermodynamics: time t flows in the same direction as the degradation of gradients $d\iota/dr$ occurs, i.e. when the entropy S of the universe increases ($\Delta S_{\text{universe}} > 0$ for spontaneous, irreversible processes); a reversible process, on the other hand, is infinitely slow and the entropy of the universe remains constant ($\Delta S_{\text{universe}} = 0$); the spontaneous enhancement of a new gradient is not possible, as the total entropy of the universe would decrease in that way ($\Delta S_{\text{universe}} < 0$).

In mathematical terms, adding up the total entropy changes of a sub-system corresponds to the summation of all single energetic flows, i.e. individual line integrals $\int \iota \times \delta\varepsilon$ (such as force × distance, pressure × volume, electrochemical potential × number of charged particles, heat) multiplied by the corresponding $1/T$ value; this operation must be computed both for all the relevant sub-systems as well as for the environment and then added up, yielding a positive result for a spontaneous process:

$$\Delta S_{\text{universe}} = \Delta S_{\text{environment}} + \sum \Delta S_{(\text{sub})\text{system(s)}} = \left[(1/T) \left(\int F\,ds + \int P\,dV + \int \mu\,dn + \int \phi\,dq + \int \delta Q \right) \right]_{\text{environment}} + \Sigma \left[(1/T) \left(\int F\,ds + \int P\,dV + \int \mu\,dn + \int \phi\,dq + \int dQ \right) \right]_{(\text{sub})\text{system(s)}}$$

If the (sub)system(s) is (are) in thermal and mechanical equilibrium with the environment (i.e. no T, P, or F gradients), the free enthalpy difference of a process ΔG results from the total entropy change of the universe $\Delta S_{universe}$ as follows:

$$\Delta G = -T \Delta S_{universe} = -T \left[\Delta S_{environment} + \Sigma \Delta S_{(sub)system(s)}\right] = \Sigma \left[\Delta H - T \Delta S\right]_{(sub)system(s)}$$
$$= \Sigma \left[\Delta G\right]_{(sub)system(s)}$$

Recalling Maxwell's demon and modern statistical and information physics, building up a gradient di/dr requires gain of information (i.e. reduction of entropy), which is only possible if coupled with a larger loss of information (i.e. increase of entropy) in a second process. Here is where the laws of thermodynamics are at work: They ensure the maximization of entropy (or the minimization of information) by regulating the flux of heat, particles, charges, volume, and distance.

In chemical technology as well as engineering sciences and chemical industry, one rather uses the term exergy. Exergy by definition is the maximum utilizable amount of energy. The non-utilizable part is called anergy. The sum of exergy and anergy is the Gibbs free energy. This can easily be demonstrated with the example of a pumped storage power plant, here with the one in Lichtenberg, Saxony, Germany. The mural crown's altitude is 497 meters above mean sea level (MAMSL). The height of fall of water is 42.8 m. This is also the steepest part in the entire system between reservoir and the North Sea (Figure 16.1).

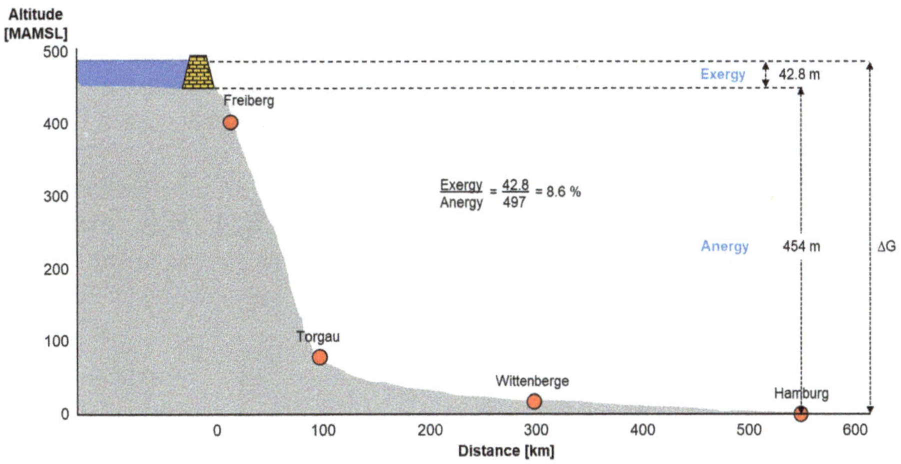

Figure 16.1: Schematic illustration of exergy and anergy. Only the potential energy between the maximum reservoir filling level and the valley bottom can be utilized for electricity generation. The slope in the entire distance from the reservoir's foundation sole to the mouth of the river Elbe into the sea is too flat to generate additional electricity through utilizing fall energy. Please be aware that the abscissa (distance) is in km, while the ordinate (altitude) is in the meter scale. Through the thousand-fold over increase, the slope between reservoir wall and Torgau (km 100) looks much steeper.

Only this steepest section, i.e. the highest gradient, can be used to generate electricity by utilizing fall energy. This height difference of 42.8 m corresponds to the exergy. The quotient of fall height and maximum altitude of the reservoir level is 0.086, i.e. only 8.6% of the potential energy can be utilized at all. Calculating the efficiency of pumped storage power plants therefore always must consider this small difference in altitude. The situation is only different in those cases where there is a lower reservoir from which the water can be pumped back to refill the upper reservoir. One has to be aware, though, that the anergy fraction of the overall energy is not unusable. With a run-of-river power plant, additional electricity could be generated, thus extending the exergy fraction at the expense of anergy. However, since the difference in altitude, Δh, is less, i.e. since the gradient is smaller, the amount of electricity produced is less, too. As a matter of course the difference between upper level and lower level, which is ΔG in our model, remains unaltered. Only the fraction of exergy and anergy may vary, while the sum of both always remains the same.

Taking these aspects into account, it is clear that low concentrations (e.g. CO_2 in the atmosphere) and enthalpically stable compounds (such as H_2O and CO_2) constitute deep thermodynamic sinks. For instance, production of methane from atmospheric carbon dioxide (which is enthalpically very stable and only available at 0.042 vol-%) implies the buildup of a higher chemical potential (concentration enhancement) and an increase of enthalpy (reduction to methane and water), thus implying a vast jump in free enthalpy (Gibbs free energy G), along with additional losses (by friction, overpotentials, activation barriers, *etc.*); hence, this process needs to be coupled to the degradation of an even vaster source of free energy (such as gravitational potential difference by means of a hydroelectric power plant or conversion of mass into heat in a nuclear power plant).

These fundamental aspects must be taken into account for a proper context in which mining, concentration, chemical transformation, purification, controlled technical dilution (for defined applications), and recycling need to be interpreted and assessed.

16.2 Resources

Nature delivers the elements and compounds we need for all kinds of chemical transformations and device construction (the previous chapters impressively demonstrate the multitude of materials we use in daily life) [1–3]. Until the invention of the steam engine in the eighteenth century, only few elements were exploited; due to scientific and engineering progress, technically exploited elements literally exploded. Today, in fact, almost the entire PSE (till Americium) is in use. However, raw material recovery is complex. While crop farming simply requires a certain area of arable land, the situation with mineral ores is completely different. Both topics have in common that crop harvesting or mineral extraction on a technical scale only makes sense when the target product is available as a concentrate. What in agriculture is a field, is a deposit in geology. In order to get a grip on mineral resources, they must be present in enriched form. The necessity for such mineral concentration sites, i.e. deposits, aggravates technical exploitation of nature's wealth. Surely, any element is sufficiently abundant, yet mostly insufficiently concentrated. The ability to extract minerals from ore deposits is a matter of gradients (cf. 16.1). Nobody would extract gold from the backyard, although being present in some 4 grams per 1000 tons of soil (4 ppb). If, however, an enrichment is given, exploitation gets into reach. Where the ambient factors are favorable such as in Freiberg (Saxony, Germany), where silver ore was found in 1168 on-surface, mining becomes commercially viable and effective. This can be the case even if sub-surface mining is required. The situation may be adverse, however, if local factors stand against an immediate exploitation. An example is the Salar de Uyuni in Bolivia. Besides being a unique ecosystem with very slow recovery rates, the world's richest lithium deposit suffers from being located at 3,653 m above sea level with poor supplies of fresh water and an insufficient infrastructure.

We have seen that mineral concentrations are coupled to certain geological conditions, which are rare, and also some anthropogenic factors, such as infrastructure, which in sum is even rarer. Once a decision has been made, mineral ore deposits are exploited as open pit mines, underground mines, or salt lakes. It is well understood that exploitation of such deposits will inevitably interfere with nature. This interference in itself is typically nothing really harmful for the environment, especially when compared to the damages that a local mudslide or a weather event may cause. Nowadays, this clearly is an issue, since global overpopulation is getting even more serious every year. To get in impression: already in the 1950s, there were concerns of an overpopulated planet Earth, with roughly 2,500 million people at that time. In 2021 it reached 7,900 million (Source: United Nations), corresponding to more than three times the value of 70 years earlier [4]. The impact of our species is the basis of the term Anthropocene, while presently the biomass of human beings and domesticated livestock exceeds 20 times the mass of all other mammalian species on Earth! [5] Until 2050, the UN expect mankind to grow by another 2,000 to 2,500 million people if not even up to 12,000 million in 2060. We are adding ourselves the entire world population of 1950 or even more in only 30 years, not bearing in mind that these human beings strive for the same living standard encountered in the industrialized countries (or at least a fraction of it). One may be inclined to regard this as the mother of all conflict situations, and in the case of raw material supply this view certainly may hold true (i.e. competition for resources in the broadest sense).

Societal entities such as national economies will therefore have to invest considerable efforts to get a grip on raw materials. This fact has been widely understood, with the exception of the industrialized countries where the dream of solving problems by recycling is dreamt, while being deeply convinced that mining is dirty.

Both is not true. Modern mining technology in fact hardly affects on-surface ecosystems – if done properly. One such example is fluorspar mining in the local community "Niederschlag" in the Saxonian ore mountains, where trucks enter and leave the mine like a road tunnel. Pre-beneficiation is done subsurface in the mine, and the run-of-mine concentrate is beneficiated in a nearby industrial site. Recycling, however, is limited by the gradient issue again. Although a smartphone, which for whatever reason constitutes the ideograph for recycling, contains several elements in relatively concentrated form, they are far from being pure. They often are present as alloys, from which minor by-components must be separated in order to re-establish primary material quality. In the vast majority of cases, the overall anti-entropic effort exceeds the intrinsic value of the target material. This is the reason why recycling truly needs to be specified as down-cycling (i.e. with losses of quality). Real recycling in its truest sense, i.e. primary product quality, is economically viable for a small set of raw-materials only: noble metals, Cu, In, Ga, Sn, Li_2CO_3, and some rare earth elements, such as Nd, Gd or Sm. Even if the qualities after down-cycling were sufficient, industrialized countries are export-oriented national economies. Hence, one has to accept the fact that recyclable products hardly will find their way back from final-consumer countries back to the EU or the US. Thus, for such industrialized nations, the so-called "circular economies" would run empty within months if not regularly replenished with fresh material in sufficiently high quality provided by mining activities. Hence, for the industrialized countries, which to a large extent expel vital industry to Asia for the sake of being CO_2-neutral, the access to high-tech-related primary raw materials is an existential question. By the way, the atmosphere (and neither local nor global climate) does not care for where CO_2 is emitted or if its production is reduced by 1–2% by complete shut-down of an economy representing a minor fraction of the entire global population (e.g., the German economy and industries). And, last not least, a growing world population unquestionably will consume growing amounts of raw materials, particularly considering that the poorest fraction of the population representing the vast majority of humanity will steadily strive for an enhancement of live standards – a fact that will be barely influenced by ascetic efforts from the developed nations intending to establish a moral example after two centuries of high-living standards enabled by industrialization.

If now political and ideological incentives enter the market and claim truth for themselves, the necessity of raw material supply and raw material access get out of sight. Just to give an example, battery-driven electric vehicles (BEV) by far cannot be produced in the amounts "forecasted" by politics, simply because the sheer number of units greatly exceeds the available supplies, such as in the case of cobalt. In many cases, unstable countries control the access to raw materials; in contrast to expectations of industrialized economies, these countries rightfully do to not intend to negotiate their grip on wealth.

Although some of the elements occur in elemental form in nature, they are highly diluted in the ores. Typical examples are gold deposits with contents of around 10 g with one ton of accompanying rock. Grinding, flotation, and cyanide leaching necessitate energy and large amounts of chemicals. In the case of the platinum metals, they are always socialized (also highly diluted in accompanying rock) and require highly sophisticated separation processes, in order to obtain the six elements in pure form. The high dilution of the noble metals in the natural resources forces the buildup of a high gradient to enable element recovery (and this automatically implies the need for energy sources). A third example is elemental sulfur that requires large amounts of overheated water vapor in the Frasch process (see Chapter 12.2), and is nowadays produced by the Claus process instead.

Many elements have high oxo- and thiophilicity, thus dominating oxide and sulfide formation. Nature works with lattice energy and already produced thermo-dynamically stable compounds – binary oxides besides silicates, aluminates, and other intricate multinary oxides. The sulfides are mainly transition metal based. Their decomposition for element production needs the reverse process, thus requiring large amounts of energy!

Besides the delicate work-up procedures, in several cases, element extraction already produces hazardous by-products during mining. One of the striking processes is rare earth element production. The rare-earth-rich deposits naturally contain significant amounts of accompanying radioactive minerals, particularly due to the presence of thorium. A striking example is the Bayan Obo mine in China [6]. Further examples concern lithium production (destruction of the Atacama Desert or the Salar de Uyuni), cobalt and tantalum mining (critical human mining conditions), bauxite mining (red mud waste) or illegal gold digging (mercury and amalgam waste).

The final energy-consuming step of element production is purification. The higher the purity of an element (i.e. the higher its chemical potential), the higher is the external gradient degradation required (especially if thermal or frictional losses along the process are considered, *vide infra* 16.3). Typical final purification steps concern refining by electrolysis, distillation, or sublimation. In this context, one of the most energy-consuming procedures concerns the production of semiconducting materials, e.g. silicon production in ultra-pure form through the silico-chloroform distillation process along with repeated zone-melting refining.

16.3 Processes

All chemical processes in which gradients are built up require a coupled destruction of a primary gradient that overcompensates for frictional losses (e.g. electrochemical overpotentials, corrosion, and other dissipative processes). Overall, the entropy of the universe increases and the quality of usable energy is diminished. Such simple energetic considerations hold true for all large scale chemo-technical processes – gradients are the crucial parameters.

An important issue of large-scale chemical synthesis concerns elements or compounds that are inevitably needed for synthesis, but that are not visible in the product. This is underlined by three examples: (i) NaOH for the Bayer process [1, 2], i.e. for aluminum/iron separation found in bauxite, (ii) Cl_2 for silicon and titanium production (and all other carbochlorination processes) and (iii) NH_3 from the Haber-Bosch synthesis as a key chemical for a multitude of further chemical reactions and food production (i.e. as a fertilizer precursor for agriculture). Of the first two processes, NaOH and Cl_2 production are coupled in the chloralkali electrolysis. Such a chlorine electrolysis plant consumes about 2.2–2.6 MWh per electrochemical unit (ECU = 1 t Cl_2, 1.1 t caustic, 0.03 t H_2). In total, the chloralkali electrolysis annually consumes about 10% of the global electricity production [7]. The Haber-Bosch process, on the other hand, requires the synthesis of pure H_2 and N_2 as educts – two distinctly endergonic processes.

The next steps concern reduction chemistry. Due to the oxophilic and thiophilic nature of many elements, they need to be reduced in order to get the pure elements. Metallothermal reductions with Na, Mg, or Al play an extraordinary role in large-scale chemical synthesis. The striking problem concerns the reduction agents themselves, since they are synthesized *via* highly endergonic salt flux electrolysis delivering one, two or three electrons to Na^+, Mg^{2+} and Al^{3+}, respectively. This vast amount of energy needs to be applied before any of the metallothermal reductions, with all the inevitable dissipative losses associated in every step (including the metallothermal reduction itself).

Many other processes require carbon as a reducing agent: the blast furnace process (1); the carbothermal reduction of silicon dioxide (2), the carbochlorination of rutile (3) for titanium production, the Søderberg electrodes for molten flux electrolysis of aluminum, and the production of phosphorus (4).

$$FeO + C \rightarrow Fe + CO \tag{16.1}$$

$$SiO_2 + 2C \rightarrow Si + 2CO \tag{16.2}$$

$$TiO_2 + 2Cl_2 + 2C \rightarrow TiCl_4 + 2CO \tag{16.3}$$

$$Ca_3(PO_4)_2 + 3SiO_2 + 5C \rightarrow 3CaSiO_3 + \tfrac{1}{2}P_4 + 5CO \tag{16.4}$$

As a consequence of the Boudouard equilibrium, these high-temperature carbothermal reductions release carbon monoxide, which is used for further combustion in thermal power plants for the generation of process steam, heat, or electricity.

In almost all cases, hydrogen is no suitable substitute as reducing agent. Hydrogen needs to be synthesized from fossil fuels or through electrolysis (water splitting or chloralkali electrolysis) with additional energy losses due to electrochemical overpotentials and undesired hydride formation as an additional obstacle involving the desired element!

Summing up, many large-scale chemical processes lead to massive reduction of the quality of energy due to the formation of heat (phonons), which in most cases cannot economically be utilized (converted). Striking examples are the overload during any electrolysis of aqueous solution (>20% of the applied energy for water electrolysis) or the waste heat of molten flux electrolysis.

16.4 Products

The previous discussion also holds for all kind of chemical products, materials, and functional structures, since gradients are utilized for their synthesis, purification, morphology optimization, and structuring. In general, the required energy to build up suitable concentration, temperature, pressure, or field gradients increases with the complexity of the product aimed at. This statement shall be underlined by two examples, viz. the separation of air into numerous air products and the production/ structuring of semiconductors.

Dry air at sea level is composed of a number of inert gases and oxygen, while diluted oxygen at ambient pressure is rather inert too.

Consequently, the separation of air is not performed by chemical reactions but by employing physical gradients, which are built up by temperature and pressure (cryogenic air separation) or by adsorption/desorption processes (pressure-swing-adsorption). Facilities for the production of up to several thousand tons per day and consuming 40 MW are being operated worldwide, since the air products N_2, O_2, and Ar are widely applied in large quantities, e.g. for the production of (petro) chemicals and steel [9]. Therefore, air separation is mainly done to gain O_2, N_2 and Ar, while these products consume a substantial amount of the globally produced primary energy. For instance, air separation units as a single industrial equipment item accounts for a considerable proportion of China's total national power consumption, viz. about 5% [10].

The other air components are present in such low concentrations that separation is extremely challenging. While Ne, Kr and Xe are enriched up to 500 ppm during the abovementioned first step, they have to be further processed to pure gases in additional cryogenic facilities; due to their extremely low relative atmospheric concentration, the production of CO_2 and He is not achieved by air separation. It is economically much more advisably to separate He from natural gas as some CH_4 sources comprise between 2 to 10% He. These sources are thus regarded as the best choice for an economically sustainable He production. CO_2 can be extracted from geological carbonate reservoirs, mostly in areas with active volcanism or hot, CO_2-rich springs. Moreover, CO_2 is obtained from combustion processes, since its volume fraction in exhaust gas is much higher than in air and thus separation requires much lower physicochemical gradients.

While CO_2 separation from air is a tremendous challenge for technology and nature (mainly due to its low relative concentration in the atmosphere), another problem for the biosphere is the reduction of CO_2 as a thermodynamic sink to obtain carbon in lower oxidation states. CO_2 is the carbon source for all carbohydrates and finally for all secondary biomolecules. Autotrophic organisms are able to take-up and to reduce CO_2 by the so-called dark reaction, despite a low atmospheric partial pressure. A low CO_2 concentration requires the complete opening of the stomata and thus the diffusion-driven CO_2 uptake goes hand in hand with significant water loss. To survive in an environment with declining CO_2 partial pressure and low water availability, Crassulaceae Acid Mechanism (CAM) plants have evolved, which have a temporal separation of CO_2 take-up from the light reaction to reduce water loss over daytime [11]. In fact, one of the best ways to increase the yields of greenhouse-based farming is the use of a CO_2 enriched atmosphere. However, the dark reaction, i.e. the Calvin cycle, requires a large amount of free energy (ATP) and reductive power (NADPH) to convert CO_2 into sugars, where solar energy is the energy source for the photochemical driven water splitting, which delivers energy (ATP) and reductive power (NADPH). Consequently, solar energy is responsible for the buildup of the electrochemical gradients over the thylakoid membrane located in chloroplasts and thus in turn for the production of energy-rich organic molecules.

In summary, the effort to separate an air component as listed in Table 16.1 mainly depends on its concentration in the respective source.

The second example concerns one of the biggest challenges of present technology, which is the synthesis of semiconducting materials, viz. element and compound semiconductors, and their subsequent 2D or 3D structuring to obtain respective active electronic components. This requires first of all the activation of thermodynamic

Table 16.1: Components of dry air and their boiling point at 1 bar [8], *Global Monitoring Laboratory, Mauna Loa, HI, updated at July 6th,2021.

Component	Volume fraction [%]	Boiling point [K]
Nitrogen N_2	78.08	77
Oxygen O_2	20.95	90
Argon Ar	0.93	87
Carbon dioxide CO_2	0.0419 (419 ppm)*	195 (sublimation)
Hydrogen H_2	5×10^{-5}	20
Neon Ne	1.82×10^{-3}	27
Helium He	5.2×10^{-4}	4.2
Krypton Kr	1.14×10^{-4}	120
Xenon Xe	8.7×10^{-4}	165

sinks such as SiO_2, Al_2O_3, Ga_2O_3, In_2O_3, N_2, $Ca_3(PO_4)_2$, MgO, and borax. The respective (semi)metals are mostly obtained by the reduction of these oxides either by carbon or by electrolysis as mentioned above.

Metallurgic-grade silicon is then highly purified (with $HSiCl_3$ as intermediate for distillation purification) and molten to obtain Si single crystals, which is a very energy and time-consuming process (zone refining). These crystals are cut into wavers, e.g. 6" or 12" in diameter, and then doped by B or P to obtain p- and n-type silicon. The main challenge is then the generation of nanoscale patterns by photolithography upon using masks, photoresists and chemicals for etching. Photolithography at the nano-scale utilizes very coherent excimer DUV or VUV lasers, e.g. krypton fluoride (KrF*) emitting at 248 nm or argon fluoride (ArF*) emitting at 193 nm, where the latter allows the generation of structures down to 50 nm and lower by special techniques. Due to the ongoing trend of miniaturization of ICs, Extreme UltraViolet (EUV) lithography working at 13.5 nm is the next evolutionary step in Si patterning. The demand in terms of energy and investments is tremendous, expressed by the low efficiency of <0.5% for the conversion of EUV in-band power emitted from the 0.25 mm diameter Sn plasma into 2π solid angle radiation and by the value and mass of a typical EUV tool: A single machine costs 120 million € and weighs 180 tons [12, 13]. In summary, silicon chip manufacturing requires a lot of energy and is enabled at the cost of the reduction of several gradients, such as needed for the generation of coherent radiation, which in turn becomes incoherent during the process by scattering, absorption, and re-emission.

While silicon is structured upon using respective wavers, III/V compound semicon-ductors require the conversion of the above-mentioned elements, as well as N_2, P_4, As_4 into highly pure volatile compounds, such as hydrides or metal-organic (MO) com-pounds. These gaseous compounds, viz. NH_3, PH_3, AsH_3, $Al(CH_3)_3$, $Ga(CH_3)_3$, $In(CH_3)_3$, $(C_5H_5)_2Mg$, and SiH_4, are deposited from a nitrogen stream onto a VUV radiation-purified single-crystalline waver, mostly Al_2O_3 according to the following reactions:

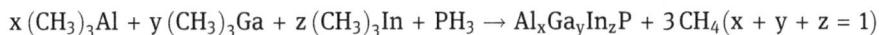

$$x\,(CH_3)_3Al + y\,(CH_3)_3Ga + z\,(CH_3)_3In + NH_3 \rightarrow Al_xGa_yIn_zN + 3\,CH_4\,(x + y + z = 1)$$

$$x\,(CH_3)_3Al + y\,(CH_3)_3Ga + z\,(CH_3)_3In + PH_3 \rightarrow Al_xGa_yIn_zP + 3\,CH_4\,(x + y + z = 1)$$

For all kind of compositions, very cost-intensive MO-CVD machines are utilized to build up a concentration and temperature gradient for the epitaxial deposition of the compound semiconductors at the top of the sapphire wafer. The layer structure of the stack determines the function, while the defect density governs the final device effi-ciency. Then, the wafer with the functional layers on top of is cut into typically 1 mm^2 chips by laser cutting. Most of these semiconductor chips are used in LEDs, sensors, or diode array detectors. As far as LEDs are concerned, the chip is solely the primary radi-ation source, i.e. it must be contacted, wired, placed into secondary optics, and equipped with an electronic driver to obtain a solid-state light source, e.g. an LED lamp (see Chapter 8.1). However, it must be mentioned that such LED lamps, as many other electronic devices, comprise precious elements (here Ga and In) in a final high dilution with respect to the device as a whole, i.e. the LED lamp. Therefore, the large concentration gradient that has to be overcome during recycling in order to enrich these precious elements again towards an economically viable content is energy- and time-consuming.

16.5 Recycling

Mining and recycling have similar boundary conditions. If the element to be refined is present in high dilution (in the mineral (ore, rock) or within a product such as a device), the concentration gradient buildup is the determining and energy-consuming step. Again, gold production is the typical example – high dilution of gold within the accompanying rock *vs.* highly diluted recyclable minor gold amounts within a used mobile phone (or other highly diluted valuable elements in most consumer electronics). On the other hand, good recycling rates are possible with almost pure starting materials. One of the best examples is the lead acid battery, a device with very high recycling quota [14–16]. It should be kept in mind that every production-use-recycling step is always associated with losses of quantity and quality of material, along with the dissipation of free energy gradients.

Pure functional materials that are (highly) diluted in further use are hardly recyclable. To give an example, the amount of titanium used as the white pigment TiO_2 (see Chapter 1.4) is completely lost (dissipated) in rubble and domestic waste without any recycling potential. Other functional materials find application in devices in parallel with many other compounds. A typical example is the lithium-ion battery. The energy demand for separation of the many and chemically different components is higher than primary production from mining educts [17].

A final example concerns screens, used billions of times as human-machine interface for electronic devices. Critical elements are indium, that is used in form of indium-tin-oxide (ITO, ca. 3 g of indium for a 32" display) [18, 19] in LCD screens, as well as iridium from Ir(III) complexes [20, 21] in OLED displays (70.000 oz = 2 t in 2020 for all consumer device applications, 1 g Ir(III) complex ~ 0.3 g Ir for 3000 smartphone displays (average screen size is 5.5" or almost 100 cm^2), ~ 1 µg Ir/cm^2, ~ 1 ppm in a smartphone (weighing ~ 100 g). The sales of OLED displays will be 42 billion USD in 2020 [22]. Both elements show strong dilution and dissipation; however, the huge absolute amount that is requested in future makes recycling indispensable – which is inefficient from an energetic point of view but without alternative, unless the technology evolves towards abundant-element-based approaches.

Summing up, the best resource management is reached for energy and resources that are not consumed at all. One of the striking topics in this sense is digitalization. Meanwhile, we have all information available worldwide 24/7. This involves vast amounts of mobile devices, constructed by using valuable materials and tremendous amounts of energy. The World Wide Web network demands growing server clusters and satellites; meanwhile, it is projected to consume 21% of the global electricity in 2030 – a significant fraction of the primary energy production [23].

References

[1] Bertau M, Müller A, Fröhlich P, Katzberg M. Industrielle Anorganische Chemie, 4. Auflage, Wiley-VCH, Weinheim, Germany, 2013.

[2] Holleman AF, Wiberg N. Anorganische Chemie, 103. Auflage, De Gruyter, Berlin, 2016. ISBN 978-3-11-051854-2.

[3] Fröhlich P, Lorenz T, Martin G, Brett B, Bertau M. Angew Chem Int Ed, 2017, 56, 2544–80.

[4] https://population.un.org/wpp/Graphs/DemographicProfiles/Line/900, last accessed 01.12.2021

[5] Bar-On YM, Phillips R, Milo R. PNAS, 2018, 115, 6506–11.

[6] Lorenz T, Bertau M, Möckel R. Rare-earth minerals and rare-earth mining. In: Pöttgen R, Jüstel T, Strassert CA. (Eds). Rare Earth Chemistry – Basics for Master and PhD Students, De Gruyter, Berlin, 2020, 15–35. ISBN: 978-3-11-065360-1.

[7] Li K, Fan Q, Chuai H, Liu H, Zhang S, Ma X. Trans Tianjin Univ, 2021, 27, 202–16.

[8] Hausen H, Linde H. Tieftemperaturtechnik, 2nd edition, Springer, Heidelberg, 1985.

[9] 1×1 der Gase, 3rd Edition, Air Liquide Deutschland GmbH, 2007.

[10] He X, Liu Y, Rehman A, Wang L. Appl Energy, 2021, 281, 115976.

[11] Berg JM, Stryer L, Tymoczko J, Gatto G. Biochemistry, Elsevier, Amsterdam, 2019.

[12] Stamm U, Ahmad I, Borisov VM, Flohrer F, Gaebel K, Goetze S, Ivanov AS, Khristoforov OB, Kloepfel D, Koehler P, Kleinschmidt J, Korobotchko V, Ringling J, Schriever G, Vinokhodov AY. Proc SPIE, 2002, 4688, 122–33.

[13] https://www.asml.com/en/products/euv-lithography-systems, accessed at July 22nd, 2021.

[14] Sun Z, Cao H, Zhang X, Lin X, Zheng W, Cao G, Sun Y, Zhang Y. Waste Manage, 2017, 64, 190–201.

[15] Zakiyya H, Distya YD, Ellen R. IOP Conf Series Mater Sci Eng, 2018, 288, 012074. DOI: 10.1088/1757-899X/288/1/012074.

[16] Ballantyne AD, Hallett JP, Jason Riley D, Shah N, Payne DJ. R Soc Opensci, 2018, 5, 171368. DOI: http://dx.doi.org/10.1098/rsos.171368.

[17] Piątek J, Afyon S, Budnyak TM, Budnyk S, Sipponen MH, Slabon A. Adv Energy Mater, 2021, 11, 2003456.

[18] Ciacci L, Werner TT, Vassura I, Passarini F. J Indust Ecol, 2019, 23, 426–37.

[19] Schoch K, Liedtke C, Bieng K. Resources, 2021, 10, 5.

[20] Zysman-Colman E (Ed). Iridium(III) in Optoelectronic and Photonics Applications, John Wiley & Sons, Chichester, UK, 2017. DOI: 10.1002/9781119007166.

[21] Minke C, Suermann M, Bensmann B, Hanke-Rauschenbach R. Int J Hydrog Energy, 2021, 46, 23581–90.

[22] Tremblay J-F. The rise of OLED displays, Chem Eng News, 2016, 94(28).

[23] Jones N. Nature, 2018, 561, 163–66.

Subject index

480 — Subject index

AZO 58
azurite 359

backlight LEDs 18
bacterial proliferation 343
ball bearings 199
ball clays 178
ballistic protection 199
ballistic protection ceramics 231
ballistics 236
band gap engineering 10
barium sulfate 399
barium titanate 238
Barkhausen jump 91
barrels 315
basic refractories 255
battery recycling 432
bauxite 272
bauxite bricks 255
beam deflectors 203
bearings 266
beta particles 83
Betaisodona® 354
biaxial alignment 32
binder burnout 199
bioactive coating 224
bioceramics 200, 238
biocompatibility 404, 412
biofilm 342
biogas plants 308
bioglass 406
biomedical implants 199
biomembranes 337
biosensors 199, 394
biphasic calcium phosphate (BCP) 224
biscuit firing 167, 177
bismuth citrate 397
bismuth potassium titanate 218
bismuth salts 397
bismuth subsalicylate 397
bismuth telluride 423
bis-sulfur-chelate complex 449
bisulfites 315
bleaching 313
blende 11
Bloch walls 91, 219
block copolymers 154
blood plasma 456
body color 165

boehmite 274
Bohr radius 12
bolometer 70
bolus alba 396
bone 405
bone china 198
bone china tableware 165
bone grafts 199
Bordeaux mixture 360
borides 198, 260
boron 155
boron carbide 239
boron oxide 276
bouillie bordelaise 360
boundary potential formed 45
Bragg reflection 2
brilliance 18
broadcast satellites 215
bronchioles 456
bronze process 29
brown fused alumina 272
BSCCO 32
burst nucleation 14

CAD/CAM 409
cadmium 442
cadmium minerals 446
cadmium telluride 423
calamine 395
calcium phosphate 405
calcium phosphate cement 406
calcium polysulfide 314
calcium sulfate 307
calcium sulfite 307
calcium thiosulfate 314
Calvin cycle 474
Campylobacter 337
cancer chemotherapy 393
capacitance 157
carbazole 154
carbides 198, 260
carbo medicinalis 398
carbon 45, 272, 275
carbon dioxide 307, 453
carbon disulfide 314
carbon membranes 185
carbon monoxide 41, 452
carbon nanotubes 185, 295
carbonitride 261

Formula index

https://doi.org/10.1515/9783110733471-034

www.ingramcontent.com/pod-product-compliance
Lightning Source LLC
Chambersburg PA
CBHW060956210326

41598CB00031B/4843